Computational Structural Analysis and Finite Element Methods

A. Kaveh

Computational Structural Analysis and Finite Element Methods

Springer

A. Kaveh
Centre of Excellence for Fundamental Studies in
Structural Engineering
School of Civil Engineering
Iran University of Science and Technology
Tehran
Iran

ISBN 978-3-319-02963-4 ISBN 978-3-319-02964-1 (eBook)
DOI 10.1007/978-3-319-02964-1
Springer Cham Heidelberg New York Dordrecht London

Library of Congress Control Number: 2013956541

Printed on acid-free paper

Springer is part of Springer Science+Business Media (www.springer.com)

Preface

Recent advances in structural technology require greater accuracy, efficiency and speed in the analysis of structural systems. It is therefore not surprising that new methods have been developed for the analysis of structures with complex configurations and large number of elements.

The requirement of accuracy in analysis has been brought about by the need for demonstrating structural safety. Consequently, accurate methods of analysis had to be developed, since conventional methods, although perfectly satisfactory when used on simple structures, have been found inadequate when applied to complex and large-scale structures. Another reason why higher speed is required results from the need to have optimal design, where analysis is repeated hundred or even thousands of times.

This book can be considered as an application of discrete mathematics rather than the more usual calculus-based methods of analysis of structures and finite element methods. The subject of graph theory has become important in science and engineering through its strong links with matrix algebra and computer science. At first glance, it seems extraordinary that such abstract material should have quite practical applications. However, as the author makes clear, the early relationship between graph theory and skeletal structures and finite element models is now obvious: the structure of the mathematics is well suited to the structure of the physical problem. In fact, could there be any other way of dealing with this structural problem? The engineer studying these applications of structural analysis has either to apply the computer programs as a black box, or to become involved in graph theory, matrix algebra and sparse matrix technology. This book is addressed to those scientists and engineers, and their students, who wish to understand the theory.

The methods of analysis in this book employ matrix algebra and graph theory, which are ideally suited for modern computational mechanics. Although this text deals primarily with analysis of structural engineering systems, it should be recognised that these methods are also applicable to other types of systems such as hydraulic and electrical networks.

The author has been involved in various developments and applications of graph theory in the last four decades. The present book contains part of this research suitable for various aspects of matrix structural analysis and finite element methods, with particular attention to the finite element force method.

In Chap. 1, the most important concepts and theorems of structures and theory of graphs are briefly presented. Chapter 2 contains different efficient approaches for determining the degree of static indeterminacy of structures and provides systematic methods for studying the connectivity properties of structural models. In this chapter, force method of analysis for skeletal structures is described mostly based on the author's algorithms. Chapter 3 provides simple and efficient methods for construction of stiffness matrices. These methods are especially suitable for the formation of well-conditioned stiffness matrices. In Chaps. 4 and 5, banded, variable banded and frontal methods are investigated. Efficient methods are presented for both node and element ordering. Many new graphs are introduced for transforming the connectivity properties of finite element models onto graph models. Chapters 6 and 7 include powerful graph theory and algebraic graph theory methods for the force method of finite element meshes of low order and high order, respectively. These new methods use different graphs of the models and algebraic approaches. In Chap. 8, several partitioning algorithms are developed for solution of multi-member systems, which can be categorized as graph theory methods and algebraic graph theory approaches. In Chap. 9, an efficient method is presented for the analysis of near-regular structures which are obtained by addition or removal of some members to regular structural models. In Chap. 10, energy formulation based on the force method is derived and a new optimization algorithm called SCSS is applied to the analysis procedure. Then, using the SCSS and prescribed stress ratios, structures are analyzed and designed. In all the chapters, many examples are included to make the text easier to be understood.

I would like to take this opportunity to acknowledge a deep sense of gratitude to a number of colleagues and friends who in different ways have helped in the preparation of this book. Mr. J. C. de C. Henderson, formerly of Imperial College of Science and Technology, first introduced me to the subject with most stimulating discussions on various aspects of topology and combinatorial mathematics. Professor F. Ziegler and Prof. Ch. Bucher encouraged and supported me to write this book. My special thanks are due to Mrs. Silvia Schilgerius, the senior editor of the Applied Sciences of Springer, for her constructive comments, editing and unfailing kindness in the course of the preparation of this book. My sincere appreciation is extended to our Springer colleagues Ms. Beate Siek and Ms. G. Ramya Prakash.

I would like to thank my former Ph.D. and M.Sc. students, Dr. H. Rahami, Dr. M. S. Massoudi, Dr. K. Koohestani, Dr. P. Sharafi, Mr. M. J. Tolou Kian, Dr. A. Mokhtar-zadeh, Mr. G. R. Roosta, Ms. E. Ebrahimi, Mr. M. Ardalan, and Mr. B. Ahmadi for using our joint papers and for their help in various stages of writing this book. I would like to thank the publishers who permitted some of our papers to be utilized in the preparation of this book, consisting of Springer-Verlag, John Wiley and Sons, and Elsevier.

My warmest gratitude is due to my family and in particular my wife, Mrs. Leopoldine Kaveh, for her continued support in the course of preparing this book.

Every effort has been made to render the book error free. However, the author would appreciate any remaining errors being brought to his attention through his email-address: alikaveh@iust.ac.ir.

Tehran A. Kaveh
December 2013

Contents

Chapter 1
Basic Definitions and Concepts of Structural Mechanics and Theory of Graphs

1.1 Introduction

This chapter consists of two parts. In the first part, basic definitions, concepts and theorems of structural mechanics are presented. These theorems are employed in the following chapters and are very important for their understanding. For determination of the distribution of internal forces and displacements, under prescribed external loading, a solution to the basic equations of the theory of structures should be obtained, satisfying the boundary conditions. In the matrix methods of structural analysis, one must also use these basic equations. In order to provide a ready reference for the development of the general theory of matrix structural analysis, the most important basic theorems are introduced in this chapter, and illustrated through simple examples.

In the second part, basic concepts and definitions of graph theory are presented. Since some of the readers may be unfamiliar with the theory of graphs, simple examples are included to make it easier to understand the presented concepts.

1.1.1 Definitions

A *structure* can be defined as a body that resists external effects such as loads, temperature changes, and support settlements, without undue deformation. Building frames, industrial building, bridges, halls, towers, dams, reservoirs, tanks, retaining walls, channels, pavements are typical structures of interest to civil engineers.

A structure can be considered as an assemblage of members and nodes. Structures with clearly defined members are known as *skeletal structures*. Planar and space trusses, planar and space frames, single and double-layer grids are examples of skeletal structures, Fig. 1.1.

A. Kaveh, *Computational Structural Analysis and Finite Element Methods*,
DOI 10.1007/978-3-319-02964-1_1,

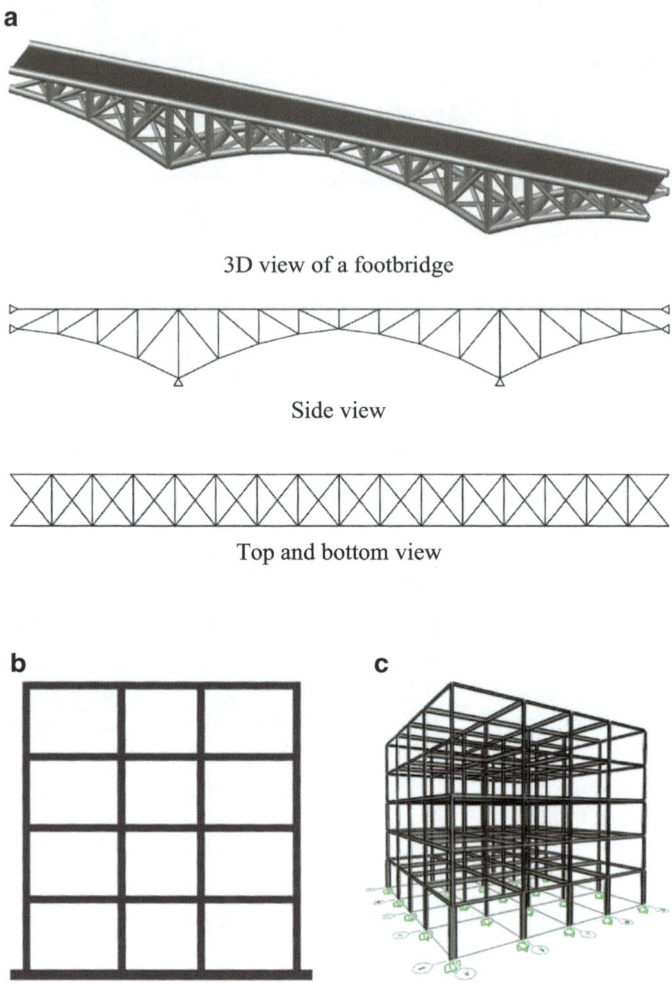

Fig. 1.1 (continued)

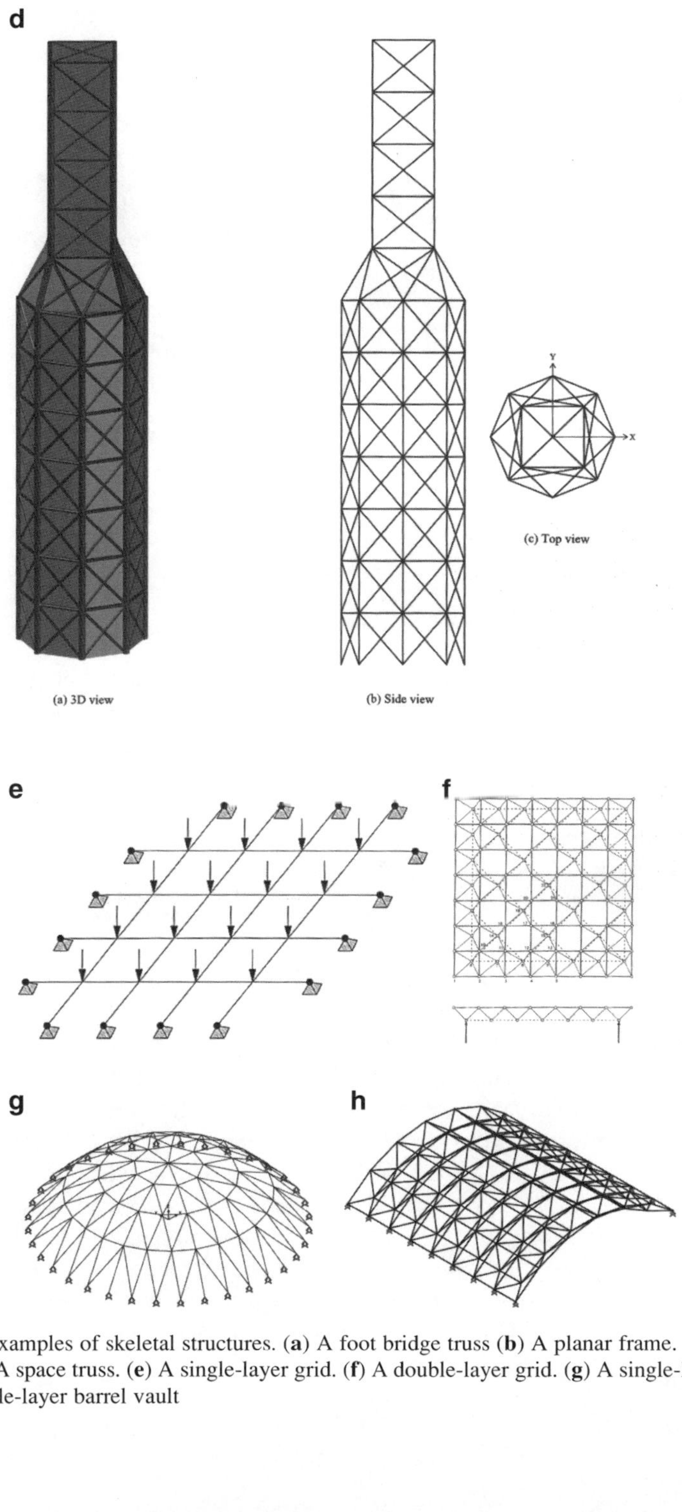

Fig. 1.1 Examples of skeletal structures. (**a**) A foot bridge truss (**b**) A planar frame. (**c**) A space frame. (**d**) A space truss. (**e**) A single-layer grid. (**f**) A double-layer grid. (**g**) A single-layer dome. (**h**) A double-layer barrel vault

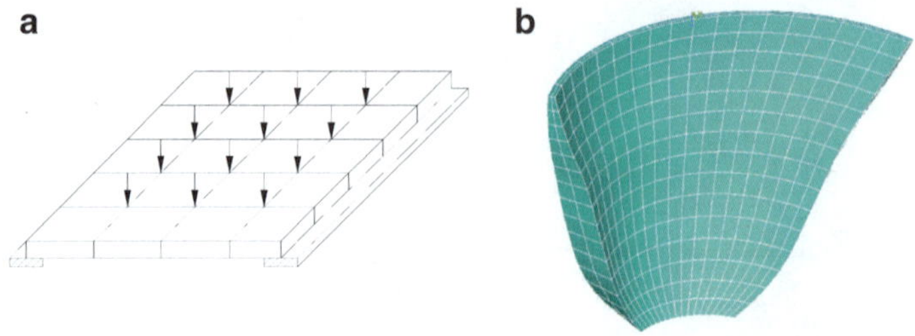

Fig. 1.2 Examples of continua. (**a**) A plate. (**b**) A dam

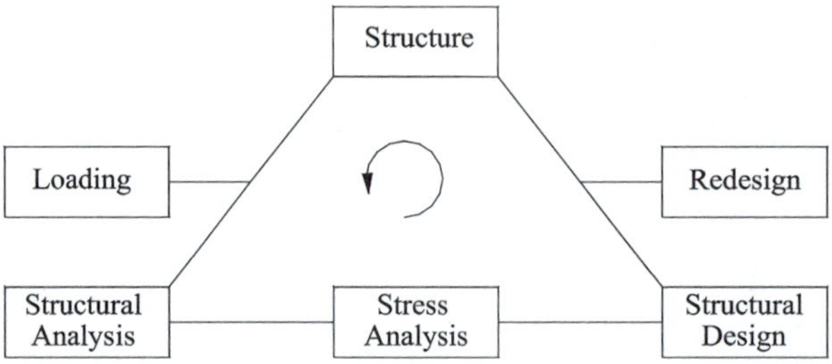

Fig. 1.3 The cycle of analysis and design of a structure

Structures which must artificially be divided into members (elements) are called *continua*. Concrete dams, plates, and pavements are examples of continua, Fig. 1.2.

The underlying principles for the analysis of other structures are more or less the same. Airplane, missile and satellite structures are of interest to the aviation engineer. The analysis and design of a ship is interesting for a naval architect. A machine engineer should be able to design machine parts. However, in this book only structures of interest to structural engineers are studied.

1.1.2 Structural Analysis and Design

Structural analysis is the determination of the response of a structure to external effects such as loading, temperature changes and support settlements. *Structural design* is the selection of a suitable arrangement of members, and a selection of materials and member sections, to withstand the stress resultants (internal forces) by a specified set of loads, and satisfy the stress and displacement constraints, and other requirements specified by the utilized code of practice. The diagram shown in Fig. 1.3 is a simple illustration for the cycle of structural analysis and design.

In optimal design of structures this cycle should be repeated hundred and sometime thousands of times to reduce the weight or cost of the structure.

Structural theories may be classified from different points of view as follows:

Static versus dynamic;
Planar versus space;

Linear versus non-linear;
Skeletal versus continua;
Statically determinate versus statically indeterminate.

In this book, static analyses of linear structures are mainly discussed for the statically determinate and indeterminate cases. Here, both planar and space skeletal structures and continua models are of interest.

1.2 General Concepts of Structural Analysis

1.2.1 Main Steps of Structural Analysis

A correct solution of a structure should satisfy the following requirements:

1. **Equilibrium**: The external forces applied to a structure and the internal forces induced in its members should be in equilibrium at each node.
2. **Compatibility**: The members should deform so that they all fit together.
3. **Force-displacement relationship**: The internal forces and deformations satisfy the stress–strain relationships of the members.

For structural analysis two basic methods are in use:

Force method: In this method, some of the internal forces and/or reactions are taken as primary unknowns, called redundants. Then the stress–strain relationship is used to express the deformations of the members in terms of external and redundant forces. Finally, by applying the compatibility conditions that the deformed members must fit together, a set of linear equations yield the values of the redundant forces. The stress resultants in the members are then calculated and the displacements at the nodes in the direction of external forces are found. This method is also known as the *flexibility method* and *compatibility approach*.

Displacement method: In this method, the displacements of the nodes necessary to describe the deformed state of the structure are taken as unknowns. The deformations of the members are then calculated in terms of these displacements, and by use of the stress–strain relationship, the internal forces are related to them. Finally, by applying the equilibrium equations at each node, a set of linear equations is obtained, the solution of which results in the unknown nodal displacements. This method is also known as the *stiffness method* and *equilibrium approach*.

For choosing the most suitable method for a particular structure, the number of unknowns is one of the main criteria. A comparison for the force and displacement methods can be made, by calculating the degree of static indeterminacy and kinematic indeterminacy. As an example, for the truss structure shown in Fig. 1.4a, the number of redundants is 2 in the force method, while the number of unknown displacements is 13 for the displacement approach. For the 3×3 planar frame shown in Fig. 1.4b, the static indeterminacy and the kinematic indeterminacy are 27 and 36, respectively. For the simple six-bar planar truss of Fig. 1.4c, the

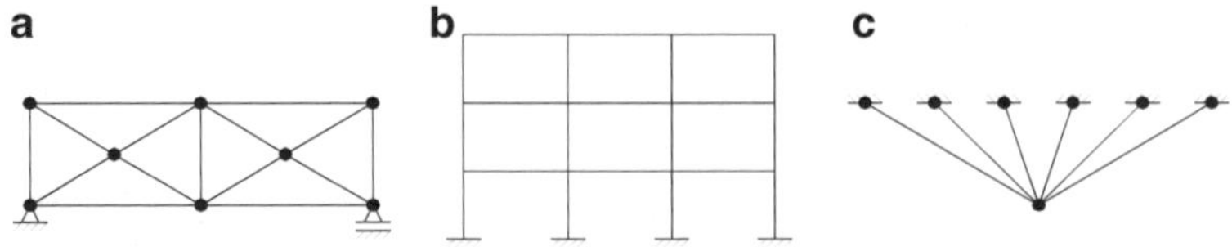

Fig. 1.4 Some simple structures. (**a**) A planar truss. (**b**) A planar frame. (**c**) A simple planar truss

number of unknowns for the force and displacement methods is 4 and 2, respectively. Efficient methods for calculating the indeterminacies are discussed in Chap. 2. The number of unknowns is not the only consideration: another criterion for choosing the most suitable method is the conditioning of the flexibility and stiffness matrices, which are discussed in Kaveh [1, 2].

1.2.2 *Member Forces and Displacements*

A structure can be considered as an assembly of its members, subjected to external effects. These effects will be considered as external loads applied at nodes, since any other effect can be reduced to such equivalent nodal loads. The state of stress in a member (internal forces) is defined by a vector,

$$\mathbf{r}_m = \left\{ r_1^k \, r_2^k \, r_3^k \ldots r_n^k \right\}^t, \tag{1.1}$$

and the associated member deformation (distortion) is designated by a vector,

$$\mathbf{u}_m = \left\{ u_1^k \, u_2^k \, u_3^k \ldots u_n^k \right\}^t, \tag{1.2}$$

where n is the number of force or displacement components of the kth member (element), and t shows the transposition of the vector. Some simple examples of typical elements, common in structural mechanics, are shown in Fig. 1.5.

The relationship between member forces and displacements can be written as:

$$\mathbf{r}_m = \mathbf{k}_m \mathbf{u}_m \text{ or } \mathbf{u}_m = \mathbf{f}_m \mathbf{r}_m, \tag{1.3}$$

where $\mathbf{k}_m$ and $\mathbf{f}_m$ are called *member stiffness* and *member flexibility matrices*, respectively. Obviously, $\mathbf{k}_m$ and $\mathbf{f}_m$ are related as:

$$\mathbf{k}_m \mathbf{f}_m = \mathbf{I}. \tag{1.4}$$

Flexibility matrices can be written only for members supported in a stable manner, because rigid body motion of the undefined amplitude would otherwise result from application of applied loads. These matrices can be written in as many ways as there are stable and statically determinate support conditions.

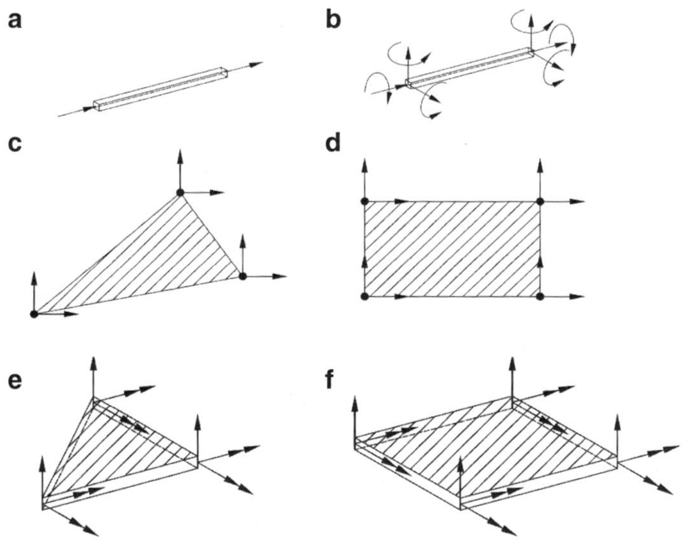

Fig. 1.5 Some simple elements. (**a**) Bar element. (**b**) Beam element. (**c**) Triangular plane stress element. (**d**) Rectangular plane stress element. (**e**) Triangular plate bending element. (**f**) Rectangular plate bending element

The stiffness and flexibility matrices can be derived using different approaches. For simple members like bar elements and beam elements, methods based on the principles of strength of materials or classical theory of structures will be sufficient. However, for more complicated elements the principle of virtual work or alternatively variational methods can be employed. In this section, only simple members are studied, and further considerations will be presented in Chaps. 2, 6, and 7.

1.2.3 Member Flexibility and Stiffness Matrices

Consider a bar element as shown in Fig. 1.6 which carries only axial forces, and has two components of member forces. From the equilibrium,

$$\mathrm{N}_{\mathrm{m}}^{\mathrm{L}} + \mathrm{N}_{\mathrm{m}}^{\mathrm{R}} = 0, \tag{1.5}$$

then only one end force need be specified in order to determine the state of stress throughout the member. The corresponding deformation of the member is simply the elongation, and hence:

$$\mathrm{r}_{\mathrm{m}}^{1} = \mathrm{N}_{\mathrm{m}}^{\mathrm{R}}, \ \text{ and } \ \mathrm{u}_{\mathrm{m}}^{1} = \delta_{\mathrm{m}}^{\mathrm{R}}. \tag{1.6}$$

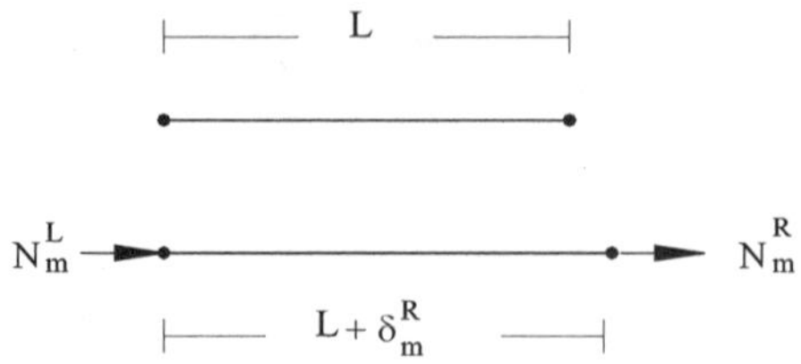

Fig. 1.6 Internal forces and deformation of a bar element

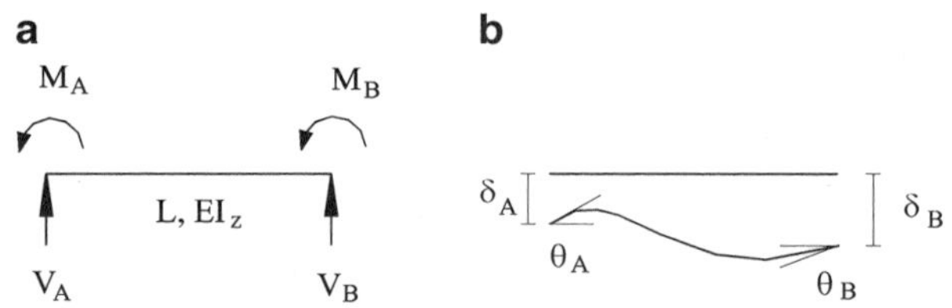

Fig. 1.7 End forces and deflected shape of a beam element

From Hooke's law $N_m^R = \frac{EA}{L}\delta_m^R$, and therefore:

$$\mathbf{f}_m = \frac{L}{EA} \text{ and } \mathbf{k}_m = \frac{EA}{L}. \tag{1.7}$$

Now consider a prismatic beam of a planar frame with length L and bending stiffness EI. The internal forces are shown in Fig. 1.7.

This element is assumed to be subjected to four end forces, as shown in Fig. 1.7a, and the deflected shape and position is illustrated in Fig. 1.7b. Four end forces are related by the following two equilibrium equations:

$$\begin{aligned} V_A + V_B &= 0, \\ M_A + M_B + V_B L &= 0. \end{aligned} \tag{1.8}$$

Therefore, only two end-force components should be specified as internal forces. Some possible choices for $\mathbf{r}_m$ are $\{M_A, M_B\}$, $\{V_B, M_B\}$ and $\{V_A, M_A\}$.

Using classical formulae, such as those of the strength of materials or slope-deflection equations of the theory of structures, the force-displacement relationships can be established. As an example, the flexibility matrix for a prismatic beam supported as a cantilever is obtained using the differential equation of the elastic deformation curve as follows:

$$\frac{d^2v}{dx^2} = \frac{M_z}{EI_z} = \frac{1}{EI_z}[V_B(L - x) + M_B].$$

Integrating the above equation leads to,

$$\frac{dv}{dx} = \frac{1}{EI_z}\left[V_B\left(Lx - \frac{1}{2}x^2\right) + M_B x\right] + C_1,$$

and integrating again results in:

$$v = \frac{1}{EI_z}\left[V_B\left(\frac{1}{2}Lx^2 - \frac{1}{6}x^3\right) + \frac{1}{2}M_B x^2\right] + C_1 x + C_2.$$

Using the boundary conditions at A as,

$$\left[\frac{dv}{dx}\right]_{x=0} = 0 \text{ and } [v]_{x=0} = 0,$$

results in:

$$C_1 = 0 \text{ and } C_2 = 0.$$

Substituting these constants leads to:

$$v = \frac{1}{EI_z}\left[V_B\left(\frac{1}{2}Lx^2 - \frac{1}{6}x^3\right) + \frac{1}{2}M_B x^2\right],$$
$$\frac{dv}{dx} = \frac{1}{EI_z}\left[V_B\left(Lx - \frac{1}{2}x^2\right) + M_B x\right].$$

For $x = L$, the displacement and rotation of end B are obtained as,

$$\delta_B = \frac{V_B L^3}{3EI_z} + \frac{M_B L^2}{2EI_z} \text{ and } \theta_B = \frac{V_B L^2}{2EI_z} + \frac{M_B L}{EI_z},$$

using $I_z = I$, the above relationships in matrix form become,

$$\begin{bmatrix} \delta_B \\ \theta_B \end{bmatrix} = \begin{bmatrix} u_m^1 \\ u_m^2 \end{bmatrix} = \begin{bmatrix} \dfrac{L^3}{3EI} & \dfrac{L^2}{2EI} \\ \dfrac{L^2}{2EI} & \dfrac{L}{EI} \end{bmatrix} \begin{bmatrix} V_B \\ M_B \end{bmatrix},$$

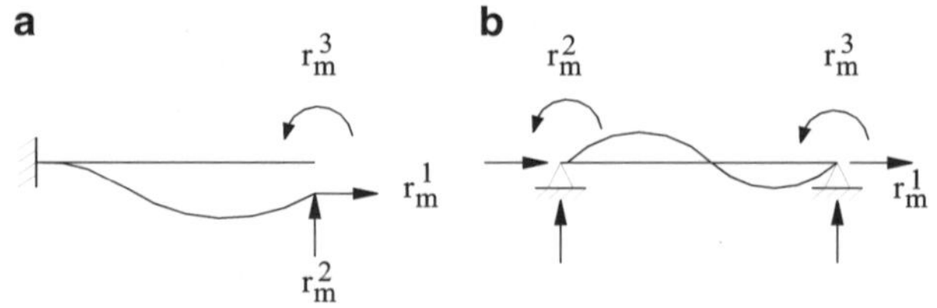

Fig. 1.8 Two sets of end forces and displacements for a beam element

or

$$\mathbf{f}_m = \frac{L^2}{6EI}\begin{bmatrix} 2L & 3 \\ 3 & 6/L \end{bmatrix}. \tag{1.9}$$

Using a similar method, for a simply supported beam with two moments acting at the two ends, we have:

$$\mathbf{f}_m = \begin{bmatrix} \frac{L}{3EI} & -\frac{L}{6EI} \\ -\frac{L}{6EI} & \frac{L}{3EI} \end{bmatrix} = \frac{L}{6EI}\begin{bmatrix} 2 & -1 \\ -1 & 2 \end{bmatrix}. \tag{1.10}$$

If the axial forces are also included as member forces, then $\mathbf{r}_m^t = [N_B\ V_B\ M_B]$ and $\mathbf{r}_m^t = [N_B\ M_A\ M_B]$, as shown in Fig. 1.8. The above matrices become:

$$\mathbf{f}_m = \begin{bmatrix} \frac{L}{EA} & 0 & 0 \\ 0 & \frac{L^3}{3EI} & \frac{L^2}{2EI} \\ 0 & \frac{L^2}{2EI} & \frac{L}{EI} \end{bmatrix} \text{ and } \mathbf{f}_m = \begin{bmatrix} \frac{L}{EA} & 0 & 0 \\ 0 & \frac{L}{3EI} & -\frac{L}{6EI} \\ 0 & -\frac{L}{6EI} & \frac{L}{3EI} \end{bmatrix}. \tag{1.11}$$

The corresponding stiffness matrices are:

$$\mathbf{k}_m = \begin{bmatrix} \frac{EA}{L} & 0 & 0 \\ 0 & \frac{12EI}{L^3} & -\frac{6EI}{L^2} \\ 0 & -\frac{6EI}{L^2} & \frac{4EI}{L} \end{bmatrix} \text{ and } \mathbf{k}_m = \begin{bmatrix} \frac{EA}{L} & 0 & 0 \\ 0 & \frac{4EI}{L} & \frac{2EI}{L} \\ 0 & \frac{2EI}{L} & \frac{4EI}{L} \end{bmatrix}. \tag{1.12}$$

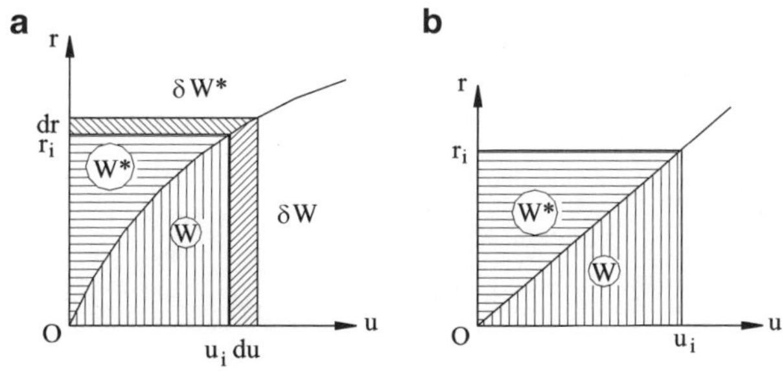

Fig. 1.9 Force-displacement relationships. (**a**) A non-linear relationship. (**b**) A linear relationship

It should be mentioned that both flexibility and stiffness matrices are symmetric, on account of the Maxwell-Betti reciprocal work theorem proven in the next section. More general methods for the derivation of member flexibility and stiffness matrices will be studied in Chaps. 2, 3, 6, and 7.

1.3 Important Structural Theorems

1.3.1 Work and Energy

The work, δW, of a force **r** acting through a change in displacement d**u** in the direction of that force is the product **rdu**.

Consider a general load–displacement relationship as shown in Fig. 1.9a. The area under this curve represents the work done, denoted by W. The area above this curve is the complementary work designated by W*.

For a total displacement of u_1, the total work is given by,

$$W = \int_0^{u_1} \mathbf{r}d\mathbf{u}, \tag{1.13}$$

and the complementary work is:

$$W^* = \int_0^{r_1} \mathbf{u}d\mathbf{r}. \tag{1.14}$$

For a linear case, as shown in Fig. 1.9b, we have:

$$W = W^*. \tag{1.15}$$

In this book, it is assumed that the loads are applied to a structure in a gradual manner, and attention is limited to linear behaviour. Thus the load–displacement relationship is as shown in Fig. 1.9b, and the relation can be expressed as,

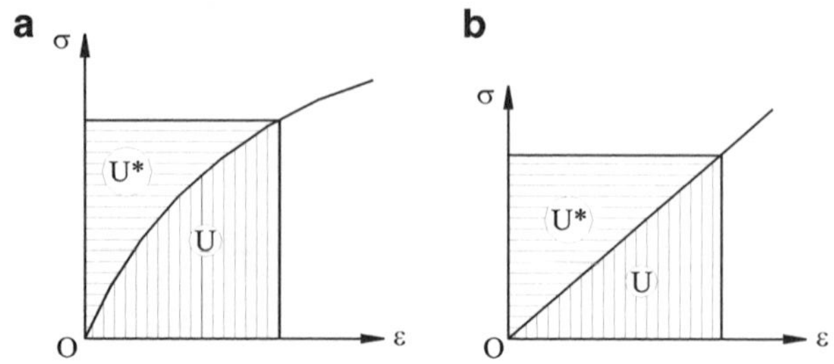

Fig. 1.10 Stress-strain relationships. (**a**) A general stress-strain relationship. (**b**) Linear stress–strain relationship

$$\mathbf{r} = \mathbf{k}\mathbf{u}, \tag{1.16}$$

where **k** is a constant. The work in Fig. 1.9b can be written as:

$$W = \frac{1}{2} r_i u_i. \tag{1.17}$$

Forces and displacements at a point are both represented by vectors, and their work is represented as a dot product. In matrix notations, however, the work can be written as:

$$W = \frac{1}{2} \mathbf{r}^t \mathbf{u}. \tag{1.18}$$

Using Eq. 1.3,

$$W = \frac{1}{2} \mathbf{u}^t \mathbf{k}^t \mathbf{u} = \frac{1}{2} \mathbf{u}^t \mathbf{k} \mathbf{u}. \tag{1.19}$$

Similarly, W* can be calculated as:

$$W^* = \frac{1}{2} \mathbf{r}^t \mathbf{f} \mathbf{r}. \tag{1.20}$$

Consider the stress–strain relationship as illustrated in Fig. 1.10a. The area under this curve represents the density of the strain energy, and when integrated over the volume of the member (or structure) results in the strain energy U. The area to the left of the stress–strain curve is the density of the complementary strain energy, and by integration over the member (or structure) the complementary energy U* is obtained. For the linear stress–strain relationship as shown in Fig. 1.10b, U = U*.

Since the work done by external actions on an elastic system is equal to the strain energy stored internally in the system (work-energy law), therefore:

$$W = U \text{ and } W^* = U^*. \tag{1.21}$$

1.3.2 Castigliano's Theorems

Consider the force-displacement curve in Fig. 1.9a, and suppose an imaginary displacement δu_i is imposed on the system. The work done, δW, under the action of r_i in moving through δu_i is equal to:

$$\delta W = r_i \delta u_i. \tag{1.22}$$

Using Eq. 1.21, and taking limit, leads to the first theorem of Castigliano as,

$$\frac{\partial U}{\partial q_i} = r_i, \tag{1.23}$$

which can be stated as follows [3]:

> The partial derivative of the strain energy with respect to a displacement, is equal to the force applied at the point and along the considered displacement.

Similarly, if the system is subjected to an imaginary force δr_i along the displacement u_i, then the complementary work done δW^* is equal to,

$$\delta W^* = u_i \delta r_i = \delta U^*, \tag{1.24}$$

and in the limit, the second theorem of Castigliano is obtained as:

$$\frac{\partial U^*}{\partial r_i} = u_i. \tag{1.25}$$

> The partial derivative of the complementary strain energy with respect to a force is equal to the displacement at the point where the force is applied and directed along the action of the force.

For the linear case, $U^* = U$ and therefore Eq. 1.25 becomes as:

$$\frac{\partial U}{\partial r_i} = u_i. \tag{1.26}$$

1.3.3 Principle of Virtual Work

The *principle of virtual work* is a very powerful means for deducing the conditions of compatibility and equilibrium [4], and it can be stated as follows:

> The work done by a set of external forces **P** acting on a structure, in moving through the associated displacements **v**, is equal to the work done by some other set of forces **R**, which is statically equivalent to **P**, moving through associated displacements **u**, which is compatible with **v**. Associated forces and displacements have the same lines of actions.

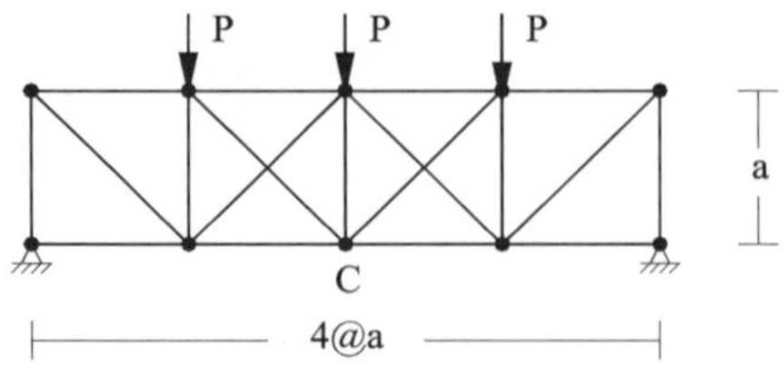

Fig. 1.11 A planar truss

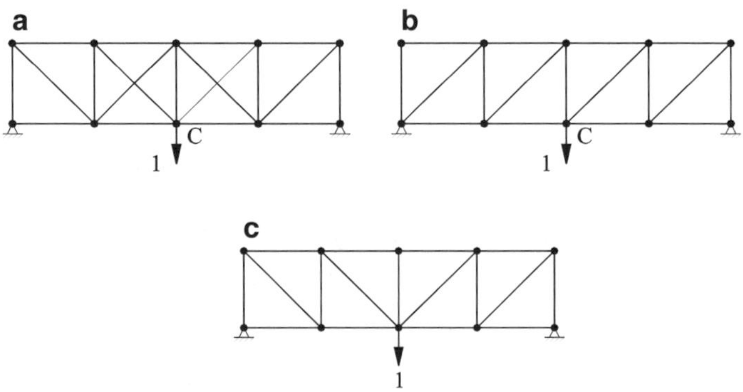

Fig. 1.12 Three different systems capable of supporting the dummy load

Using a statically admissible set of forces and the work equation, the compatibility relations between the deformations and displacements can be derived. Alternatively, employing a compatible set of displacements and the work equation, one obtains the equations of equilibrium between the forces. These approaches are elegant and practical.

Dummy Load Theorem. This theorem can be used to determine the conditions of compatibility. Suppose that the deformed shape of each member of a structure is known, then it is possible to find the deflection of the structure at any point by using the principle of virtual work. For this purpose a dummy load (usually unit load) is applied at the point and in the direction of required displacement, which is why it is also known as the *unit load method*. The dummy load theorem can be stated as:

$$\left\{\begin{array}{l}\text{applied}\\\text{dummy}\\\text{load}\end{array}\right\} \times \left\{\begin{array}{l}\text{actual displacement}\\\text{of structure where external}\\\text{dummy load is applied}\end{array}\right\}$$

$$= \left\{\begin{array}{l}\text{internal forces}\\\text{statically equivalent to}\\\text{the applied dummy load}\end{array}\right\} \times \left\{\begin{array}{l}\text{actual}\\\text{deformation}\\\text{of elements}\end{array}\right\}$$

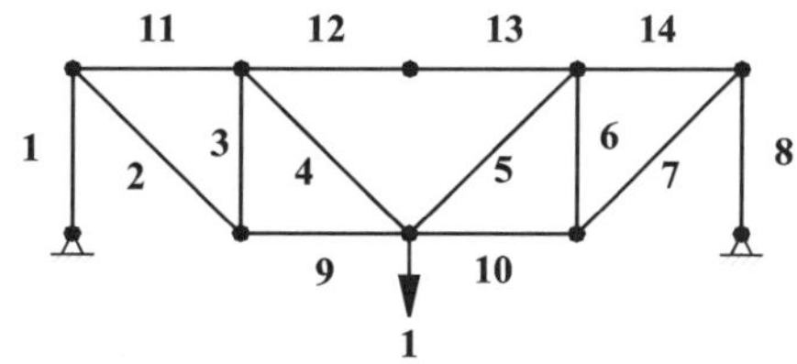

Fig. 1.13 Internal forces equivalent to unit dummy load

It should be noted that the dummy load theorem is a condition on the geometry of the structure. In fact, once the deformations of elements are known, one can draw the deflected shape of the structure, and the results obtained for the deflections will agree with those of the dummy load theorem.

Example 1. Consider a truss as shown in Fig. 1.11. It is desired to measure the vertical deflection at node C, when the structure is subjected to a certain loading.

A unit load is applied at C, and a set of internal forces statically equivalent to the unit load is chosen. However, for such equivalent internal forces, there exists a wide choice of systems, since there are several numbers of structural possibilities which can sustain the load at C. Three examples of such systems are shown in Fig. 1.12a–c.

Obviously, system (a) will need more calculation because of being statically indeterminate.

System (c) is used here, since it has a smaller number of members than (b), and symmetry is also preserved. Internal forces of the members in this system shown in Fig. 1.13 are:

$$\mathbf{r}=\left\{-1/2,\sqrt{2}/2,-1/2,\sqrt{2}/2,\sqrt{2}/2,-1/2,\sqrt{2}/2,-1/2,1/2,1/2,-1/2,-1,-1,-1/2\right\}^t.$$

Measuring the elongation in members of this system containing 14 bars, and using the dummy-load theorem, results in:

$$\left(\frac{1}{2}\right)(0)+(1)(\mathrm{v_c})+\left(\frac{1}{2}\right)(0)=\mathrm{v_c}=-\frac{1}{2}\mathrm{e}_1+\frac{\sqrt{2}}{2}\mathrm{e}_2+-\frac{1}{2}\mathrm{e}_3+\frac{\sqrt{2}}{2}\mathrm{e}_4+\frac{\sqrt{2}}{2}\mathrm{e}_5-\frac{1}{2}\mathrm{e}_6$$
$$+\frac{\sqrt{2}}{2}\mathrm{e}_7-\frac{1}{2}\mathrm{e}_8+\frac{1}{2}\mathrm{e}_9+\frac{1}{2}\mathrm{e}_{10}-\frac{1}{2}\mathrm{e}_{11}-\mathrm{e}_{12}-\mathrm{e}_{13}-\frac{1}{2}\mathrm{e}_{14}.$$

Dummy Displacement Theorem. This method is usually used to find the applied external forces when the internal forces are known. In order to obtain the external force at a particular point, one subjects the structure to a unit displacement at that point in the direction of the force and chooses any set of deformations compatible with the unit displacement. Then from the principle of work, the dummy displacement theorem can be stated as:

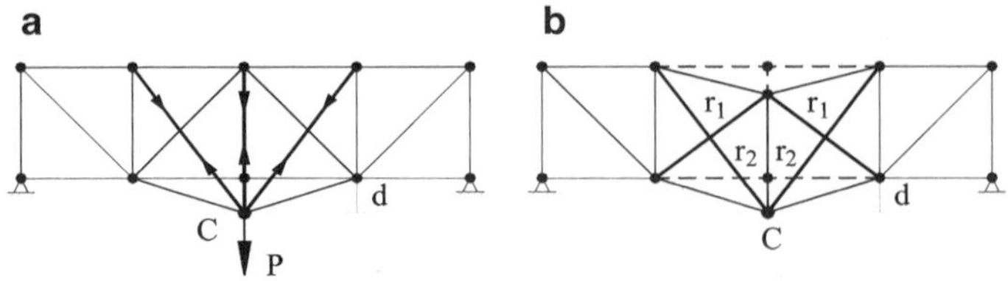

Fig. 1.14 Element deformations equivalent to unit dummy displacement

$$\left\{\begin{array}{l}\text{dummy displacement applied}\\ \text{in the direction of unknowns}\\ \text{actual external forces}\end{array}\right\} \times \left\{\begin{array}{l}\text{actual}\\ \text{external}\\ \text{forces}\end{array}\right\}$$
$$= \left\{\begin{array}{l}\text{deformation of elements}\\ \text{compatible with}\\ \text{dummy displacement}\end{array}\right\} \times \left\{\begin{array}{l}\text{actual}\\ \text{internal}\\ \text{forces}\end{array}\right\}$$

This method is also known as the *unit displacement method.*

Example 2. For the truss studied in Example 1, it is required to find the magnitude of P by measuring the internal forces in the members of the truss.

Again, many systems can be chosen; two of which are illustrated in Fig. 1.14a, b. In these systems, the internal forces to be measured are shown in bold lines. Due to the symmetry, in both cases only two measurements are needed. Applying the dummy-displacement theorem to system (a) yields:

$$\mathrm{Pd} = \mathrm{r}_1 \mathrm{d}\frac{\sqrt{2}}{2} + \mathrm{r}_2 \mathrm{d} + \mathrm{r}_1 \mathrm{d}\frac{\sqrt{2}}{2} = \mathrm{d}\left(\sqrt{2}\mathrm{r}_1 + \mathrm{r}_2\right).$$

1.3.4 Contragradient Principle

Consider two statically equivalent force systems **R** and **P**, related by a linear transformation as:

$$\mathbf{R} = \mathbf{BP}, \tag{1.27}$$

R is considered to have more entries than **P**, i.e. there are solutions to **R** for which **P** is zero. Associated with **R** and **P** let there be two sets of displacements **v** and **u**, respectively. These are compatible displacements and therefore the work done in each system is the same, i.e.

$$\mathbf{P}^{\mathrm{t}}\mathbf{u} = \mathbf{R}^{\mathrm{t}}\mathbf{v}. \tag{1.28}$$

From Eq. 1.27,

$$\mathbf{R}^{t} = \mathbf{P}^{t}\mathbf{B}^{t}. \tag{1.29}$$

Therefore:

$$\mathbf{P}^{t}\mathbf{u} = \mathbf{P}^{t}\mathbf{B}^{t}\mathbf{v}. \tag{1.30}$$

Since **P** is arbitrary, hence:

$$\mathbf{u} = \mathbf{B}^{t}\mathbf{v}. \tag{1.31}$$

Equations 1.27 and 1.31 will be used in the formulation of the force method.

In a general structure, if member forces **R** are related to external nodal loads **P**, similar to Eq. 1.27, then according to the contragradient principle [4], the member distortions **v** and nodal displacement **u** will be related by an equation similar to Eq. 1.31.

If two displacement systems **u** and **v** are related by a linear transformation as,

$$\mathbf{v} = \mathbf{C}\mathbf{u}, \tag{1.32}$$

and **R** and **P** are statically equivalent forces, then equating the work done for compatible displacements results in:

$$\mathbf{P}^{t}\mathbf{u} = \mathbf{R}^{t}\mathbf{v} = \mathbf{R}^{t}\mathbf{C}\mathbf{u}. \tag{1.33}$$

Again **u** is arbitrary and:

$$\mathbf{P} = \mathbf{C}^{t}\mathbf{R}. \tag{1.34}$$

Equations 1.32 and 1.34 are employed in the formulation of the displacement method.

For a statically determinate structure,

$$\mathbf{P} = \mathbf{B}^{-1}\mathbf{R}, \tag{1.35}$$

and therefore:

$$\mathbf{C}^{t} = \mathbf{B}^{-1}. \tag{1.36}$$

1.3.5 Reciprocal Work Theorem

Consider a structure as shown in Fig. 1.15a subjected to a set of loads, $\{P_1, P_2, \ldots, P_m\}$. The same structure is considered under the action of a second set of loads $\{Q_1, Q_2, \ldots, Q_n\}$, Fig. 1.15b. The reciprocal work theorem can be stated as:

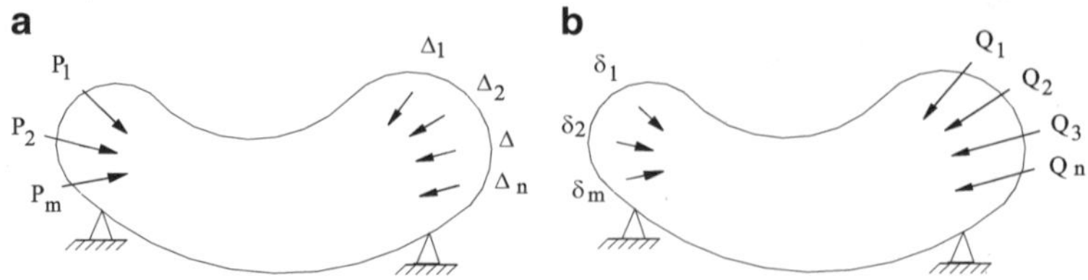

Fig. 1.15 A structure subjected to two sets of loads

The work done by $\{P_1, P_2, \ldots, P_m\}$ through displacements $\{\delta_1, \delta_2, \ldots, \delta_m\}$ produced by $\{Q_1, Q_2, \ldots, Q_n\}$, is the same as the work done by $\{Q_1, Q_2, \ldots, Q_n\}$ through displacements $\{\Delta_1, \Delta_2, \ldots, \Delta_n\}$ produced by $\{P_1, P_2, \ldots, P_m\}$; i.e.

$$\sum_{i=1}^{m} P_i\delta_i = \sum_{j=1}^{n} Q_j\Delta_j. \tag{1.37}$$

When single loads P and Q are considered, Eq. 1.37 reduces to,

$$P\delta_i = Q\Delta_j, \tag{1.38}$$

and for the case where P = Q, one obtains:

$$\delta_i = \Delta_j. \tag{1.39}$$

Equation 1.39 is known as Betti's law, and can be stated as follows:

The deflection at point i due to a load at point j is the same as deflection at j when the same load is applied at i.

The proof of the reciprocal work theorem is constructed by equating the strain energy of the structure in two different loading sequences [5]. In the first sequence, both sets of loads are applied simultaneously, while in the second sequence, loads $\{P_1, P_2, \ldots, P_m\}$ are applied first, followed by the application of the second set of loads $\{Q_1, Q_2, \ldots, Q_n\}$.

1.4 Basic Concepts and Definitions of Graph Theory

Some of the uses of the theory of graphs in the context of civil engineering are as follows: A graph can be a model of a structure, a hydraulic network, a traffic network, a transportation system, a construction system, or a resource allocation system, for example. In this book, the theory of graphs is used as the model of a skeletal structure, and it is employed also as a way of transforming the connectivity properties of finite element meshes to those of graphs. Many such graphs are previously defined in [6], and employed throughout the combinatorial optimisations performed for optimal analysis of skeletal structures and finite element models. This part of the chapter will also enable the readers to develop their own ideas and methods in the light of the principles of graph theory. For further definitions and proofs, the reader may refer to Harary [7], Berge [8], and West [9].

1.4.1 Basic Definitions

The performance of a structure depends not only on the characteristics of its components, but also on their relative location. On the other hand, in a structure, if the properties of one member are altered, the overall behaviour may be changed. This indicates that the performance of a structure depends on the detailed characteristics of its members. On the other hand, if the location of a member is altered, the properties of the structure may again be different. Therefore, the connectivity (topology) of the structure influences the performance of the whole structure and is as important as the mechanical properties of its members. Hence, it is important to represent a structure so that its topology can be understood clearly. The graph model of a structure provides a powerful means for this purpose.

1.4.2 Definition of a Graph

A *graph* S consists of a non-empty set N(S) of elements called *nodes* (vertices or points) and a set M(S) of elements called *members* (edges or arcs), together with a relation of *incidence* which associates each member with a pair of nodes, called its *ends*.

Two or more members joining the same pair of nodes are collectively known as a *multiple member*, and a member joining a node to itself is called a *loop*. A graph with no loops and multiple members is called a *simple graph*. If N(S) and M(S) are countable sets, then the corresponding graph S is *finite*. In this book, only finite graphs are needed, which are referred to as *graphs*.

The above definitions correspond to abstract graphs; however, a graph may be visualised as a set of points connected by line segments in Euclidean space; the nodes of a graph are identified with points, and its members are identified as line segments without their end points. Such a configuration is known as a *topological graph*. These definitions are illustrated in Fig. 1.16.

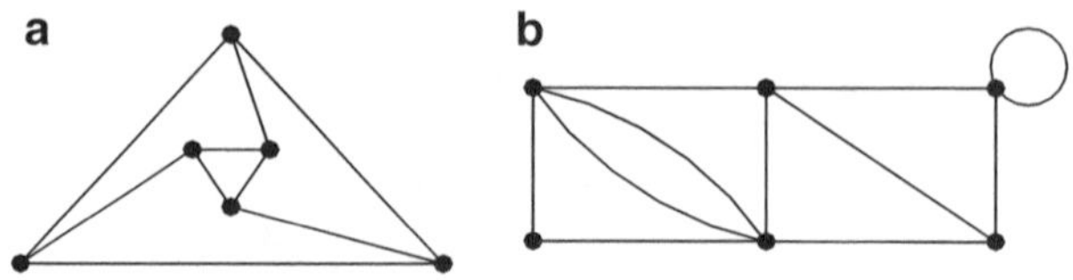

Fig. 1.16 Simple and non-simple graphs. (**a**) A simple graph. (**b**) A graph with loop and multiple members

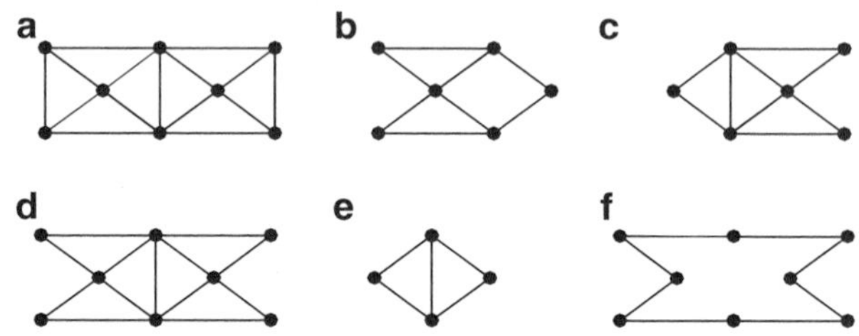

Fig. 1.17 A graph, two of its subgraphs, their union, intersection and ring sum. (**a**) S. (**b**) S_i. (**c**) S_j. (**d**) $S_i \cup S_j$. (**e**) $S_i \cap S_j$. (**f**) $S_i \oplus S_j$

1.4.3 Adjacency and Incidence

Two nodes of a graph are called *adjacent* if these nodes are the end nodes of a member. A member is called *incident with a node* if this node is an end node of the member. Two members are called *incident* if they have a common end node. The *degree* (valency) of a node n_i of a graph, denoted by $\deg(n_i)$, is the number of members incident with that node. Since each member has two end nodes, the sum of node-degrees of a graph is twice the number of its members.

1.4.4 Graph Operations

A *subgraph* S_i of S is a graph for which $N(S_i) \subseteq N(S)$ and $M(S_i) \subseteq M(S)$, and each member of S_i has the same ends as in S.

The *union* of subgraphs $S_1, S_2, \ldots, S_k$ of S, denoted by $S^k = \bigcup_{i=1}^{k} S_i = S_1 \cup S_2 \cup \ldots \cup S_k$, is a subgraph of S with $N(S^k) = \bigcup_{i=1}^{k} N(S_i)$ and $M(S^k) = \bigcup_{i=1}^{k} M(S_i)$. The *intersection* of two subgraphs S_i and S_j is similarly defined using intersections of node-sets and member-sets of the two subgraphs. The intersection of two subgraphs does not need to consist only of nodes, but it is usually considered to do so in the substructuring technique of structural analysis. The *ring sum* of two subgraphs $S_i \oplus S_j$ is a subgraph that contains the nodes and members of S_i and S_j except those elements common to S_i and S_j. These definitions are illustrated in Fig. 1.17.

Two graphs S and K are called *homeomorphic* if one can obtain K from S, by suppressing or inserting nodes of degree 2 in the members.

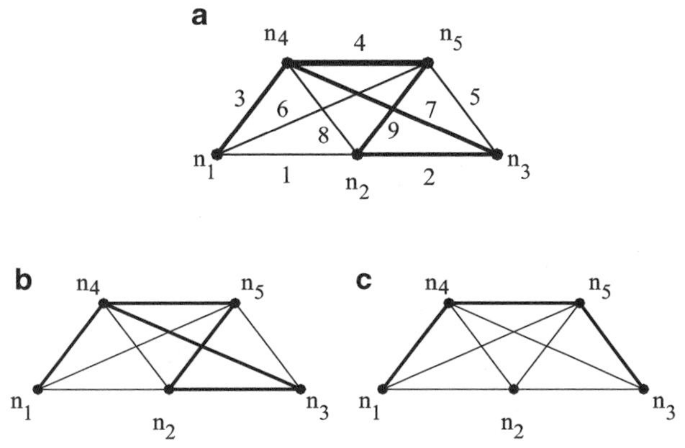

Fig. 1.18 A walk, a trail and a path in S. (**a**) A walk w in S. (**b**) A trail t in S. (**c**) A path P in S

1.4.5 Walks, Trails and Paths

A *walk* w of S is a finite sequence $w = \{n_0, m_1, n_1, \ldots, m_p, n_p\}$ whose terms are alternately nodes n_i and members m_i of S for $1 \leq i \leq p$, and n_{i-1} and n_i are the two ends of m_i. A *trail* t in S is a walk in which no member of S appears more than once. A *path* P is a trail in which no node appears more than once. The *length* of a path P_i, denoted by $L(P_i)$, is taken as the number of its members. P_i is called the *shortest path* between the two nodes n_0 and n_p, if for any other path P_j between these nodes, $L(P_i) \leq L(P_j)$. The *distance* between two nodes of a graph is defined as the number of the members of a shortest path between these nodes.

As an example, in Fig. 1.18,

$$w = (n_1,\ m_3,\ \mathbf{n_4},\ \mathbf{m_4},\ \mathbf{n_5},\ m_9,\ n_2,\ m_2,\ n_3,\ m_7,\ \mathbf{n_4},\ \mathbf{m_4},\ \mathbf{n_5})$$

is a walk between n_1 and n_5 in which member m_4 and nodes n_4 and n_5 are repeated twice.

$$t = (n_1,\ m_3,\ \mathbf{n_4},\ m_4,\ n_5,\ m_9,\ n_2,\ m_2,\ n_3,\ m_7,\ \mathbf{n_4})$$

is a trail between n_1 and n_4 in which node n_5 is repeated twice.

$$P = (n_1, m_3, n_4, m_4, n_5, m_5, n_3)$$

is a path of length 3 in which no node and no member is repeated.

The path $(n_1, m_6, n_5, m_5, n_3)$ is a shortest path of length 2 between the two nodes n_1 and n_3, where the length of each member is taken as unity.

Two nodes n_i and n_j are said to be *connected* in S if there exists a path between these nodes. A graph S is called *connected* if all pairs of its nodes are connected. A *component* of a graph S is a maximal connected subgraph, i.e. it is not a subgraph of any other connected subgraph of S.

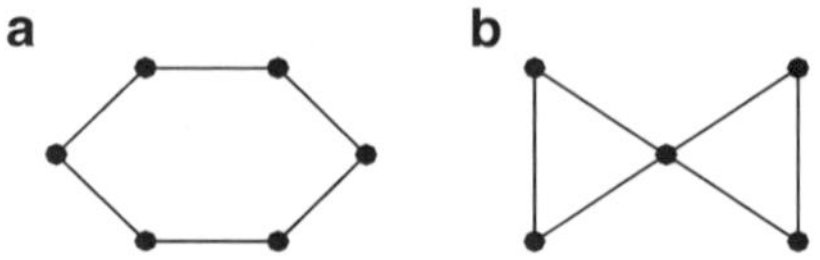

Fig. 1.19 Cycles of S.
(**a**) A cycle of S.
(**b**) A hinged cycle of S

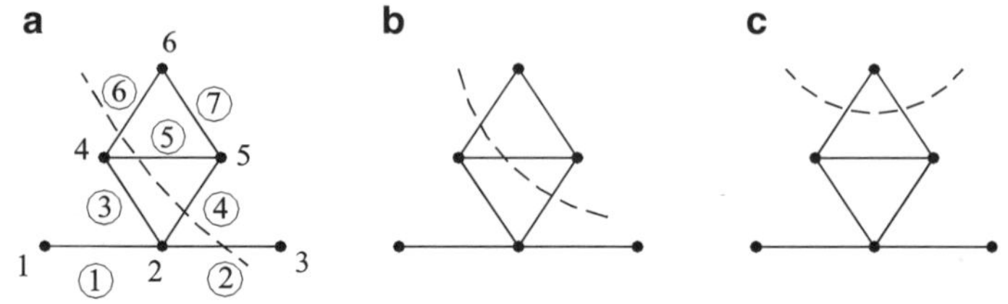

Fig. 1.20 Cutsets of S. (**a**) A cutset of S. (**b**) A prime cutest. (**c**) A cocycle of S

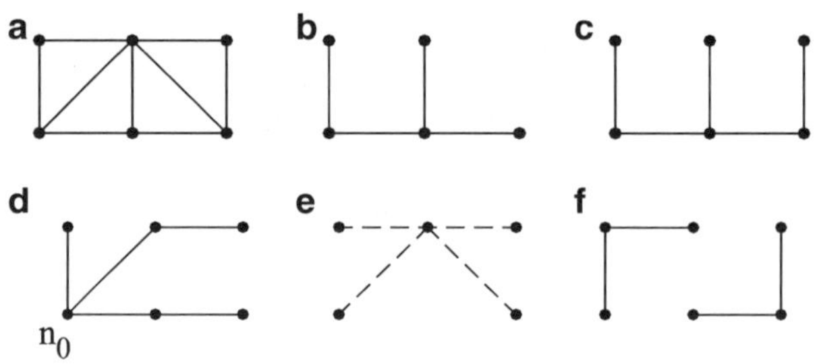

Fig. 1.21 Different trees, a cotree and a forest of S. (**a**) A graph S. (**b**) A tree of S. (**c**) A spanning tree of S. (**d**) An SRT rooted from n_0. (**e**) The cotree of (**c**). (**f**) A forest with two trees

1.4.6 Cycles and Cutsets

A *cycle* is a path (n_0, m_1, n_1, . . ., m_p, n_p) for which $n_0 = n_p$ and $p \geq 1$; i.e. a cycle is a closed path. Similarly, a *closed trail (hinged cycle)* and a *closed walk* can be defined, Fig. 1.19.

A *cutset* is a collection of members whose removal from the graph increases the number of its components. If a cutset results in two disjoint subgraphs S_1 and S_2, then it is called a *prime cutset*. Notice that no proper subsets of a cutset have this property. A *link* is a member which has its ends in S_1 and S_2. Each S_1 and S_2 may or may not be connected. If both are connected, the cutset is called *prime*. If one of S_1 or S_2 consists of a single node, the cutset is called a *cocycle*. These definitions are illustrated in Fig. 1.20.

1.4.7 *Trees, Spanning Trees and Shortest Route Trees*

A *tree* T of S is a connected subgraph of S which contains no cycle. A set of trees of S forms a *forest*. Obviously a forest with k trees contains N(S) − k members. If a tree contains all the nodes of S, it is called a *spanning tree* of S. Henceforth, for simplicity it will be referred to as a *tree*.

A *shortest route tree* (SRT) rooted at a specified node n_0 of S, is a tree for which the distance between every node n_j of T and n_0 is a minimum. An SRT of a graph can be generated by the following simple algorithm:

Algorithm. Label the selected root n_0 as "0" and the adjacent nodes as "1". Record the members incident to "0" as tree members. Repeat the process of labelling with "2" the unnumbered ends of all the members incident with nodes labelled as "1", again recording the tree members. This process terminates when each node of S is labelled and all the tree members are recorded. This algorithm has many applications in engineering, and it is called a *breadth-first-search* algorithm.

A graph is called *acyclic* if it has no cycle. A tree is a connected acyclic graph. Any graph without cycles is a *forest*, thus the components of a forest are trees.

The above definitions are illustrated in Fig. 1.21.

It is easy to prove that, for a tree T,

$$M(T) = N(T) - 1, \tag{1.40}$$

where M(T) and N(T) are the numbers of members and nodes of T, respectively.

The complement of T in S is called a *cotree*, denoted by T*. The members of T are known as *branches* and those of T* are called *chords*. For a connected graph S, the number of chords is given by:

$$M(T^*) = M(S) - M(T). \tag{1.41}$$

Since N(T) = N(S), hence,

$$M(T^*) = M(S) - N(S) + 1, \tag{1.42}$$

where M(S) and N(S) are the numbers of members and nodes of S, respectively. Notice that for a set and its cardinality the same notation is used and the difference should be obvious from the context.

1.4.8 *Different Types of Graphs*

In order to simplify the study of the properties of graphs, different types of graphs have been defined. Some important ones are as follows:

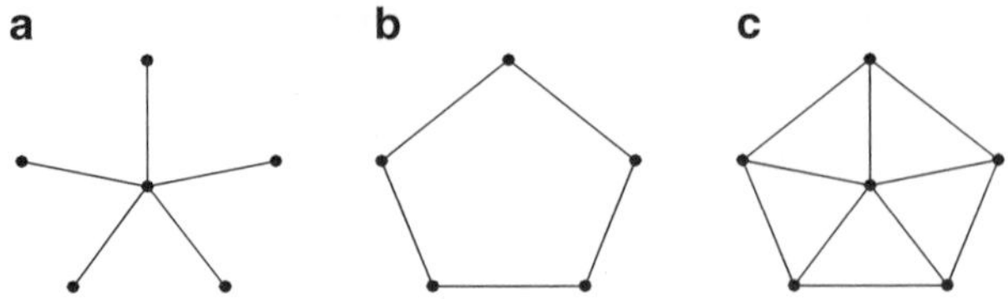

Fig. 1.22 Wheel graph W_6. (**a**) Star graph S_6. (**b**) Cycle graph C_5. (**c**) Wheel graph W_6

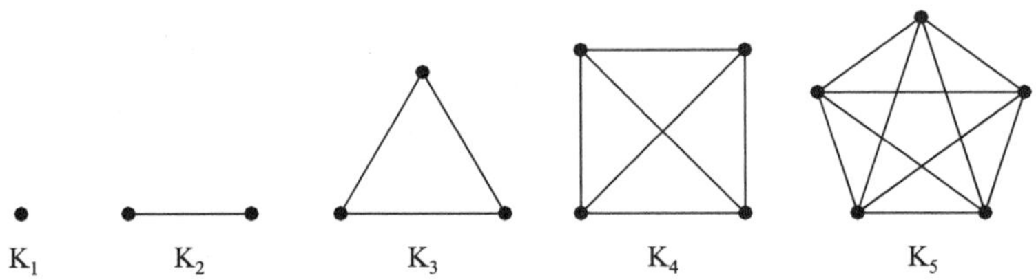

Fig. 1.23 Five complete graphs

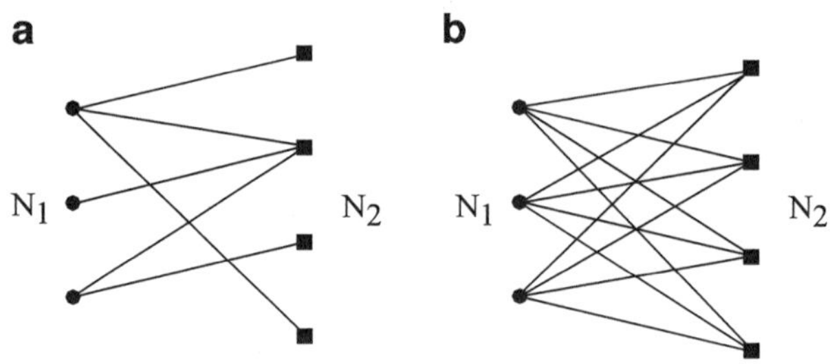

Fig. 1.24 Two bipartite graphs. (**a**) A bipartite graph. (**b**)A complete bipartite graph $K_{3,\ 4}$

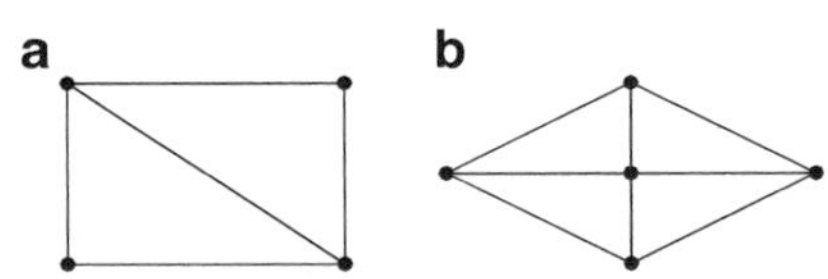

Fig. 1.25 A simple graph and its line graph. (**a**) A graph S. (**b**) The line graph L(S) of S

A *null graph* is a graph that contains no members. Thus, N_k is a graph containing k isolated nodes.

A *cycle graph* is a graph consisting of a single cycle. Therefore, C_k is a polygon with k members.

A *path graph* is a graph consisting of a single path. Hence, P_k is a path with k nodes and (k−1) members.

A *wheel graph* W_k is defined as the union of a star graph with k−1 members and a cycle graph C_{k-1}, connected as shown in Fig. 1.22, for k = 6. Alternatively, a wheel graph W_k can be obtained from the cycle graph C_{k-1} by adding a node O and members (spokes) joining O to each node of C_{k-1}.

A *complete graph* is a graph in which every two distinct nodes are connected by exactly one member, Fig. 1.23. A complete graph with N nodes is denoted by K_N. It is easy to prove that a complete graph with N nodes has N(N−1)/2 members.

A graph is called *bipartite* if the corresponding node set can be split into two sets N_1 and N_2 in such a way that each member of S joins a node of N_1 to a node of N_2. This graph is denoted by B(S) = (N_1, M, N_2). A *complete bipartite* graph is a bipartite graph in which each node N_1 is joined to each node of N_2 by exactly one member. If the numbers of nodes in N_1 and N_2 are denoted by r and s, respectively, then a complete bipartite graph is denoted by $K_{r,s}$. Examples of bipartite and complete bipartite graphs are shown in Fig. 1.24.

A graph S is called *regular* if all of its nodes have the same degree. If this degree is k, then S is *k-regular graph*. As an example, a triangle graph is 2-regular and a cubic graph is 3-regular.

Consider the set M of members of a graph S as a family of 2-node subsets of N (S). The *line graph* L(S) of S has its vertices in a one-to-one correspondence with members of S, and two vertices are connected by an edge if the corresponding members in S are incident. Thus the vertices of L(S) are the members of S, with two vertices of L(S) being adjacent when the corresponding members of S are incident. As an example, the line graph of Fig. 1.25a is illustrated in Fig. 1.25b.

For the original graph S, the terms nodes and members are used, and for the line graph L(S), the terms vertices and edges are employed. In this book, many new graphs are defined and employed for transforming the connectivity properties of the original models to those of the induced new graphs.

1.5 Vector Spaces Associated with a Graph

A vector space can be associated with a graph by defining a vector, the field and the binary operations as follows:

Any subset of the M(S) members of a graph S can be represented by a vector **x** whose M(S) components are elements of the field of integer modulo 2, where component $x_i = 1$ when the *i*th member is an element of the subset, and $x_i = 0$ otherwise. The sum of two subset vectors **x** and **y** is a vector **z** with entries defined by $z_i = x_i + y_i$, representing the symmetric difference of the original subsets. The scalar product of **x** and **y** defined by $\Sigma x_i y_i$ is 0 or 1 according as the original subsets have an even or an odd number of members in common. Although this vector space can be constructed over an arbitrary field, for simplicity the field of integer modulo 2 is considered, in which 1 + 1 = 0.

As an example, consider $\mathbf{x} = \{0, 0, 0, 1, 1, 1, 0\}^t$ and $\mathbf{y} = \{0, 0, 1, 1, 1, 0, 0\}^t$ representing two subgraphs of S. Then, their symmetric difference is obtained as $\mathbf{z} = \{0, 0, 1, 0, 0, 1, 0\}^t$, and the scalar product $\Sigma x_i y_i = 0$ (mod 2), since these subgraphs have two members in common.

Two important subspaces of the above vector space of a graph S are the cycle subspace and cutset subspace, known as the *cycle space* and the *cutset space* of S.

1.5.1 Cycle Space

Let a cycle set of members of a graph be defined as a set of members which form a cycle or form several cycles having no common member, but perhaps common nodes. The null set is also defined as a cycle set. A vector representing a cycle set is called a *cycle set vector*. It can be shown that the sum of two cycle set vectors of a graph is also a cycle set vector. Thus, the cycle set vectors of a graph form a vector space over the field of integer modulo 2. The dimension of a cycle space is given by:

$$\text{nullity (S)} = \nu(S) = b_1(S) = M(S) - N(S) + b_0(S), \tag{1.43}$$

where $b_1(S)$ and $b_0(S)$ are the first and zero Betti numbers of S, respectively. As an example, the nullity of the graph S in Fig. 1.16a is $\nu(S) = 9 - 6 + 1 = 4$.

1.5.2 Cutset Space

Consider a cutset vector similar to that of a cycle vector. Let the null set be also defined as a cutset. It can be shown that the sum of two cutset vectors of a graph is also a cutset vector. Therefore the cutset vectors of a graph form a vector space, the dimension of which is given by:

$$\text{rank (S)} = \rho(S) = N(S) - b_0(S). \tag{1.44}$$

As an example, the rank of S in Fig. 1.16a is $\rho(S) = 6 - 1 = 5$.

1.5.3 Orthogonality Property

Two vectors are called *orthogonal* if their scalar product is zero. It can be shown that a vector is a cycle set (cutset) vector, if and only if it is orthogonal to every vector of a cutset (cycle set) basis. Since the cycle set and cutset spaces of a graph S containing M(S) members are both subspaces of the M(S)-dimensional space of all vectors which represent subsets of the members, therefore the cycle set and cutset spaces are *orthogonal components* of each other.

1.5.4 Fundamental Cycle Bases

A maximal set of independent cycles of a graph is known as its *cycle basis*. The cardinality of a cycle basis is the same as the first Betti number $b_1(S)$. A special basis known as a fundamental cycle basis can easily be constructed corresponding to a tree T of S. In a connected S, a chord of T together with T contains a cycle known as a *fundamental cycle* of S. Moreover, the fundamental cycles obtained by adding the chords to T, one at a time, are independent, because each cycle has a member which is not in the others. Also, every cycle C_i depends on the set of fundamental cycles obtained by the above process, for C_i is the symmetric difference of the cycles determined by the chords of T which lie in C_i. Thus the cycle rank (cyclomatic number, first Betti number, nullity) of graph S, which is the number of cycles in a basis of the cycle space of S, is given by,

$$b_1(S) = M(S) - N(S) + 1, \tag{1.45}$$

and if S contains $b_0(S)$ components, then:

$$b_1(S) = M(S) - N(S) + b_0(S). \tag{1.46}$$

As an example, the selected tree and three fundamental cycles of S are illustrated in Fig. 1.26.

1.5.5 Fundamental Cutset Bases

A basis can be constructed for the cutset space of a graph S. Consider the tree T and its cotree T*. The subgraph of S consisting of T* and any member of T (branch) contains exactly one cutset known as a *fundamental cutset*. The set of cutsets obtained by adding branches of T to T*, one at a time, forms a basis for the cutset space of S, known as a *fundamental cutset basis* of S. The cutset rank (rank of S) is the number of cutsets in a basis for the cutset space of S, and it can be obtained by a similar reasoning to that of the cycle basis as,

$$\rho(S) = N(S) - 1, \tag{1.47}$$

and for a graph with $b_0(S)$ components:

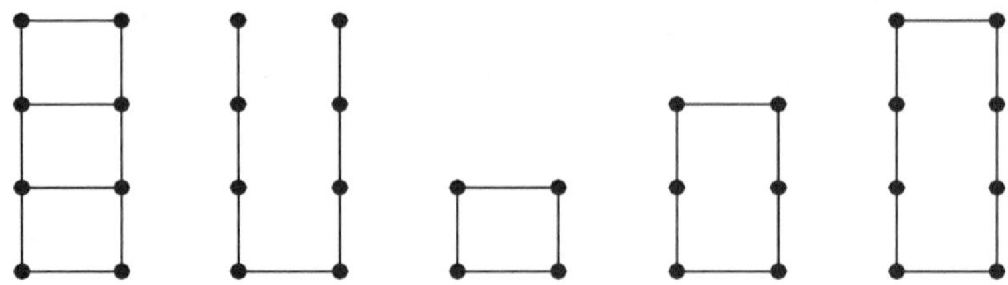

Fig. 1.26 A graph S and a fundamental cycle basis of S

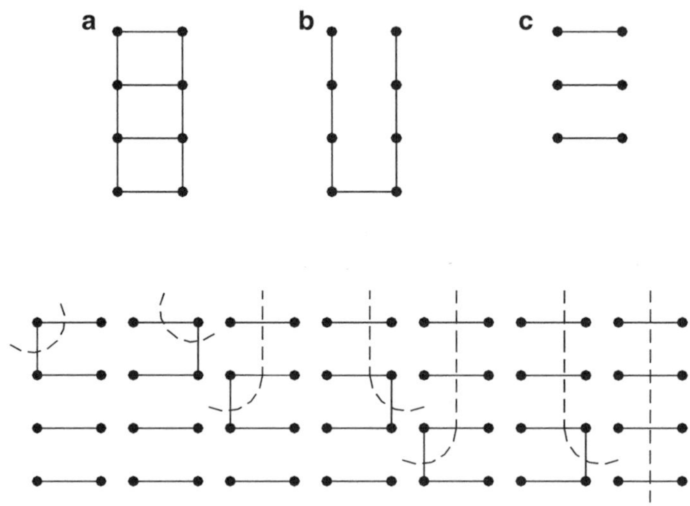

Fig. 1.27 A graph S and a fundamental cutset basis of S. (**a**) A graph S. (**b**) A tree T of S. (**c**) Cotree T* of T

$$\rho(S) = N(S) - b_0(S). \tag{1.48}$$

A graph S and a fundamental cutset basis of S are shown in Fig. 1.27.

1.6 Matrices Associated with a Graph

Matrices play a dominant role in the theory of graphs and especially in applications to structural analysis. Some of these matrices conveniently describe the connectivity properties of a graph and others provide useful information about the patterns of the structural matrices, and some reveal additional information about transformations such as those of equilibrium and compatibility equations.

In this section various matrices are studied which reflect the properties of the corresponding graphs. For simplicity, all graphs are assumed to be connected, since

$$\mathbf{A} = \begin{bmatrix} 0 & 1 & 1 & 1 & 0 \\ 1 & 0 & 1 & 1 & 0 \\ 1 & 1 & 0 & 0 & 1 \\ 1 & 1 & 0 & 0 & 1 \\ 0 & 0 & 1 & 1 & 0 \end{bmatrix}. \quad \text{(A-10)}$$

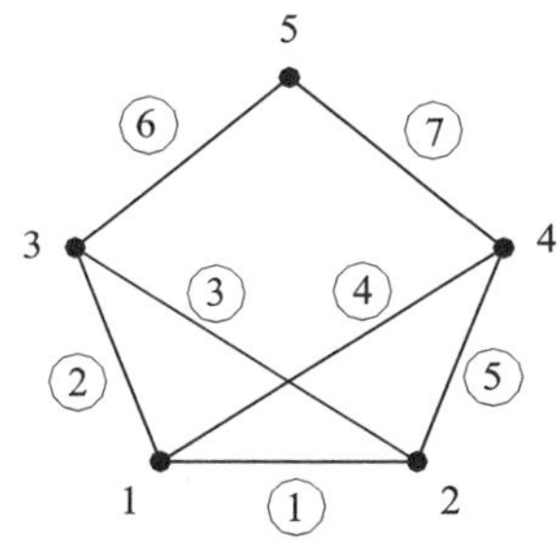

Fig. 1.28 A graph S

the generalisation to non-connected graphs is trivial and consists of considering the direct sum of the matrices for their components.

1.6.1 Matrix Representation of a Graph

A graph can be represented in various forms. Some of these representations are of theoretical importance, others are useful from the programming point of view when applied to realistic problems. In this section six different representations of a graph are described.

Node Adjacency Matrix. Let S be a graph with N nodes. The *adjacency matrix* **A** is an N × N matrix in which the entry in row i and column j is 1 if node n_i is adjacent to n_j, and is 0 otherwise. This matrix is symmetric, and the row sums of **A** are the degrees of the nodes of S.

The adjacency matrix of the graph S, shown in Fig. 1.28, is a 5 × 5 matrix as:

It can be noted that **A** is a symmetric matrix of trace zero. The (i, j)th entry of $\mathbf{A}^2$ shows the number of walks of length 2 with n_i and n_j as end nodes. Similarly, the entry in the (i, j) position of $\mathbf{A}^k$ is equal to the number of walks of length k with n_i and n_j as end nodes. The polynomial,

$$f(\lambda) = \det(\mathbf{I}\lambda - \mathbf{A}), \tag{1.49}$$

is called the *characteristic polynomial* of S. The collection of N(S) eigenvalues of **A** is known as the *spectrum* of S. Since **A** is symmetric, the spectrum of S consists of N(S) real numbers. The sum of eigenvalues of **A** is equal to zero.

Node-Member Incidence Matrix. Let S be a graph with M members and N nodes. The *node-member incidence matrix* $\overline{\mathbf{B}}$ is an N × M matrix in which the entry in row i and column j is 1 if node n_i is incident with member m_j, and is 0 otherwise. As an example, the node-member incidence matrix of the graph in Fig. 1.28 is a 5 × 7 matrix of the form:

$$\overline{\mathbf{B}} = \begin{bmatrix} 1 & 1 & 0 & 1 & 0 & 0 & 0 \\ 1 & 0 & 1 & 0 & 1 & 0 & 0 \\ 0 & 1 & 1 & 0 & 0 & 1 & 0 \\ 0 & 0 & 0 & 1 & 1 & 0 & 1 \\ 0 & 0 & 0 & 0 & 0 & 1 & 1 \end{bmatrix}. \tag{1.50}$$

Obviously, the pattern of an incidence matrix depends on the particular way in which its nodes and members are labelled. One incidence matrix can be obtained from another by simply interchanging rows (corresponding to re-labelling the nodes) and columns (corresponding to re-labelling the members).

The incidence matrix $\overline{\mathbf{B}}$ and the adjacency matrix $\mathbf{A}$ of a graph S are related by,

$$\overline{\mathbf{B}}\overline{\mathbf{B}}^{t} = \mathbf{A} + \mathbf{V}, \tag{1.51}$$

where $\mathbf{V}$ is a diagonal matrix of order N(S) whose typical entry v_i is the valency of the node n_i of S for $i = 1, \ldots, N(S)$. For the example of Fig. 1.28, Eq. 1.51 becomes:

$$\overline{\mathbf{B}}\overline{\mathbf{B}}^{t} = \begin{bmatrix} 0 & 1 & 1 & 1 & 0 \\ 1 & 0 & 1 & 1 & 0 \\ 1 & 1 & 0 & 0 & 1 \\ 1 & 1 & 0 & 0 & 1 \\ 0 & 0 & 1 & 1 & 0 \end{bmatrix} + \begin{bmatrix} 3 & & & & \\ & 3 & & & \\ & & 3 & & \\ & & & 3 & \\ & & & & 2 \end{bmatrix}. \tag{1.52}$$

The rows of $\overline{\mathbf{B}}$ are dependent, and one row can arbitrarily be deleted to ensure the independence of the rest of the rows. The node corresponding to the deleted row is called a *datum (reference) node*. The matrix obtained after deleting a dependent row is called an *incidence matrix* of S, and is denoted by $\mathbf{B}$.

Although $\mathbf{A}$ and $\mathbf{B}$ are of great theoretical value, the storage requirements for these matrices are high and proportional to $N \times N$ and $M \times N$, respectively. In fact, a large number of unnecessary zeros is stored in these matrices. In practice, one can use different approaches to reduce the storage required, some of which are described in the following.

Member List: This type of representation is a common approach in structural mechanics. A member list consists of two rows (or columns) and M columns (or rows). Each column (or row) contains the labels of the two end nodes of each member, in which members are arranged sequentially. For example, the member list of S in Fig. 1.28 is:

$$\mathbf{ML} = \begin{array}{c} \begin{matrix} m_1 & m_2 & m_3 & m_4 & m_5 & m_6 & m_7 \end{matrix} \\ \begin{bmatrix} 1 & 1 & 2 & 1 & 2 & 3 & 4 \\ 2 & 3 & 3 & 4 & 4 & 5 & 5 \end{bmatrix} \end{array}. \tag{1.53}$$

It should be noted that a member list can also represent orientations on members. The storage required for this representation is 2 × M. Some engineers prefer to add a third row containing the member's labels, for easy addressing. In this case, the storage is increased to 3 × M.

A different way of preparing a member list is to use a vector containing the end nodes of members sequentially; e.g. for the previous example this vector becomes:

$$(1,2;1,3;2,3;1,4;2,4;3,5;4,5). \tag{1.54}$$

This is a compact description of a graph; however, it is impractical because of the extra search required for its use in various algorithms.

Adjacency List. This list consists of N rows and D columns, where D is the maximum degree of the nodes of S. The *i*th row contains the labels of the nodes adjacent to node i of S. For the graph S shown in Fig. 1.28, the adjacency list is:

$$\mathbf{AL} = \begin{matrix} n_1 \\ n_2 \\ n_3 \\ n_4 \\ n_5 \end{matrix} \begin{bmatrix} 2 & 3 & 4 \\ 1 & 3 & 4 \\ 1 & 2 & 5 \\ 1 & 2 & 5 \\ 3 & 4 & \end{bmatrix}_{N \times D} \tag{1.55}$$

The storage needed for an adjacency list is N × D.

Compact Adjacency List. In this list, the rows of **AL** are continually arranged in a row vector **R**, and an additional vector of pointers **P** is considered. For example, the compact adjacency list of Fig. 1.28 can be written as:

$$\begin{aligned} \mathbf{R} &= (2,3,4,1,3,4,1,2,5,1,2,5,3,4), \\ \mathbf{P} &= (1,4,7,10,13,15). \end{aligned} \tag{1.56}$$

P is a vector (p_1, p_2, p_3, ...) which helps to list the nodes adjacent to each node. For node n_i, one should start reading **R** at entry p_i and finish at $p_{i+1}-1$.

An additional restriction can be put on **R**, by ordering the nodes adjacent to each node n_i in ascending order of their degrees. This ordering can be of some advantage, an example of which is nodal ordering for bandwidth optimisation. The storage required for this list is 2M + N + 1.

1.6.2 Cycle Bases Matrices

The cycle-member incidence matrix $\overline{\mathbf{C}}$ of a graph S, has a row for each cycle or hinged cycle and a column for each member. An entry c_{ij} of $\overline{\mathbf{C}}$ is 1 if cycle C_i contains member m_j, and it is 0 otherwise. In contrast to the node adjacency and node-member incidence matrices, the cycle-member incidence matrix does not determine a graph up to isomorphism; i.e. two totally different graphs may have the same cycle-member incidence matrix.

For a graph S, there exist $2^{b_1(S)} - 1$ cycles or hinged cycles. Thus $\overline{\mathbf{C}}$ is a $(2^{b_1(S)} - 1) \times M$ matrix. However, one does not need all the cycles of S, and the elements of a cycle basis are sufficient. For a cycle basis, a cycle-member incidence matrix becomes a $b_1(S) \times M$ matrix, denoted by $\mathbf{C}$, known as the *cycle basis incidence matrix* of S. As an example, matrix $\mathbf{C}$ for the graph shown in Fig. 1.28, for the following cycle basis,

$$C_1 = (m_1,\ m_2,\ m_3)$$
$$C_2 = (m_1,\ m_4,\ m_5)$$
$$C_3 = (m_2,\ m_4,\ m_6,\ m_7)$$

is given by:

$$\mathbf{C} = \begin{matrix} C_1 \\ C_2 \\ C_3 \end{matrix} \begin{bmatrix} 1 & 1 & 1 & 0 & 0 & 0 & 0 \\ 1 & 0 & 0 & 1 & 1 & 0 & 0 \\ 0 & 1 & 0 & 1 & 0 & 1 & 1 \end{bmatrix}. \tag{1.57}$$

The *cycle adjacency matrix* $\mathbf{D}$ is a $b_1(S) \times b_1(S)$ matrix, each entry d_{ij} of which is 1 if C_i and C_j have at least one member in common and it is 0 otherwise. This matrix is related to the cycle-member incidence matrix by the following relationship,

$$\mathbf{CC}^t = \mathbf{D} + \mathbf{W}, \tag{1.58}$$

where $\mathbf{W}$ is a diagonal matrix with w_{ii} being the length of the *i*th cycle, and its trace being equal to the total length of the cycles of the basis.

For the above example:

$$\mathbf{CC}^t = \begin{bmatrix} 0 & 1 & 1 \\ 1 & 0 & 1 \\ 1 & 1 & 0 \end{bmatrix} + \begin{bmatrix} 3 & 0 & 0 \\ 0 & 3 & 0 \\ 0 & 0 & 3 \end{bmatrix}. \tag{1.59}$$

An important theorem can now be stated which is based on the orthogonality property studied in Sect. 1.5.3.

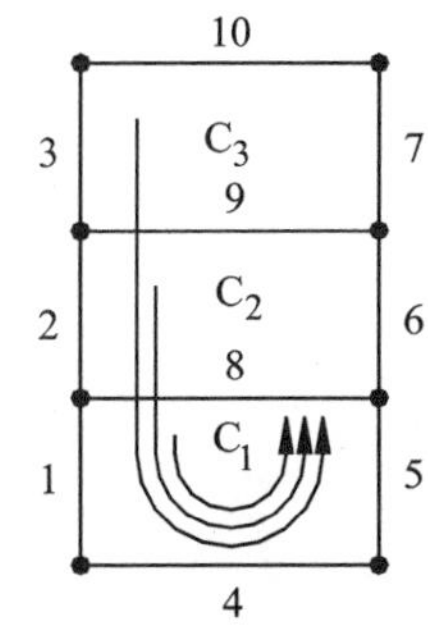

Fig. 1.29 A graph with oriented members and cycles

Theorem. Let S have an incidence matrix **B** and a cycle basis incidence matrix **C**. Then:

$$\mathbf{C}\mathbf{B}^t = \mathbf{0}(\mathrm{mod}\ 2). \tag{1.60}$$

A simple proof of this theorem can be found in Kaveh [10]. Notice that Eq. 1.60 holds due to the orthogonality property discussed in Sect. 1.5.3. In fact, the above relation holds even if the cutsets or cycles do not form bases, or the matrices contain additional cutsets and/or cycle vectors.

1.6.3 Special Patterns for Fundamental Cycle Bases

Matrix C for a fundamental cycle basis, with special labels for its tree members and chords, finds a particular pattern. Let S have a tree T whose members are M (T) = $(m_1, m_2, \ldots, m_p)$ and a cotree for which $M(T^*) = (m_{p+1}, m_{p+2}, \ldots, m_{M(S)})$. Then there is a unique fundamental cycle C_i in $S - M(T^*) + m_i$, $p + 1 \le i \le M(S)$, and this set of cycles forms a basis for the cycle space of S. As an example, for the graph S of Fig. 1.27a whose members are labelled as shown in Fig. 1.29, the fundamental cycle basis consists of,

$$C_1 = (m_1,\ m_4,\ m_5,\ m_8),$$
$$C_2 = (m_2,\ m_1,\ m_4,\ m_5,\ m_6,\ m_9),$$
$$C_3 = (m_3,\ m_2,\ m_1,\ m_4,\ m_5,\ m_6,\ m_7,\ m_{10}),$$

given by:

$$\mathbf{C} = \begin{matrix} C_1 \\ C_2 \\ C_3 \end{matrix}\left[\begin{array}{ccccccc|ccc} 1 & 0 & 0 & 1 & 1 & 0 & 0 & 1 & 0 & 0 \\ 1 & 1 & 0 & 1 & 1 & 1 & 0 & 0 & 1 & 0 \\ 1 & 1 & 1 & 1 & 1 & 1 & 1 & 0 & 0 & 1 \end{array}\right] = [\mathbf{C}_T|\mathbf{I}]. \tag{1.61}$$
$$\qquad\qquad M(T) \qquad\qquad\qquad M(T^*)$$

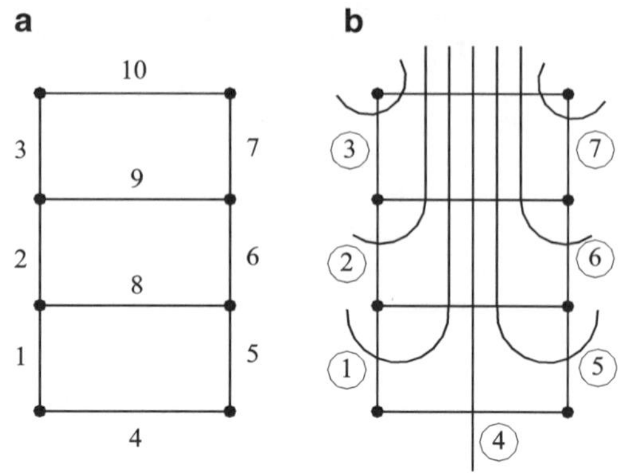

Fig. 1.30 A graph with oriented members and a cutset basis

1.6.4 Cutset Bases Matrices

The *cutset-member incidence matrix* $\overline{\mathbf{C}}^*$ for a graph S, has a row for each cutset of S and a column for each member. An entry $\overline{\mathbf{c}}^*_{ij}$ of $\overline{\mathbf{C}}^*$ is 1 if cutset $\mathbf{C}^*_i$ contains member m_j, and it is 0 otherwise. This matrix, like $\overline{\mathbf{C}}$, does not determine a graph completely.

Independent rows of $\overline{\mathbf{C}}^*$ for a cutset basis, denoted by **C***, form a matrix known as a *cutset basis incidence matrix*, which is a η(S) × M matrix, η(S) being the rank of graph S. As an example, **C*** for the cutsets of Fig. 1.27 with members labelled as in Fig. 1.30a, is given below:

$$\mathbf{C}^* = \begin{bmatrix} 0 & 0 & 1 & 0 & 0 & 0 & 0 & 0 & 0 & 1 \\ 0 & 0 & 0 & 0 & 0 & 0 & 1 & 0 & 0 & 1 \\ 0 & 1 & 0 & 0 & 0 & 0 & 0 & 0 & 1 & 1 \\ 0 & 0 & 0 & 0 & 0 & 1 & 0 & 0 & 1 & 1 \\ 0 & 0 & 0 & 1 & 0 & 0 & 0 & 1 & 1 & 1 \\ 1 & 0 & 0 & 0 & 0 & 0 & 0 & 1 & 1 & 1 \\ 0 & 0 & 0 & 0 & 1 & 0 & 0 & 1 & 1 & 1 \end{bmatrix}. \tag{1.62}$$

The *cutset adjacency matrix* **D*** is a η(S) × η(S) matrix defined analogously to cycle adjacency matrix **D**.

1.6.5 Special Patterns for Fundamental Cutset Bases

For a fundamental cutset basis with appropriate labelling of the members in T and T*, as illustrated in Fig. 1.30b, if the cutsets are taken in the order of their generators (tree members), the matrix **C*** will have a particular pattern as:

$$\mathbf{C}_0^* = \left[\begin{array}{ccccccc|ccc} 1 & 0 & 0 & 0 & 0 & 0 & 0 & 1 & 1 & 1 \\ 0 & 1 & 0 & 0 & 0 & 0 & 0 & 0 & 1 & 1 \\ 0 & 0 & 1 & 0 & 0 & 0 & 0 & 0 & 0 & 1 \\ 0 & 0 & 0 & 1 & 0 & 0 & 0 & 1 & 1 & 1 \\ 0 & 0 & 0 & 0 & 1 & 0 & 0 & 1 & 1 & 1 \\ 0 & 0 & 0 & 0 & 0 & 1 & 0 & 0 & 1 & 1 \\ 0 & 0 & 0 & 0 & 0 & 0 & 1 & 0 & 0 & 1 \end{array}\right] = \left[\mathbf{I}\middle|\mathbf{C}_c^*\right]. \tag{1.63}$$

From the orthogonality condition, $\mathbf{C}_0\mathbf{C}_0^{*t} = \mathbf{0}$; i.e.

$$[\mathbf{C}_T \quad \mathbf{I}]\begin{bmatrix} \mathbf{I} \\ \mathbf{C}_c^{*t} \end{bmatrix} = \mathbf{0}. \tag{1.64}$$

$$\text{Hence } \mathbf{C}_T + \mathbf{C}_c^{*t} = \mathbf{0}(\text{mod } 2), \text{ and :}$$

$$\mathbf{C}_T = \mathbf{C}_c^{*t}. \tag{1.65}$$

Therefore, for a graph having $\mathbf{C}_0$, one can construct $\mathbf{C}_0^*$ and vice versa.

There exists a very simple basis for the cutset space of a graph which consists of N−1 cocycles of S. As an example, for the graph of Fig. 1.28, considering n_5 as a datum node, we have,

$$\mathbf{C}^* = \begin{bmatrix} 1 & 1 & 0 & 1 & 0 & 0 & 0 \\ 1 & 0 & 1 & 0 & 1 & 0 & 0 \\ 0 & 1 & 1 & 0 & 0 & 1 & 0 \\ 0 & 0 & 0 & 1 & 1 & 0 & 1 \end{bmatrix}, \tag{1.66}$$

which is the same as the incidence matrix **B** of S. The simplicity of the displacement method of structural analysis is due to the existence of such a simple basis.

1.7 Directed Graphs and Their Matrices

An *oriented* or *directed* graph is a graph in which each member is assigned an orientation. A member is oriented from its initial node (*tail*) to its final node (*head*). The initial node is said to be positively incident on the member, and the final node negatively incident, as shown in Fig. 1.31a.

The choice of orientation of members of a graph is arbitrary; however, once it is chosen, it must be retained. Cycles and cutsets can also be oriented as shown in Fig. 1.31b.

As an example, m_4 is positively oriented in cycle C_i, and m_7 is negatively oriented in cutset $\mathbf{C}_i^*$.

All the matrices $\overline{\mathbf{B}}$, **B**, **C** and **C*** can be defined as before, with the difference of having +1, −1 and 0 as entries, according to whether the member is positively, negatively or zero incident with a cutset or a cycle.

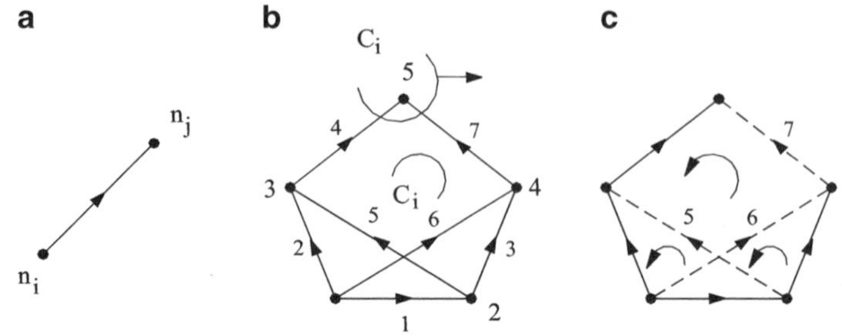

Fig. 1.31 An oriented member, a directed graph, and a directed tree (with chords shown in *dashed lines*)

As an example, for graph S in Fig. 1.31b, the matrix **B** with n_1 as the datum node is formed:

$$\mathbf{B} = \begin{matrix} n_1 \\ n_2 \\ n_3 \\ n_4 \end{matrix} \begin{bmatrix} -1 & 0 & 1 & 0 & 1 & 0 & 0 \\ 0 & -1 & 0 & 0 & -1 & 0 & 0 \\ 0 & 0 & -1 & 0 & 0 & -1 & 1 \\ 0 & 0 & 0 & -1 & 0 & 0 & -1 \end{bmatrix}. \tag{1.67}$$

Consider a tree as shown in continuous lines, Fig. 1.31c. When the directions of the cycles are taken as those of their corresponding chords (dashed lines), the fundamental cycle basis incidence matrix can be written as:

$$\mathbf{C} = \begin{matrix} C_1 \\ C_2 \\ C_3 \end{matrix} \left[\begin{array}{cccc|ccc} 1 & -1 & 0 & 0 & 1 & 0 & 0 \\ 1 & 0 & 1 & 0 & 0 & 1 & 0 \\ 1 & -1 & 1 & -1 & 0 & 0 & 1 \end{array}\right]. \tag{1.68}$$

It should be noted that the tree members are numbered first, followed by the chords of the cycles in the same sequence as their generation.

Obviously,

$$\mathbf{BC}^t = \mathbf{CB}^t = \mathbf{0}(\text{mod } 2), \tag{1.69}$$

with a proof similar to that of the non-oriented case.

A cuset basis incidence matrix is similarly obtained as:

$$\mathbf{C}^* = \left[\begin{array}{cccc|ccc} 1 & 0 & 0 & 0 & -1 & 1 & -1 \\ 0 & 1 & 0 & 0 & 1 & 0 & 1 \\ 0 & 0 & 1 & 0 & 0 & 1 & -1 \\ 0 & 0 & 0 & 1 & 0 & 0 & 1 \end{array}\right], \qquad \begin{matrix} \mathbf{C}_T^* & \mathbf{C}_c^* \end{matrix} \tag{1.70}$$

where the direction of a cutset is taken as the orientation of its generator (the corresponding tree member).

It can easily be proven that:

$$\mathbf{C}_{\mathrm{T}} = -\mathbf{C}_{\mathrm{c}}^{*\mathrm{t}}. \tag{1.71}$$

For a directed graph, Eq. 1.51 becomes:

$$\mathbf{B}\mathbf{B}^{\mathrm{t}} = \mathbf{A} - \mathbf{V}, \tag{1.72}$$

Similarly, Eq. 1.59 for the directed case becomes:

$$\mathbf{C}\mathbf{C}^{\mathrm{t}} = \mathbf{D} - \mathbf{W}. \tag{1.73}$$

References

1. Kaveh A (2004) Structural mechanics: graph and matrix methods, 3rd edn. Research Studies Press, Baldock, Herdfordshire, UK
2. Kaveh A (1979) Optimal structural analysis, 1st edn. Research Studies Press, Chicherster, Herdfordshire, UK
3. Castigliano A (1879) Théorie de l'équilibre des cutset élastiques et ses applications. AF Negro, Turin
4. Argyris JH, Kelsey S (1960) Energy theorems and structural analysis. Butterworth, London
5. Timoshenko S, Young DH (1945) Theory of structures. McGraw-Hill, New York
6. Kaveh A, Roosta GR (1998) Comparative study of finite element nodal ordering methods. Eng Struct 20(1–2):86–96
7. Harary F (1969) Graph theory. Addison-Wesley, Reading
8. Berge C (1973) Graphs and hypergraphs. North-Holland Publishing, Amsterdam
9. West DB (1996) Introduction to graph theory. Prentice-Hall, Upper Saddle River, NJ, USA
10. Kaveh A (1974) Application of topology and matroid theory to the analysis of structures. Ph.D. thesis, Imperial College, London University

Chapter 2
Optimal Force Method: Analysis of Skeletal Structures

2.1 Introduction

This chapter starts with presenting simple and general methods for calculating the degree of static indeterminacy of different types of skeletal structures, such as rigid-jointed planar and space frames, pin-jointed planar trusses and ball-jointed space trusses.

Then the progress made in the force method of structural analysis in recent years is presented, and the state of art is summarized. Efficient methods are developed leading to highly sparse flexibility matrices. The methods are mainly developed for frame structures, however, extensions are made to general skeletal structures.

The force method of structural analysis, in which the member forces are used as unknowns, is appealing to engineers, since the properties of members of a structure most often depend on the member forces rather than joint displacements. This method was used extensively until 1960. After this, the advent of the digital computer and the amenability of the displacement method for computation attracted most researchers. As a result, the force method and some of the advantages it offers in optimisation and non-linear analysis, have been neglected.

Six different approaches are adopted for the force method of structural analysis, which will be classified as follows:

1. Topological force methods,
2. Combinatorial force methods,
3. Algebraic force methods,
4. Mixed algebraic-combinatorial force methods,
5. Integrated force method, and
6. Metaheuristic based methods.

Topological methods have been developed by Henderson [1], Maunder [2] and Kaveh [3]. Combinatorial force method is mainly developed by Kaveh [3] using different graph theoretical algorithms. Algebraic topology is employed extensively in the work of Langefors [4]. Algebraic methods have been developed by Denke [5],

A. Kaveh, *Computational Structural Analysis and Finite Element Methods*,

DOI 10.1007/978-3-319-02964-1_2,

Robinson [6], Topçu [7], and Kaneko et al. [8], and mixed algebraic-topological methods have been used by Gilbert et al. [9], Coleman and Pothen [10]. The integrated force method has been developed by Patnaik [11]. Meta-heuristic based methods are also developed for the formation of null basis in the work of Kaveh and Jahamshahi [12] and Kaveh and Daei [13].

2.2 Static Indeterminacy of Structures

Skeletal structures are the most common type of structures encountered in civil engineering practice. These structures sustain the applied loads mainly by virtue of their topology, i.e. the way members are connected to each other (connectivity). Therefore, topology plays a vital role in their design. The first step in design of such structures is to provide sufficient rigidity and make it reliable, but this depends in part on the degrees of static indeterminacy of the structures. One way to calculate the degree of static indeterminacy is to use classical formulae such as those given in Timoshenko and Young [14]; however, the application of these usually provides only a small part of the necessary topological properties. The methods presented in this chapter provide powerful means for understanding the distribution of the indeterminacy within a structure. The concepts presented are efficient in both the optimal force method of structural analysis, as will be discussed in the second part of this chapter.

In the analysis of skeletal structures, three different properties are encountered, which are classified as topological, geometrical and material. Separate study of these properties results in a considerable simplification in understanding the structural behaviour leading to methods for efficient analysis. This chapter is confined to a study of those topological properties of skeletal structures needed to study force and displacement methods. The number of equations to be solved in the two methods may differ widely for the same structure. This number depends on the size of the flexibility and the stiffness matrices. The orders of this matrix are the same as the degree of static indeterminacy and the degree of kinematic indeterminacy of a structure, respectively. Obviously, the method that leads to the required results with the least amount of computational time and storage should be used for the analysis of a given structure. Thus, the comparison of the degree of static indeterminacy and the degree of kinematic indeterminacy may be the main criterion for selecting the method of analysis.

The degree of kinematic indeterminacy of a structure, also known as its total number of degrees of freedom, can easily be obtained by summing up the degrees of freedom of its nodes. A node of planar and space trusses has two and three degrees of freedom, respectively. For planar and space frames, these numbers are 3 and 6, respectively. Single-layer grids have also three degrees of freedom for each node.

For determining the degree of static indeterminacy of structures, numerous formulae depending on the kinds of members or types of joints have been given, e.g. Ref. [14]. The use of these classical formulae, in general, requires counting the number of members and joints, which becomes a tedious process for multi-member and/or complex pattern structures; moreover, the count provides no additional information about connectivity.

Henderson and Bickley [1] related the degree of static indeterminacy of a rigid-jointed frame to the first Betti number of its graph model S. Generalising the Betti's number to a linear function and using an expansion process, Kaveh [15] developed a general method for determining the degree of static indeterminacy and degree of kinematic indeterminacy of different types of skeletal structures. Special methods have also been developed to transform the topological properties of space structures to those of their planar drawings, in order to simplify the calculation of their degrees of static indeterminacy, Ref. [16].

It should be noted that various methods for determining the degree of static indeterminacy of structures are a by-product of the general methods developed by Kaveh [15]. The method of expansion and its control at each step, using the intersection theorem presented in this chapter, provides a powerful tool for further studies in the field of structural analysis.

2.2.1 Mathematical Model of a Skeletal Structure

The mathematical model of a structure is considered to be a finite, connected graph S. There is a one-to-one correspondence between the elements of the structure and the members (edges) of S. There is also a one-to-one correspondence between the joints of the structure and the nodes of S, except for the support joints of some models.

For frame structures, shown in Fig. 2.1(a1) and (a2), two graph models can be considered. For the first model, all the support joints are identified as a single node called a *ground node*, as shown in Fig. 2.1(b1) and (b2). For the second model, all the joints are connected by an artificial arbitrary spanning tree, termed *ground tree*, Fig. 2.1(c1) and (c2).

Truss structures shown in Fig. 2.2(a1) and (a2) are assumed to be supported in a statically determinate fashion (Fig. 2.2(b1) and (b2)), and the effect of additional supports can easily be included in calculating the degree of static indeterminacy (DSI) of the corresponding structures. Alternatively artificial members can be added as shown in Fig. 2.2(c1) and (c2) to model the components of the corresponding supports. For a fixed support, two members and three members are considered for planar and space trusses, respectively, and one member is used for representing a roller.

The skeletal structures are considered to be in perfect condition; i.e. planar and space trusses have pin and ball joints only. Obviously, the effect of extra constraints or releases can be taken into account in determining their degrees of static indeterminacy and also in their analysis, Mauch and Fenves [17].

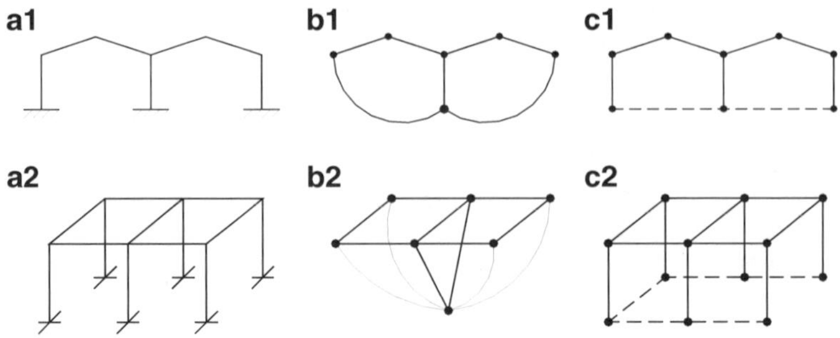

Fig. 2.1 Frame structures and their mathematical models. (**a1**) A plane frame. (**b1**) First model with a ground node. (**c1**) Second model with a ground tree. (**a2**) A space frame. (**b2**) First model with a ground node. (**c2**) Second model with a ground tree

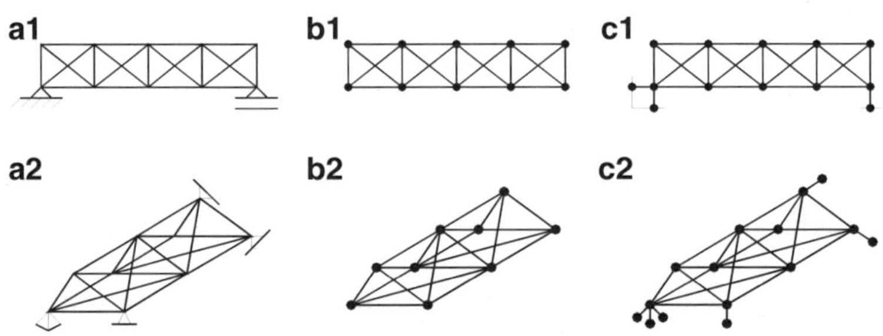

Fig. 2.2 Trusses and their graph models. (**a1**) A plane truss. (**b1**) First model without added members. (**c1**) Second model with replaced members. (**a2**) A space truss. (**b2**) First model without added members. (**c2**) Second model with replaced members

2.2.2 Expansion Process for Determining the Degree of Static Indeterminacy

The *degree of kinematic indeterminacy* of a structure is the number of independent displacement components (translations and rotations) required for describing a general state of deformation of the structure. The degree of kinematic indeterminacy is also referred to as the *total degrees of freedom* of the structure. On the other hand, the *degree of static indeterminacy* (redundancy) of a structure is the number of independent force components (forces and moments) required for describing a general equilibrium state of the structure. The DSI of a structure can be obtained by subtracting the number of independent equilibrium equations from the number of its unknown forces.

2.2.2.1 Classical Formulae

Formulae for calculating the DSI of various skeletal structures can be found in textbooks on structural mechanics, e.g. the DSI of a planar truss, denoted by $\gamma(S)$, can be calculated from,

$$\gamma(S) = M(S) - 2N(S) + 3, \tag{2.1}$$

where S is supported in a statically determinate fashion (internal indeterminacy). For extra supports (external indeterminacy), $\gamma(S)$ should be further increased by the number of additional unknown reactions.

A similar formula holds for space trusses:

$$\gamma(S) = M(S) - 3N(S) + 6. \tag{2.2}$$

For planar and space frames, the classical formulae is given as,

$$\gamma(S) = \alpha[M(S) - N(S) + 1], \tag{2.3}$$

where all supports are modelled as a datum (ground) node, and $\alpha = 3$ or 6 for planar and space frames, respectively.

All these formulae require counting a great number of members and nodes, which makes their application impractical for multi-member and complex pattern structures. These numbers provide only a limited amount of information about the connectivity properties of structures. In order to obtain additional information, the methods developed in the following sections will be utilised:

2.2.2.2 A Unifying Function

All the existing formulae for determining DSI have a common property, namely their linearity with respect to M(S) and N(S). Therefore, a general unifying function can be defined as,

$$\gamma(S) = aM(S) + bN(S) + c\gamma_0(S), \tag{2.4}$$

where M(S), N(S) and $\gamma_0(S)$ are the numbers of members, nodes and components of S, respectively. The coefficients a, b and c are integer numbers depending on both the type of the corresponding structure and the property which the function is expected to represent. For example, $\gamma(S)$ with appropriate values for a, b and c may describe the DSI of certain types of skeletal structures, Table 2.1. For $a = 1$, $b = -1$ and $c = 1$, $\gamma(S)$ becomes the first Betti number $b_1(S)$ of S, as described in Sect. 1.5.1.

Table 2.1 Coefficients of $\gamma(S)$ for different types of structures

Type of structure	a	b	c
Plane truss	+1	−2	+3
Space truss	+1	−3	+6
Plane frame	+3	−3	+3
Space frame	+6	−6	+6

2.2.2.3 An Expansion Process

An expansion process, in its simplest form, has been used by Müller-Breslau [18] for re-forming structural models, such as simple planar and space trusses. In his expansion process, the properties of typical subgraphs, selected in each step to be joined to the previously expanded subgraph, guarantee the determinacy of the simple truss. These subgraphs consist of two and three concurrent bars for planar and space trusses, respectively.

The idea can be extended to other types of structure, and more general subgraphs can be considered for addition at each step of the expansion process. A cycle, a planar subgraph, and a subgraph with prescribed connectivity properties are examples of these, which will be employed in this book. For example, the planar truss of Fig. 2.3a can be formed in four steps, joining a substructure S_i with $\gamma(S_i) = 1$ as shown in Fig. 2.3b, sequentially, as illustrated in Fig. 2.3c.

2.2.2.4 An Intersection Theorem

In a general expansion process, a subgraph S_i may be joined to another subgraph S_j in an arbitrary manner. For example, $\gamma(S_i)$ or $\gamma(S_j)$ may have any arbitrary value and the union $S_i \cup S_j$ may be a connected or a disjoint subgraph. The intersection $S_i \cup S_j$ may also be connected or disjoint. It is important to find the properties of $S_1 \cup S_2$ having the properties of S_1, S_2 and $S_1 \cap S_2$. The following theorem provides a correct calculation of the properties of $S_i \cup S_j$. In order to have the formula in its general form, q subgraphs are considered in place of two subgraphs.

Theorem (Kaveh [15]). Let S be the union of q subgraphs $S_1, S_2, S_3, \ldots, S_q$ with the following functions being defined:

$$\begin{aligned}
\gamma(S) &= aM(S) + bN(S) + c\gamma_0(S), \\
\gamma(S_i) &= aM(S_i) + bN(S_i) + c\gamma_0(S_i) \quad i = 1, 2, \ldots, q, \\
\gamma(A_i) &= aM(A_i) + bN(A_i) + c\gamma_0(A_i) \quad i = 2, 3, \ldots, q,
\end{aligned}$$

where $A_i = S^{i-1} \cap S_i$ and $S^i = S_1 \cup S_2 \cup \ldots \cup S_i$. Then:

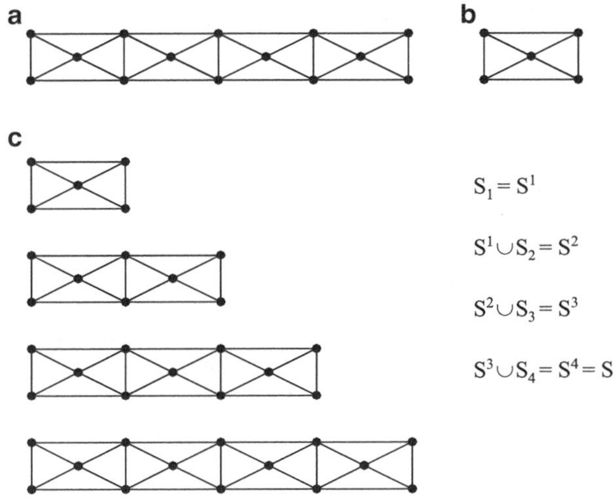

Fig. 2.3 Process for the formation of a planar truss. (**a**) A planar truss. (**b**) Selected unit. (**c**) The process of expansion as $S_1 = S^1 \rightarrow S^2 \rightarrow S^3 \rightarrow S^4 = S$

$$[\gamma(S) - c\gamma_0(S)] = \sum_{i=1}^{q} [\gamma(S_i) - c\gamma_0(S_i)] - \sum_{i=2}^{q} [\gamma(A_i) - c\gamma_0(A_i)] \tag{2.5}$$

For proof, the interested reader may refer to Kaveh [19].

Special Case. If S and each of its subgraphs considered for expansion (S_i for $i = 1, \ldots, q$) are non-disjoint (connected), then Eq. 2.5 can be simplified as:

$$\gamma(S) = \sum_{i=1}^{q} \gamma(S_i) - \sum_{i=2}^{q} \overline{\gamma}(A_i), \tag{2.6}$$

where $\overline{\gamma}(A_i) = aM(A_i) + bN(A_i) + c$.

For calculating the DSI of a multi-member structure, one normally selects a repeated unit of the structure and joins these units sequentially in a connected form. Therefore, Eq. 2.6 can be applied in place of Eq. 2.5 to obtain the overall property of the structure.

2.2.2.5 A Method for Determining the DSI of Structures

Let S be the union of its repeated and/or simple pattern subgraphs S_i ($i = 1, \ldots, q$). Calculate the DSI of each subgraph, using the appropriate coefficients from Table 2.1. Now perform the union–intersection method with the following steps:

Step 1: Join S_1 to S_2 to form $S^2 = S_1 \cup S_2$, and calculate the DSI of their intersection $A_2 = S_1 \cap S_2$. The value of $\gamma(S^2)$ can be found using Eq. 2.5 or Eq. 2.6, as appropriate.
Step 2: Join S_3 to S^2 to obtain $S^3 = S^2 \cup S_3$, and determine the DSI of $A_3 = S^2 \cap S_3$. Similarly to Step 1, calculate $\gamma(S^3)$.
Step k: Subsequently join S_{k+1} to S^k, calculating the DSI of $A_{k+1} = S^k \cap S_{k+1}$ and evaluating the magnitude of $\gamma(S^{k+1})$.

Repeat Step k until the entire structural model $S = \bigcup_{i=1}^{q} S_i$ has been reformed and its DSI determined.

In the above expansion process, the value of q depends on the properties of the substructures (subgraphs) which are considered for reforming S. These subgraphs have either simple patterns for which $\gamma(S_i)$ can easily be calculated, or the DSIs of which are already known.

In the process of expansion, if an intersection A_i itself has a complex pattern, further refinement is also possible; i.e. the intersection can be considered as the union of simpler subgraphs.

2.3 Formulation of the Force Method

In this section, a matrix formulation using the basic tools of structural analysis—equilibrium, compatibility and load–displacement relationships—is described. The notations are chosen from those most commonly utilized in structural mechanics.

2.3.1 Equilibrium Equations

Consider a structure S with M members and N nodes, which is $\gamma(S)$ times statically indeterminate. Select $\gamma(S)$ independent unknown forces as redundants. These unknown forces can be selected from external reactions and/or internal forces of the structure. Denote these redundants by:

$$\mathbf{q} = \left\{q_1, q_2, \ldots, q_{\gamma(S)}\right\}^t. \tag{2.7}$$

Remove the constraints corresponding to redundants, in order to obtain the corresponding statically determinate structure, known as the *basic* (*released* or *primary*) *structure* of S. Obviously, a basic structure should be rigid. Consider the joint loads as,

$$\mathbf{p} = \{p_1, p_2, \ldots, p_n\}^t, \tag{2.8}$$

where n is the number of components for applied nodal loads.

Now the stress resultant distribution **r**, due to the given load **p**, for a linear analysis by the force method can be written as,

$$\mathbf{r} = \mathbf{B}_0\mathbf{p} + \mathbf{B}_1\mathbf{q}, \tag{2.9}$$

where $\mathbf{B}_0$ and $\mathbf{B}_1$ are rectangular matrices each having m rows, and n and $\gamma(S)$ columns, respectively, m being the number of independent components for member forces. $\mathbf{B}_0\mathbf{p}$ is known as a *particular solution,* which satisfies equilibrium with the imposed load, and $\mathbf{B}_1\mathbf{q}$ is a *complementary solution*, formed from a maximal set of independent self-equilibrating stress systems (S.E.Ss), known as a *statical basis.*

Example 1. Consider a planar truss, as shown in Fig. 2.4a, which is two times statically indeterminate. EA is taken to be the same for all the members.

One member force and one component of a reaction may be taken as redundants. Alternatively, two member forces can also be selected as unknowns, as shown in Fig. 2.4b. Selecting the latter choice, the corresponding $\mathbf{B}_0$ and $\mathbf{B}_1$ matrices can now be obtained by applying unit values of p_i (i = 1, 2) and q_j (j = 1, 2), respectively:

$$\mathbf{B}_0^t = \begin{bmatrix} -1 & 0 & 0 & 0 & \sqrt{2} & 0 & -1 & 0 & 0 & 0 \\ -2 & -1 & +1 & 0 & \sqrt{2} & 0 & -1 & \sqrt{2} & 0 & -1 \end{bmatrix},$$

and

$$\mathbf{B}_1^t = \begin{bmatrix} -1/\sqrt{2} & 0 & -1/\sqrt{2} & 0 & +1 & +1 & -1/\sqrt{2} & 0 & 0 & 0 \\ 0 & -1/\sqrt{2} & 0 & -1/\sqrt{2} & 0 & 0 & -1/\sqrt{2} & +1 & +1 & -1/\sqrt{2} \end{bmatrix}.$$

The columns of $\mathbf{B}_1$ (rows of $\mathbf{B}_1^t$) form a statical basis of S. The underlying subgraph of a typical self-equilibrating stress system (for $q_2 = 1$) is shown in bold lines, Fig. 2.4b.

Example 2. Consider a portal frame shown in Fig. 2.5a, which is three times statically indeterminate.

This structure is made statically determinate by an imaginary cut at the middle of its beam. The unit value of external load p_1 and each of the bi-actions q_i (i = 1, 2, 3) lead to the formation of $\mathbf{B}_0$ and $\mathbf{B}_1$ matrices, in which the two end bending moments (M_i, M_j) of a member are taken as its member forces. Using the sign convention introduced in Chap. 1, $\mathbf{B}_0$ and $\mathbf{B}_1$ matrices are formed as:

$$\mathbf{B}_0^t = [+4 \quad 0 \quad 0 \quad 0 \quad 0 \quad 0],$$

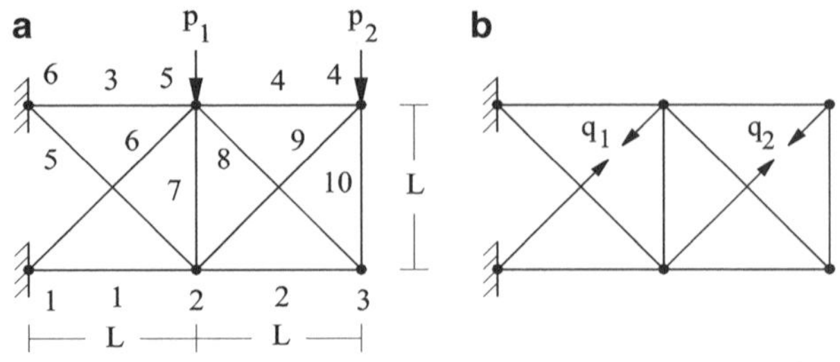

Fig. 2.4 A statically indeterminate planar truss. (**a**) A planar truss. (**b**) The selected unknown forces

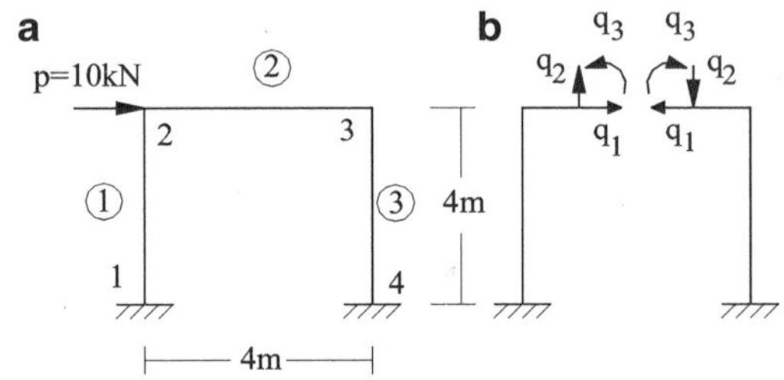

Fig. 2.5 A statically indeterminate frame. (**a**) A portal frame S. (**b**) The basic structure of S

and

$$\mathbf{B}_1^t = \begin{bmatrix} +4 & 0 & 0 & 0 & 0 & -4 \\ -2 & +2 & -2 & -2 & +2 & -2 \\ -1 & +1 & -1 & +1 & -1 & +1 \end{bmatrix}.$$

The columns of $\mathbf{B}_1$ form a statical basis of S, and the underlying subgraph of each self-equilibrating stress system is a cycle, as illustrated in bold lines, Fig. 2.5b. Notice that three self-equilibrating stress systems can be formed on each cycle of a planar frame.

In both of the above examples, particular and complementary solutions are obtained from the same basic structure. However, this is not a necessary requirement, as imagined by some authors. In fact a particular solution is any solution satisfying equilibrium with the applied loads, and a complementary solution is any maximal set of independent self-equilibrating systems. The latter is a basis of a vector space over the field of real numbers, known as a *complementary solution space*, Henderson and Maunder [20].

Using the same basic structure is equivalent to searching for a cycle basis of a graph, but restricting the search to fundamental cycles only, which is convenient but not efficient when the structure is complex or cycle bases with specific properties are needed.

As an example, consider a three-storey frame as shown in Fig. 2.6a. A cut system as shown in Fig. 2.6b corresponds to a statical basis, containing three self-equilibrating stress systems formed on each element of the cycle basis shown in Fig. 2.6b. However, the same particular solution can be employed with a statical basis formed on the cycles of the basis shown in Fig. 2.6c.

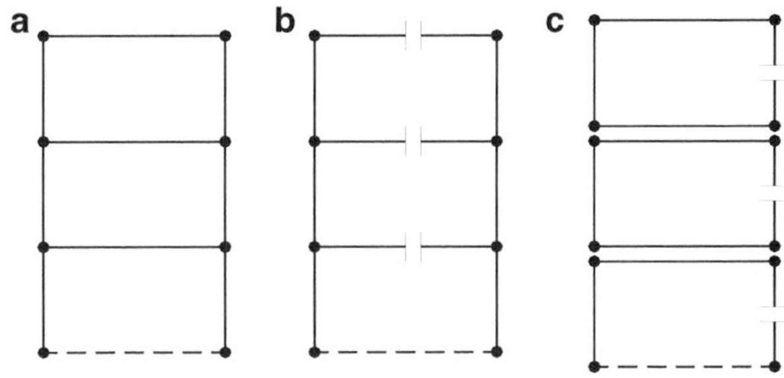

Fig. 2.6 A three-storey frame with different cut systems

A basic structure need not be selected as a determinate one. For a redundant basic structure, one may obtain the necessary data either by analysing it first for the loads **p** and bi-actions $q_i = 1(i = 1, 2, \ldots, \gamma(S))$, or by using existing information.

2.3.2 Member Flexibility Matrices

In the force method of analysis, the determination of the member flexibility matrix is an important step. A number of alternative methods are available for the formation of displacement-force relationships describing the flexibility properties of the members. Four such approaches are:

1. Inversion of the force-displacement relationship;
2. Unit load method;
3. Castigliano's theorem;
4. Solution of differential equations for member displacements.

In the following, the unit load method is briefly described for the formation of the flexibility matrices:

Consider a general element with n member forces,

$$\mathbf{r}_m^t = \{r_1, r_2, \ldots, r_n\}, \tag{2.10}$$

and member displacements:

$$\mathbf{u}_m^t = \{u_1, u_2, \ldots, u_n\}. \tag{2.11}$$

A typical component of the displacement u_i can be found using the unit load method as:

$$u_i = \iiint_V \bar{\sigma}_i^t \varepsilon dV, \tag{2.12}$$

where $\bar{\sigma}_i$ represents the matrix of statically equivalent stresses due to a unit load in the direction of r_i, and $\boldsymbol{\varepsilon}$ is the exact strain matrix due to all applied forces $\mathbf{r}_m$.

The unit loads can be used in turn for all the points where member force are applied, and therefore,

$$\mathrm{u_m} = \iiint_V \overline{\boldsymbol{\sigma}}^t \varepsilon \mathrm{dV}, \tag{2.13}$$

where,

$$\overline{\boldsymbol{\sigma}} = \{\overline{\boldsymbol{\sigma}}_1 \overline{\boldsymbol{\sigma}}_2 \ldots \overline{\boldsymbol{\sigma}}_n\}^t. \tag{2.14}$$

For a linear system,

$$\boldsymbol{\sigma} = \mathbf{c}\mathbf{r}_m, \tag{2.15}$$

where $\mathbf{c}$ is the stress distribution due to unit forces $\mathbf{r}_m$.

The stress-strain relationship can be written as:

$$\boldsymbol{\varepsilon} = \boldsymbol{\phi}\boldsymbol{\sigma} = \boldsymbol{\phi}\mathbf{c}\mathbf{r}_m. \tag{2.16}$$

Substituting in Eq. 2.13 leads to,

$$\mathbf{u}_m = \iiint_V \overline{\boldsymbol{\sigma}}^t \boldsymbol{\phi}\mathbf{c}\mathrm{dV}\,\mathbf{r}_m \tag{2.17}$$

or,

$$\mathbf{u}_m = \mathbf{f}_m \mathbf{r}_m, \tag{2.18}$$

where,

$$\mathbf{f}_m = \iiint_V \overline{\boldsymbol{\sigma}}^t \boldsymbol{\phi}\mathbf{c}\mathrm{dV}, \tag{2.19}$$

represents the element flexibility matrix.

The evaluation of $\overline{\boldsymbol{\sigma}}$ representing the exact stress distribution due to the forces $\mathbf{r}_m$, may not be possible, and hence an approximate relationship should be used. Usually the matrix $\mathbf{c}$ is selected such that it will satisfy at least the equations of equilibrium. Denoting this approximate matrix by $\overline{\mathbf{c}}$, and using $\overline{\boldsymbol{\sigma}} = \overline{\mathbf{c}}$:

$$\mathbf{f}_m = \iiint_V \overline{\mathbf{c}}^t \boldsymbol{\phi}\overline{\mathbf{c}}\mathrm{dV}. \tag{2.20}$$

This equation will be used for the derivation of the flexibility matrices of some finite elements in the proceeding sections.

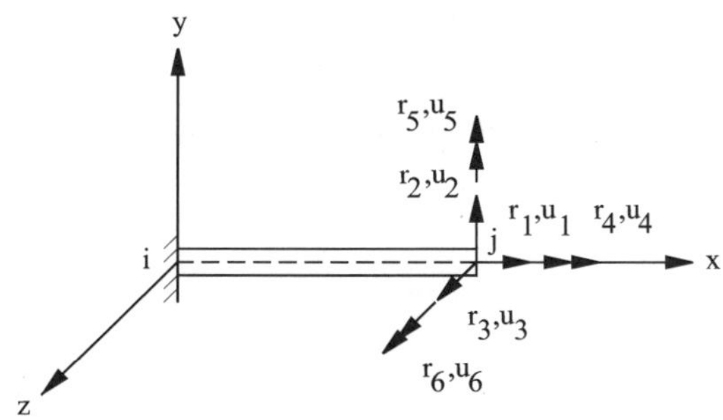

Fig. 2.7 A beam element and selected independent member forces

For a bar element of a space truss, however, the flexibility matrix can easily be obtained using Hooke's law as already discussed in Chap. 1. For a beam element ij of a space frame, y and z axes are taken as the principal axes of the beams cross sections, Fig. 2.7. The forces of end j are selected as a set of independent member forces, and the element is considered to be supported at point i. The axial, torsional, and flexural behaviour in respective planes are uncoupled, and therefore, one needs only to consider the flexibility relationships for four separate members:

1. An axial force member (along x axis);
2. A pure torsional member (about x axis);
3. A beam bent about y axis;
4. A beam bent about z axis.

Direct adaptation of the flexibility relationships derived in Chap. 1, gives the following 6×6 flexibility matrix,

$$\mathbf{f}_m = \begin{bmatrix} \frac{L}{EA} & & & & & \\ 0 & \frac{L^3}{3EI_z} & & & \text{sym.} & \\ 0 & 0 & \frac{L^3}{3EI_y} & & & \\ 0 & 0 & 0 & \frac{L}{GJ} & & \\ 0 & 0 & -\frac{L^2}{2EI_y} & 0 & \frac{L}{EI_y} & \\ 0 & \frac{L^2}{2EI_z} & 0 & 0 & 0 & \frac{L}{EI_z} \end{bmatrix}, \tag{2.21}$$

where G is the shear modulus, I_y and I_z are the moments of inertia with respect to y and z axes, respectively. J is the Saint-Venant torsion constant of the cross section.

2.3.3 *Explicit Method for Imposing Compatibility*

The compatibility equations in the actual structure will now be derived. Using the displacement-load relationship for each member, and collecting them in the diagonal of the unassembled flexibility matrix $\mathbf{F}_{\mathrm{m}}$, one can write member distortions as:

$$\mathbf{u} = \mathbf{F}_{\mathrm{m}}\mathbf{r} = \mathbf{F}_{\mathrm{m}}\mathbf{B}_0\mathbf{p} + \mathbf{F}_{\mathrm{m}}\mathbf{B}_1\mathbf{q}. \tag{2.22}$$

In matrix form:

$$[\mathbf{u}] = [\mathbf{F}_{\mathrm{m}}]\,[\mathbf{B}_0\ \ \mathbf{B}_1]\begin{bmatrix}\mathbf{p}\\ \mathbf{q}\end{bmatrix}. \tag{2.23}$$

From the contragradient principle of Chap. 1,

$$[\mathbf{v}] = \begin{bmatrix}\mathbf{B}_0^{\mathrm{t}}\\ \mathbf{B}_1^{\mathrm{t}}\end{bmatrix}[\mathbf{u}]. \tag{2.24}$$

Combining Eqs. 2.23 and 2.24 results in,

$$\begin{bmatrix}\mathbf{v}_0\\ \mathbf{v}_{\mathrm{c}}\end{bmatrix} = \begin{bmatrix}\mathbf{B}_0^{\mathrm{t}}\\ \mathbf{B}_1^{\mathrm{t}}\end{bmatrix}[\mathbf{F}_{\mathrm{m}}]\,[\mathbf{B}_0\ \mathbf{B}_1]\begin{bmatrix}\mathbf{p}\\ \mathbf{q}\end{bmatrix}, \tag{2.25}$$

in which $\mathbf{v}_0$ contains the displacements corresponding to the force components of $\mathbf{p}$, and $\mathbf{v}_{\mathrm{c}}$ denotes the relative displacements of the cuts for the basic structure. Performing the multiplication,

$$\begin{bmatrix}\mathbf{v}_0\\ \mathbf{v}_{\mathrm{c}}\end{bmatrix} = \begin{bmatrix}\mathbf{B}_0^{\mathrm{t}}\mathbf{F}_{\mathrm{m}}\mathbf{B}_0 & \mathbf{B}_0^{\mathrm{t}}\mathbf{F}_{\mathrm{m}}\mathbf{B}_1\\ \mathbf{B}_1^{\mathrm{t}}\mathbf{F}_{\mathrm{m}}\mathbf{B}_0 & \mathbf{B}_1^{\mathrm{t}}\mathbf{F}_{\mathrm{m}}\mathbf{B}_1\end{bmatrix}\begin{bmatrix}\mathbf{p}\\ \mathbf{q}\end{bmatrix}. \tag{2.26}$$

Defining:

$$\begin{aligned}\mathbf{D}_{00} &= \mathbf{B}_0^{\mathrm{t}}\mathbf{F}_{\mathrm{m}}\mathbf{B}_0, & \mathbf{D}_{10} &= \mathbf{B}_0^{\mathrm{t}}\mathbf{F}_{\mathrm{m}}\mathbf{B}_1,\\ \mathbf{D}_{01} &= \mathbf{B}_1^{\mathrm{t}}\mathbf{F}_{\mathrm{m}}\mathbf{B}_0, & \mathbf{D}_{11} &= \mathbf{B}_1^{\mathrm{t}}\mathbf{F}_{\mathrm{m}}\mathbf{B}_1,\end{aligned} \tag{2.27}$$

the expansion of Eq. 2.14 leads to:

$$\mathbf{v}_0 = \mathbf{D}_{00}\mathbf{p} + \mathbf{D}_{01}\mathbf{q}, \tag{2.28}$$

and

$$\mathbf{v}_{\mathrm{c}} = \mathbf{D}_{10}\mathbf{p} + \mathbf{D}_{11}\mathbf{q}. \tag{2.29}$$

Consider now the compatibility conditions as:

$$\mathbf{v}_{\mathrm{c}} = 0. \tag{2.30}$$

Equation 2.30 together with Eq. 2.29 leads to:

$$\mathbf{q} = -\mathbf{D}_{11}^{-1}\mathbf{D}_{10}\mathbf{p} = \mathbf{F}\mathbf{p}. \tag{2.31}$$

Substituting in Eq. 2.22 yields,

$$\mathbf{v}_0 = \left[\mathbf{D}_{00} - \mathbf{D}_{01}\mathbf{D}_{11}^{-1}\mathbf{D}_{10}\right]\mathbf{p}, \tag{2.32}$$

and the stress resultant in a structure can be obtained as:

$$\mathbf{r} = \left[\mathbf{B}_0 - \mathbf{B}_1\mathbf{D}_{11}^{-1}\mathbf{D}_{10}\right]\mathbf{p}. \tag{2.33}$$

2.3.4 Implicit Approach for Imposing Compatibility

A direct application of the work principle of Chap. 1, can also be used to impose the compatibility conditions in an implicit form as follows:

Since the structure is considered to be linearly elastic, a linear relation exists between the unknown forces $\mathbf{q}$ and the applied forces $\mathbf{p}$; i.e.

$$\mathbf{q} = \mathbf{Q}\mathbf{p}, \tag{2.34}$$

where $\mathbf{Q}$ is a transformation matrix which is still unknown.

Equation 2.9 can now be written as:

$$\mathbf{r} = \mathbf{B}_0\mathbf{p} + \mathbf{B}_1\mathbf{Q}\mathbf{p} = (\mathbf{B}_0 + \mathbf{B}_1\mathbf{Q})\mathbf{p} = \mathbf{B}\mathbf{p}. \tag{2.35}$$

Using the work theorem:

$$\mathbf{P}^{t}\mathbf{v} = \mathbf{r}^{t}\mathbf{u} = \mathbf{p}^{t}\mathbf{B}^{t}\mathbf{u}. \tag{2.36}$$

Now a set of suitable internal forces, $\mathbf{r}^*$, is considered which is statically equivalent to the external loads. From work principle:

$$\mathbf{p}^{t}\mathbf{v} = \mathbf{r}^{*t}\mathbf{u}, \tag{2.37}$$

or

$$\mathbf{p}^{t}\mathbf{v} = \mathbf{p}^{t}\mathbf{B}_0^{t}\mathbf{u}. \tag{2.38}$$

Comparison of the above two equations leads to:

$$\mathbf{p}^t\mathbf{B}^t\mathbf{u} = \mathbf{p}^t\mathbf{B}_0^t\mathbf{u}. \tag{2.39}$$

Substituting $\mathbf{u} = \mathbf{F}_m\mathbf{B}\mathbf{P}$ in the above equation:

$$\mathbf{p}^t\mathbf{B}^t\mathbf{F}_m\mathbf{B}\mathbf{p} = \mathbf{p}^t\mathbf{B}_0^t\mathbf{F}_m\mathbf{B}\mathbf{p}. \tag{2.40}$$

This holds for any $\mathbf{p}$, and therefore:

$$\mathbf{B}^t\mathbf{F}_m\mathbf{B} = \mathbf{B}_0^t\mathbf{F}_m\mathbf{B}. \tag{2.41}$$

From Eq. 2.35 by transposition,

$$\mathbf{B}^t = \mathbf{B}_0^t + \mathbf{Q}^t\mathbf{B}_1^t, \tag{2.42}$$

therefore,

$$\left(\mathbf{B}_0^t + \mathbf{Q}^t\mathbf{B}_0^t\right)\mathbf{F}_m\mathbf{B} = \mathbf{B}_0^t\mathbf{F}_m\mathbf{B}, \tag{2.43}$$

or

$$\mathbf{Q}^t\mathbf{B}_0^t\mathbf{F}_m(\mathbf{B}_0 + \mathbf{B}_1\mathbf{Q}) = \mathbf{0}, \tag{2.44}$$

or

$$\mathbf{Q}^t\left(\mathbf{B}_1^t\mathbf{F}_m\mathbf{B}_0 + \mathbf{B}_1^t\mathbf{F}_m\mathbf{B}_1\mathbf{Q}\right) = \mathbf{0}. \tag{2.45}$$

Using the notation introduced in Eq. 2.15 leads to,

$$\mathbf{Q}^t(\mathbf{D}_{10} + \mathbf{D}_{11}\mathbf{Q}) = \mathbf{0}, \tag{2.46}$$

or

$$\mathbf{D}_{10} + \mathbf{D}_{11}\mathbf{Q} = \mathbf{0}. \tag{2.47}$$

Therefore,

$$\mathbf{Q} = -\mathbf{D}_{11}^{-1}\mathbf{D}_{10}, \tag{2.48}$$

and

$$\mathbf{q} = -\mathbf{D}_{11}^{-1}\mathbf{D}_{10}\mathbf{p}, \tag{2.49}$$

and Eq. 2.20 is obtained as in the previous approach.

2.3.5 Structural Flexibility Matrices

The overall flexibility matrix of a structure can be expressed as:

$$\mathbf{v} = \mathbf{F}\mathbf{p}. \tag{2.50}$$

Pre-multiplying the above equation by $\mathbf{p}^{\mathrm{t}}$, we have:

$$\mathbf{p}^{\mathrm{t}}\mathbf{F}\mathbf{p} = \mathbf{p}^{\mathrm{t}}\mathbf{B}_0^{\mathrm{t}}\mathbf{F}_{\mathrm{m}}\mathbf{B}\mathbf{p}. \tag{2.51}$$

Since $\mathbf{p}$ is arbitrary,

$$\mathbf{F} = \mathbf{B}_0^{\mathrm{t}}\mathbf{F}_{\mathrm{m}}\mathbf{B}, \tag{2.52}$$

or

$$\mathbf{F} = \mathbf{B}_0^{\mathrm{t}}\mathbf{F}_{\mathrm{m}}(\mathbf{B}_0 + \mathbf{B}_1\mathbf{Q}), \tag{2.53}$$

or

$$\mathbf{F} = \mathbf{B}_0^{\mathrm{t}}\mathbf{F}_{\mathrm{m}}\mathbf{B}_0 - \mathbf{B}_0^{\mathrm{t}}\mathbf{F}_{\mathrm{m}}\mathbf{B}_1\mathbf{D}_{11}^{-1}\mathbf{D}_{10}. \tag{2.54}$$

Since $\mathbf{F}_{\mathrm{m}}$ is symmetric, it follows that:

$$\mathbf{D}_{10}^{\mathrm{t}} = \mathbf{B}_0^{\mathrm{t}}\mathbf{F}_{\mathrm{m}}\mathbf{B}_1 = \mathbf{B}_0^{\mathrm{t}}\mathbf{F}_{\mathrm{m}}^{\mathrm{t}}\mathbf{B}_1. \tag{2.55}$$

Therefore, the *overall flexibility matrix* (known also as influence matrix) of the structure is obtained as,

$$\mathbf{F} = \mathbf{D}_{00} - \mathbf{D}_{10}^{\mathrm{t}}\mathbf{D}_{11}^{-1}\mathbf{D}_{10}, \tag{2.56}$$

and $\mathbf{D}_{11} = \mathbf{B}_1^{\mathrm{t}}\mathbf{F}_{\mathrm{m}}\mathbf{B}_1 = \mathbf{G}$ is also referred to as the *flexibility matrix* of the structure. In this book, properties of $\mathbf{G}$ will be studied, since its pattern is the most important factor in optimal analysis of the structure by the force method.

Equation 2.34 can now be used to calculate the nodal displacements.

2.3.6 Computational Procedure

The sequence of computational steps for the force method can be summarized as:

1. Construct $\mathbf{B}_0$ and obtain $\mathbf{B}_0^{\mathrm{t}}$.
2. Construct $\mathbf{B}_1$ and obtain $\mathbf{B}_1^{\mathrm{t}}$.
3. Form unassembled flexibility matrix $\mathbf{F}_{\mathrm{m}}$.

4. Form $\mathbf{F}_m\mathbf{B}_0$ followed by $\mathbf{F}_m\mathbf{B}_1$.
5. Calculate $\mathbf{D}_{00}$, $\mathbf{D}_{10}^t$, $\mathbf{D}_{10}$ and $\mathbf{D}_{11}$, sequentially.
6. Compute $-\mathbf{D}_{11}^{-1}$.
7. Calculate $\mathbf{Q} = -\mathbf{D}_{11}^{-1}\mathbf{D}_{10}$.
8. Form $\mathbf{B}_1\mathbf{Q}$ and find $\mathbf{B} = \mathbf{B}_0 + \mathbf{B}_1\mathbf{Q}$.
9. Form $\mathbf{D}_{10}^t\mathbf{Q}$ and find $\mathbf{D}_{00} + \mathbf{D}_{10}^t\mathbf{Q}$.
10. Compute the internal forces as $\mathbf{r} = \mathbf{Bp}$.
11. Compute nodal displacements as $\mathbf{v}_0 = \mathbf{Fp}$.

Example 3. In this example, the complete analysis of the truss of Example 1 will be given.

$\mathbf{B}_0$ and $\mathbf{B}_1$ matrices are already formed in Example 1 of Sect. 2.2.1. The unassembled flexibility matrix can be constructed as:

$$\mathbf{F}_m = \frac{L}{EA}\begin{bmatrix} 1 & & & & & & & & & \\ & 1 & & & & & & & & \\ & & 1 & & & & & \mathbf{0} & & \\ & & & 1 & & & & & & \\ & & & & \sqrt{2} & & & & & \\ & & & & & \sqrt{2} & & & & \\ & & & & & & 1 & & & \\ & & \mathbf{0} & & & & & \sqrt{2} & & \\ & & & & & & & & \sqrt{2} & \\ & & & & & & & & & 1 \end{bmatrix}.$$

Using the above matrix and the matrices from Example 1, leads to:

$$\mathbf{D}_{11} = \frac{L}{EA}\begin{bmatrix} 2\sqrt{2}+3/2 & 1/2 \\ 1/2 & 2\sqrt{2}+2 \end{bmatrix},$$

and

$$\mathbf{D}_{10} = \frac{L}{EA}\begin{bmatrix} 2+2/\sqrt{2} & 2+2/\sqrt{2} \\ 1/\sqrt{2} & 2+3/\sqrt{2} \end{bmatrix}.$$

Substituting in Eq. 2.25, results in:

$$\begin{bmatrix} q_1 \\ q_2 \end{bmatrix} = -\begin{bmatrix} 2\sqrt{2}+3/2 & 1/2 \\ 1/2 & 2\sqrt{2}+2 \end{bmatrix}^{-1}\begin{bmatrix} 2+2/\sqrt{2} & 2+2/\sqrt{2} \\ 1/\sqrt{2} & 2+3/\sqrt{2} \end{bmatrix}\begin{bmatrix} p_1 \\ p_2 \end{bmatrix}.$$

Taking $p_1 = p_2 = P$ for simplicity, and solving the above equations gives:

$$q_1 = -1.43P \text{ and } q_2 = -1.17P.$$

Equation 2.3 is then used to calculate the member forces as:

$$\mathbf{r}=\{r_1\ r_2\ r_3\ r_4\ r_5\ r_6\ r_7\ r_8\ r_9\ r_{10}\}^t$$
$$=\{-1.95P\ -0.17P\ 2.05P\ 0.83P\ 1.36P\ -1.44P\ -0.12P\ 0.24P\ -1.17P\ -0.17P\}^t.$$

Nodal displacements can be found using Eq. 2.25.

Example 4. In this example, the complete analysis of the frame in Example 2 is given.

$\mathbf{B}_0$ and $\mathbf{B}_1$ matrices are already formed in Example 2 of Sect. 2.2.1. The unassembled flexibility matrix of the structure, using the sign convention introduced in Chap. 1, is formed as:

$$\mathbf{F}_m = \frac{L}{6EI}\begin{bmatrix} 2 & -1 & & & & \\ -1 & 2 & & & & \\ & & 2 & -1 & & \\ & & -1 & 2 & & \\ & & & & 2 & -1 \\ & & & & -1 & 2 \end{bmatrix}.$$

Substituting in Eq. 2.21 leads to:

$$\mathbf{D}_{11} = \frac{L}{6EI}\begin{bmatrix} 64 & 0 & -24 \\ 0 & 56 & 0 \\ -24 & 0 & 18 \end{bmatrix},$$

and

$$\mathbf{D}_{10} = \frac{L}{6EI}\begin{bmatrix} 32 \\ -24 \\ -12 \end{bmatrix}.$$

The inverse of $\mathbf{D}_{11}$ is computed as,

$$\mathbf{D}_{11}^{-1} = \frac{6EI}{L}\begin{bmatrix} 18/576 & 0 & 3/72 \\ 0 & 576 & 0 \\ 3/72 & 0 & 1/9 \end{bmatrix},$$

and $\mathbf{Q}$ can be obtained as:

$$\mathbf{Q} = -\mathbf{D}_{11}^{-1}\mathbf{D}_{10} = \begin{bmatrix} -1/2 \\ +3/7 \\ 0 \end{bmatrix}.$$

Matrix $\mathbf{B}$ is now computed as,

$$\mathbf{B} = \begin{bmatrix} 4 \\ 0 \\ 0 \\ 0 \\ 0 \\ 0 \end{bmatrix} + \begin{bmatrix} +4 & -2 & -1 \\ 0 & +2 & +1 \\ 0 & -2 & -1 \\ 0 & -2 & +1 \\ 0 & +2 & -1 \\ -4 & -2 & +1 \end{bmatrix} \begin{bmatrix} -1/2 \\ +3/7 \\ 0 \end{bmatrix},$$

and finally by using Eq. 2.23 the member forces are obtained as:

$$\mathbf{r} = \{+11.43 \quad +8.57 \quad -8.57 \quad -8.57 \quad +8.57 \quad +11.43\}^{\mathrm{t}}.$$

General Loading. When members are loaded in a general form, then it must be replaced by an equivalent loading. Such a loading can be found as the superposition of two cases; case 1 consists of the given loading but the ends of the member are fixed. The fixed end forces (actions), denoted by FEA, can be found using tables from books on strength of materials. Case 2 is the given structure subjected to the reverse of the fixed end actions only. Obviously, the sum of the loads and reactions of case 1 and case 2 will be the same effect as that of the given loading. This superposition process is illustrated in the following example:

Example 5. A two-span beam is considered as shown in Fig. 2.8a. The fixed end actions are provided in b, and the equivalent forces are illustrated in Fig. 2.8c. The structure is twice indeterminate, and the primary structure is obtained by introducing two hinges as shown in d. The applied nodal forces and redundants are depicted in Fig. 2.8e, f, respectively.

$\mathbf{B}_0$ and $\mathbf{B}_1$ matrices are formed as,

$$\mathbf{B}_0 = \begin{bmatrix} -1 & 0 & 0 \\ 0 & +1 & 0 \\ 0 & 0 & 0 \\ 0 & 0 & +1 \end{bmatrix} \quad \text{and} \quad \mathbf{B}_1 = \begin{bmatrix} -1 & 0 \\ 0 & +1 \\ 0 & -1 \\ 0 & 0 \end{bmatrix},$$

and the unassembled flexibility matrix of the structure is constructed as:

$$\mathbf{F}_{\mathrm{m}} = \frac{\mathrm{L}}{6\mathrm{EI}} \begin{bmatrix} 2 & -1 & & \\ -1 & 2 & & \\ & & 2 & -1 \\ & & -1 & 2 \end{bmatrix}.$$

Substituting in Eq. 2.27 leads:

$$\mathbf{D}_{11} = \frac{\mathrm{L}}{6\mathrm{EI}} \begin{bmatrix} 2 & 1 \\ 1 & 4 \end{bmatrix},$$

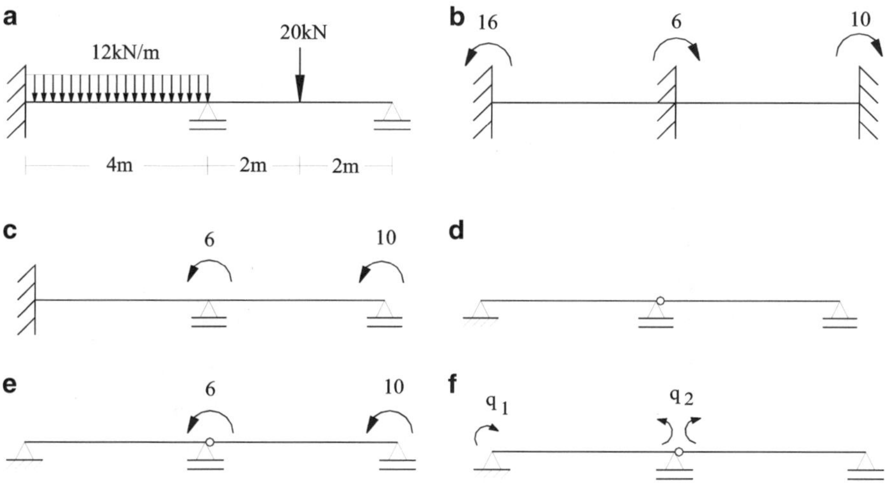

Fig. 2.8 A two-span beam with general loading. (**a**) A two-span beam. (**b**) Fixed end actions. (**c**) The equivalent loading. (**d**) The selected primary structure. (**e**) Applied force on primary structure. (**f**) Redundants on primary structure

and

$$\mathbf{D}_{10} = \frac{L}{6EI}\begin{bmatrix} 2 & 1 & 0 \\ 1 & 2 & 1 \end{bmatrix}.$$

The inverse of $\mathbf{D}_{11}$ is computed as,

$$\mathbf{D}_{11}^{-1} = -\frac{1}{7} \times \frac{6EI}{L}\begin{bmatrix} 4 & -1 \\ -1 & 2 \end{bmatrix},$$

and $\mathbf{Q}$ can be obtained as:

$$\mathbf{Q} = -\mathbf{D}_{11}^{-1}\mathbf{D}_{10} = -\frac{1}{7}\begin{bmatrix} 4 & -1 \\ -1 & 2 \end{bmatrix}\begin{bmatrix} 2 & 1 & 0 \\ 1 & 2 & 1 \end{bmatrix} = -\frac{1}{7}\begin{bmatrix} 7 & 2 & -1 \\ 0 & 3 & 2 \end{bmatrix}.$$

Now $\mathbf{r}$ is computed as,

$$\mathbf{r}' = \left(\begin{bmatrix} -1 & 0 & 0 \\ 0 & 1 & 0 \\ 0 & 0 & 0 \\ 0 & 0 & 1 \end{bmatrix} + \begin{bmatrix} 1 & 2/7 & -1/7 \\ 0 & -3/7 & -2/7 \\ 0 & 3/7 & 2/7 \\ 0 & 0 & 0 \end{bmatrix}\right)\begin{bmatrix} 0 \\ 6 \\ 10 \end{bmatrix} = \begin{bmatrix} 0.285 \\ 0.572 \\ 5.428 \\ 10.00 \end{bmatrix},$$

àdding the fixed end reaction, the final member forces are obtained as:

$$\mathbf{r} = \{\, 16.285 \quad -15.428 \quad 15.428 \quad 0.000 \,\}^t.$$

2.3.7 Optimal Force Method

For an efficient force method, the matrix **G** should be:

(a) Sparse;
(b) Well conditioned;
(c) Properly structured, i.e. narrowly banded.

In order to provide the properties (a) and (b) for **G**, the structure of $\mathbf{B}_1$ should be carefully designed, since the pattern of $\mathbf{F}_m$ for a given discretization is unchanged; i.e. a suitable statical basis should be selected. This problem is treated in different forms by various methods. In the following, graph theoretical methods are described for the formation of appropriate statical bases of different types of skeletal structures. The property (c) above has a totally combinatorial nature and is studied in Chaps. 5 and 6.

Pattern Equivalence. Matrix $\mathbf{B}_1$ containing a statical basis, in partitioned form, is pattern equivalent to $\mathbf{C}^t$, where **C** is the cycle-member incidence matrix. Similarly, $\mathbf{B}_1^t\mathbf{F}_m\mathbf{B}_1$ is pattern equivalent to $\mathbf{CIC}^t$ or $\mathbf{CC}^t$. This correspondence transforms some structural problems associated with the characterization of $\mathbf{G} = \mathbf{B}_1^t\mathbf{F}_m\mathbf{B}_1$ into combinatorial problems of dealing with $\mathbf{CC}^t$.

As an example, if a sparse matrix **G** is required, this can be achieved by increasing the sparsity of $\mathbf{CC}^t$. Similarly for a banded **G**, instead of ordering the elements of a statical basis (self-equilibrating stress systems), one can order the corresponding cycles. This transformation has many advantages, such as:

1. The dimension of $\mathbf{CC}^t$ is often smaller than that of **G**. For example, for a space frame the dimension of $\mathbf{CC}^t$ is six-fold and for a planar frame three-fold smaller than that of **G**. Therefore, the optimisation process becomes much simpler when combinatorial properties are used.
2. The entries of **C** and $\mathbf{CC}^t$ are elements of Z_2 and therefore easier to operate on, compared to $\mathbf{B}_1$ and **G** which have real numbers as their entries.
3. The advances made in combinatorial mathematics and graph theory become directly applicable to structural problems.
4. A correspondence between algebraic and graph theoretical methods becomes established.

2.4 Force Method for the Analysis of Frame Structures

In this section, frame structures are considered in their perfect conditions; i.e. the joints of a frame are assumed to be rigid, and connected to each other by elastic members and supported by a rigid foundation.

For this type of skeletal structure, a statical basis can be generated on a cycle basis of its graph model. The function representing the degree of static indeterminacy, $\gamma(S)$, of a rigid-jointed structure is directly related to the first Betti number $b_1(S)$ of its graph model,

$$\gamma(S) = \alpha b_1(S) = \alpha[M(S) - N(S) + b_0(S)], \tag{2.57}$$

where $\alpha = 3$ or 6 depending on whether the structure is either a planar or a space frame.

For a frame structure, matrix $\mathbf{B}_0$ can easily be generated using a shortest route tree of its model, and $\mathbf{B}_1$ can be formed by constructing 3 or 6 self-equilibrating stress systems on each element of a cycle basis of S.

In order to obtain a flexibility matrix of maximal sparsity, special cycle bases should be selected as defined in the next section. Methods for the formation of a cycle basis can be divided into two groups, namely

(a) Topological methods, (b) graph theoretical approaches.

Topological methods useful for the formation of cycle bases by hand, were developed by Henderson and Maunder [20] and a complete description of these methods is presented in Kaveh [3]. Graph-theoretical methods suitable for computer applications were developed by Kaveh [21].

2.4.1 Minimal and Optimal Cycle Bases

A matrix is called *sparse* if many of its entries are zero. The interest in sparsity arises because its exploitation can lead to enormous computational saving, and because many large matrices that occur in the analysis of practical structures, can be made sparse if they are not already so. A matrix can therefore be considered sparse, if there is an advantage in exploiting its zero entries.

The *sparsity coefficient* χ of a matrix is defined to be its number of non-zero entries. A cycle basis $C = \{C_1, C_2, C_3, \ldots, C_{b_1(S)}\}$ is called *minimal*, if it corresponds to a minimum value of:

$$L(C) = \sum_{i=1}^{b_1(S)} L(C_i). \tag{2.58}$$

Obviously, $\chi(C) = L(C)$ and a minimal cycle basis can be defined as a basis which corresponds to minimum $\chi(C)$. A cycle basis for which L(C) is near minimum is called a *subminimal* cycle basis of S.

A cycle basis corresponding to maximal sparsity of the $\mathbf{CC}^t$ is called an *optimal* cycle basis of S. If $\chi(\mathbf{CC}^t)$ does not differ considerably from its minimum value, then the corresponding basis is termed *suboptimal*.

The matrix intersection coefficient $\sigma_i(\mathbf{C})$ of row i of cycle member incidence matrix $\mathbf{C}$ is the number of row j such that:

(a) $j \in \{i + 1, i + 2, \ldots, b_1(S)\}$,
(b) $C_i \cap C_j \neq \emptyset$, i.e. there is at least one k such that the column k of both cycles C_i and C_j (rows i and j) contain non-zero entries.

Now it can be shown that:

$$\chi(\mathbf{D}) = b_1(S) + 2 \sum_{i=1}^{b_1(S)-1} \sigma_i(\mathbf{C}). \tag{2.59}$$

This relationship shows the correspondence of a cycle member incidence matrix **C** and that of its cycle basis adjacency matrix. In order to minimize $\chi(\mathbf{CC}^t)$, the value of $\sum_{i=1}^{b_1(S)-1} \sigma_i(\mathbf{C})$ should be minimized, since $b_1(S)$ is a constant for a given structure S, i.e. γ-cycles with a minimum number of overlaps should be selected.

In the force method, an optimal cycle basis is needed corresponding to the maximum sparsity of $\mathbf{CC}^t$ matrix. However, because of the complexity of this problem, most of the research has been concentrated on minimal cycle basis selection, except those of Ref. [22], which minimize the overlaps of the cycles rather than only their length.

2.4.2 Selection of Minimal and Subminimal Cycle Bases

Cycle bases of graphs have many applications in various fields of engineering. The amount of work in these applications depends on the cycle basis chosen. A basis with shorter cycles reduces the time and storage required for some applications; i.e. it is ideal to select a minimal cycle basis, and for some other applications minimal overlaps of cycles are needed; i.e. optimal cycle bases are preferred. In this section, the formation of minimal and subminimal cycle bases is first discussed. Then the possibility of selecting optimal and suboptimal cycle bases is investigated.

Minimal cycle bases were considered first by Stepanec [23] and improved by Zykov [24]. Many practical algorithms for selecting subminimal cycle bases have been developed by Kaveh [15].

In this section, the merits of the algorithms developed by different authors are discussed; a method is given for selection of minimal cycle bases, and efficient approaches are presented for the generation of subminimal cycle bases.

Formation of a Minimal Cycle on a Member. A minimal length cycle C_i on a member m_j, called its *generator*, can be formed by using the shortest route tree algorithm as follows:

Start the formation of two SRTs rooted at the two end nodes n_s and n_t of m_j, and terminate the process as soon as the SRTs intersect each other (not through m_j itself) at say n_c. The shortest paths between n_s and n_c, and n_t and n_c, together with m_j, form

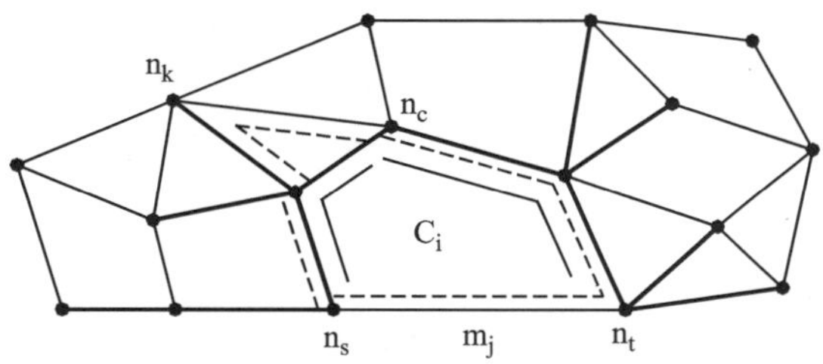

Fig. 2.9 A minimal cycle on a member

a minimal cycle C_i on m_j. Using this algorithm, cycles of prescribed lengths can also be generated.

As an example, C_i is a minimal cycle on m_j in Fig. 2.9. The SRTs are shown in bold lines. The generation of SRTs is terminated as soon as n_c has been found.

A minimal cycle on a member m_j passing through a specified node n_k can similarly be generated. An SRT rooted at n_k is formed and as soon as it hits the end nodes of m_j, the shortest paths are found by backtracking between n_k and n_s, and n_k and n_t. These paths together with m_j form the required cycle. As an example, a minimal cycle on m_j containing n_k, is illustrated by dashed lines in Fig. 2.9.

Different Cycle Sets for Selecting a Cycle Basis. It is obvious that a general cycle can be decomposed into its simple cycles. Therefore, it is natural to confine the considered set to only simple cycles of S. Even such a cycle set, which forms a subspace of the cycle space of the graph, has many elements and is therefore uneconomical for practical purposes.

In order to overcome the above difficulty, Kaveh [15] used an expansion process, selecting the smallest admissible (independent with additional restriction) cycles, one at a time, until $b_1(S)$ cycles forming a basis had been obtained. In this approach, a very limited number of cycles were checked for being an element of a basis. As an example, the expansion process for selecting a cycle basis of S is illustrated in Fig. 2.10.

Hubicka and Sysløø [25] employed a similar approach, without the restriction of selecting one cycle at each step of expansion. In their method, when a cycle has been added to the previously selected cycles, increasing the first Betti number of the expanded part by "p", then p created cycles have been formed. As an example, in this method, Steps 4 and 5 will be combined into a single step, and addition of cycle 5 will require immediate formation of the cycle 4. The above method is modified, and an efficient algorithm is developed for the formation of cycle bases by Kaveh and Roosta [26],

Finally, Horton [27] proved that the elements of a minimal cycle basis lie in between a cycle set consisting of the minimal cycles on each member of S which passes through each node of S, i.e. each member is taken in turn and all cycles of minimal length on such a member passing through all the nodes of S are generated. Obviously, $M(S) \times M(S)$ such cycles will be generated.

Independence Control. Each cycle of a graph can be considered as a column vector of its cycle-member incidence matrix. An algebraic method such as the

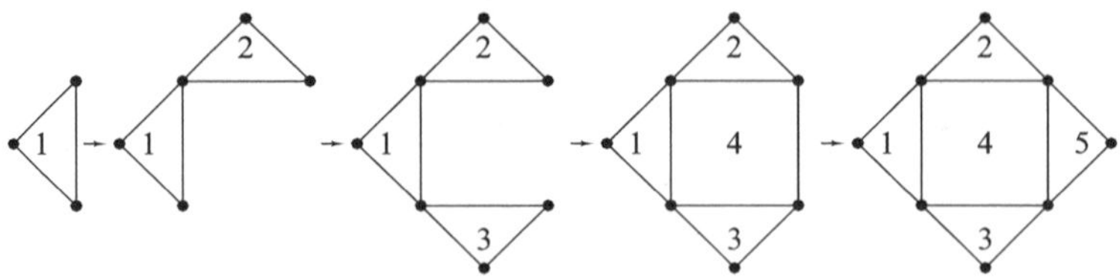

Fig. 2.10 A graph S and selected cycles

Gaussian elimination may then be used for checking the independence of a cycle with respect to the previously selected sub-basis. However, although this method is general and reduces the order dependency of the cycle selection algorithms, like many other algebraic approaches its application requires a considerable amount of storage space.

The most natural graph theoretical approach is to employ a spanning tree of S, and form its fundamental cycles. This method is very simple; however, in general its use leads to long cycles. The method can be improved by allowing the inclusion of each used chord in the branch set of the selected tree. Further reduction in length may be achieved by generating an SRT from a centre node of a graph, and the use of its chords in ascending order of distance from the centre node, Kaveh [21].

A third method, which is also graph-theoretical, consists of using admissible cycles. Consider the following expansion process, with S being a 2-connected graph,

$$C_1 = C^1 \rightarrow C^2 \rightarrow C^3 \rightarrow \ldots \rightarrow C^{b_1(S)} = S,$$

where $C^k = \bigcup_{i=1}^{k} C_i$. A cycle C_{k+1} is called an *admissible* cycle, if for $C^{k+1} = C^k \cup C_{k+1}$:

$$b_1\left(C^{k+1}\right) = b_1\left(C^k \cup C_{k+1}\right) = b_1\left(C^k\right) + 1. \tag{2.60}$$

It can easily be proved that, the above admissibility condition is satisfied if any of the following conditions hold:

1. $A_{k+1} = C^k \cap C_{k+1} = \varnothing$, where $\varnothing$ is an empty intersection;
2. $\overline{b}_1(A_{k+1}) = r - s$, where r and s are the numbers of components of C^{k+1} and C^k, respectively;
3. $\overline{b}_1(A_{k+1}) = 0$ when C^k and C^{k+1} are connected ($r = s$).

In the above relations, $\overline{b}_1(A_i) = \overline{M}_i - \overline{N}_i + 1$, where $\overline{M}_i$ and $\overline{N}_i$ are the numbers of members and nodes of A_i, respectively.

As an example, the sequence of cycle selection in Fig. 2.11 will be as specified by their numbers.

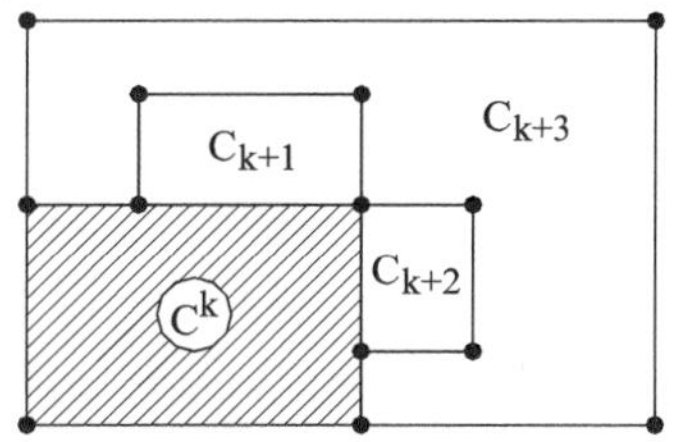

Fig. 2.11 A cycle and its bounded cycles

A different approach suggested by Hubicka and Syslø, in which,

$$b_1\left(C^{k+1}\right) = b_1\left(C^k\right) + p, \tag{2.61}$$

is considered to be permissible. However, a completion is performed for $p > 1$. As an example, when C_3 is added to C^k, its first Betti number is increased by 3 and therefore, cycles C_1 and C_2 must also be selected at that stage, before further expansion.

Having discussed the mathematical concepts involved in a cycle basis selection, three different algorithms are now described.

Algorithm 1 (Kaveh [15])

Step 1: Select a pseudo-centre node of maximal degree O. Such a node can be selected manually or automatically using the graph or algebraic graph theoretical methods discussed in Chap. 5.

Step 2: Generate an SRT rooted at O, form the set of its chords and order them according to their distance from O.

Step 3: Form one minimal cycle on each chord in turn, starting with the chord nearest to the root node. A corresponding simple path is chosen which contains members of the tree and the previously used chords, hence providing the admissibility of the selected cycle.

This method selects subminimal cycle bases, using the chords of an SRT. The nodes and members of the tree and consequently the cycles are partially ordered according to their distance from O. This is the combinatorial version of the Turn Back method to be discussed in the section on algebraic force methods.

Algorithm 2 (Kaveh [15])

Step 1: Select a centre or seudo-centre node of maximal degree O.

Step 2: Use any member incident with O as the generator of the first minimal cycle. Take any member not used in C_1 and incident with O, and generate on it the second minimal cycle. Continue this process until all the members incident with O are used as the members of the selected cycles. The cycles selected so far are admissible, since the intersection of each cycle with the previously selected cycles is a simple path (or a single node) resulting in an increase of the first Betti number by unity for each cycle.

Step 3: Choose a starting node O′, adjacent to O, which has the highest degree. Repeat a step similar to Step 2, testing each selected cycle for admissibility.

If the cycle formed on a generator m_k fails the test, then examine the other minimal cycles on m_k if any such cycle exists. If no admissible minimal cycle can be found on m_k, then,

Form admissible minimal cycles on the other members incident with O′. If m_k does not belong to one of these subsequent cycles, then:

Search for an admissible minimal cycle on m_k, since the formation of cycles on other previous members may now have altered the admissibility of this cycle. If no such cycle can be found, leave m_k unused. In this step more than one member may be left unused.

Step 4: Repeat Step 3 using as starting nodes a node adjacent to O and/or O′, having the highest degree. Continue the formation of cycles until all the nodes of S have been tested for cycle selection. If all the members have not been used, select the shortest admissible cycle available for an unused member as generator. Then test the minimal cycles on the other unused members, in case the formation of the longer cycle has altered the admissibility. Each time a minimal cycle is found to be admissible, add to C^i and test all the minimal cycles on the other unused members again. Repeat this process, forming other shortest admissible cycles on unused members as generators, until S is re-formed and a subminimal cycle basis has been obtained.

Both of the above two algorithms are order-dependent, and various starting nodes may alter the result. The following algorithm is more flexible and less order-dependent, and in general leads to the formation of shorter cycle bases.

Algorithm 3 (Kaveh [21])

Step 1: Generate as many admissible cycles of length 3 as possible. Denote the union of the selected cycles by C^n.

Step 2: Select an admissible cycle of length 4 on an unused member. Once such a cycle C_{n+1} is found, check the other unused members for possible admissible cycles of length 3. Again select an admissible cycle of length 4 followed by the formation of possible 3-sided cycles. This process is repeated until no admissible cycles of length 3 and 4 can be formed. Denote the generated cycles by C^m.

Step 3: Select an admissible cycle of length 5 on an unused member. Then check the unused members for the formation of 3-sided admissible cycles. Repeat Step 2 until no cycle of length 3 or 4 can be generated. Repeat Step 3 until no cycle of length 3, 4 or 5 can be found.

Step 4: Repeat similar steps to Step 3, considering higher-length cycles, until $b_1(S)$ admissible cycles forming a subminimal cycle basis are generated.

Remark. The cycle basis C formed by Algorithms 1–3 can further be improved by exchanging the elements of the selected basis. In each step of this process, a shortest cycle C_i' independent of the cycles of $C \backslash C_i$ is replaced by C_i if $L(C_i') < L(C_i)$. This process is repeated for $i = 1, 2, \ldots, b_1(S)$.

This additional operation increases the computational time and storage, and its use is recommended only when the formation of minimal cycle basis is required.

Algorithm 4 (Horton [27])

Step 1: Find a minimum path $P(n_i, n_j)$ between each pair of nodes n_i and n_j.

Step 2: For each node n_k and member $m_l = (n_i, n_j)$, generate the cycle having m_l and n_k as $P(n_k,n_i) + P(n_k,n_j) + (n_i,n_j)$ and calculate its length. Degenerate cases in which $P(n_k, n_i)$ and $P(n_k, n_j)$ have nodes other than n_k in common, can be omitted.

Step 3: Order the cycles by their weight (or length).

Step 4: Use the Greedy Algorithm, to find a minimal cycle basis from this set of cycles. This algorithm is given in in Kaveh [15, 20].

A simplified version of the above Algorithm can be designed as follows:

Step 1: Form a spanning tree of S rooted from an arbitrary node, and select its chords.

Step 2: Take the first chord and form $N(S) - 2$ minimal cycles, each being formed on the specified chord containing a node of S (except the two end nodes of this chord).

Step 3: Repeat Step 2 for the other chords, in turn, until $[M(S) - N(S) + 1] \times [N(S) - 2]$ cycles are generated. Repeated and degenerate cycles should be discarded.

Step 4: Order the cycles in ascending magnitude of their lengths.

Step 5: Using the above set of cycles, employ the Greedy Algorithm to form a minimal cycle basis of S.

The main contribution of Horton's Algorithm is the limit imposed on the elements of the cycle-set used in the Greedy Algorithm. The use of matroids and the Greedy Algorithm, has been suggested by Kaveh [15], and they have been employed by Lawler [28] and Kolasinska [29].

2.4.3 *Examples*

Example 1. Consider a planar graph S, as shown in Fig. 2.12, for which $b_1(S) = 18-11 + 1 = 8$. Using Algorithm 3, the selected basis consists of four cycles of length 3, three cycles of length 4 and one cycle of length 5, as follows:

$$C_1 = (1,2,3), C_2 = (1,8,9), C_3 = (2,6,3), C_4 = (2,5,6), C_5 = (1,4,5,2),$$
$$C_6 = (1,7,5,2), C_7 = (8,6,2,1), C_8 = (10,8,6,3,11)$$

The total length of the selected basis is $L(C) = 29$, which is a counter example for minimality of a mesh basis, since, for any such basis of S, $L(C) > 29$.

Example 2. In this example, S is the model of a space frame, considered as $S = \bigcup_{i=1}^{27} S_i$, where a typical S_i is depicted in Fig. 2.13a. For S_i there are 12 members joining eight corner nodes, and a central node joined to these corner nodes.

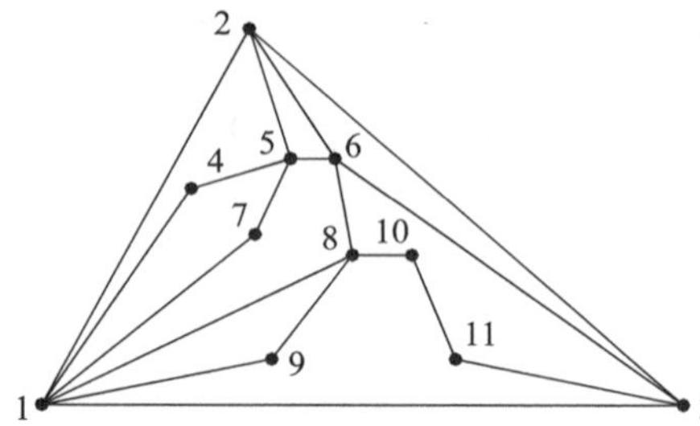

Fig. 2.12 A planar graph S

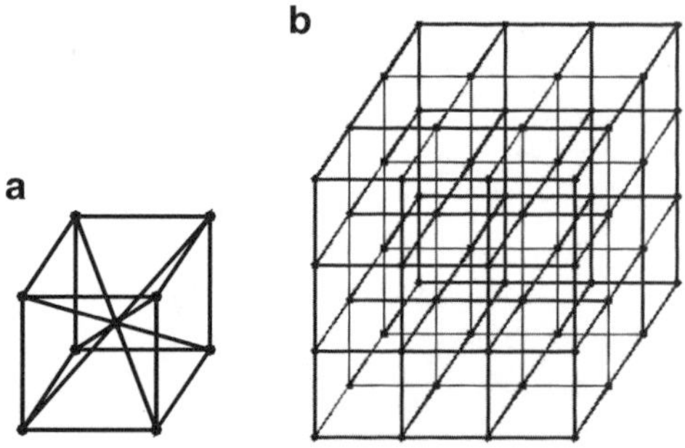

Fig. 2.13 A space frame S. (**a**) A typical S_i (i = 1, . . ., 27). (**b**) S with some omitted members

The model S is shown in Fig. 2.13b, in which some of the members are omitted for clarity in the diagram. For this graph, $b_1(S) = 270$.

The selected cycle basis using any of the algorithms consists of 270 cycles of length 3, forming a minimal cycle basis of S. For Algorithm 3, the use of different starting nodes leads to a minimal cycle basis, showing the capability of this method.

Example 3. S is a planar graph with $b_1(S) = 9$, as shown in Fig. 2.14. The application of Algorithm 3 results in the formation of a cycle of length 3 followed by the selection of five cycles of length 4. Then member {1, 6} is used as the generator of a six-sided cycle $C_7 = (1,2,3,4,5,6,1)$. Member {2, 10} is then employed to form a seven-sided cycle $C_8 = (2,11,12,13,14,15,10,2)$, followed by the selection of a five-sided cycle $C_9 = (10,5,4,3,2,10)$. The selected cycle basis has a total length of $L(C) = 41$, and is not a minimal cycle basis. A shorter cycle basis can be found by Algorithm 4 consisting of one three-sided and five four-sided cycles, together with the following cycles,

$$C_7 = (1, 2, 10, 5, 6, 1), C_8 = (2, 3, 4, 5, 10, 2) \text{ and}$$

$$C_9 = (2, 11, 12, 13, 14, 15, 10, 2),$$

forming a basis with the total length of 40. However, the computation time and storage for Algorithm 3 is far less than that of Algorithm 4, as compared in Ref. [30].

Fig. 2.14 A planar graph S

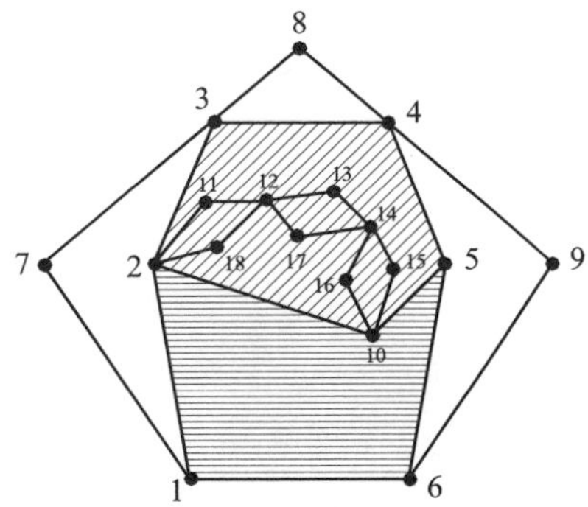

2.4.4 *Optimal and Suboptimal Cycle Bases*

In what follows, a direct method and an indirect approach, which often lead to the formation of optimal cycle bases, are presented. Much work is needed before the selection of an optimal cycle basis of a graph becomes feasible.

2.4.4.1 Suboptimal Cycle Bases; A Direct Approach

Definition 1. An *elementary contraction* of a graph S is obtained by replacing a path containing all nodes of degree 2 with a new member. A graph S contracted to a graph S' is obtained by a sequence of elementary contractions. Since in each elementary contraction k nodes and k members are reduced, the first Betti number does not change in a contraction, i.e. $b_1(S) = b_1(S')$. The graph S is called homeomorphic to S′, Fig. 2.15.

This operation is performed in order to reduce the size of the graph and also because the number of members in an intersection of two cycles is unimportant; a single member is enough to render $C_i \cap C_j$ nonempty, and hence to produce a non-zero entry in $\mathbf{CC}^t$.

Definition 2. Consider a member m_i of a graph S. On this member, p minimal cycles of length q can be generated. P is called the *incidence number* and q is defined as the *cycle length number* of m_i. In fact, p and q are measures assigned to a member to indicate its potential as a member in the elements of a cycle basis. In the process of expansion for cycle selection, an artificial increase in p results in the exclusion of this element from a minimal cycle, keeping the number of overlaps as small as possible.

Space graphs need special treatment. For these graphs, when a member has $p = 1$, then the next shortest length cycles with $q' = q + 1$ (1 being the next smallest possible integer) are also considered. Denoting the number of such cycles by p', the incidence number and cycle length number for this type of member are taken as,

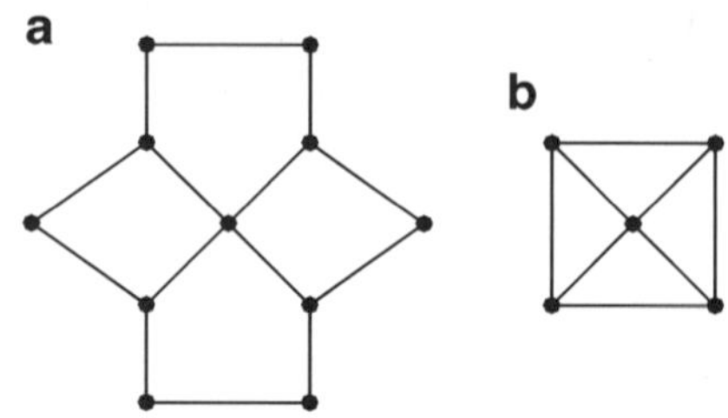

Fig. 2.15 S and its contracted graph S'. (**a**) S. (**b**) S'

$$I_{jk} = p' + 1 \quad \text{and} \quad I^{c}_{jk} = \left(q + p'q'\right)/\left(1 + p'\right), \tag{2.62}$$

respectively. The end nodes of the considered member are j and k.

Definition 3. The *weight* of a cycle is defined as the sum of the incidence numbers of its members.

Algorithm A

Step 1: Contract S into S', and calculate the incidence number (IN) and cycle length number (CLN) of all its members.

Step 2: Start with a member of the least CLN and generate a minimal weight cycle on this member. For members with equal CLNs, the one with the smallest IN should be selected. A member with these two properties will be referred to as "a member of the least CLN with the smallest IN".

Step 3: On the next unused member of the least CLN with the smallest IN, generate an admissible minimal weight cycle. In the case when a cycle of minimal weight is rejected due to inadmissibility, the next unused member should be considered. This process is continued as far as the generation of admissible minimal weight cycles is possible. After a member has been used as many times as its IN, before each extra usage, increase the IN of such a member by unity.

Step 4: On an unused member of the least CLN, generate one admissible cycle of the smallest weight. This cycle is not a minimal weight cycle, otherwise it would have been selected at Step 3. Such a cycle is called a *subminimal weight cycle*. Again, update the incidence numbers for each extra usage. Now repeat Step 3, since the formation of the new subminimal weight cycle may have altered the admissibility condition of the other cycles, and selection of further minimal weight cycles may now have become possible.

Step 5: Repeat Step 4, selecting admissible minimal and subminimal weight cycles, until $b_1(S')$ of these cycles are generated.

Step 6: A reverse process to that of the contraction of Step 1, transforms the selected cycles of S' into those of S.

This algorithm leads to the formation of a suboptimal cycle basis, and for many models encountered in practice, the selected bases have been optimal.

2.4.4.2 Suboptimal Cycle Bases; an Indirect Approach

Definition 1. The *weight* of a member in the following algorithm is taken as the sum of the degrees of its end nodes.

Algorithm B

Step 1: Order the members of S in ascending order of weight. In all the subsequent steps use this ordered member set.

Step 2: Generate as many admissible cycles of length α as possible, where α is the length of the shortest cycle of S. Denote the union of the selected cycles by C^m. When α is not specified, use the value $\alpha = 3$.

Step 3: Select an admissible cycle of length α+1 on an unused member (use the ordered member set). Once such a cycle C_{m+1} is found, control the other unused members for possible admissible cycles of length α. Again select an admissible cycle of length α+1 followed by the formation of possible α-sided cycles. This process is repeated until no admissible cycles of length α and α+1 can be found. Denote the generated cycles by C^n.

Step 4: Select an admissible cycle C_{n+1} of length α+2 on an unused member. Then check the unused members for the formation of α-sided cycles. Repeat Step 2 until no cycle of length α or α+1 can be generated. Repeat Step 3 until no cycles of length α, α+1 or α+2 can be found.

Step 5: Take an unused member and generate an admissible cycle of minimal length on this member. Repeat Steps 1, 2 and 3.

Step 6: Repeat steps similar to that of Step 4 until $b_1(S)$ admissible cycles, forming a suboptimal cycle basis, are generated.

Using the ordered member set affects the selection process in two ways:

1. Generators are selected in ascending weight order, hence increasing the possibility of forming cycles from the dense part of the graph. This increases the chance of cycles with smaller overlaps being selected.
2. From cycles of equal length formed on a generator, the one with smallest total weight (sum of the weights of the members of a cycle) is selected.

The cycle bases generated by this algorithm are suboptimal; however, the results are inferior to those of the direct method A.

Remark. Once a cycle basis C is formed by Algorithm A or Algorithm B, it can further be improved by exchanging the elements of C. In each step of this process, a cycle C_k is controlled for the possibility of being exchanged by ring sum of C_k and a combination of the cycles of $C \backslash C_k$, in order to reduce the overlap of the cycles. The process is repeated until no improvement can be achieved. This additional operation increases the computational time and storage, and should only be used when the corresponding effort is justifiable, e.g. this may be the case when a non-linear analysis or a design optimisation is performed using a fixed cycle basis.

2.4.5 Examples

In this section, examples of planar and space frames are studied. The cycle bases selected by Algorithms A and B are compared with those developed for generating minimal cycle bases (Algorithms 1–4). Simple examples are chosen, in order to illustrate clearly the process of the methods presented. The models, however, can be extended to those containing a greater number of members and nodes of high degree, to show the considerable improvements to the sparsity of matrix $\mathbf{CC}^t$.

Example 1. Consider a space frame as shown in Fig. 2.16a with the corresponding graph model S as illustrated in Fig. 2.16b. For this graph $b_1(S) = 12$, and therefore 12 independent cycles should be selected as a basis. Algorithm B selects a minimal cycle basis containing the following cycles,

$$C_1 = (1,2,3), C_2 = (1,2,5), C_3 = (1,3,4), C_4 = (1,5,4), C_5 = (2,3,6,7),$$
$$C_6 = (3,4,7,8), C_7 = (4,5,8,9), C_8 = (6,7,8,9), C_9 = (7,8,11,12),$$
$$C_{10} = (6,7,10,11), C_{12} = (9,8,12,13), C_{12} = (10,11,12,13)$$

which corresponds to:

$$\chi(\mathbf{C}) = 4 \times 3 + 8 \times 4 = 44,$$

and

$$\chi(\mathbf{CC}^t) = 12 + 2 \times 23 = 58.$$

Using Algorithm A leads to the formation of a similar basis, with the difference that $C_8' = (6,9,10,13)$ is generated in place of $C_8 = (6, 7, 8, 9)$, corresponding to:

$$\chi\left(\mathbf{C}'\right) = 4 \times 3 + 8 \times 4 = 44,$$
$$\chi\left(\mathbf{C}'\mathbf{C}'^t\right) = 12 + 2 \times 20 = 52.$$

The CLNs and Ins of the members used in this algorithm are illustrated in Fig. 2.16b.

Example 2. In this example, S is a space structure with $b_1(S) = 33$, as shown in Fig. 2.17a. Both Algorithms 3 and A select 33 cycles of length 4, i.e. a minimal cycle basis with $\chi(\mathbf{C}) = 4 \times 33 = 132$ is obtained.

The basis selected by Algorithm 3 contains (in the worst case) all four-sided cycles of S except those which are shaded in Fig. 2.17a, with $\chi(\mathbf{CC}^t) = 233$.

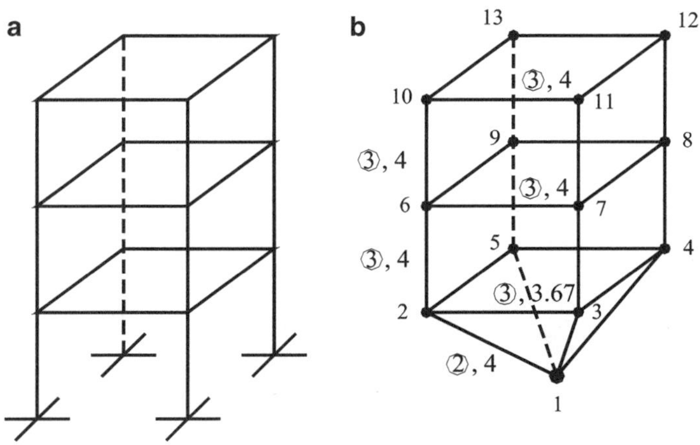

Fig. 2.16 A space frame, and CLNs and Ins of its members. (**a**) A space structure. (**b**) The graph model S of the structure

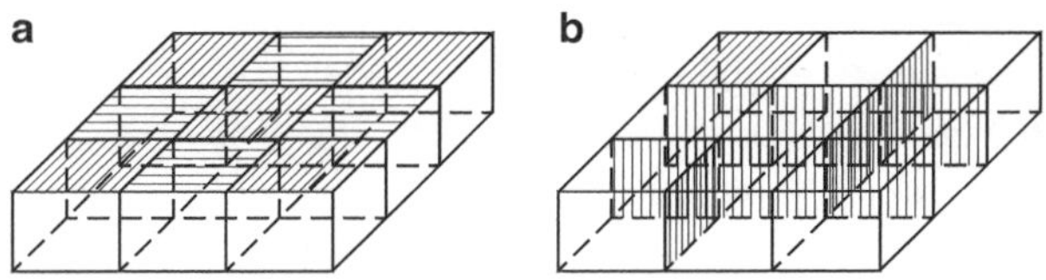

Fig. 2.17 Minimal and suboptimal cycle bases of S. (**a**) A minimal cycle basis. (**b**) A suboptimal cycle basis

Algorithm A selects all three-sided cycles of S except those shaded in Fig. 2.17b, with $\chi(\mathbf{CC}^t) = 190$. It will be noticed that, for structures containing nodes of higher degrees, considerable improvement is obtained by the use of Algorithm A.

Example 3. Consider a space frame as shown in Fig. 2.18, for which $b_1(S) = 10$. The minimal cycle basis selected by Algorithm 3 consists of the following cycles,

$$
\begin{aligned}
&C_1 = (1,2,3), C_2 = (4,5,6), C_3 = (7,8,9), C_4 = (10,11,12),\\
&C_5 = (1,2,5,4), C_6 = (2,3,6,5), C_7 = (4,5,8,7), C_8 = (5,6,9,8),\\
&C_9 = (7,8,11,10), C_{10} = (8,9,12,11),
\end{aligned}
$$

corresponding to $\chi(\mathbf{C}) = 4 \times 3 + 6 \times 4 = 36$ and $\chi(\mathbf{CC}^t) = 10 + 2\,[0 + 0 + 0 + 2 + 3 + 3 + 4 + 3 + 4] = 10 + 2 \times 19 = 48$.

However, the following non-minimal cycle basis has a higher $\chi(\mathbf{C})$, and leads to a more sparse $\mathbf{CC}^t$ matrix. The selected cycles are as follows,

Fig. 2.18 A space frame S

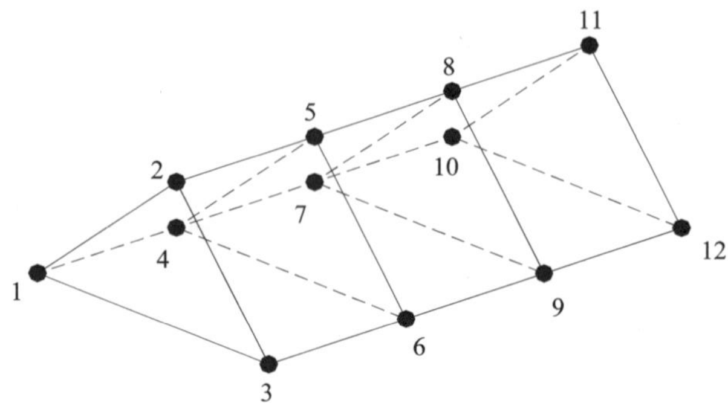

$$C_1 = (1,2,3), C_2 = (1,2,5,4), C_3 = (2,3,6,5), C_4 = (1,3,6,4),$$
$$C_5 = (4,5,8,7), C_6 = (5,6,9,8), C_7 = (4,6,9,7), C_8 = (7,8,11,10),$$
$$C_9 = (8,9,12,11), C_{10} = (10,11,12),$$

for which $\chi(\mathbf{C}') = 2 \times 3 + 8 \times 4 = 38$ corresponding to $\chi(\mathbf{C}'\mathbf{C}'^{t}) = 10 + 2$ $[1 + 2 + 3 + 1 + 2 + 3 + 1 + 2 + 2] = 10 + 2 \times 16 = 42$.

Therefore, the idea of having an optimal cycle basis in between minimal cycle bases is incorrect.

Example 4. Consider the skeleton of a structure S, comprising of six flipped flags, as shown in Fig. 2.19a, for which $b_1(S) = 6$. After contraction, S′ is obtained as illustrated in Fig. 2.19b. Obviously, this is a planar graph. The CLNs for the members are 3 and IN for member (1, 2) is 6 and for the remaining members it is equal to 1., Algorithm 3 selects a minimal cycle basis for S′, consists of six 3-sided cycles, corresponding to:

$$\chi(\mathbf{C}) = 6 \times 3 = 18 \text{ and } \chi(\mathbf{C}\mathbf{C}^{t}) = 6 + 2[0 + 1 + 2 + 3 + 4 + 5] = 6 + 2 \times 15 = 36$$

However, the following non-minimal cycle basis has a higher $\chi(S')$, and leads to a lower sparsity, $\chi(\mathbf{C}'\mathbf{C}'^{t})$:

$$C_1 = (1,3,2,4), C_2 = (1,4,2,5), C_3 = (1,2,3), C_4 = (1,2,6),$$

$$C_5 = (1,6,2,7), C_6 = (1,7,2,8).$$

For this basis, $\chi(\mathbf{C}') = 4 \times 4 + 2 \times 3 = 22$, corresponding to $\chi(\mathbf{C}'\mathbf{C}'^{t}) = 6 + 2$ $[0 + 1 + 1 + 1 + 1 + 1] = 6 + 2 \times 5 = 16$. After the back transformation from S′ to S, we have $\chi(\mathbf{C}) = 4 \times 6 + 2 \times 4 = 32$, corresponding to $\chi(\mathbf{C}\mathbf{C}') = 6 + 2$ $[0 + 1 + 1 + 1 + 1 + 1] = 16$.

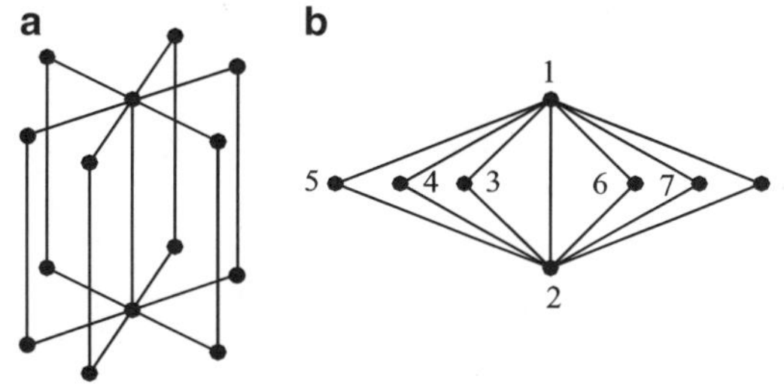

Fig. 2.19 A flipped flag before and after contraction. (**a**) S. (**b**) S′

2.4.6 An Improved Turn Back Method for the Formation of Cycle Bases

In this section, the combinatorial Turn-back method of Kaveh [15] is improved to obtain shorter cycle bases. This method covers all the counter examples, known for the minimality of the selected cycle bases.

Step 1: Generate an SRT rooted from an arbitrary node O. Identify its chords, and order them according to their distance numbers from O.

Step 2: Select the shortest length cycle of the graph on a chord and add this chord (generator) to the tree members. Repeat this process to all the chords, forming cycles of the least length containing the tree members and the previously used chords only. The selected cycles are all admissible, i.e. the addition of each cycle increases the first Betti number of the expanded part of the graph by unity. Store these cycles in C.

Step 3: Form all the new cycles of the same length on the remaining chords, allowing the use of more than one unused chords in their formation.

Step 4: Control the cycles formed in Step 3 to find only one cycle having a generator, which is in none of the other connected cycles formed in Step 3. When such a chord is found, add the corresponding cycle to C and include its generator in the tree members. Repeat this control until no such a cycle can be found.

Step 5: Select a cycle of the next higher length in the graph containing only one chord. Add the selected cycle to C and its generator to the tree members.

Step 6: Control the cycles formed in Step 3 to find a cycle containing only one unused chord. Add such a cycle to C and add its chord to the tree members. Repeat this control until no cycle of this property can be found.

Step 7: Repeat Step 4.

Step 8: Repeat Steps 5 and 6 and continue this repetition with the same length until no cycle in Step 5 can be found.

Step 9: Repeat Steps 3 to 8, until $b_1(S)$ cycles forming a cycle basis is included in C.

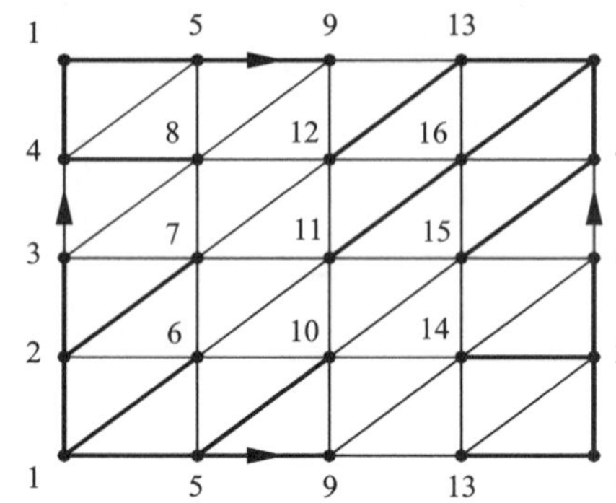

Fig. 2.20 Graph S and the selected SRT

2.4.7 Examples

Example 1. A graph is considered in the form of the 1-skeleton of a torus-type structure, Fig. 2.20. An SRT is selected, as shown in bold lines. The cycles selected in Step 2 are given in the following:

$$\begin{aligned} C = \{&(1,2,6),(1,4,5),(1,5,6),(1,2,13),(1,4,16),(1,13,16),(2,3,7),\\ &(2,6,7),(2,3,14),(2,13,14),(4,5,8),(4,15,16),(5,6,10),\\ &(5,9,10),(5,8,9),(12,13,16),(11,12,16),(11,15,16)\}. \end{aligned}$$

The execution of Step 3 results in the following cycles:

$$\begin{aligned} &(3,7,8),(3,4,8),(7,11,12),(7,8,12),(8,9,12),(9,13,14),(9,10,14),\\ &(10,14,15),(10,11,15),(9,12,13),(3,14,15),(3,4,15). \end{aligned}$$

Twelve cycles are generated, increasing the first Betti number by twelve. The control of Step 4, leads to generators {10, 11} and {7, 11} corresponding to the cycles (10, 11, 15) and (7, 11, 12), respectively. Thus no cycle is selected.

In Step 5, a cycle of length 4 containing an unused chord is formed. On {3, 4}, cycle (1, 2, 3, 4) is generated and added to C. Then in Step 6, the following cycles are added to C:

$$(3,4,8) \text{ for } \{3,8\}, (3,7,8) \text{ for } \{7,8\}, (3,4,15) \text{ for } \{3,15\}, (3,14,15) \text{ for } \{14,15\}.$$

In Step 7 no cycle is found, but in Step 8, the execution of Step 5 leads to cycle (1, 5, 9, 13) on {9, 13}, and Step 6 leads to the following cycles completing C, and forming a minimal cycle basis of S:

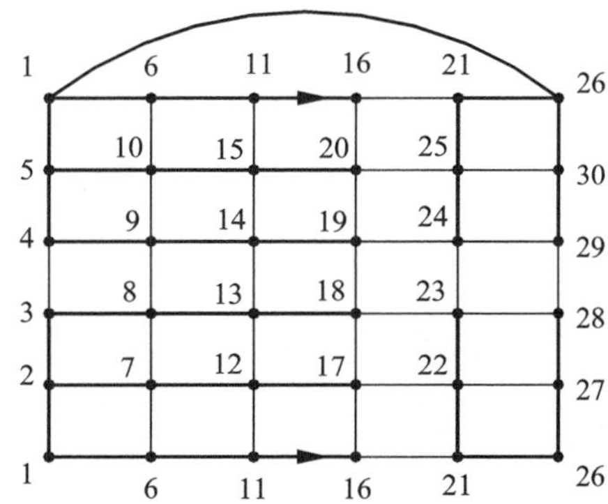

Fig. 2.21 A space graph and the selected SRT

(9, 12, 13) for {9, 12}, (9, 13, 14) for {9, 14}, (8, 9, 12) for {8, 12}, (7, 8, 12) for {7, 12}, (7, 11, 12)for {7, 11}, (9, 10, 14) for {10, 14}, (10, 14, 15) for {10, 15}, and (10, 11, 15) for {10, 11}.

Example 2. A space graph is considered as illustrated in Fig. 2.21. An SRT is selected as shown in bold lines. The application of Step 2, leads to the following cycle set:

$$C = \{(1,2,6,7),(1,5,6,10),(2,3,7,8),(4,5,9,10),(6,7,11,12),(6,10,11,15),\\ (7,8,12,13),(9,10,14,15),(11,12,16,17),(11,15,16,20),(12,13,17,18),\\ (14,15,19,20),(21,22,26,27),(21,25,26,30),(22,23,27,28),(24,25,29,30)\}.$$

In Step 3, the following cycles are generated:

$$(3,4,8,9),(8,9,13,14),(13,14,18,19),(16,17,21,22),(17,18,22,23),\\ (18,19,22,23),(18,19,23,24),(19,20,24,25),(16,20,21,25),(23,24,28,29).$$

These cycles contain 11 unused chords. The control of Step 4 shows that {3, 4} and {28, 29} are included in one cycle, and therefore all the chords remain unused. In the next step, a cycle of length 5 including an unused chord is generated and added to C. Only with chord {3, 4}, the 5-sided cycle (1, 2, 3, 4, 5) is generated, and in Step 6 the following three-sided cycles are selected:

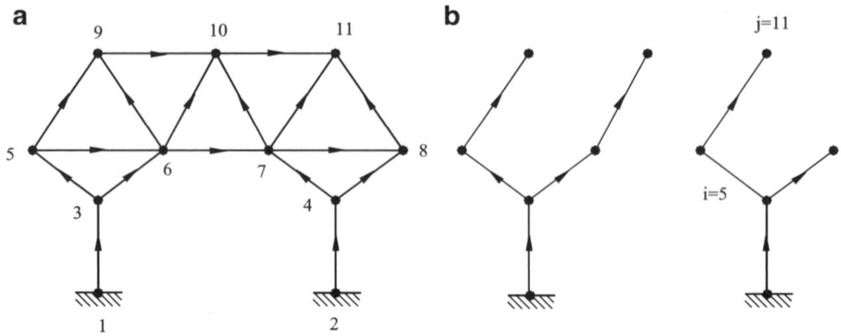

Fig. 2.22 S and two of its SR subtrees

$$(3, 4, 8, 9), (8, 9, 13, 14), \text{and } (13, 14, 18, 19).$$

Step 7 is carried out and cycle (23, 24, 28, 29) on {28, 29} is found, repetition of this control leads to cycle (18, 19, 23, 24) on {23, 24}. In the next step, no cycle is selected. The execution of Steps 3 and 4 in Step 9 results in no cycle.

The execution of Step 5 in Step 9, forms cycle (1, 6, 11, 16, 21, 26) on chord {16, 21}, and the execution of Step 6 leads to the following cycles,

$$(16, 20, 21, 25) \text{ for } \{20, 25\}, (19, 20, 24, 25) \text{ for } \{19, 24\}, (16, 17, 21, 22) \text{ for } \{17, 22\}, \text{and } (17, 18, 22, 23) \text{ for } \{18, 23\}.$$

The selected cycles form a minimal cycle basis.

2.4.8 *Formation of $\mathbf{B}_0$ and $\mathbf{B}_1$ Matrices*

In order to generate the elements of a $\mathbf{B}_0$ matrix, a basic structure of S should be selected. For this purpose a spanning forest consisting of NG(S) SRTs is used, where NG(S) is the number of ground (support) nodes of S. As an example, for S shown in Fig. 2.22a, two SR subtrees are generated, Fig. 2.22b.

The orientation assigned to each member of S is from the lower numbered node to its higher numbered end. For each SR subtree, the orientation is given in the direction of its growth from its support node.

MATRIX $\mathbf{B}_0$: This is a 6M(S) × 6NL(S) matrix, where M(S) and NL(S) are the numbers of members and loaded nodes of S, respectively. If all the free nodes are loaded, then

$$\text{NL(S)} = \text{N(S)} - \text{NG(S)},$$

where NG(S) is the number of support nodes.

For a member, the internal forces are represented by the components at the lower numbered end. Obviously the components at the other end can be obtained by considering the equilibrium of the member.

The coefficients of $\mathbf{B}_0$ can be obtained by considering the transformation of each joint load to the ground node of the corresponding subtree. $[\mathbf{B}_0]_{ij}$ for member i and node j is given by a 6 × 6 submatrix as,

$$[\mathbf{B}_0]_{ij} = \alpha_{ij} \begin{bmatrix} 1 & 0 & 0 & 0 & 0 & 0 \\ 0 & 1 & 0 & 0 & 0 & 0 \\ 0 & 0 & 1 & 0 & 0 & 0 \\ 0 & -\Delta z & \Delta y & 1 & 0 & 0 \\ \Delta z & 0 & -\Delta x & 0 & 1 & 0 \\ -\Delta y & \Delta x & 0 & 0 & 0 & 1 \end{bmatrix}, \tag{2.63}$$

in which Δx, Δy and Δz are the differences of the coordinates of node j with respect to the lower numbered end of member i, in the selected global coordinate system, and α_{ij} is the orientation coefficient defined as:

$$\alpha_{ij} = \begin{cases} +1 \text{ if member is positively oriented in the tree containing j,} \\ -1 \text{ if member is negatively oriented in the tree containing j,} \\ \;\;0 \text{ if member is not in the tree containing node j.} \end{cases}$$

The $\mathbf{B}_0$ matrix can be obtained by assembling the $[\mathbf{B}_0]_{ij}$ submatrices as shown schematically in the following:

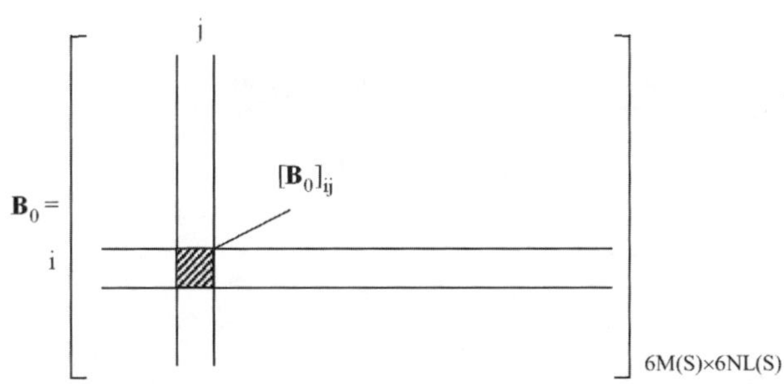

(2.64)

MATRIX $\mathbf{B}_1$: This is a $6M(S) \times 6b_1(S)$ matrix, which can be formed using the elements of a selected cycle basis. For a space structure, six self-equilibrating stress systems can be formed on each cycle. Consider C_j and take a member of this cycle as its generator. Cut the generator in the neighbourhood of its beginning node and apply six bi-actions as illustrated in Fig. 2.23.

The internal forces under the application of each bi-action are a self-equilibrating stress system As for the matrix $\mathbf{B}_0$, a submatrix $[\mathbf{B}_1]_{ij}$ of $\mathbf{B}_1$ is a 6 × 6 submatrix, the columns of which show the internal forces at the lower numbered end of member i under the application of six bi-actions at the cut of the generator j,

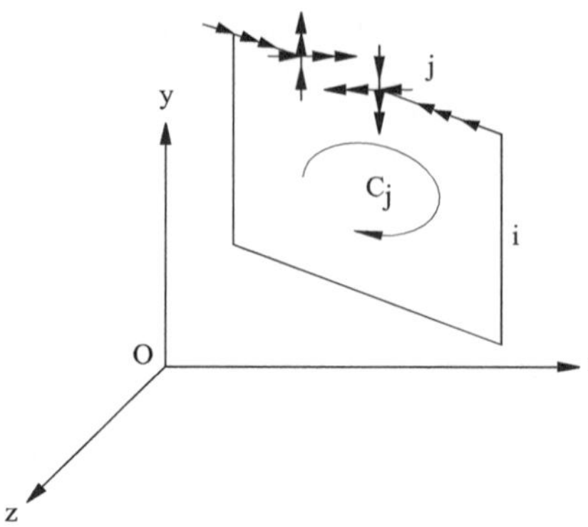

Fig. 2.23 A cycle and the considered bi-action at a cut

$$[\mathbf{B}_1]_{ij} = \beta_{ij} \begin{bmatrix} 1 & 0 & 0 & 0 & 0 & 0 \\ 0 & 1 & 0 & 0 & 0 & 0 \\ 0 & 0 & 1 & 0 & 0 & 0 \\ 0 & -\Delta z & \Delta y & 1 & 0 & 0 \\ \Delta z & 0 & -\Delta x & 0 & 1 & 0 \\ -\Delta y & \Delta x & 0 & 0 & 0 & 1 \end{bmatrix}, \tag{2.65}$$

in which Δx, Δy and Δz are the differences of the coordinates x, y and z of the beginning node of the generator j and the beginning node of the member i. The orientation coefficient β_{ij} is defined as:

$$\beta_{ij} = \begin{cases} +1 & \text{if member i has the same orientation of the cycle generated on j,} \\ -1 & \text{if member i has the reverse orientation of the cycle generated on j,} \\ 0 & \text{if member is not in the cycle whose generator is j.} \end{cases}$$

The pattern of $\mathbf{B}_1$ containing $[\mathbf{B}_1]_{ij}$ submatrices is shown in the following:

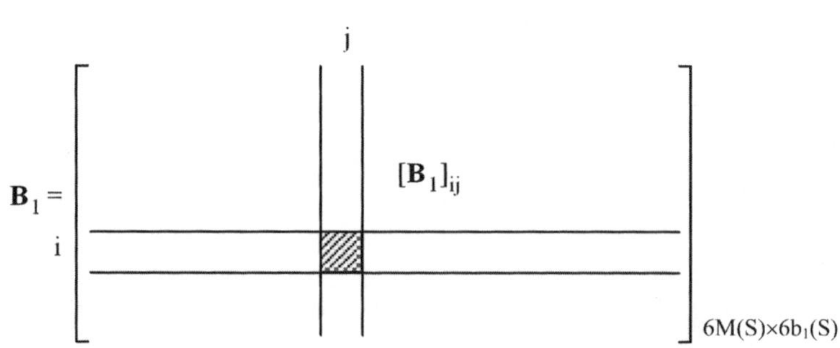

(2.66)

Subroutines for the formation of $\mathbf{B}_0$ and $\mathbf{B}_1$ matrices are included in the program presented in Ref. [19].

Fig. 2.24 A four by four planar frame S

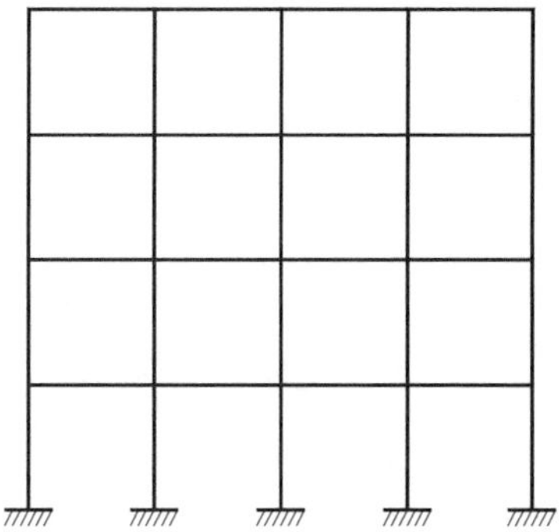

Fig. 2.25 Patterns of $\mathbf{B}_1$ and $\mathbf{B}_1^t\mathbf{B}_1$ matrices for S. (a) Pattern of $\mathbf{B}_1$. (b) Pattern of $\mathbf{B}_1^t\mathbf{B}_1$

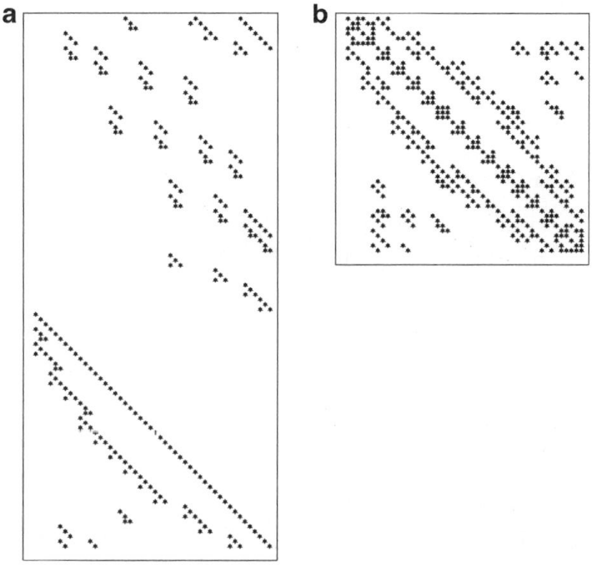

Fig. 2.26 A simple space frame S

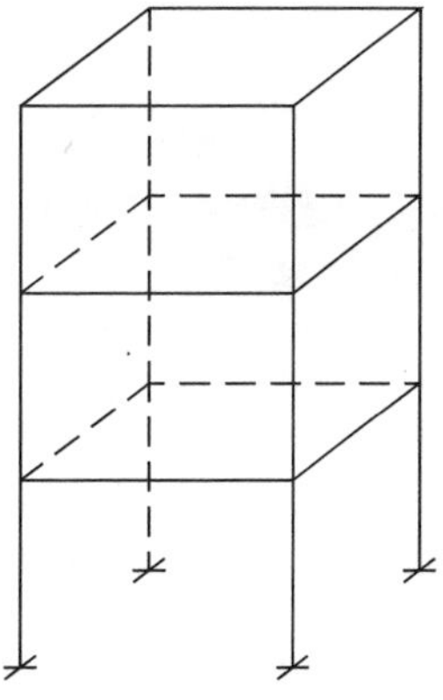

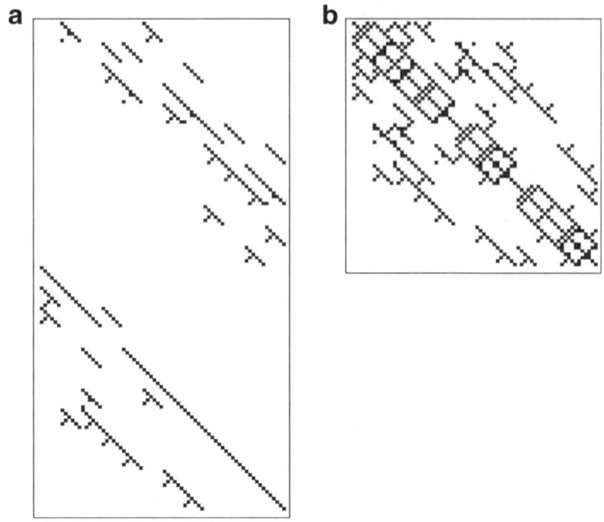

Fig. 2.27 Patterns of $\mathbf{B}_1$ and $\mathbf{B}_1^t\mathbf{B}_1$ matrices for S. (**a**) Pattern of $\mathbf{B}_1$. (**b**) Pattern of $\mathbf{B}_1^t\mathbf{B}_1$

Example 1. A four by four planar frame is considered as shown in Fig. 2.24.

The patterns of $\mathbf{B}_1$ and $\mathbf{B}_1^t\mathbf{B}_1$ formed on the elements of the cycle basis selected by any of the methods of the previous section are depicted in Fig. 2.25, corresponding to $\chi(\mathbf{B}_1) = 241$ and $\chi(\mathbf{B}_1^t\mathbf{B}_1) = 388$.

Example 2. A one-bay three-storey frame is considered as shown in Fig. 2.26.

The patterns of $\mathbf{B}_1$ and $\mathbf{B}_1^t\mathbf{B}_1$ matrices formed on the elements of the cycle basis selected by any of the graph theoretical algorithms of the previous Section are shown in Fig. 2.27, corresponding to $\chi(\mathbf{B}_1) = 310$ and $\chi(\mathbf{B}_1^t\mathbf{B}_1) = 562$.

Once $\mathbf{B}_0$ and $\mathbf{B}_1$ are computed, the remaining steps of the analysis are the same as those presented in Sect. 2.3.6. The interested reader may also refer to standard textbooks such as those of McGuire and Gallagher [31], Przemieniecki [32], or Pestel and Leckie [33] for further information.

2.5 Generalized Cycle Bases of a Graph

In this section, S is considered to be a connected graph. For $\gamma(S) = aM(S) + bN(S) + c\gamma_0(S)$, the coefficients b and c are assumed to be integer multiples of the coefficient $a > 0$. Only those coefficients given in Table 2.1 are of interest.

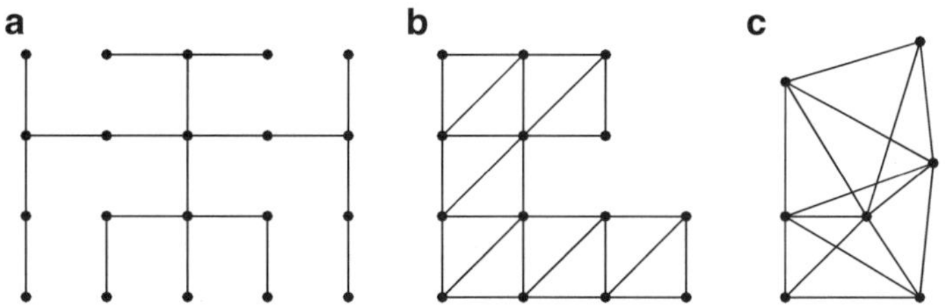

Fig. 2.28 Examples of γ-trees (**a**) $\gamma(S) = 3M - 3N + 3$. (**b**) $\gamma(S) = M - 2N + 3$. (**c**) $\gamma(S) = M - 3N + 6$

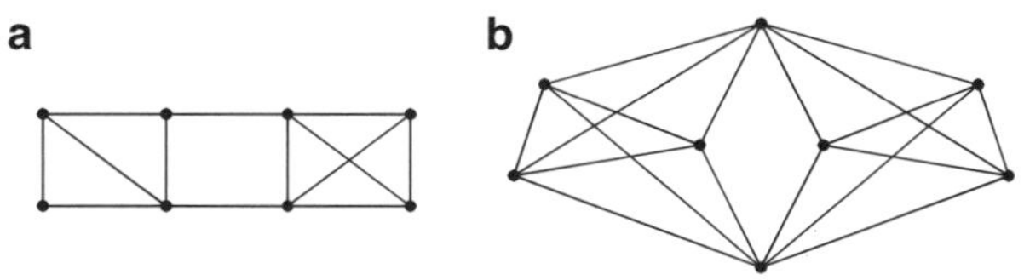

Fig. 2.29 Structures satisfying $\gamma(T) = 0$ which are not rigid. (**a**) $\gamma(S) = M - 2N + 3$. (**b**) $\gamma(S) = M - 3N + 6$

2.5.1 Definitions

Definition 1. A subgraph S_i is called an *elementary subgraph* if it does not contain a subgraph $S_i' \subseteq S_i$ with $\gamma(S_i') > 0$. A connected rigid subgraph T of S containing all the nodes of S is called a *γ-tree* if $\gamma(T) = 0$. For $\gamma(S_i) = b_1(S_i)$, a γ-tree becomes a tree in graph theory.

Obviously a structure whose model is a γ-tree is statically determinate when γ(S) describes the degree of static indeterminacy of the structure. The ensuing stress resultants can uniquely be determined everywhere in the structure by equilibrium only. Examples of γ-trees are shown in Fig. 2.28.

Notice that $\gamma(T) = 0$ does not guarantee the rigidity of a γ-tree. For example, the graphs models depicted in Fig. 2.29 both satisfy $\gamma(T) = 0$; however, neither is rigid.

Definition 2. A member of $S - T$ is called a *γ-chord* of T. The collection of all γ-chords of a γ-tree is called the *γ-cotree* of S.

Definition 3. A *removable subgraph* S_j of a graph S_i, is the elementary subgraph for which $\gamma(S_i - S_j) = \gamma(S_i)$, i.e. the removal of S_j from S_i does not alter its DSI. A γ-tree of S containing two chosen nodes, which has no removable subgraph is called a *γ-path* between these two nodes.

As an example, the graphs shown in Fig. 2.30 are γ-paths between the specified nodes n_s and n_t.

Definition 4. A connected rigid subgraph of S with $\gamma(C_k) = a$, which has no removable subgraph is termed a *γ-cycle* of S. The total number of members of

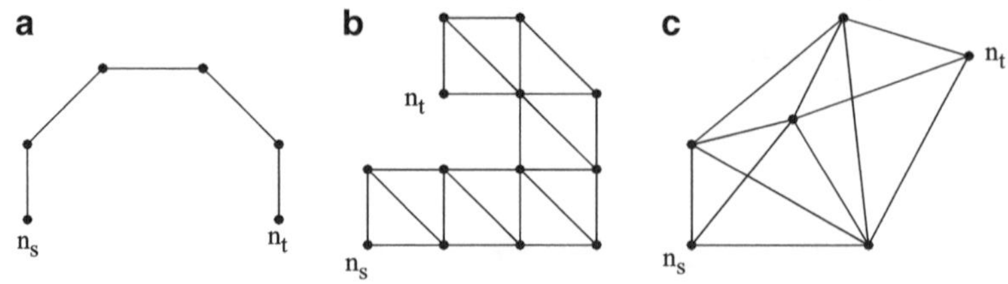

Fig. 2.30 Examples of γ-paths. (**a**) γ(S) = α(M − N + 1). (**b**) γ(S) = M − 2N + 3. (**c**) γ(S) = M − 3N + 6

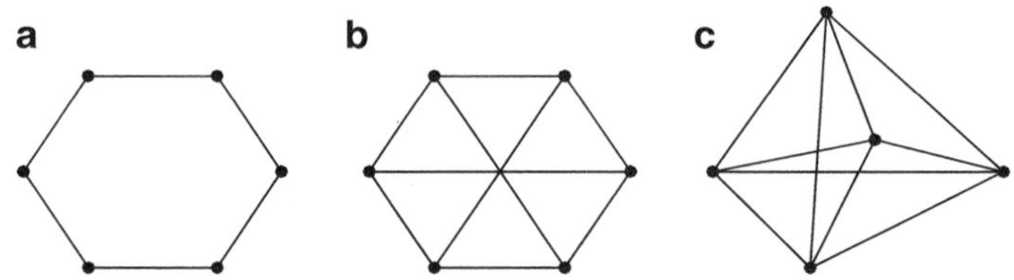

Fig. 2.31 Examples of γ-cycles. (**a**) γ(S) = α(M − N + 1). (**b**) γ(S) = M − 2N + 3. (**c**) γ(S) = M − 3N + 6

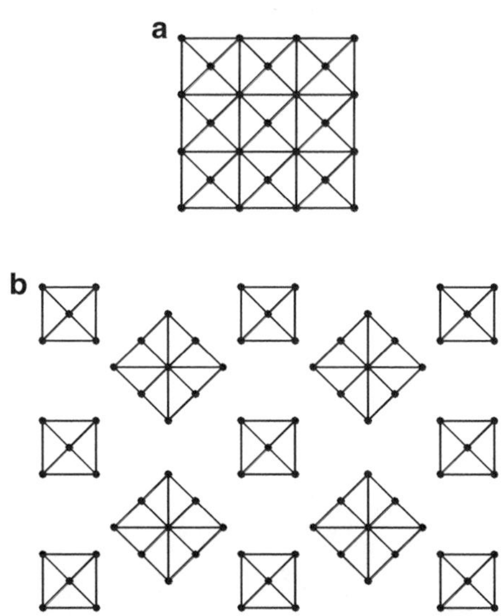

Fig. 2.32 A planar truss S, and the elements of a GCB of S. (**a**) A planar truss S. (**b**) A generalized cycle basis of S

C_k, denoted by $L(C_k)$, is called the *length* of C_k. Examples of γ-cycles are shown in Fig. 2.31.

Definition 5. Let m_i be a γ-chord of T. Then $T \cup m_i$ contains a γ-cycle C_i which is defined as a *fundamental γ-cycle* of S with respect to T. Using the Intersection Theorem of Sect. 2.2.2, it can easily be shown that,

$$\gamma(T \cup m_i) = 0 + (a + 2b + c) - (2b + c) = a,$$

indicating the existence of a γ-cycle in $T \cup m_i$. For a rigid T, the corresponding fundamental γ-cycle is also rigid, since the addition of an extra member between the existing nodes of a graph cannot destroy the rigidity. A fundamental γ-cycle can be obtained by omitting all the removable subgraphs of $T \cup m_i$.

Definition 6. A maximal set of independent γ-cycles of S is defined as a *generalized cycle basis* (GCB) of S. A maximal set of independent fundamental γ-cycles is termed a *fundamental generalized cycle basis* of S. The dimension of such a basis is given be $\eta(S) = \gamma(S)/a$.

As an example, a generalized cycle basis of a planar truss is illustrated in Fig. 2.32.

Definition 7. A *generalized cycle basis-member incidence matrix* **C** is an $\eta(S) \times M$ matrix with entries -1, 0 and +1, where $c_{ij} = 1$ (or -1) if γ-cycle C_i contains positively (or negatively) oriented member m_j, and $c_{ij} = 0$ otherwise. The *generalized cycle adjacency matrix* is defined as **D** which is an $\eta(S) \times \eta(S)$ matrix when undirected γ-cycles are considered; then the negative entries of **C** become positive.

2.5.2 *Minimal and Optimal Generalized Cycle Bases*

A generalized cycle basis $C = \{C_1, C_2, \ldots, C_{\eta(S)}\}$ is called *minimal* if it corresponds to a minimum value of:

$$L(C) = \sum_{i=1}^{\eta(S)} L(C_i). \tag{2.67}$$

Obviously, $\chi(\mathbf{C}) = L(C)$ and a minimal GCB can be defined as a basis which corresponds to minimum $\chi(\mathbf{C})$. A GCB for which L(C) is near minimum is called a *subminimal* GCB of S.

A GCB corresponding to maximal sparsity of the GCB adjacency matrix is called an *optimal* generalized cycle basis of S. If $\chi(\mathbf{CC}^t)$ does not differ considerably from its minimum value, then the corresponding basis is termed *suboptimal.*

The matrix intersection coefficient $\sigma_i(\mathbf{C})$ of row i of GCB incidence matrix **C** is the number of row j such that:

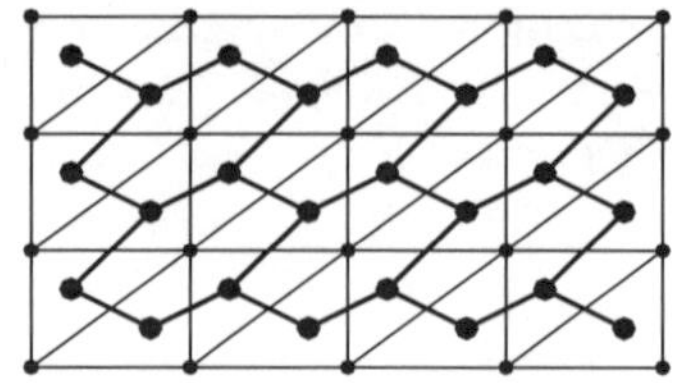

Fig. 2.33 A planar truss S and its associate graph A(S)

(a) $j \in \{i + 1, i + 2, \ldots, \eta(S)\}$,
(b) $C_i \cap C_j \neq \varnothing$, i.e. there is at least one k such that the column k of both γ-cycles C_i and C_j (rows i and j) contain non-zero entries.

Now it can be shown that:

$$\chi(\mathbf{CC}^t) = \eta(S) + 2\sum_{i=1}^{\eta(S)-1} \sigma_j(C). \tag{2.68}$$

This relationship shows the correspondence of a GCB incidence matrix **C** and that of its GCB adjacency matrix. In order to minimize $\chi(\mathbf{CC}^t)$, the value of $\sum_{i=1}^{\eta(S)-1} \sigma_j(C)$ should be minimized, since $\eta(S)$ is a constant for a given structure S, i.e. γ-cycles with a minimum number of overlaps should be selected.

2.6 Force Method for the Analysis of Pin-Jointed Planar Trusses

The methods described in Sect. 2.5 are applicable to the selection of generalized cycle bases for different types of skeletal structures. However, the use of these algorithms for trusses engenders some problems, which are discussed in Ref. [34]. In this section, two methods are developed for selecting suitable GCBs for planar trusses. In both methods, special graphs are constructed for the original graph model S of a truss, containing all the connectivity properties required for selecting a suboptimal GCB of S.

2.6.1 Associate Graphs for Selection of a Suboptimal GCB

Let S be the model of a planar truss with triangulated panels, as shown in Fig. 2.33. The associate graph of S, denoted by A(S), is a graph whose nodes are in a one-to-one correspondence with triangular panels of S, and two nodes of A(S) are connected by a member if the corresponding panels have a common member in S.

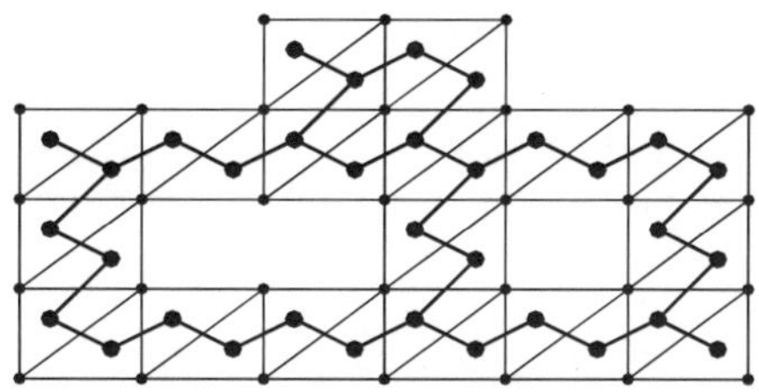

Fig. 2.34 S with two cut-outs and its A(S)

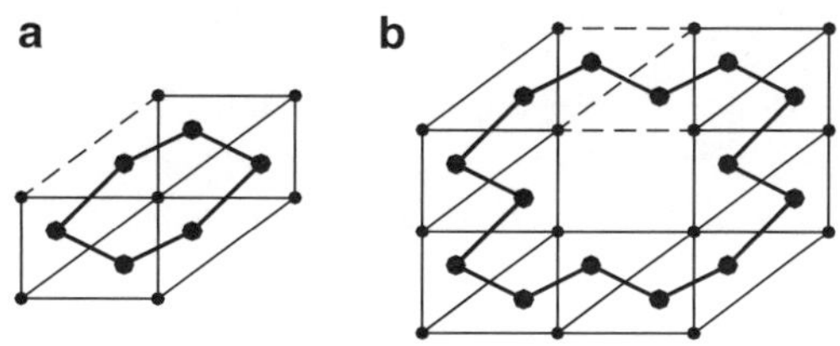

Fig. 2.35 Two different types of cycles. (**a**) A type C_I cycle. (**b**) A type C_{III} cycle

If S has some cut-outs, as shown in Fig. 2.34, then its associate graph can still be formed, provided that each cut-out is surrounded by triangulated panels.

For trusses containing adjacent cut-outs, a cut-out with cut-nodes in its boundary, or any other form violating the above-mentioned condition, extra members can be added to S. The effect of such members should then be included in the process of generating its self-equilibrating stress systems.

Theorem A. For a fully triangulated truss (except for the exterior boundary), as in Fig. 2.33, the dimension of a statical basis γ(S) is equal to the number of its internal nodes, which is the same as the first Betti number of its associate graph, i.e.

$$\gamma(S) = N_i(S) = b_1[A(S)]. \tag{2.69}$$

Proof. Let M′ and N′ be the numbers of members and nodes of A(S), respectively. By definition,

$$N' = R(S) - 1,$$

and $M' = M_i(S) = M(S) - M_e(S) = M(S) - N_e(S) = M(S) - [N(S) - N_i(S)]$.

Thus: $b_1[A(S)] = M' - N' + 1 = M(S) - [N(S) - N_i(S)] - R(S) + 1 + 1$ $= 2 - R(S) + M(S) - N(S) + N_i(S)$.

By Euler's polyhedron formula, we have:

$$2 - R(S) + M(S) - N(S) = 0.$$

Therefore:

For trusses which are not fully triangulated, we have:

$$\gamma(S) = N_i(S) - M_c(S).$$

A Cycle of A(S) and the Corresponding γ-Cycle of S. In Fig. 2.35a, a triangulated truss and its associate graph, which is a cycle, are shown for which

$$\gamma(S_i) = N_i = 1 = b_1[A(S)].$$

Since C_1 of A(S) corresponds to one γ-cycle of S, it is called a *type I cycle*, denoted by C_I. A γ-cycle of S is shown by continuous lines, and its γ-chords are depicted in dashed lines.

Figure 2.35b shows a truss unit with one cut-out. In general, if a cut-out is an m-gon, then the completion of the triangulation requires m−3 members. Instead, m internal nodes will be created, increasing the DSI by m. Hence Eq. 2.68 yields,

$$\gamma(S) = m - (m - 3) = 3,$$

while: $b_1[A(S)] = 1$.

However, in this case S contains three γ-cycles. A γ-path P and three γ-chords (dashed lines) are depicted in Fig. 2.35b. Obviously $P \cup m_i$ (i = 1, 2, 3) form three γ-cycles which correspond to a cycle of type C_{III} of A(S). Thus two types of cycles C_I and C_{III} should be recognized in A(S) and an appropriate number of γ-cycles will then be generated.

Algorithm AA

Step 1: Construct the associate graph A(S) of S.

Step 2: Select a mesh basis of A(S), using an appropriate cycle selection algorithm. For fully triangulated S, Algorithms 1–3 generate cycle bases with three-sided elements.

Step 3: Select the γ-cycles of S corresponding to the cycles of A(S). One γ- cycle for each cycle of type C_I, and three γ-cycles for each cycle of type C_{III} should be chosen.

Once a GCB is selected, on each γ-cycle one self-equilibrating stress system can easily be formed. Therefore, a statical basis with localized self-equilibrating stress systems will be obtained.

Example. Let S be the graph model of a planar truss, as shown in Fig. 2.34, for which γ(S) = 12. For A(S), six cycles of length 6 of type C_I and two cycles of lengths 18 and 26 of type C_{III} are selected. Therefore, the total of 6 + 3 × 2 = 12 γ-cycles of S are obtained. On each γ-cycle one self-equilibrating stress system is constructed, and a statical basis consisting of localized self-equilibrating stress systems is thus obtained.

2.6.2 Minimal GCB of a Graph

Theoretically a minimal GCB of a graph can be found using the Greedy Algorithm developed for matroids. This will be discussed in Kaveh [15, 20] after matroids have been introduced, and here only the algorithm is briefly outlined.

Consider the graph model of a structure, and select all of its γ-cycles. Order the selected γ-cycles in ascending order of length. Denote these cycles by a set C. Then perform the following steps:

Step 1: Choose a γ-cycle C_1 of the smallest length, i.e. $L(C_1) < L(C_i)$ for all $C_i \in C$
Step 2: Select the second γ-cycle C_2 from $C - \{C_1\}$ which is independent of C_1 and $L(C_2) \leq L(C_i)$ for all γ-cycles of $C - \{C_1\}$.
Step k: Subsequently choose a γ-cycle C_k from $C - \{C_1, C_2, \ldots, C_{k-1}\}$ which is independent of $C_1, C_2, \ldots, C_{k-1}$ and $L(C_k) \leq L(C_i)$ for all $C_i \in C - \{C_1, C_2, \ldots, C_{k-1}\}$.

After η(S) steps, a minimal GCB will be selected by this process, a proof of which can be found in Kaveh [19].

2.6.3 Selection of a Subminimal GCB: Practical Methods

In practice, three main difficulties are encountered in an efficient implementation of the Greedy Algorithm. These difficulties are briefly mentioned in the following:

1. Selection of some of the γ-cycles for some γ(S) functions.
2. Formation of all of the γ-cycles of S.
3. Checking the independence of γ-cycles.

In order to overcome the above difficulties, various methods are developed. The bases selected by these approaches correspond to very sparse GCB adjacency matrices, although these bases are not always minimal.

Method 1. This is a natural generalization of the method for finding a fundamental cycle basis of a graph, and consists of the following steps:

Step 1: Select an arbitrary γ-tree of S, and find its γ-chords.
Step 2: Add one γ-chord at a time to the selected γ-tree to form fundamental γ-cycles of S with respect to the selected γ-tree.

The main advantage of this method is the fact that the independence of γ-cycles is guaranteed by using a γ-tree. However, the selected γ-cycles are often quite long, corresponding to highly populated CCB adjacency matrices.

Method 2. This is an improved version of Method 1, in which a special γ-tree has been employed and each γ-chord is added to γ-tree members after being used for formation of a fundamental γ-cycle.

Step 1. Select the centre "O" of the given graph. Methods for selecting such a node will be discussed in Chap. 5.

Step 2: Generate a shortest route γ-tree rooted at the selected node O, and order its γ-chords according to their distance from O. The *distance* of a member is taken as the sum of the shortest paths between its end nodes and O.

Step 3: Form a γ-cycle on the γ-chord of the smallest distance number, and add the used γ-chord to the tree members, i.e. form $T \cup m_1$.

Step 4: Form the second γ-cycle on the next nearest γ-chord to O, by finding a γ-path in $T \cup m_1$ (not through m_2). Then add the second used γ-chord m_2 to $T \cup m_1$ obtaining $T \cup m_1 \cup m_2$.

Step 5: Subsequently form the *k*th γ-cycle on the next unused γ-chord nearest to O, by finding a γ-path in the $T \cup m_1 \cup m_2 \cup \ldots \cup m_{k-1}$ (not through m_k). Such a γ-path together with m_k forms a γ-cycle.

Step 6: Repeat Step 5 until η(S) of γ-cycles are selected.

Addition of the used γ-chords to the γ-tree members leads to a considerable reduction in the length of the selected γ-cycles, while maintaining the simplicity of the independence check.

In this algorithm, the use of an SRT, orders the nodes and members of the graph. Such an ordering leads to fairly banded member-node incidence matrices. Considering the columns corresponding to tree members as independent columns, a base is effectively selected for the cycle matroid of the graph, Kaveh [34].

Method 3. This method uses an expansion process, at each step of which one independent γ-cycle is selected and added to the previously selected ones. The independence is secured using an admissibility condition defined as follows.

A γ-cycle C_{k+1} added to the previous selected γ-cycles $C^k = C_1 \cup C_2 \cup \ldots \cup C_k$ is called *admissible* if,

$$\gamma\left(C^k \cup C_{k+1}\right) = \gamma\left(C^k\right) + a, \tag{2.70}$$

where "a" is the coefficient defined in Table 2.1. The algorithm can now be described as follows.

Step 1: Select the first γ-cycle of minimal length C_1.

Step 2: Select the second γ-cycle of minimal length C_2 which is independent of C_1, i.e. select the second admissible γ-cycle of minimal length.

Step k: Subsequently, find the *k*th admissible γ-cycle of minimal length. Continue this process until η(S) independent γ-cycles forming a subminimal GCB are obtained.

A γ-cycle of minimal length can be generated on an arbitrary member by adding a γ-path of minimal length between the two end nodes of the member (not through the member itself). The main advantage of this algorithm is avoiding the formation of all γ-cycles of S and also the independence control, which becomes feasible by graph theoretical methods.

The above methods are elaborated for specific $\gamma(S)$ functions in subsequent sections, and examples are included to illustrate their simplicity and efficiency.

2.7 Algebraic Force Methods of Analysis

Combinatorial methods for the force method of structural analysis have been presented in previous sections. These methods are very efficient for skeletal structures and in particular for rigid-jointed frames. For general structures, the underlying graph of self-equilibrating stress systems will be discussed in Chaps. 6 and 7. Algebraic methods can be formulated in a more general form to cover different types of structures such as skeletal structures and finite element models. The main drawbacks of pure algebraic methods are the larger storage requirements, and the higher number of operations.

2.7.1 Algebraic Methods

Consider a discrete or discretized structure S, which is assumed to be statically indeterminate. Let $\mathbf{r}$ denote the m-dimensional vector of generalized independent element (member) forces, and $\mathbf{p}$ the n-vector of nodal loads. The equilibrium conditions of the structure can then be expressed as,

$$\mathbf{Ar} = \mathbf{p}, \tag{2.71}$$

where $\mathbf{A}$ is an $n \times m$ *equilibrium matrix*. The structure is assumed to be rigid, and therefore, $\mathbf{A}$ has a full rank, i.e. $t = m-n > 0$, and rank $\mathbf{A} = n$.

The member forces can be written as,

$$\mathbf{r} = \mathbf{B}_0\mathbf{p} + \mathbf{B}_1\mathbf{q}, \tag{2.72}$$

where $\mathbf{B}_0$ is an $m \times n$ matrix such that $\mathbf{AB}_0$ is an $n \times n$ identity matrix, and $\mathbf{B}_1$ is an $m \times t$ matrix such that $\mathbf{AB}_1$ is an $n \times t$ zero matrix. $\mathbf{B}_0$ and $\mathbf{B}_1$ always exist for a structure and in fact many of them can be found for a structure. $\mathbf{B}_1$ is called a *self-stress matrix* as well as *null basis matrix*. Each column of $\mathbf{B}_1$ is known as a *null vector*. Notice that the null space, null basis and null vectors correspond to complementary solution space, statical basis and self-equilibrating stress systems, respectively, when S is taken as a general structure.

Minimizing the potential energy requires that $\mathbf{r}$ minimize the quadratic form,

$$\frac{1}{2}\mathbf{r}^t\mathbf{F}_m\mathbf{r}, \tag{2.73}$$

subject to the constraint as in Eq. 2.71. $\mathbf{F}_m$ is an m × m block diagonal element flexibility matrix. Using Eq. 2.72, it can be seen that $\mathbf{q}$ must satisfy the following equation.

$$\left(\mathbf{B}_1^t\mathbf{F}_m\mathbf{B}_1\right)\mathbf{q} = -\mathbf{B}_1^t\mathbf{F}_m\mathbf{B}_0\mathbf{p}, \tag{2.74}$$

where $\mathbf{B}_1^t\mathbf{F}_m\mathbf{B}_1 = \mathbf{G}$ is the *overall flexibility matrix of the structure*. Computing the redundant forces $\mathbf{q}$ from Eq. 2.49, $\mathbf{r}$ can be found using Eq. 2.9. The structure of $\mathbf{G}$ is again important and its sparsity, bandwidth and conditioning govern the efficiency of the force method. For the sparsity of $\mathbf{G}$ one can search for a sparse $\mathbf{B}_1$ matrix, which is often referred to as the sparse null basis problem.

Many algorithms exist for computing a null basis $\mathbf{B}_1$ of a matrix $\mathbf{A}$. For the moment let $\mathbf{A}$ be partitioned so that,

$$\mathbf{AP} = [\mathbf{A}_1, \mathbf{A}_2], \tag{2.75}$$

where $\mathbf{A}_1$ is n × n and non-singular, and $\mathbf{P}$ is a permutation matrix that may be required in order to ensure that $\mathbf{A}_1$ is non-singular. One can write:

$$\mathbf{B}_1 = \mathbf{P}\begin{bmatrix} -\mathbf{A}_1^{-1}\mathbf{A}_2 \\ \mathbf{I} \end{bmatrix}. \tag{2.76}$$

By simple multiplication it becomes obvious that:

$$\mathbf{AB}_1 = [\mathbf{A}_1 \quad \mathbf{A}_2]\begin{bmatrix} -\mathbf{A}_1^{-1}\mathbf{A}_2 \\ \mathbf{I} \end{bmatrix} = \mathbf{0}.$$

A permutation $\mathbf{P}$ that yields a non-singular $\mathbf{A}_1$ matrix can be chosen purely symbolically, but this says nothing about the possible numerical conditioning of $\mathbf{A}_1$ and the resulting $\mathbf{B}_1$.

In order to control the numerical conditioning, pivoting must be employed. There are many such methods based on various matrix factorizations, including the Gauss-Jordan elimination, **QR**, **LU**, **LQ** and Turn-back method. Some of these methods are briefly studied in the following:

Gauss-Jordan Elimination Method. In this approach one creates an n × n identity matrix $\mathbf{I}$ in the first columns of $\mathbf{A}$ by column changes and a sequence of n pivots. This procedure can be expressed as,

$$\mathbf{G}_n\mathbf{G}_{n-1}\ldots\mathbf{G}_2\mathbf{G}_1\mathbf{AP} = [\mathbf{I}, \mathbf{M}], \tag{2.77}$$

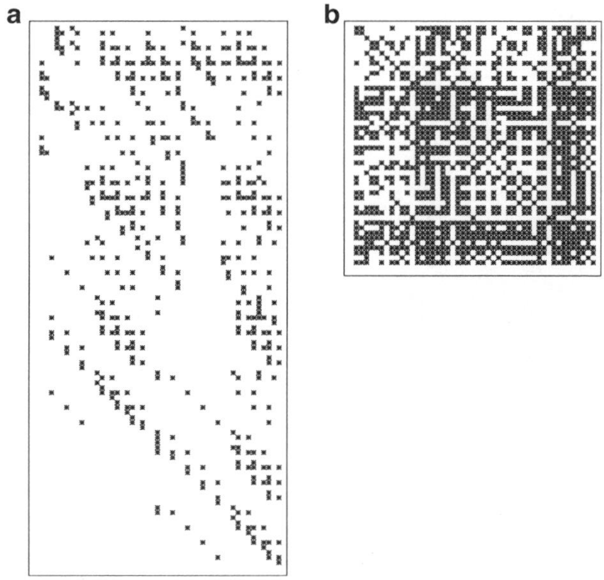

Fig. 2.36 Patterns of B_1 and $\mathbf{B}_1^t\mathbf{B}_1$ matrices for S using Gauss-Jordan elimination method, Example 1. (**a**) Pattern of B_1. (**b**) Pattern of $\mathbf{B}_1^t\mathbf{B}_1$

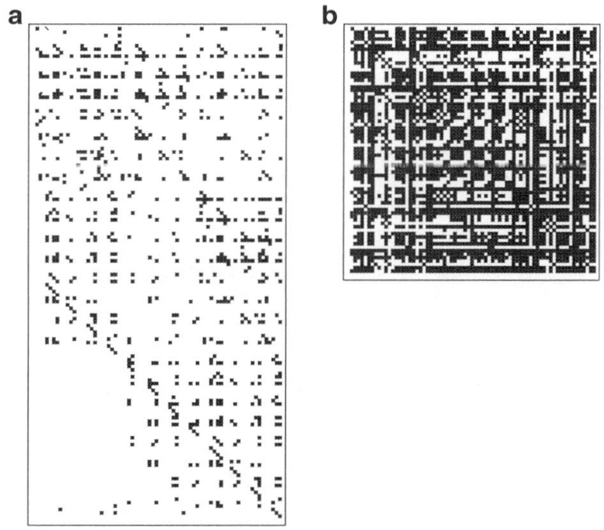

Fig. 2.37 Patterns of B_1 and $\mathbf{B}_1^t\mathbf{B}_1$ matrices for S using Gauss-Jordan elimination method, Example 2. (**a**) Pattern of B_1. (**b**) Pattern of $\mathbf{B}_1^t\mathbf{B}_1$

where $\mathbf{G}_i$ is the *i*th pivot matrix and $\mathbf{P}$ is an m $\times$ m column permutation matrix (so $\mathbf{P}^t = \mathbf{P}$) and $\mathbf{I}$ is an n$\times$n identity matrix, and $\mathbf{M}$ is an n $\times$ t matrix. Denoting $\mathbf{G}_n\mathbf{G}_{n-1} \ldots \mathbf{G}_2\mathbf{G}_1$ by $\mathbf{G}$ we have,

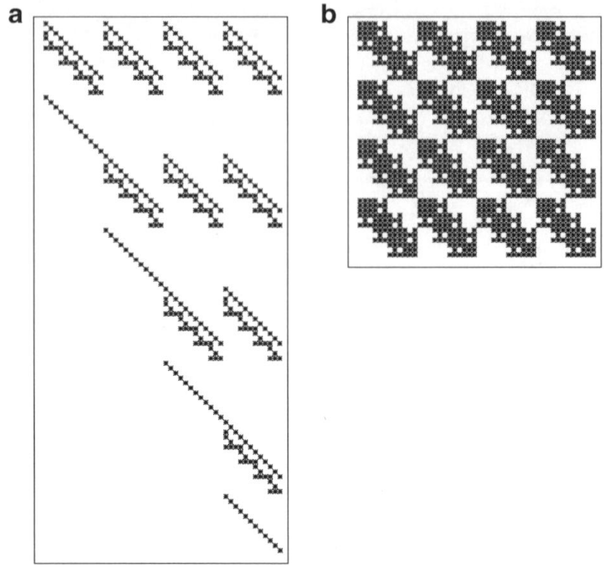

Fig. 2.38 Patterns of B_1 and $\mathbf{B}_1^t\mathbf{B}_1$ matrices for S using LU decomposition method, Example 1. (**a**) Pattern of B_1. (**b**) Pattern of $\mathbf{B}_1^t\mathbf{B}_1$

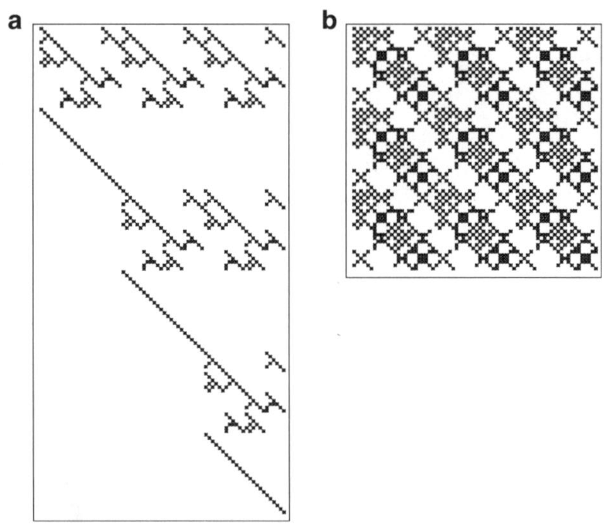

Fig. 2.39 Patterns of B_1 and $\mathbf{B}_1^t\mathbf{B}_1$ matrices for S using LU decomposition method, Example 2. (**a**) Pattern of B_1. (**b**) Pattern of $\mathbf{B}_1^t\mathbf{B}_1$

$$\mathbf{GAP} = [\mathbf{I}, \mathbf{M}], \tag{2.78}$$

or
$$\mathbf{AP} = \mathbf{G}^{-1}[\mathbf{I}, \mathbf{M}] = \left[\mathbf{G}^{-1}, \mathbf{G}^{-1}\mathbf{M}\right], \tag{2.79}$$

which can be regarded as Gauss-Jordan factorization of **A**, and:

$$\mathbf{B}_0 = \overline{\mathbf{P}}\begin{bmatrix}\mathbf{G}\\ \mathbf{0}\end{bmatrix} \quad \text{and} \quad \mathbf{B}_1 = \overline{\mathbf{P}}\begin{bmatrix}-\mathbf{M}\\ \mathbf{I}\end{bmatrix} \tag{2.80}$$

Example 1. The four by four planar frame of Fig. 2.24 is reconsidered. The patterns of $\mathbf{B}_1$ and $\mathbf{B}_1^t\mathbf{B}_1$ formed by the Gauss-Jordan elimination method are depicted in Fig. 2.36, corresponding to $\chi(\mathbf{B}_1) = 491$ and $\chi(\mathbf{B}_1^t\mathbf{B}_1) = 1342$.

Example 2. The three-story frame of Fig. 2.24 is re-considered, and the Gauss-Jordan elimination method is used. The patterns of $\mathbf{B}_1$ and $\mathbf{B}_1^t\mathbf{B}_1$ matrices formed are shown in Fig. 2.37, corresponding to $\chi(\mathbf{B}_1) = 483$ and $\chi(\mathbf{B}_1^t\mathbf{B}_1) = 1592$.

LU Decomposition Method. Using the **LU** decomposition method, one obtains the **LU** factorization of **A** as,

$$\mathbf{PA} = \mathbf{LU} \quad \text{and} \quad \mathbf{U}\overline{\mathbf{P}} = [\mathbf{U}_1, \mathbf{U}_2], \tag{2.81}$$

P and $\overline{\mathbf{P}}$ are again permutation matrices of order n × n and m × m, respectively. Now $\mathbf{B}_0$ and $\mathbf{B}_1$ can be written as:

$$\mathbf{B}_0 = \overline{\mathbf{P}}\begin{bmatrix}\mathbf{U}_1^{-1}\mathbf{L}^{-1}\mathbf{P}\\ \mathbf{0}\end{bmatrix} \quad \text{and} \quad \mathbf{B}_1 = \overline{\mathbf{P}}\begin{bmatrix}-\mathbf{U}_1^{-1}\mathbf{U}_2\\ \mathbf{I}\end{bmatrix}. \tag{2.82}$$

Example 1. The four by four planar frame of Fig. 2.24 is re-considered. The patterns of $\mathbf{B}_1$ and $\mathbf{B}_1^t\mathbf{B}_1$ formed by the **LU** factorization method are depicted in Fig. 2.38. The sparsity for the corresponding matrices are $\chi(\mathbf{B}_1) = 408$ and $\chi(\mathbf{B}_1^t\mathbf{B}_1) = 1248$.

Example 2. The three-storey frame of Fig. 2.24 is re-considered, and the **LU** factorization method is used. The patterns of $\mathbf{B}_1$ and $\mathbf{B}_1^t\mathbf{B}_1$ matrices formed are shown in Fig. 2.39, corresponding to $\chi(\mathbf{B}_1) = 504$ and $\chi(\mathbf{B}_1^t\mathbf{B}_1) = 1530$.

QR Decomposition Method. Using a **QR** factorization algorithm with column pivoting yields,

$$\mathbf{AP} = \mathbf{Q}[\mathbf{R}_1, \mathbf{R}_2], \tag{2.83}$$

where **P** is again a permutation matrix, and $\mathbf{R}_1$ is an upper triangular matrix of order n. $\mathbf{B}_1$ can be obtained as:

$$\mathbf{B}_1 = \mathbf{P}\begin{bmatrix}-\mathbf{R}_1^{-1}\mathbf{R}_2\\ \mathbf{I}\end{bmatrix}. \tag{2.84}$$

Turn-back LU Decomposition Method. Topçu developed a method, the so-called Turn-back **LU** procedure, which is based on **LU** factorization and often results in highly sparse and banded $\mathbf{B}_1$ matrices. Heath et al. [35] adopted this method for use with **QR** factorization. Due to the efficiency of this method, a brief description of their approach will be presented in the following.

Fig. 2.40 Patterns of B_1 and $\mathbf{B}_1^t\mathbf{B}_1$ matrices for S using Turn-back LU decomposition method, Example 1. (**a**) Pattern of B_1. (**b**) Pattern of $\mathbf{B}_1^t\mathbf{B}_1$

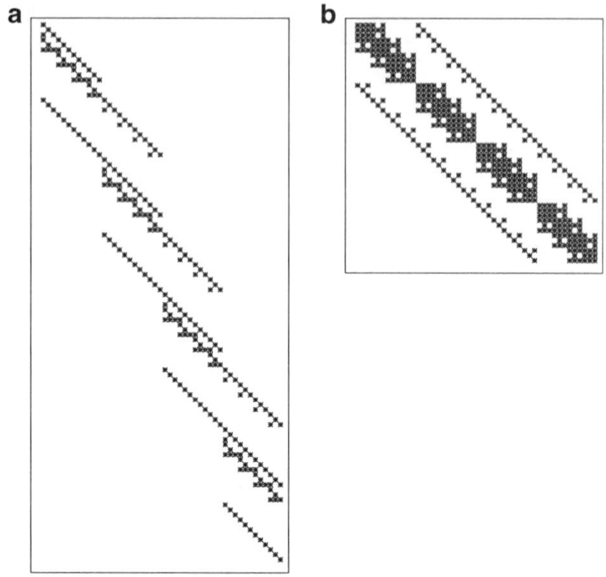

Fig. 2.41 Patterns of B_1 and $\mathbf{B}_1^t\mathbf{B}_1$ matrices for S using Turn-back LU decomposition method, Example 2. (**a**) Pattern of B_1. (**b**) Pattern of $\mathbf{B}_1^t\mathbf{B}_1$

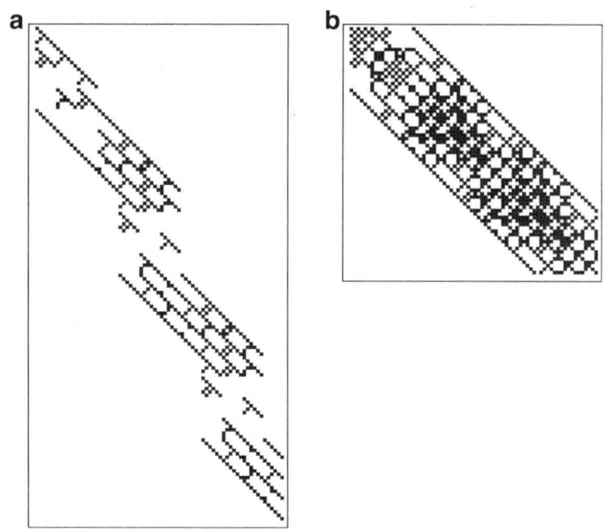

Write the matrix $\mathbf{A} = (a_1, a_2, \ldots, a_n)$ by columns. A *start column* is a column such that the ranks of $(a_1, a_2, \ldots, a_{s-1})$ and $(a_1, a_2, \ldots, a_s)$ are equal. Equivalently, a_s is a start column if it is linearly dependent on lower-numbered columns. The coefficients of this linear dependency give a null vector whose highest numbered non-zero is in position s. It is easy to see that, the number of start columns is $m - n = t$, the dimension of the null space of $\mathbf{A}$.

The start column can be found by performing a **QR** factorization of **A**, using orthogonal transformations to annihilate the subdiagonal non-zeros. Suppose that in carrying out the **QR** factorization we do not perform column interchanges but simply skip over any columns that are already zero on and below the diagonal. The result will then be a factorization of the form:

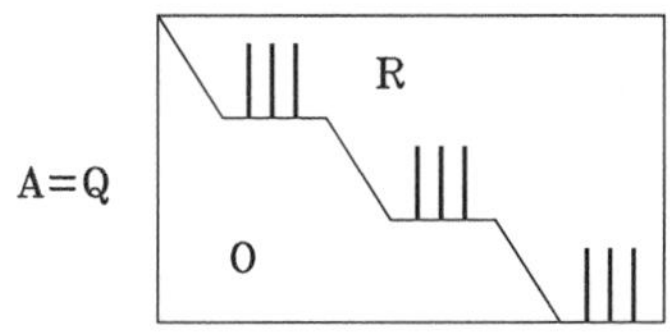

(2.85)

The start columns are those columns where the upper triangular structure jogs to the right; that is, a_s is a start column if the highest non-zero position in column s of **R**, is no larger than the highest non-zero position in earlier columns of **R**.

The Turn-back method finds one null vector for each start column a_s, by "turning back" from column s to find the smallest k for which columns $a_s, a_{s-1}, \ldots, a_{s-k}$ are linearly dependent. The null vector has a non-zero only in position s-k throughs. Thus if k is small for most of the start columns, then the null basis will have a small profile. Notice that the turn-back operates on **A**, and not on **R**. The initial **QR** factorization of **A** is used only to determine the start columns, and then discarded.

The null vector that Turn-back finds from start column as may not be non-zero in position s. Therefore, Turn-back needs to have some way to guarantee that its null vectors are linearly independent. This can be accomplished by forbidding the left-most column of the dependency for each null vector from participating in any later dependencies. Thus, if the null vector for start column a_s has its first non-zero in position s-k, every null vector for a start column to the right of a_s will be zero in position s-k.

Although the term "Turn-back" is introduced in Ref. [7], the basic idea had also been used in Refs. [36]. Since this correspondence simplifies the understanding of the Turn-back method, it is briefly described in the following.

For the Algorithm 1 of Sect. 2.3, the use of an SRT orders the nodes and members of the graph simultaneously, resulting in a fairly banded member-node incidence matrix **B**. Considering the columns of **B** corresponding to tree members as independent columns, effectively a cycle is formed on each ordered chord (start column) by turning back in **B** and establishing a minimal dependency, using the tree members and previously used chords. The cycle basis selected by this process forms a base for the cycle matroid of the graph, as it is described in Kaveh [37]. Therefore, the idea used in Algorithm 1 and its generalization for the formation of a generalized cycle bases in Ref. [38] seems to constitute a similar idea to that of the algebraic Turn-back method.

Example 1. The four by four planar frame of Fig. 2.24 is re-considered. The patterns of $\mathbf{B}_1$ and $\mathbf{B}_1^t\mathbf{B}_1$ formed by the Turn-back LU factorization method are depicted in Fig. 2.40, corresponding to $\chi(\mathbf{B}_1) = 240$ and $\chi(\mathbf{B}_1^t\mathbf{B}_1) = 408$.

Example 2. The four by four planar frame of Fig. 2.24 is re-considered, and the Turn-back LU factorization method is used. The patterns of $\mathbf{B}_1$ and $\mathbf{B}_1^t\mathbf{B}_1$ matrices formed are shown in Fig. 2.41, corresponding to $\chi(\mathbf{B}_1) = 476$ and $\chi(\mathbf{B}_1^t\mathbf{B}_1) = 984$.

A comparative study of various force methods has been made in Ref. [30].

Many algorithms have been developed for selection of null bases, and the interested reader may refer to Refs. [38, 39].

References

1. Henderson JC de C, Bickley WG (1955) Statical indeterminacy of a structure. Aircr Eng 27:400–402
2. Maunder EWA (1971) Topological and linear analysis of skeletal structures. Ph.D. thesis, London University, IC
3. Kaveh A (1992) Recent developments in the force method of structural analysis. Appl Mech Rev 45:401–418
4. Langefors B (1961) Algebraic topology and elastic networks, SAAB TN49. Linköping
5. Denke PH (1962) A general digital computer analysis of statically indeterminate structures, NASA-TD-D-1666
6. Robinson J (1973) Integrated theory of finite element methods. Wiley, New York
7. Topçu A (1979) A contribution to the systematic analysis of finite element structures using the force method (in German). Doctoral dissertation, Essen University
8. Kaneko I, Lawo M, Thierauf G (1982) On computational procedures for the force methods. Int J Numer Method Eng 18:1469–1495
9. Gilbert JR, Heath MT (1987) Computing a sparse basis for the null space. SIAM J Algebra Discr Method 8:446–459
10. Coleman TF, Pothen A (1987) The null space problem II; algorithms. SIAM J Algebra Discr Method 8:544–561
11. Patnaik SN (1986) Integrated force method versus the standard force method. Comput Struct 22:151–164
12. Kaveh A, Jahanshahi M (2006) An efficient program for cycle basis selection and bandwidth optimization. Asian J Civil Eng 7(1):95–109
13. Kaveh A, Daei M (2010) Suboptimal cycle bases of graphs using an ant colony system algorithm. Eng Comput 27(4):485–494
14. Timoshenko S, Young DH (1945) Theory of structures. McGraw-Hill, New York
15. Kaveh A (1974) Application of topology and matroid theory to the analysis of structures. Ph.D. thesis, London University, IC
16. Kaveh A (1988) Topological properties of skeletal structures. Comput Struct 29:403–411
17. Mauch SP, Fenves SJ (1967) Release and constraints in structural networks. J Struct Div ASCE 93:401–417
18. Müller-Breslau H (1912) Die graphische Statik der Baukonstruktionen. Alfred Kröner Verlag, 1907, und Leipzig
19. Kaveh A (2004) Structural mechanics: graph and matrix methods, 3rd edn. Research Studies Press, Baldock
20. Henderson JC de C, Maunder EWA (1969) A problem in applied topology. J Inst Math Appl 5:254–269
21. Kaveh A (1976) Improved cycle bases for the flexibility analysis of structures. Comput Method Appl Mech Eng 9:267–272
22. Kaveh A (1988) Suboptimal cycle bases of graphs for the flexibility analysis of skeletal structures. Comput Method Appl Mech Eng 71:259–271

23. Stepanec GF (1964) Basis systems of vector cycles with extremal properties in graphs. Uspekhi Mat Nauk 19:171–175 (in Russian)
24. Zykov AA (1969) Theory of finite graphs. Nuaka, Novosibirsk (in Russian)
25. Hubicka E, Syslø MM (1975) Minimal bases of cycles of a graph. In: Fiedler M (ed) Recent advances in graph theory. Academia Praha, Prague, pp 283–293
26. Kaveh A, Roosta GR (1994) Revised Greedy algorithm for the formation of minimal cycle basis of a graph. Commun Numer Method Eng 10:523–530
27. Horton JD (1987) A polynomial time algorithm to find the shortest cycle basis of a graph. SIAM J Comput 16:358–366
28. Lawler EL (1976) Combinatorial optimization; networks and matroids. Holt, Rinehart and Winston, New York
29. Kolasinska E (1980) On a minimum cycle basis of a graph. Zastos Math 16:631–639
30. Kaveh A, Mokhtar-zadeh A (1993) A comparative study of the combinatorial and algebraic force methods. In: Proceedings of the Civil-Comp93, Edinburgh, pp 21–30
31. Brusa L, Riccio F (1989) A frontal technique for vector computers. Int J Numer Method Eng 28:1635–1644
32. Prezemieniecki JS (1968) Theory of matrix structural analysis. McGraw-Hill, New York
33. Pestel EC, Leckie FA (1963) Matrix methods in elastomechanics. McGraw-Hill, New York
34. Kaveh A (1993) Matroids applied to the force method of structural analysis. Z Angew Math Mech 73:T355–T357
35. Heath MT, Plemmons RJ, Ward RC (1984) Sparse orthogonal schemes for structural optimization using the force method. SIAM J Sci Stat Comput 5:514–532
36. Cassell AC (1976) An alternative method for finite element analysis; a combinatorial approach to the flexibility method. Proc R Soc Lond A352:73–89
37. Kaveh A (1979) A combinatorial optimization problem; optimal generalized cycle bases. Comput Method Appl Mech Eng 20:39–52
38. Coleman TF, Pothen A (1986) The null space problem I; complexity. SIAM J Algebra Disc Method 7:527–537
39. Plemmons RJ, White RE (1990) Substructuring methods for computing the null space of equilibrium matrices. SIAM J Matrix Anal Appl 11:1–22

Chapter 3
Optimal Displacement Method of Structural Analysis

3.1 Introduction

In this chapter, the principles introduced in Chap. 1 are used for the formulation of the general displacement method of structural analysis. Computational aspects are discussed and many worked examples are included to illustrate the concepts and principles being used. In order to show the generality of the methods introduced for the formation of the element stiffness matrices, the stiffness matrix of a simple finite element is also derived.

Special attention is paid to the graph theory aspects of the displacement method for rigid jointed structures, where the pattern equivalence of structural and graph theory matrices is used. The standard displacement method employs cocycle bases of structural graph models; however, for general solutions a cutset basis of the model should be employed. This becomes vital, when solutions leading to well conditioned stiffness matrices are required. Methods for the selection of such cutset bases are described in this chapter.

In the last half-century, considerable progress has been made in the matrix analysis of structures; see for example, Argyris and Kelsey [1], Livesley [2], McGuire and Gallagher [3], Przemieniecki [4], Zienkiewicz [5], and Kaveh [6, 7]. The topic has been generalized to finite elements, and extended to the stability, non-linear and dynamic analysis of structures. This progress is due to the simplicity, modularity and flexibility of matrix methods.

3.2 Formulation

In this section, a matrix formulation using the basic tools of structural analysis—equilibrium of forces, compatibility of displacements, and force-displacement relationships—is provided. The notations are chosen from those most often encountered versions in structural mechanics.

DOI 10.1007/978-3-319-02964-1_3,

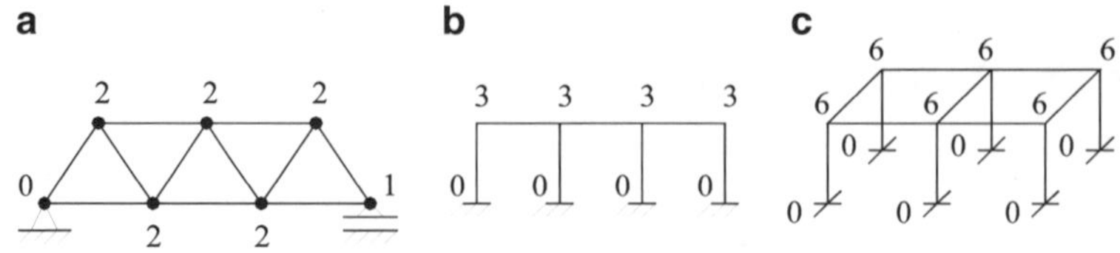

Fig. 3.1 The degrees of freedom of the joints for three structures. (**a**) A planar truss. (**b**) A planar frame. (**c**) A space frame

3.2.1 Coordinate Systems Transformation

Consider a structure S with M members and N nodes; each node having α degrees of freedom (DOF). The degree of kinematic indeterminacy (DKI) of S may then be determined as,

$$\eta(S) = \alpha N - \beta, \tag{3.1}$$

where β is the number of constraints due to the support conditions. As an example, η(S) for the planar truss S depicted in Fig. 3.1a is given by $\eta(S) = 7 \times 2 - 3 = 11$, for the plane frame illustrated in Fig. 3.1b, it is calculated as $\eta(S) = 8 \times 3 - 4 \times 3 = 12$, and for the space frame shown in Fig. 3.1c, it is calculated as $\eta(S) = 12 \times 6 - 6 \times 6 = 36$.

One can also calculate η(S) by simple addition of the degrees of freedom of the joints of the structure, i.e. for the truss S, $\eta(S) = 2 + 2 + 2 + 2 + 2 + 1 = 11$, and for the planar frame $\eta(S) = 4 \times 3 = 12$, and for the space frame $\eta(S) = 6 \times 6 = 36$.

For a structure, the stiffness matrices of the elements should be prepared in a single coordinate system known as the *global coordinate system,* in order to be able to perform the assembling process. However, the stiffness matrices of individual members are usually written first in coordinate systems attached to the members, known as *local coordinate systems*. Therefore a transformation is needed, before the assembling process. Typical local and global coordinate systems are illustrated in Fig. 3.2.

A global coordinate system can be selected arbitrarily, however, it may be advantageous to select this system such that the structure falls in the first quadrant of the plane, in order to have positive coordinates for the nodes of the structure. On the other hand, a local coordinate system of a member is so chosen that it has one of its axes along the member, the second axis lies in its plane of symmetry (if it has one) and the third axis is chosen such that it results in a right handed coordinate system.

The transformation from a local coordinate to a global coordinate system can be performed as illustrated in Fig. 3.3, in which x, y, z is the global system and x_2, y_2, z_2, often denoted by $\overline{xyz}$, is the local system.

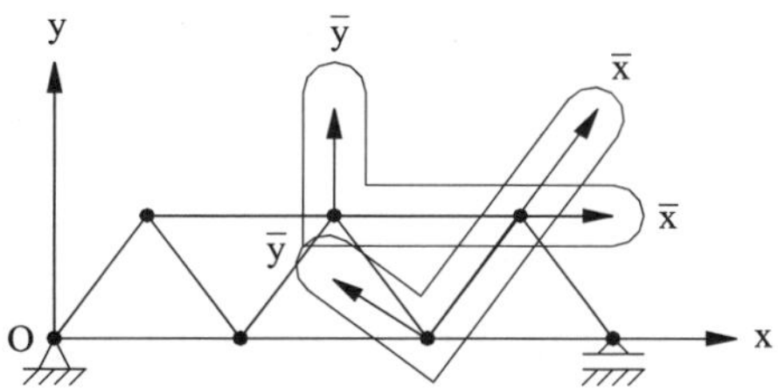

Fig. 3.2 Local $\bar{x}, \bar{y}$ and global coordinate x, y systems

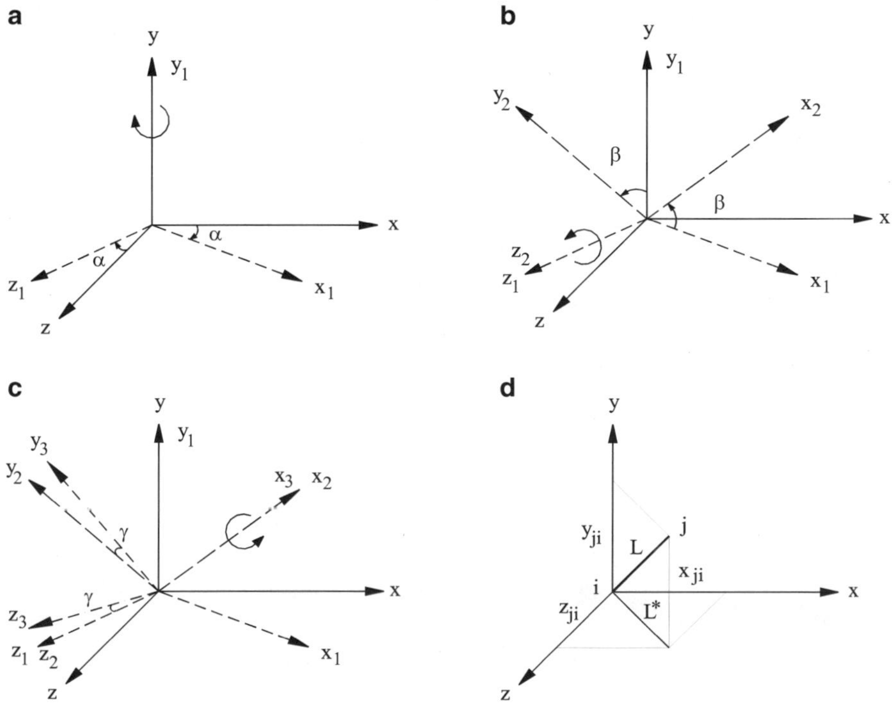

Fig. 3.3 Transformation from local coordinate system to global coordinate system

For rotation about the y axis the relation between x_1, y_1, z_1 and x, y, z can be expressed as:

$$\begin{bmatrix} x_1 \\ y_1 \\ z_1 \end{bmatrix} = \begin{bmatrix} \cos\alpha & 0 & \sin\alpha \\ 0 & 1 & 0 \\ -\sin\alpha & 0 & \cos\alpha \end{bmatrix} \begin{bmatrix} x \\ y \\ z \end{bmatrix}. \tag{3.2}$$

Similarly, for rotation about the z_1 axis x_2, y_2, z_2 and x_1, y_1, z_1 are related by,

$$\begin{bmatrix} x_2 \\ y_2 \\ z_2 \end{bmatrix} = \begin{bmatrix} \cos\beta & \sin\beta & 0 \\ -\sin\beta & \cos\beta & 0 \\ 0 & 0 & 1 \end{bmatrix} \begin{bmatrix} x_1 \\ y_1 \\ z_1 \end{bmatrix} \tag{3.3}$$

and for rotation about the x_2 axis x_3, y_3, z_3 and x_2, y_2, z_2 are related as:

$$\begin{bmatrix} x_3 \\ y_3 \\ z_3 \end{bmatrix} = \begin{bmatrix} 1 & 0 & 0 \\ 0 & \cos\gamma & \sin\gamma \\ 0 & -\sin\gamma & \cos\gamma \end{bmatrix} \begin{bmatrix} x_2 \\ y_2 \\ z_2 \end{bmatrix} \tag{3.4}$$

Combining the above transformations, results in:

$$\mathbf{T} = \begin{bmatrix} (\cos\alpha\cos\beta) & (\sin\beta) & (\cos\beta\sin\alpha) \\ -(\sin\alpha\sin\gamma + \cos\alpha\sin\beta\cos\gamma) & (\cos\beta\cos\gamma) & (\sin\gamma\cos\alpha - \sin\alpha\sin\beta\cos\gamma) \\ -(\sin\alpha\cos\gamma - \cos\alpha\sin\beta\sin\gamma) & (-\cos\beta\sin\gamma) & (\cos\alpha\cos\gamma + \sin\alpha\sin\beta\sin\gamma) \end{bmatrix}. \tag{3.5}$$

where :

$$\begin{bmatrix} x_3 \\ y_3 \\ z_3 \end{bmatrix} = [\mathbf{T}] \begin{bmatrix} x \\ y \\ z \end{bmatrix}. \tag{3.6}$$

The representation of a vector in the local coordinate system $\overline{\boldsymbol{\Gamma}}$ and the global coordinate system $\boldsymbol{\Gamma}$ are related by:

$$\overline{\boldsymbol{\Gamma}} = \mathbf{T}\,\boldsymbol{\Gamma}. \tag{3.7}$$

It can easily be proved that **T** is an orthogonal matrix, i.e.

$$[\mathbf{T}]^{-1} = [\mathbf{T}]^{t}. \tag{3.8}$$

In the above transformation, γ represents the tilt of the member, which is quite often zero. Thus, **T** can be simplified as,

$$\mathbf{T} = \begin{bmatrix} \cos\alpha\cos\beta & \sin\beta & \sin\alpha\cos\beta \\ -\cos\alpha\sin\beta & \cos\beta & -\sin\alpha\sin\beta \\ -\sin\alpha & 0 & \cos\alpha \end{bmatrix}. \tag{3.9}$$

and for the two dimensional case and "α equal to zero", **T** reduces to:

$$\mathbf{T} = \begin{bmatrix} \cos\beta & \sin\beta \\ -\sin\beta & \cos\beta \end{bmatrix} \tag{3.10}$$

Equation 3.9 can easily be written in terms of the coordinates of the two ends of a vector. Considering Fig. 3.3b and using simple trigonometry, Eq. 3.9 becomes,

$$\mathbf{T} = \begin{bmatrix} x_{ji}/L & y_{ji}/L & z_{ji}/L \\ -x_{ji}y_{ji}/L*L & L*/L & y_{ji}z_{ji}/L*L \\ -z_{ji}/L* & 0 & x_{ji}/L* \end{bmatrix}, \tag{3.11}$$

where:

$$x_{ji} = x_j - x_i \quad y_{ji} = y_j - y_i \quad z_{ji} = z_j - z_i$$

$$L^* = \left(z_{ji}^2 + x_{ji}^2\right)^{\frac{1}{2}} \text{and } L = \left(z_{ji}^2 + y_{ji}^2 + x_{ji}^2\right)^{\frac{1}{2}}. \tag{3.12}$$

Notice that $\mathbf{T}$ transforms a 3-dimensional vector from a global to a local coordinate system and $\mathbf{T}^t$ performs the reverse transformation. However, if the element forces or element displacements (distortions) consist of p vectors, the block diagonal matrix with p submatrices should be used. As an example, for a beam element of a space frame, with each node having six degrees of freedom, the transformation matrix is a 12 × 12 matrix of the form:

$$\mathbf{T} = \begin{bmatrix} \mathbf{T} & & & \\ & \mathbf{T} & & \\ & & \mathbf{T} & \\ & & & \mathbf{T} \end{bmatrix}. \tag{3.13}$$

3.2.2 Element Stiffness Matrix Using Unit Displacement Method

Consider a general element, as shown in Fig. 3.4, with n member forces,

$$\mathbf{r}_m = \{r_1\, r_2 \ldots r_n\}^t, \tag{3.14}$$

and n member displacements:

$$\mathbf{u}_m = \{u_1\, u_2 \ldots u_n\}^t. \tag{3.15}$$

A typical force component r_i can be found by using the unit displacement method to be,

$$r_i = \iiint_V \widehat{\varepsilon}_i^t \sigma dV, \tag{3.16}$$

where $\widehat{\varepsilon}_i$ represents the matrix of compatible strains due to a unit displacement in the direction of r_i, and $\boldsymbol{\sigma}$ is the exact stress matrix due to the applied forces $\mathbf{r}_m$. The unit displacements can be used in turn for all the points where member forces are applied, and therefore,

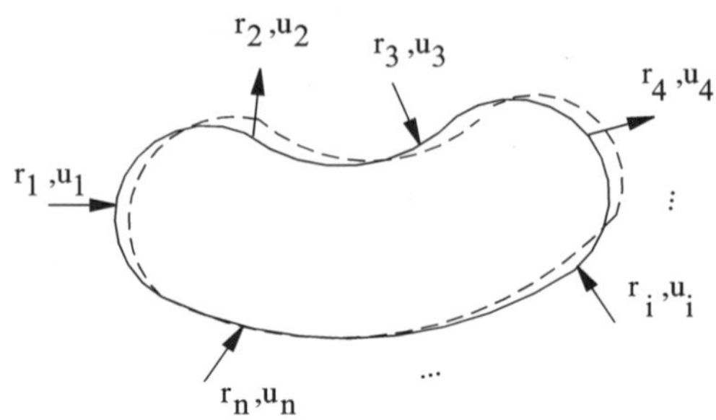

Fig. 3.4 A general element with its nodal loads and nodal displacements

$$\mathbf{r}_m = \iiint_V \widehat{\varepsilon}^t \sigma dV, \tag{3.17}$$

where:

$$\widehat{\varepsilon} = \left\{ \widehat{\varepsilon}_1 \; \widehat{\varepsilon}_2 \ldots \widehat{\varepsilon}_n \right\}^t. \tag{3.18}$$

For a linear system the total strain,

$$\mathbf{e} = \left\{ e_{xx} \; e_{yy} \; e_{zz} \; e_{xy} \; e_{yz} \; e_{xz} \right\}^t. \tag{3.19}$$

can be expressed as,

$$\mathbf{e}=\mathbf{bu}, \tag{3.20}$$

where **b** is the exact strain due to the unit displacement **u**.

The stress-strain relationship can be written as,

$$\boldsymbol{\sigma} = \chi \mathbf{bu}, \tag{3.21}$$

where:

$$\boldsymbol{\chi} = \frac{E}{(1+v)(1-2v)} \begin{bmatrix} 1-v & v & v & & & \\ v & 1-v & v & & & \\ v & v & 1-v & & & \\ & & & \frac{1-2v}{2} & 0 & 0 \\ & & & 0 & \frac{1-2v}{2} & 0 \\ & & & 0 & 0 & \frac{1-2v}{2} \end{bmatrix} \tag{3.22}$$

Substituting in Eq. 3.17 leads to,

$$\mathbf{r}_m = \iiint_V \widehat{\boldsymbol{\varepsilon}}^t \boldsymbol{\chi} \mathbf{b} dV \mathbf{u}_m, \tag{3.23}$$

or

$$\mathbf{r}_m = \mathbf{k}_m \mathbf{u}_m, \tag{3.24}$$

where:

$$\mathbf{k}_m = \iiint_V \widehat{\varepsilon}^t \chi b dV, \tag{3.25}$$

represents the element stiffness matrix.

The evaluation of the matrix **b**, representing the exact strain distributions can often be difficult, if not impossible. Hence in case there is no exact distribution, an approximate relationship may be used. Usually the matrix **b** is selected such that it will satisfy the equations of compatibility at least. Denoting this approximate matrix by $\widehat{\varepsilon}$ and using $\widehat{\boldsymbol{\varepsilon}} = \widehat{\mathbf{b}}$ results in:

$$\mathbf{k}_m = \iiint_V \widehat{b}^t \chi \, \widehat{b} \, dV. \tag{3.26}$$

This equation will be used for the derivation of the stiffness matrices of a finite element in Sect. 3.5.1.

As an example, consider a prismatic bar element shown in its local coordinate system, in Fig. 3.5. According to the definition of such an element, only axial forces are present.

From the theory of elasticity, the axial strain is expressed as:

$$\varepsilon_{xx} = \text{strain} = \frac{\partial u_x}{\partial x}, \tag{3.27}$$

The displacement u_x along the longitudinal axis of the bar can be expressed as:

$$u_x = A_1 x + A_2. \tag{3.28}$$

From the boundary conditions:

$$\begin{aligned} u_x &= \overline{u}_1 \ \text{at} \ x = 0, \\ u_x &= \overline{u}_4 \ \text{at} \ x = L. \end{aligned} \tag{3.29}$$

Hence:

$$A_1 = \frac{\overline{u}_4 - \overline{u}_1}{L} \quad \text{and} \quad A_2 = \overline{u}_1. \tag{3.30}$$

By substitution in Eq. 3.28:

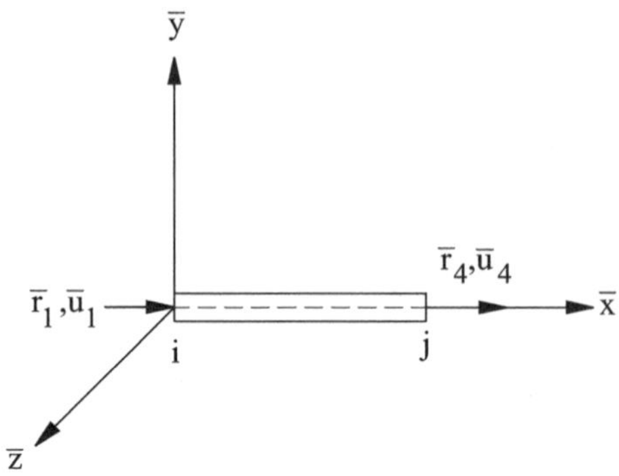

Fig. 3.5 A bar element in its local coordinate system

$$u_x = \frac{\bar{u}_4 - \bar{u}_1}{L} x + \bar{u}_1. \tag{3.31}$$

Now axial strain can be evaluated as:

$$\varepsilon_{xx} = \frac{\partial u_x}{\partial x} = \frac{1}{L}(u_2 - u_1) = \frac{1}{L}[-1 \quad +1]\begin{bmatrix} u_1 \\ u_2 \end{bmatrix}. \tag{3.32}$$

The above strain distribution is exact, and

$$\hat{\mathbf{b}} = \mathbf{b} = \frac{1}{L}[-1 \quad +1]. \tag{3.33}$$

Since a bar element is one dimensional, $\boldsymbol{\chi}$ is a 1×1 matrix defined as:

$$\boldsymbol{\chi} = E. \tag{3.34}$$

Substituting in Eq. 3.26 leads to:

$$\mathbf{k}_m = \int_0^L \frac{1}{L}\begin{bmatrix} -1 \\ 1 \end{bmatrix} \frac{E}{L}[-1 \quad 1] A dx, \tag{3.35}$$

and

$$\mathbf{k}_m = \frac{EA}{L}\begin{bmatrix} 1 & -1 \\ -1 & 1 \end{bmatrix}. \tag{3.36}$$

This method will also be used for the derivation of the finite element stiffness matrices in subsequent sections.

3.2.3 Element Stiffness Matrix Using Castigliano's Theorem

In this section, a different approach is described for the formation of element stiffness matrices, using Castigliano's theorem. Consider a general element as shown in Fig. 3.4. Suppose that loads are applied at certain points (specified as nodes) 1, 2, . . ., n. Let v_i be the displacement of node i along the applied load p_i. The loads are applied in a pseudo-static manner increasing gradually from zero. Assuming a linear behaviour, the work done by an external force $\mathbf{p} = \{p_1, p_2, \ldots, p_n\}$ through the displacement $\mathbf{v} = \{v_1, v_2, \ldots, v_n\}$ can be written as:

$$W = \frac{1}{2}(p_1 v_1 + p_2 v_2 + \ldots + p_n v_n). \tag{3.37}$$

According to the principle of the conservation of energy,

$$W = U, \tag{3.38}$$

and therefore:

$$U = \frac{1}{2}(p_1 v_1 + p_2 v_2 + \ldots + p_n v_n). \tag{3.39}$$

If a small variation is now given to v_i while keeping the other displacement components constant, then the variation of $\mathbf{v}$ with respect to v_i can be written as:

$$\frac{\partial U}{\partial v_i} = \frac{1}{2}\left[p_i + \frac{\partial p_1}{\partial v_i} v_1 + \frac{\partial p_2}{\partial v_i} v_2 + \ldots + \frac{\partial p_n}{\partial v_i} v_n\right]. \tag{3.40}$$

According to Castigliano's theorem:

$$\frac{\partial U}{\partial v_i} = p_i. \tag{3.41}$$

Thus,

$$p_i = \left[\frac{\partial p_1}{\partial v_i} v_1 + \frac{\partial p_2}{\partial v_i} v_2 + \ldots + \frac{\partial p_n}{\partial v_i} v_n\right], \tag{3.42}$$

or in a matrix form for all i = 1, . . ., n, we have:

$$\begin{bmatrix} p_1 \\ p_2 \\ \cdot \\ \cdot \\ p_n \end{bmatrix} = \begin{bmatrix} \frac{\partial p_1}{\partial v_1} & \frac{\partial p_2}{\partial v_1} & \cdot & \cdot & \frac{\partial p_n}{\partial v_1} \\ \frac{\partial p_1}{\partial v_2} & \frac{\partial p_2}{\partial v_2} & \cdot & \cdot & \frac{\partial p_n}{\partial v_2} \\ \cdot & \cdot & \cdot & \cdot & \cdot \\ \cdot & \cdot & \cdot & \cdot & \cdot \\ \frac{\partial p_1}{\partial v_n} & \frac{\partial p_2}{\partial v_n} & \cdot & \cdot & \frac{\partial p_n}{\partial v_n} \end{bmatrix} \begin{bmatrix} v_1 \\ v_2 \\ \cdot \\ \cdot \\ v_n \end{bmatrix}. \tag{3.43}$$

According to definition, the above coefficient matrix forms the stiffness matrix of the elastic body defined by its n nodes as illustrated in Fig. 3.4.

A typical element of the stiffness matrix k_{ij} is given by:

$$k_{ij} = \frac{\partial p_j}{\partial v_i}. \tag{3.44}$$

Using Castigliano's first theorem:

$$k_{ij} = \frac{\partial}{\partial v_i}\left(\frac{\partial U}{\partial v_j}\right) = \frac{\partial^2 U}{\partial v_i \partial v_j}. \tag{3.45}$$

Similarly:

$$k_{ji} = \frac{\partial p_i}{\partial v_j} = \frac{\partial^2 U}{\partial v_j \partial v_i}. \tag{3.46}$$

Since the order of differentiation should not affect the result for our problems, we have:

$$k_{ij} = k_{ji}, \tag{3.47}$$

which is a proof of the symmetry of the stiffness matrices both for a structure and for an element.

As an example, consider a prismatic bar element as shown in its local coordinate system, Fig. 3.5. According to the definition of such an element, only axial forces are present.

The strain energy of this bar can be calculated as:

$$U = \frac{1}{2}\int\int\int \sigma_{xx}\varepsilon_{xx}dxdydz = \frac{E}{2}\int\int\int \varepsilon_{xx}^2 dxdydz = \frac{EA}{2}\int \varepsilon_{xx}^2 dx. \tag{3.48}$$

On the other hand:

$$\varepsilon_{xx} = \frac{\partial u_x}{\partial x}. \tag{3.49}$$

Using Eq. 3.31, by substituting in Eq. 3.48, the strain energy of the bar is calculated to be:

$$U = \frac{EA}{2L}\left[\bar{u}_4^2 - 2\bar{u}_4\bar{u}_1 + \bar{u}_1^2\right]. \tag{3.50}$$

Hence

$$\begin{aligned} \bar{k}_{11} &= \frac{\partial^2 U}{\partial \bar{u}_1^2} = \frac{EA}{L}, \\ \bar{k}_{14} = \bar{k}_{41} &= \frac{\partial^2 U}{\partial \bar{u}_1 \partial \bar{u}_4} = -\frac{EA}{L}, \\ \bar{k}_{44} &= \frac{\partial^2 U}{\partial \bar{u}_4^2} = \frac{EA}{L}, \end{aligned} \tag{3.51}$$

$\bar{k}_{ij} = 0$ for all other components.

Therefore, the stiffness matrix of a bar element in the selected local coordinate system is obtained, and:

$$\begin{bmatrix} \bar{r}_1 \\ \bar{r}_2 \\ \bar{r}_3 \\ \bar{r}_4 \\ \bar{r}_5 \\ \bar{r}_6 \end{bmatrix} = \frac{EA}{L}\begin{bmatrix} 1 & 0 & 0 & -1 & 0 & 0 \\ 0 & 0 & 0 & 0 & 0 & 0 \\ 0 & 0 & 0 & 0 & 0 & 0 \\ -1 & 0 & 0 & 1 & 0 & 0 \\ 0 & 0 & 0 & 0 & 0 & 0 \\ 0 & 0 & 0 & 0 & 0 & 0 \end{bmatrix}\begin{bmatrix} \bar{u}_1 \\ \bar{u}_2 \\ \bar{u}_3 \\ \bar{u}_4 \\ \bar{u}_5 \\ \bar{u}_6 \end{bmatrix}. \tag{3.52}$$

3.2.4 The Stiffness Matrix of a Structure

Let **p** and **v** represent the joint loads and joint displacements of a structure. Then the force-displacement relationship for the structure can be expressed as,

$$\mathbf{p} = \mathbf{K}\mathbf{v}, \tag{3.53}$$

where **K** is a $\alpha N \times \alpha N$ symmetric matrix, known as the stiffness matrix of the structure. Expanding the ith equation of the above system, the force p_i can be expressed in terms of the displacements $\{v_1, v_2, \ldots, v_{aN}\}$ as:

$$p_i = K_{i1}v_1 + K_{i2}v_2 + \ldots + K_{i\alpha N}v_{\alpha N}. \tag{3.54}$$

A typical coefficient $\mathbf{K}_{ij}$ is the value of the force $\mathbf{p}_i$ required to be applied at the ith component of the structure in order to produce a displacement $v_j = 1$ at j and zero displacements at all the other components.

The member forces $\mathbf{r}$ can be related to nodal forces $\mathbf{p}$ by:

$$\mathbf{p} = \mathbf{B}\mathbf{r}. \tag{3.55}$$

Using the contragradient relationship, the joint displacements $\mathbf{v}$ can be related to member distortions $\mathbf{u}$ by:

$$\mathbf{u} = \mathbf{B}^t\mathbf{v}. \tag{3.56}$$

For each individual member of the structure, the member forces can be related to member distortions by an element stiffness matrix $\mathbf{k}_m$. A block diagonal matrix containing these element stiffness matrices is known as the *unassembled stiffness matrix* of the structure, denoted by $\mathbf{k}$. Obviously:

$$\mathbf{r} = \mathbf{k}\mathbf{u}. \tag{3.57}$$

This equation together with Eqs. 3.55 and 3.56 yields:

$$\mathbf{p} = \mathbf{B}\mathbf{k}\mathbf{B}^t\mathbf{v}. \tag{3.58}$$

Therefore,

$$\mathbf{K} = \mathbf{B}\mathbf{k}\mathbf{B}^t, \tag{3.59}$$

is obtained. The matrix $\mathbf{K}$ is singular since the boundary conditions of the structure are not yet applied. For an appropriately supported structure, the deletion of the rows and columns of $\mathbf{K}$ corresponding to the support constraints results in a positive definite matrix, known as the *reduced stiffness matrix* of the structure.

A symmetric matrix $\mathbf{S}$ is called *positive definite* if $\mathbf{x}^t\mathbf{S}\mathbf{x} > 0$ for every non-zero vector $\mathbf{x}$. As shown before, the stiffness matrix $\mathbf{K}$ of a structure is symmetric. This matrix is also positive definite since,

$$\mathbf{p}^t\mathbf{v} = (\mathbf{K}\mathbf{v})^t\mathbf{v} = \mathbf{v}^t\mathbf{K}^t\mathbf{v} = \mathbf{v}^t\mathbf{K}\mathbf{v} = 2W, \tag{3.60}$$

and W is always positive.

Let us illustrate the stiffness method by means of a simple example. Consider a fixed end beam with a load P applied at its mid span. This beam is discretized as two beam elements, as shown in Fig. 3.6a with two degrees of freedom for each node (axial deformation is ignored for simplicity). The components of element forces and element distortions are depicted in Fig. 3.6b and those of the entire structure are illustrated in Fig. 3.6c.

For each element such as element 1, the stiffness matrix can be written as:

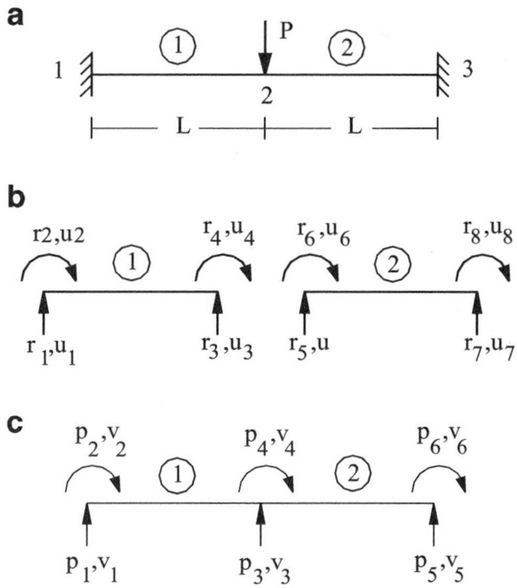

Fig. 3.6 Illustration of the analysis of a simple structure. (**a**) A fixed ended beam S. (**b**) Member forces and member distortions. (**c**) Nodal forces and nodal displacements of the entire structure

$$\begin{bmatrix} r_1 \\ r_2 \\ r_3 \\ r_4 \end{bmatrix} = \begin{bmatrix} k_{11} & k_{12} & k_{13} & k_{14} \\ k_{21} & k_{22} & k_{23} & k_{24} \\ k_{31} & k_{32} & k_{33} & k_{34} \\ k_{41} & k_{42} & k_{43} & k_{44} \end{bmatrix} \begin{bmatrix} u_1 \\ u_2 \\ u_3 \\ u_4 \end{bmatrix}, \tag{3.61}$$

and for the entire structure we have:

$$\begin{bmatrix} p_1 \\ p_2 \\ p_3 \\ p_4 \\ p_5 \\ p_6 \end{bmatrix} = \begin{bmatrix} K_{11} & K_{12} & K_{13} & K_{14} & K_{15} & K_{16} \\ K_{21} & K_{22} & K_{23} & K_{24} & K_{25} & K_{26} \\ K_{31} & K_{32} & K_{33} & K_{34} & K_{35} & K_{36} \\ K_{41} & K_{42} & K_{43} & K_{44} & K_{45} & K_{46} \\ K_{51} & K_{52} & K_{53} & K_{54} & K_{55} & K_{56} \\ K_{61} & K_{62} & K_{63} & K_{64} & K_{65} & K_{66} \end{bmatrix} \begin{bmatrix} v_1 \\ v_2 \\ v_3 \\ v_4 \\ v_5 \\ v_6 \end{bmatrix}. \tag{3.62}$$

Element stiffness matrices $\mathbf{k}_1$ and $\mathbf{k}_2$ can be easily constructed using the definition of k_{ij}. For a beam element, ignoring its axial deformation, these terms are shown in Fig. 3.7.

The structure has a uniform cross section and both elements have the same length. Therefore, using the force displacement relationship from Chap. 1:

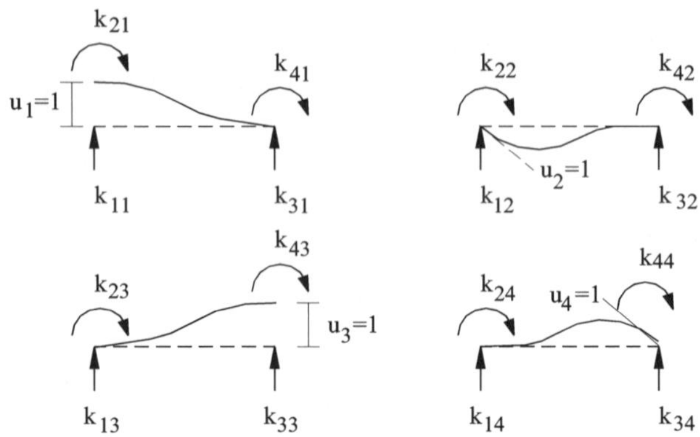

Fig. 3.7 Stiffness coefficients of a beam element ignoring its axial deformation

$$\mathbf{k}_1 = \mathbf{k}_2 = \frac{2EI}{L}\begin{bmatrix} 6/L^2 & -3/L & -6/L^2 & -3/L \\ -3/L & 2 & 3/L & 1 \\ -6/L^2 & 3/L & 6/L^2 & 3/L \\ -3/L & 1 & 3/L & 2 \end{bmatrix}. \tag{3.63}$$

The unassembled stiffness matrix is an 8 × 8 matrix of the form:

$$\mathbf{k} = \begin{bmatrix} \mathbf{k}_1 & 0 \\ 0 & \mathbf{k}_2 \end{bmatrix}. \tag{3.64}$$

Now consider the equilibrium of the joints of the structure, resulting in,

$$\begin{aligned} &p_1 = r_1 \ , \ p_2 = r_2 \ , \ p_3 = r_3 + r_5, \\ &p_4 = r_4 + r_6 \ , \ p_5 = r_7 \ , \ p_6 = r_8. \end{aligned} \tag{3.65}$$

or in a matrix form we have,

$$\begin{bmatrix} p_1 \\ p_2 \\ p_3 \\ p_4 \\ p_5 \\ p_6 \end{bmatrix} = \begin{bmatrix} 1 & \cdot & \cdot & \cdot & \cdot & \cdot & \cdot & \cdot \\ \cdot & 1 & \cdot & \cdot & \cdot & \cdot & \cdot & \cdot \\ \cdot & \cdot & 1 & \cdot & 1 & \cdot & \cdot & \cdot \\ \cdot & \cdot & \cdot & 1 & \cdot & 1 & \cdot & \cdot \\ \cdot & \cdot & \cdot & \cdot & \cdot & \cdot & 1 & \cdot \\ \cdot & \cdot & \cdot & \cdot & \cdot & \cdot & \cdot & 1 \end{bmatrix} \begin{bmatrix} r_1 \\ r_2 \\ r_3 \\ r_4 \\ r_5 \\ r_6 \\ r_7 \\ r_8 \end{bmatrix}, \tag{3.66}$$

and more compactly,

$$\mathbf{p} = \mathbf{B}\mathbf{r}, \tag{3.67}$$

where:

$$\mathbf{B} = \begin{bmatrix} 1 & \cdot & \cdot & \cdot & \cdot & \cdot & \cdot & \cdot \\ \cdot & 1 & \cdot & \cdot & \cdot & \cdot & \cdot & \cdot \\ \cdot & \cdot & 1 & \cdot & 1 & \cdot & \cdot & \cdot \\ \cdot & \cdot & \cdot & 1 & \cdot & 1 & \cdot & \cdot \\ \cdot & \cdot & \cdot & \cdot & \cdot & \cdot & 1 & \cdot \\ \cdot & \cdot & \cdot & \cdot & \cdot & \cdot & \cdot & 1 \end{bmatrix},$$

is known as the *equilibrium matrix.*

Consider now the compatibility of displacements:

$$\begin{aligned} u_1 = v_1, \quad u_2 = v_2, \quad u_3 = u_5 = v_3, \\ u_4 = u_6 = v_4, \quad u_7 = v_5, \quad u_8 = v_6. \end{aligned} \tag{3.68}$$

In a matrix form we have,

$$\begin{bmatrix} u_1 \\ u_2 \\ u_3 \\ u_4 \\ u_5 \\ u_6 \\ u_7 \\ u_8 \end{bmatrix} = \begin{bmatrix} 1 & \cdot & \cdot & \cdot & \cdot & \cdot \\ \cdot & 1 & \cdot & \cdot & \cdot & \cdot \\ \cdot & \cdot & 1 & \cdot & \cdot & \cdot \\ \cdot & \cdot & \cdot & 1 & \cdot & \cdot \\ \cdot & \cdot & 1 & \cdot & \cdot & \cdot \\ \cdot & \cdot & \cdot & 1 & \cdot & \cdot \\ \cdot & \cdot & \cdot & \cdot & 1 & \cdot \\ \cdot & \cdot & \cdot & \cdot & \cdot & 1 \end{bmatrix} \begin{bmatrix} v_1 \\ v_2 \\ v_3 \\ v_4 \\ v_5 \\ v_6 \end{bmatrix}, \tag{3.69}$$

and in compact form:

$$\mathbf{u} = \mathbf{E}\mathbf{v} = \mathbf{B}^t\mathbf{v}. \tag{3.70}$$

where:

$$\mathbf{E} = \begin{bmatrix} 1 & \cdot & \cdot & \cdot & \cdot & \cdot \\ \cdot & 1 & \cdot & \cdot & \cdot & \cdot \\ \cdot & \cdot & 1 & \cdot & \cdot & \cdot \\ \cdot & \cdot & \cdot & 1 & \cdot & \cdot \\ \cdot & \cdot & 1 & \cdot & \cdot & \cdot \\ \cdot & \cdot & \cdot & 1 & \cdot & \cdot \\ \cdot & \cdot & \cdot & \cdot & 1 & \cdot \\ \cdot & \cdot & \cdot & \cdot & \cdot & 1 \end{bmatrix},$$

is known as the *compatibility matrix.*

The reason for the matrix **E** being the transpose of the matrix **B**, has already been discussed in the previous chapter, however, by using the principle of virtual work, a simple proof can be obtained. Consider:

W = work done by external loads $= \frac{1}{2}\mathbf{v}^t\mathbf{p}$,

$U = \text{strain energy} = \frac{1}{2}\mathbf{u}^t\mathbf{r}$.

Then equating W and U, leads to $\mathbf{E} = \mathbf{B}^t$ and completes the proof. It should be mentioned that this equality holds for a general structure, and it is the result of the contragradient relationship introduced in Chap. 1.

The stiffness matrix of the entire structure is then obtained as:

$$\mathbf{K} = \frac{2EI}{L}\begin{bmatrix} 6/L^2 & -3/L & -6/L^2 & -3/L & 0 & 0 \\ -3/L & 2 & 3/L & 1 & 0 & 0 \\ -6/L^2 & 3/L & 12/L^2 & 0 & -6/L^2 & -3/L \\ -3/L & 1 & 0 & 4 & 3/L & 1 \\ 0 & 0 & -6/L^2 & 3/L & 6/L^2 & 3/L \\ 0 & 0 & -3/L & 1 & 3/L & 2 \end{bmatrix}. \tag{3.71}$$

Applying the boundary conditions,

$$v_1 = v_2 = v_5 = v_6 = 0,$$

by deleting the rows and columns corresponding to zero displacements, leads to the formation of the following reduced stiffness matrix:

$$\begin{bmatrix} p_3 \\ p_4 \end{bmatrix} = \frac{2EI}{L}\begin{bmatrix} 12/L^2 & 0 \\ 0 & 4 \end{bmatrix}\begin{bmatrix} v_3 \\ v_4 \end{bmatrix}. \tag{3.72}$$

Since $p_4 = 0$ and $p_3 = -P$, therefore $v_3 = \frac{p_3L^3}{24EI} = \frac{-PL^3}{24EI}$.

3.2.5 *Stiffness Matrix of a Structure; an Algorithmic Approach*

From the above simple example, it can be seen that the matrix **B** is a very sparse Boolean matrix and the direct formation of $\mathbf{BkB}^t$ using matrix multiplication requires a considerable amount of storage. In the following, it is shown that one can form $\mathbf{BkB}^t$ with an assembling process (known also as *planting*), as follows:

Consider an element "a" of a structure, as shown in Fig. 3.8, for which the element stiffness matrix can be written as,

$$\mathbf{k}_a = \begin{bmatrix} \mathbf{k}_{ii} & \mathbf{k}_{ij} \\ \mathbf{k}_{ji} & \mathbf{k}_{jj} \end{bmatrix}, \tag{3.73}$$

i and j are the two end nodes of member a. Pre and post multiplication in the form of $\mathbf{BkB}^t$ has the following effect on $\mathbf{k}_a$:

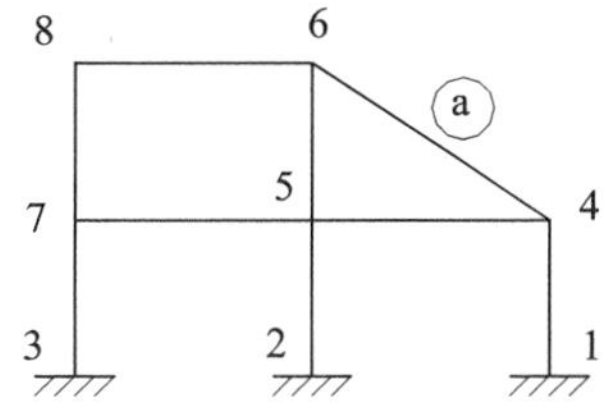

Fig. 3.8 A structural model S

$$
\begin{bmatrix} 0 & 0 \\ 0 & 0 \\ 0 & 0 \\ \mathrm{I} & 0 \\ 0 & 0 \\ 0 & \mathrm{I} \\ 0 & 0 \\ 0 & 0 \end{bmatrix}
\begin{bmatrix} \mathbf{k}_{ii} & \mathbf{k}_{ij} \\ \mathbf{k}_{ji} & \mathbf{k}_{jj} \end{bmatrix}
\begin{bmatrix} 0 & 0 & 0 & \mathrm{I} & 0 & 0 & 0 & 0 \\ 0 & 0 & 0 & 0 & 0 & \mathrm{I} & 0 & 0 \end{bmatrix}
=
\begin{bmatrix} 0 & 0 \\ 0 & 0 \\ 0 & 0 \\ \mathrm{I} & 0 \\ 0 & 0 \\ 0 & \mathrm{I} \\ 0 & 0 \\ 0 & 0 \end{bmatrix}
\begin{bmatrix} 0 & 0 & 0 & \mathbf{k}_{ii} & 0 & \mathbf{k}_{ij} & 0 & 0 \\ 0 & 0 & 0 & \mathbf{k}_{ji} & 0 & \mathbf{k}_{jj} & 0 & 0 \end{bmatrix}
\tag{3.74}
$$

$$
= \begin{array}{c} 1 \\ 2 \\ 3 \\ 4 \\ 5 \\ 6 \\ 7 \\ 8 \end{array}
\begin{bmatrix}
0 & 0 & 0 & 0 & 0 & 0 & 0 & 0 \\
0 & 0 & 0 & 0 & 0 & 0 & 0 & 0 \\
0 & 0 & 0 & 0 & 0 & 0 & 0 & 0 \\
0 & 0 & 0 & \mathbf{k}_{ii} & 0 & \mathbf{k}_{ij} & 0 & 0 \\
0 & 0 & 0 & 0 & 0 & 0 & 0 & 0 \\
0 & 0 & 0 & \mathbf{k}_{ji} & 0 & \mathbf{k}_{jj} & 0 & 0 \\
0 & 0 & 0 & 0 & 0 & 0 & 0 & 0 \\
0 & 0 & 0 & 0 & 0 & 0 & 0 & 0
\end{bmatrix}
$$

The adjacency matrix of S is also an 8×8 matrix, and the effect of node 4 being adjacent to node 6, is the existence of unit entries in the same locations as the submatrices of the element "a". One can build up the adjacency matrix of a graph by the addition of the effect of one member at a time. In the same way, one can also form the overall stiffness matrix of the structure by the addition of the contribution of every member in succession. As an example, for the graph shown in Fig. 3.8, the overall stiffness matrix has the following pattern:

$$
\begin{array}{c} 1 \\ 1 \\ 2 \\ 3 \\ 4 \\ 5 \\ 6 \\ 7 \\ 8 \end{array}
\begin{array}{c}
\begin{array}{cccccccc} 2 & 3 & 4 & 5 & 6 & 7 & 8 & \end{array} \\
\begin{bmatrix}
1 & \cdot & \cdot & 1 & \cdot & \cdot & \cdot & \cdot \\
\cdot & 1 & \cdot & \cdot & 1 & \cdot & \cdot & \cdot \\
\cdot & \cdot & 1 & \cdot & \cdot & \cdot & 1 & \cdot \\
1 & \cdot & \cdot & 1 & 1 & 1 & \cdot & \cdot \\
\cdot & 1 & \cdot & 1 & 1 & 1 & 1 & \cdot \\
\cdot & \cdot & \cdot & 1 & 1 & 1 & \cdot & 1 \\
\cdot & \cdot & 1 & \cdot & 1 & \cdot & 1 & 1 \\
\cdot & \cdot & \cdot & \cdot & \cdot & 1 & 1 & 1
\end{bmatrix}
\end{array}.
\tag{3.75}
$$

Non-zero entries are shown by "1". For a stiffness matrix each of these non-zero entries is an $\eta \times \eta$ submatrix, where η is the degrees of freedom of each node of the structure. As an example, for a planar truss $\eta = 2$, and for a space frame $\eta = 6$. The formation of the stiffness matrix by the above process is known as the *assembling* or *planting* of the stiffness matrix of a structure.

3.3 Transformation of Stiffness Matrices

Methods for the formation of element stiffness matrices have been presented in the previous section. In the following the stiffness matrices for bar and beam elements are transformed to global coordinate systems using the transformation described in Sect. 3.2.1.

From Eq. 3.7, we have:

$$\bar{\mathbf{r}} = \mathbf{T}\mathbf{r}, \tag{3.76}$$

$$\bar{\mathbf{u}} = \mathbf{T}\mathbf{u}. \tag{3.77}$$

From the definition of an element stiffness matrix in a local coordinate system:

$$\bar{\mathbf{r}} = \bar{\mathbf{k}}\bar{\mathbf{u}}. \tag{3.78}$$

By substitution of Eqs. 3.76 and 3.77 into the above equation:

$$\mathbf{r} = \mathbf{T}^{-1}\bar{\mathbf{k}}\mathbf{T}\mathbf{u} = \mathbf{T}^{t}\bar{\mathbf{k}}\mathbf{T}\mathbf{u}. \tag{3.79}$$

By definition of a stiffness matrix in a global coordinate system:

$$\mathbf{r} = \mathbf{k}\mathbf{u}. \tag{3.80}$$

Comparison of Eqs. 3.79 and 3.80 results in:

$$\mathbf{k} = \mathbf{T}^{t}\bar{\mathbf{k}}\mathbf{T}. \tag{3.81}$$

3.3.1 Stiffness Matrix of a Bar Element

Equation 3.52 provides the stiffness matrix of a bar element in its local coordinate system. This matrix in the global system, as shown in Fig. 3.9, can be written as:

$$\mathbf{k} = \begin{bmatrix} \mathbf{T} & \\ & \mathbf{T} \end{bmatrix}^{t} \left[\bar{\mathbf{k}}\right] \begin{bmatrix} \mathbf{T} & \\ & \mathbf{T} \end{bmatrix}. \tag{3.82}$$

Denoting $\mathbf{T}$ in Eq. 3.32 by,

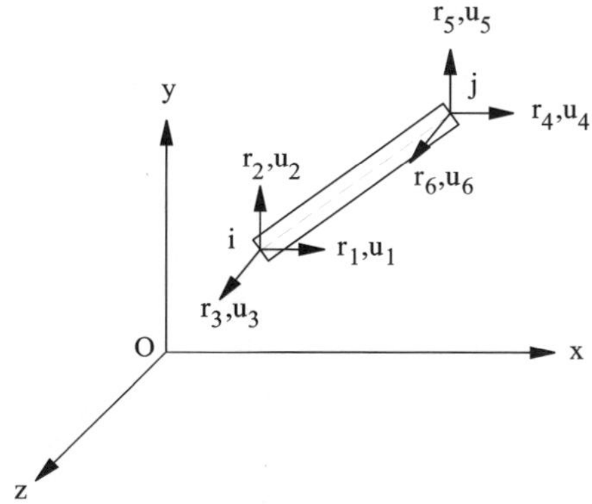

Fig. 3.9 A bar element of a space truss

$$\mathbf{T} = \begin{bmatrix} T_{11} & T_{12} & T_{13} \\ T_{21} & T_{22} & T_{23} \\ T_{31} & T_{32} & T_{33} \end{bmatrix}, \tag{3.83}$$

$\mathbf{k}_m$ can be written as,

$$\mathbf{k} = \frac{EA}{L}\begin{bmatrix} T_{11}^2 & & & & & \\ T_{11}T_{12} & T_{12}^2 & & \text{sym.} & & \\ T_{11}T_{13} & T_{12}T_{13} & T_{13}^2 & & & \\ -T_{11}^2 & -T_{11}T_{12} & -T_{11}T_{13} & T_{11}^2 & & \\ -T_{11}T_{12} & -T_{12}^2 & -T_{12}T_{13} & T_{11}T_{12} & T_{12}^2 & \\ -T_{11}T_{13} & -T_{12}T_{13} & -T_{13}^2 & T_{11}T_{13} & T_{12}T_{13} & T_{13}^2 \end{bmatrix}, \tag{3.84}$$

where "sym." denotes the symmetry of the matrix.

The entries of the above matrix can be found using the T_{ij} from Eq. 3.32. As an example, the stiffness matrix of bar 1 in the planar truss shown in Fig. 3.10 can be obtained as:

$$T_{11} = \frac{x_{21}}{(x_{12}^2 + y_{12}^2 + z_{12}^2)^{\frac{1}{2}}} = \frac{1}{\sqrt{2}} = \frac{\sqrt{2}}{2},$$

$$T_{12} = \frac{y_{21}}{(x_{12}^2 + y_{12}^2 + z_{12}^2)^{\frac{1}{2}}} = -\frac{1}{\sqrt{2}} = -\frac{\sqrt{2}}{2}.$$

Therefore:

Fig. 3.10 A planar truss and the selected global coordinate system

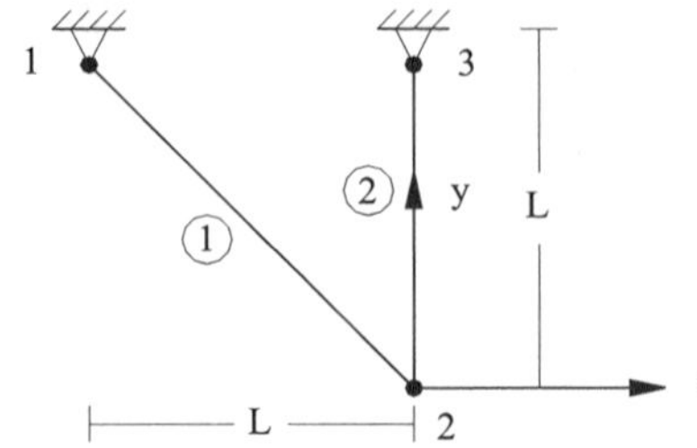

$$\mathbf{k}_1 = \frac{EA}{L\sqrt{2}} \begin{bmatrix} 0.5 & -0.5 & -0.5 & 0.5 \\ -0.5 & 0.5 & 0.5 & -0.5 \\ -0.5 & 0.5 & 0.5 & -0.5 \\ 0.5 & -0.5 & -0.5 & 0.5 \end{bmatrix}.$$

3.3.2 Stiffness Matrix of a Beam Element

Consider a prismatic beam element as shown in Fig. 3.11. The element forces and element distortions are defined by the following vectors,

$$\bar{\mathbf{r}} = \{r_1, r_2, r_3, \ldots, r_{12}\}^t,$$

and

$$\bar{\mathbf{u}} = \{u_1, u_2, u_3, \ldots, u_{12}\}^t,$$

where r_1 to r_3 are the force components at end i and r_4 to r_6 are moment components at end i. Also r_7 to r_9 are the force and r_{10} to r_{12} are the moment components, respectively at the end j, and u_i (i = 1, ..., 12) are correspondingly the translations and rotations at the ends i and j of the element.

Using one of the methods presented in Sect. 3.2.2, the stiffness matrix of the beam element, in the local coordinate system defined in Fig. 3.11, can be obtained from Eq. 3.83 as:

$$\bar{\mathbf{k}} = \frac{E}{L}\begin{bmatrix}
A & 0 & 0 & 0 & 0 & 0 & -A & 0 & 0 & 0 & 0 & 0 \\
0 & 12I_z/L^2 & 0 & 0 & 0 & 6I_z/L & 0 & 0 & -12I_z/L^2 & 0 & 0 & 6I_z/L \\
0 & 0 & 12I_y/L^2 & 0 & -6I_y/L & 0 & 0 & 0 & -12I_y/L^2 & 0 & -6I_y/L & 0 \\
0 & 0 & 0 & J/2(1+v) & 0 & 0 & 0 & 0 & 0 & -J/2(1+v) & 0 & 0 \\
0 & 0 & -6I_y/L & 0 & 4I_y & 0 & 0 & 0 & -6I_y/L & 0 & 2I_y & 0 \\
0 & 6I_z/L & 0 & 0 & 0 & 4I_z & 0 & -6I_z/L & 0 & 0 & 0 & 2I_z \\
-A & 0 & 0 & 0 & 0 & 0 & A & 0 & 0 & 0 & 0 & 0 \\
0 & -12I_z/L^2 & 0 & 0 & 0 & -6I_z/L & 0 & 12I_y/L^2 & 0 & 0 & 0 & -6I_z/L \\
0 & 0 & -12I_y/L^2 & 0 & 6I_y/L & 0 & 0 & 0 & 12I_y/L^2 & 0 & 6I_y/L & 0 \\
0 & 0 & 0 & -J/2(1+v) & 0 & 0 & 0 & 0 & 0 & J/2(1+v) & 0 & 0 \\
0 & 0 & -6I_y/L & 0 & 2I_y & 0 & 0 & 0 & 6I_y/L & 0 & 4I_y & 0 \\
0 & 6I_z/L & 0 & 0 & 0 & 2I_z & 0 & -6I_z/L & 0 & 0 & 0 & 4I_z
\end{bmatrix} \quad (3.85)$$

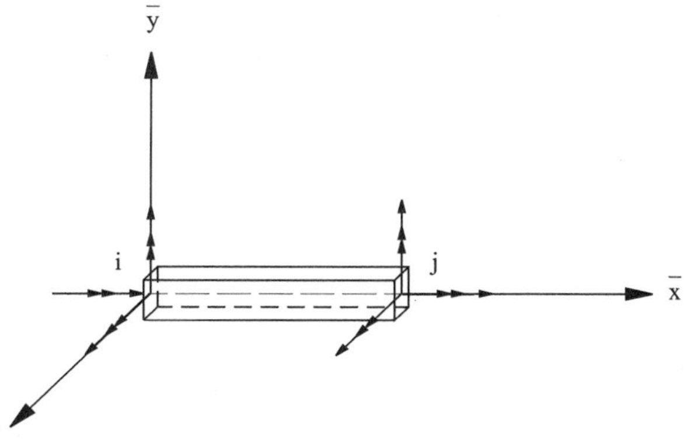

Fig. 3.11 A beam element in the local coordinate system

In this matrix, I_y, I_z and J are the moments of inertia with respect to the $\bar{y}$ and $\bar{z}$ axes and J is the polar moment of inertia of the section. E specifies the elastic modulus and v is the Poisson ratio. L denotes the length of the beam.

For the two-dimensional case, the columns and rows corresponding to the third dimension can easily be deleted, to obtain the stiffness matrix of an element of a planar frame.

The stiffness matrix in a global coordinate system can be written as:

$$\mathbf{k} = \begin{bmatrix} \mathbf{T} & & & \\ & \mathbf{T} & & \\ & & \mathbf{T} & \\ & & & \mathbf{T} \end{bmatrix}^t \boxed{\mathbf{k}} \begin{bmatrix} \mathbf{T} & & & \\ & \mathbf{T} & & \\ & & \mathbf{T} & \\ & & & \mathbf{T} \end{bmatrix}. \tag{3.86}$$

For the two-dimensional case:

$$\mathbf{k} = \begin{bmatrix} \mathbf{T} & \\ & \mathbf{T} \end{bmatrix}^t \boxed{\mathbf{k}} \begin{bmatrix} \mathbf{T} & \\ & \mathbf{T} \end{bmatrix}. \tag{3.87}$$

The entries of **k** are as follows:

$$\begin{aligned}
&k_{11} = T_{11}^2\alpha_1 + T_{21}^2\alpha_4^z \\
&k_{21} = T_{11}T_{12}\alpha_1 + T_{21}T_{22}\alpha_4^z \quad k_{22} = T_{12}^2\alpha_1 + T_{22}^2\alpha_4^z \\
&k_{31} = T_{21}\alpha_2^z \quad k_{32} = T_{22}\alpha_2^z \quad k_{33} = \alpha_3^z \\
&k_{41} = -T_{11}^2\alpha_1 + T_{21}^2\alpha_4^z \; k_{42} = -T_{21}T_{22}\alpha_4^z - T_{12}T_{11}\alpha_1 \; k_{43} = -T_{21}\alpha_2^z \; k_{44} = -T_{21}\alpha_2^z \\
&k_{51} = -T_{21}T_{22}\alpha_4^z - T_{12}T_{11}\alpha_1 \quad k_{52} = -T_{21}^2\alpha_4^z - T_{12}^2\alpha_1 \quad k_{53} = -T_{22}\alpha_2^z \\
&k_{54} = T_{21}T_{22}\alpha_4^z + T_{12}T_{11}\alpha_1 \quad k_{55} = T_{22}^2\alpha_4^z + T_{12}^2\alpha_1 \\
&k_{61} = T_{21}\alpha_2^z \; k_{62} = T_{22}\alpha_2^z \; k_{63} = \alpha_6^z \; k_{64} = -T_{21}\alpha_2^z \; k_{65} = -T_{22}\alpha_2^z \; k_{66} = \alpha_3^z.
\end{aligned} \tag{3.88}$$

in which

$$\alpha_1 = \frac{EA}{L}, \alpha_2^z = \frac{6EI_z}{L^2}, \alpha_3^z = \frac{4EI_z}{L}, \alpha_4^z = \frac{12EI_z}{L^3}, \text{ and } \alpha_6^z = \frac{2EI_z}{L}.$$

As an example, for element 1 of the planar frame, shown in Fig. 3.12, we have,

$$T_{11} = 0 \quad T_{12} = 1 \quad T_{21} = -1 \quad T_{22} = 0,$$

and the stiffness matrix of the element is obtained as:

$$\mathbf{k}_1 = 10^6 \begin{bmatrix} 1.25 & & & & & \\ 0 & 200 & & & \text{sym.} & \\ -0.75 & 0 & 6 & & & \\ -1.25 & 0 & 0.75 & 1.25 & & \\ 0 & -200 & 0 & 0 & 200 & \\ -0.75 & 0 & 3 & 0.75 & 0 & 6 \end{bmatrix}.$$

3.4 Displacement Method of Analysis

Once the stiffness matrix of an element is obtained in the selected global coordinate system, it can be planted in the specified and initialised overall stiffness matrix of the structure **K**, using the process described in Sect. 3.2.5.

Example. Let S be a planar truss with an arbitrary nodal and element numbering, as shown in Fig. 3.13. The entries of the transformation matrices of the members are calculated using Eqs. 3.32 and 3.33 as follows:

For bar 1: $T_{11} = \frac{x_2 - x_1}{2} = \frac{1-0}{2} = \frac{1}{2}$ and $T_{12} = \frac{y_2 - y_1}{2} = \frac{\sqrt{3}-0}{2} = \frac{\sqrt{3}}{2}$.
Similarly,
for bar 2: $T_{11} = \frac{1}{2}\, T_{12} = -\frac{\sqrt{3}}{2}$, and for bar 3: $T_{11} = 1$, $T_{12} = 0$.
Using the following relationship,

$$\begin{bmatrix} F_i^x \\ F_i^y \\ F_j^x \\ F_j^x \end{bmatrix} = \frac{EA}{L} \begin{bmatrix} T_{11}^2 & T_{11}T_{12} & -T_{11}^2 & -T_{11}T_{12} \\ T_{11}T_{12} & T_{12}^2 & -T_{11}T_{12} & -T_{12}^2 \\ -T_{11}^2 & -T_{11}T_{12} & T_{11}^2 & T_{11}T_{12} \\ -T_{11}T_{12} & -T_{12}^2 & T_{11}T_{12} & T_{12}^2 \end{bmatrix} \begin{bmatrix} \delta_i^x \\ \delta_i^y \\ \delta_j^x \\ \delta_j^y \end{bmatrix} \tag{3.89}$$

the stiffness matrices of the members are computed directly in the selected global coordinate system.

Fig. 3.12 A planar frame

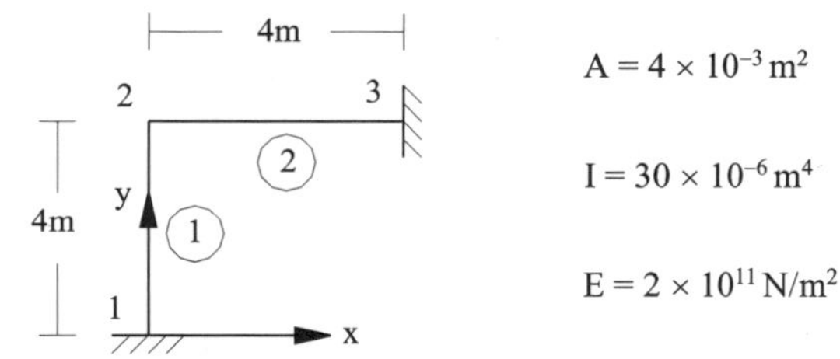

Fig. 3.13 A planar truss and the selected global coordinate system

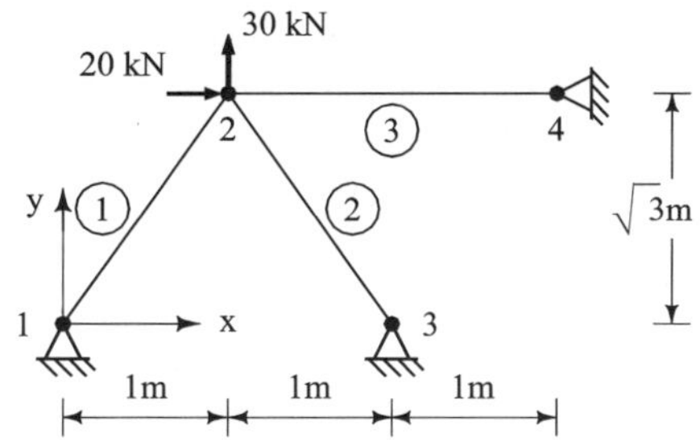

Now the stiffness matrices can be formed using Eq. 5.62:

For bar 1: $\mathbf{k}_1 = \frac{EA}{2}\begin{bmatrix} 0.25 & & & \text{sym.} \\ 0.433 & 0.75 & & \\ -0.25 & -0.433 & 0.25 & \\ -0.433 & -0.75 & 0.433 & 0.75 \end{bmatrix}$..

For bar 2: $\mathbf{k}_2 = \frac{EA}{2}\begin{bmatrix} 0.25 & & & \text{sym.} \\ -0.433 & 0.75 & & \\ -0.25 & 0.433 & 0.25 & \\ 0.433 & -0.75 & -0.433 & 0.75 \end{bmatrix}.$

For bar 3: $\mathbf{k}_3 = \frac{EA}{2}\begin{bmatrix} 1 & & & \text{sym.} \\ 0 & 0 & & \\ -1 & 0 & 1 & \\ 0 & 0 & 0 & 0 \end{bmatrix}.$

The overall stiffness matrix of the structure is an 8 × 8 matrix, which can easily be formed by planting the three member stiffness matrices as follows:

$$\mathbf{K} = \frac{EA}{2}\begin{bmatrix} 0.250 & 0.433 & -0.250 & -0.433 & 0 & 0 & 0 & 0 \\ 0.433 & 0.750 & -0.433 & -0.750 & 0 & 0 & 0 & 0 \\ -0.250 & -0.433 & 1.500 & 0 & -0.250 & 0.433 & -1.00 & 0 \\ -0.433 & -0.750 & 0 & 1.500 & 0.433 & -0.750 & 0 & 0 \\ 0 & 0 & -0.250 & 0.433 & 0.250 & -0.433 & 0 & 0 \\ 0 & 0 & 0.433 & -0.750 & -0.433 & 0.750 & 0 & 0 \\ 0 & 0 & -1.00 & 0 & 0 & 0 & 1.00 & 0 \\ 0 & 0 & 0 & 0 & 0 & 0 & 0 & 0 \end{bmatrix}.$$

Partitioning **K** into 2 × 2 submatrices, it can easily be seen that it is pattern equivalent to the node adjacency matrix of the graph model of the structure as follows:

$$\mathbf{C} * \mathbf{C}*^{t} = \begin{bmatrix} * & * & 0 & 0 \\ * & * & * & * \\ 0 & * & * & 0 \\ 0 & * & 0 & * \end{bmatrix}.$$

This pattern equivalence simplifies certain problems in structural mechanics, such as ordering the variables for bandwidth or profile reduction. Methods for increasing the sparsity, using special cutset bases, and improving the conditioning of structural matrices, are discussed in Refs. [6, 7].

3.4.1 Boundary Conditions

The matrix **K** is singular, since the boundary conditions have to be applied. Consider,

$$\mathbf{p} = \mathbf{K}\mathbf{v},$$

and partition it for free and constraint degrees of freedom as:

$$\begin{bmatrix} \mathbf{p}_{\mathrm{f}} \\ \mathbf{p}_{\mathrm{c}} \end{bmatrix} = \begin{bmatrix} \mathbf{K}_{\mathrm{ff}} & \mathbf{K}_{\mathrm{fc}} \\ \mathbf{K}_{\mathrm{cf}} & \mathbf{K}_{\mathrm{cc}} \end{bmatrix} \begin{bmatrix} \mathbf{v}_{\mathrm{f}} \\ \mathbf{v}_{\mathrm{c}} \end{bmatrix}. \tag{3.90}$$

This equation has a mixed nature; $\mathbf{p}_{\mathrm{f}}$ and $\mathbf{v}_{\mathrm{c}}$ have known values and $\mathbf{p}_{\mathrm{c}}$ and $\mathbf{v}_{\mathrm{f}}$ are unknowns. $\mathbf{K}_{\mathrm{ff}}$ is known as the *reduced stiffness matrix* of the structure, which is non-singular for a rigid structure.

For boundary conditions such as **vc** = **0**, it is easy to delete the corresponding rows and columns to obtain,

$$\mathbf{p}_{\mathrm{f}} = \mathbf{K}_{\mathrm{ff}}\mathbf{v}_{\mathrm{f}}, \tag{3.91}$$

from which $\mathbf{v}_{\mathrm{f}}$ can be obtained by solution of the above set of equations. In a computer this can be done by multiplying the diagonal entries of $\mathbf{K}_{\mathrm{cc}}$ by a large number such as 10^{20}. An alternative approach is possible by equating the diagonal entries of $\mathbf{K}_{\mathrm{cc}}$ to unity and all the other entries of these rows and columns to zero. If $\mathbf{v}_{\mathrm{c}}$ contains some specified values, $\mathbf{p}_{\mathrm{c}}$ will have corresponding $\mathbf{v}_{\mathrm{c}}$ values. A third method, which is useful when a structure has more constraint degrees of freedom (such as many supports), consists of the formation of element stiffness matrices considering the corresponding constraints, i.e. to form the reduced stiffness matrices of the elements in place of their complete matrices. This leads to some reduction in storage, and is also at the expense of additional computational effort.

As an example, the reduced stiffness matrix of the structure shown in Fig. 3.13 can be obtained from **K**, by deleting the rows and columns corresponding to the three supports 1, 3 and 4:

$$\begin{bmatrix} 20 \\ 30 \end{bmatrix} = \frac{EA}{2}\begin{bmatrix} 1.5 & 0 \\ 0 & 1.5 \end{bmatrix}\begin{bmatrix} u_{2x} \\ u_{2y} \end{bmatrix}.$$

Solving for the joint displacements, we have:

$$u_{2x} = \frac{40}{1.5EA} \text{ and } u_{2y} = \frac{40}{EA}.$$

The member distortions can easily be extracted from the displacement vector, and multiplication by the stiffness matrix of each member results in its member forces in the global coordinate system. As an example, for member 3 we have:

$$\begin{bmatrix} r_{2x} \\ r_{2y} \\ r_{4x} \\ r_{4y} \end{bmatrix} = \frac{EA}{2}\begin{bmatrix} 1 & & & \\ 0 & 0 & \text{sym.} & \\ -1 & 0 & 1 & \\ 0 & 0 & 0 & 0 \end{bmatrix}\begin{bmatrix} 40/1.5EA \\ 40/EA \\ 0 \\ 0 \end{bmatrix} = \begin{bmatrix} 13.33 \\ 0 \\ -13.33 \\ 0 \end{bmatrix}.$$

A transformation yields the member forces in the local coordinate systems, $\mathbf{r}_1 = \{-23.99 \quad 23.99\}^t, \mathbf{r}_2 = \{-10.659 \quad 10.65\}^t$ and $\mathbf{r}_3 = \{13.33 \quad -13.33\}^t$.

3.4.2 General Loading

The joint load vector of a structure can be computed in two parts. The first part comes from the external concentrated loads and/or moments, which are applied to the joints defined as the nodes of S. The components of such loads are most easily specified in a global coordinate system and can be entered into the joint load vector **p**.

The second part comes from the loads, which are applied to the spans of the members. These loads are usually defined in the local coordinate system of a member. For each member the fixed end actions (FEA) can be calculated using existing classical formulae or tables. A simple computer program can be prepared for this purpose. The fixed end actions should then be expressed in the global coordinate system using the transformation matrix given by Eq. 3.11. The FEA should then be reversed and applied to the end nodes of the members. These components can be added to **p** to form the final joint load vector. After **p** has been prepared and the boundary conditions imposed, the corresponding equations should be solved to obtain the joint displacements of the structure. Member distortions can then be extracted for each member in the reverse order to that used in assembling the **p** vector.

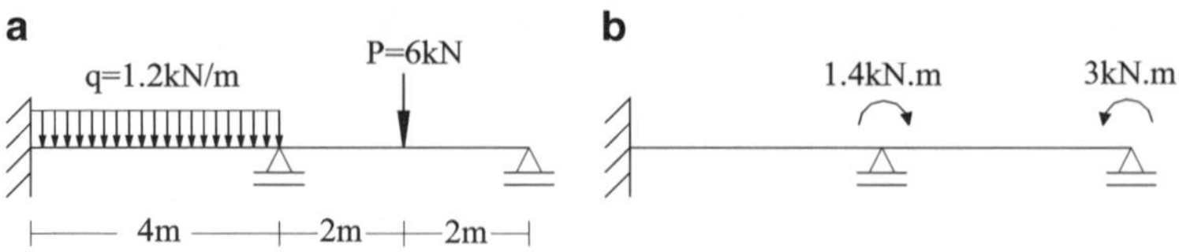

Fig. 3.14 A continuous beam and its equivalent loading

Example 1. A two span continuous beam is considered as shown in Fig. 3.14a. EI is taken to be constant along the beam.

For continuous beams, the transformation matrix **T** from local coordinate to global coordinate is identity, and therefore $\mathbf{k}_m = \overline{\mathbf{k}}_m$, i.e. no transformation is required. Ignoring the axial deformation and using Eq. 3.63, the stiffness matrices of the elements are obtained as:

$$\mathbf{k}_1 = \mathbf{k}_2 = \frac{64}{4}\begin{bmatrix} 0.75 & 1.5 & -0.75 & 1.5 \\ 1.5 & 4 & -1.5 & 2 \\ -0.75 & -1.5 & 0.75 & -1.5 \\ 1.5 & 2 & -1.5 & 4 \end{bmatrix}.$$

Assembling the overall stiffness matrix and imposing the boundary conditions, the reduced stiffness matrix of the entire beam is obtained and the force-displacement relationship for beam is written as:

$$\begin{bmatrix} -1.40 \\ 3 \end{bmatrix} = 16\begin{bmatrix} 8 & 2 \\ 2 & 4 \end{bmatrix}\begin{bmatrix} \theta_2^z \\ \theta_3^z \end{bmatrix}.$$

Solving the equations leads to:

$$\begin{bmatrix} \theta_2^z \\ \theta_3^z \end{bmatrix} = \frac{1}{448}\begin{bmatrix} 4 & -2 \\ -2 & 8 \end{bmatrix}\begin{bmatrix} -1.4 \\ 3 \end{bmatrix} = \begin{bmatrix} -0.0259 \\ 0.0598 \end{bmatrix}.$$

Member forces are calculated as:

$$\begin{bmatrix} V_1 \\ M_1 \\ V_2 \\ M_2 \end{bmatrix} = 16\begin{bmatrix} 0.75 & 1.5 & -0.75 & 1.5 \\ 1.5 & 4 & -1.5 & 2 \\ -0.75 & -1.5 & 0.75 & -1.5 \\ 1.5 & 2 & -1.5 & 4 \end{bmatrix}\begin{bmatrix} 0 \\ 0 \\ 0 \\ -0.0259 \end{bmatrix} + \begin{bmatrix} 2.4 \\ 1.6 \\ 2.4 \\ -1.6 \end{bmatrix}$$

$$= \begin{bmatrix} 1.779 \\ 0.772 \\ 3.021 \\ -3.256 \end{bmatrix},$$

and

$$\begin{bmatrix} V_2 \\ M_2 \\ V_3 \\ M_3 \end{bmatrix} = 16 \begin{bmatrix} 0.75 & 1.5 & -0.75 & 1.5 \\ 1.5 & 4 & -1.5 & 2 \\ -0.75 & -1.5 & 0.75 & -1.5 \\ 1.5 & 2 & -1.5 & 4 \end{bmatrix} \begin{bmatrix} 0 \\ -0.0259 \\ 0 \\ +0.0598 \end{bmatrix} + \begin{bmatrix} 3 \\ 3 \\ 3 \\ -3 \end{bmatrix}$$

$$= \begin{bmatrix} 3.814 \\ 3.258 \\ 2.186 \\ 0 \end{bmatrix}.$$

Example 2. A portal frame is considered as shown in Fig. 3.15. The members are made of sections with $A = 150\ cm^2$ and $I_z = 2 \times 10^4 cm^4$ and $E = 2 \times 10^4\ kN/cm^2$. Calculate the joint rotations and displacements.

The equivalent joint loads are illustrated in Fig. 3.16.

Employing Eq. 3.88, the stiffness matrices for the members are obtained as:

For member 1:

$$\mathbf{k}_1 = 10^4 \begin{bmatrix} 0.008 & & & & & \\ 0 & 0.75 & & & \text{sym.} & \\ -1.5 & 0 & 400 & & & \\ 0.008 & 0 & 1.5 & 0.008 & & \\ 0 & -0.75 & 0 & 0 & 0.75 & \\ -1.5 & 0 & 200 & 1.5 & 0 & 400 \end{bmatrix},$$

and for member 2:

$$\mathbf{k}_2 = 10^4 \begin{bmatrix} 0.6 & & & & & \\ 0 & 0.004 & & & \text{sym.} & \\ 0 & 0.96 & 320 & & & \\ -0.6 & 0 & 0 & 0.6 & & \\ 0 & -0.004 & -0.96 & 0 & 0.004 & \\ 0 & 0.96 & 160 & 0 & -0.96 & 320 \end{bmatrix}$$

For member 3:

$$\mathbf{k}_3 = 10^4 \begin{bmatrix} 0.008 & & & & & \\ 0 & 0.75 & & & \text{sym.} & \\ 1.5 & 0 & 400 & & & \\ -0.008 & 0 & -1.5 & 0.008 & & \\ 0 & -0.75 & 0 & 0 & 0.75 & \\ 1.5 & 0 & 200 & -1.5 & 0 & 400 \end{bmatrix}$$

Assembling the stiffness matrices and imposing the boundary conditions results in the following equations:

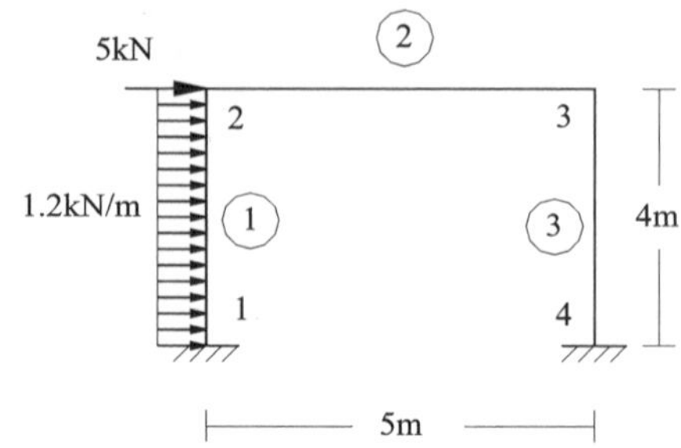

Fig. 3.15 A portal frame and its loading

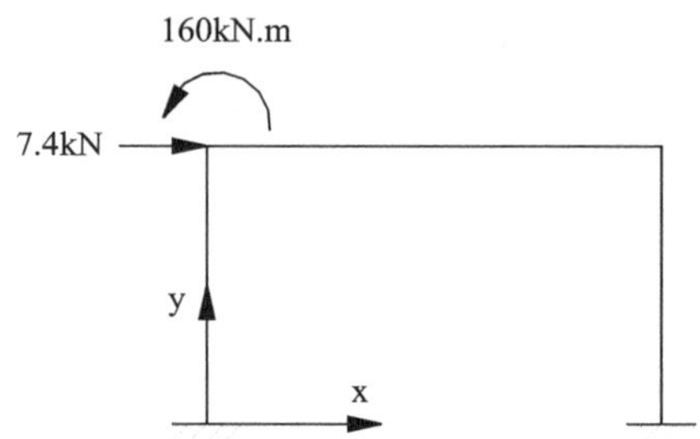

Fig. 3.16 Equivalent joint loads

$$\begin{bmatrix} 7.4 \\ 0 \\ 160 \\ 0 \\ 0 \\ 0 \end{bmatrix} = 10^4 \begin{bmatrix} 0.608 & & & & & \text{sym.} \\ 0 & 0.754 & & & & \\ 1.5 & 0.96 & 720 & & & \\ -0.6 & 0 & 0 & 0.608 & & \\ 0 & -0.004 & -0.96 & 0 & 0.754 & \\ 0 & 0.96 & 160 & 1.5 & -0.96 & 720 \end{bmatrix} \begin{bmatrix} \delta_2^x \\ \delta_2^y \\ \theta_2^z \\ \delta_3^x \\ \delta_3^y \\ \theta_3^z \end{bmatrix}$$

Solving these equations leads to:

$$\delta_2^x = 0.0659167, \quad \delta_2^y = 2.617764\text{E} - 04, \quad \theta_2^z = -8.983453\text{E} - 05,$$
$$\delta_3^x = 0.0653377, \quad \delta_3^y = -2.617704\text{E} - 04 \text{ and } \theta_3^z = -1.16855\text{E} - 04.$$

The final member forces can be found using the stiffness of the members, superimposed by the fixed end actions.

3.5 Stiffness Matrix of a Finite Element

In this section, a simple element is introduced from finite element methods, in order to show the capability of the method presented in Sect. 3.2.2, for the formation of element stiffness matrices.

3.5.1 Stiffness Matrix of a Triangular Element

For plane stress and plane strain problems, the displacements of a node can be specified by two components, and therefore for each node of the triangular element, two degrees of freedom is considered, as shown in Fig. 3.17.

Element forces and displacements are defined by the following vectors:

$$\mathbf{r}_m = \{ r_1 \quad r_2 \quad \ldots \quad r_6 \}^t \text{ and } \mathbf{u}_m = \{ u_1 \quad u_2 \quad \ldots \quad u_6 \}^t. \tag{3.92}$$

A triangular element has its boundary attached continuously to the surrounding medium, and therefore no exact stiffness matrix can be derived. Therefore an approximate solution should be sought.

The following displacement functions can be considered for the variation of the displacements,

$$u = \alpha_1 x + \alpha_2 y + \alpha_3 \quad \text{and} \quad v = \alpha_4 x + \alpha_5 y + \alpha_6, \tag{3.93}$$

where $\alpha_1, \alpha_2, \ldots, \alpha_6$ are arbitrary constants which can be found from the displacements of the three nodes of the element. From the boundary conditions,

$$\begin{aligned} &\text{at node } i(x_i, y_i), u = u_i \text{ and } v = v_i, \\ &\text{at node } j\left(x_j, y_j\right), u = u_j \text{ and } v = v_j, \\ &\text{at node } k(x_k, y_k), u = u_k \text{ and } v = v_k, \end{aligned} \tag{3.94}$$

the constants can be evaluated. Substituting in Eq. 3.93 yields:

$$\begin{aligned} u &= 1/2A\left\{\left[y_{kj}(x - x_j) - x_{kj}\left(y - y_j\right)\right]u_1 + [-y_{ki}(x - x_k) - x_{ki}(y - y_k)]u_3 + \left[y_{ji}(x - x_i) - x_{ji}(y - y_i)\right]u_5\right\}, \\ v &= 1/2A\left\{\left[y_{kj}(x - x_j) - x_{kj}\left(y - y_j\right)\right]u_2 + [-y_{ki}(x - x_k) - x_{ki}(y - y_k)]u_4 + \left[y_{ji}(x - x_i) - x_{ji}(y - y_i)\right]u_6\right\}. \end{aligned} \tag{3.95}$$

where:

$$2A = 2(\text{area of the triangle}) = x_{kj}y_{ji} - x_{ji}y_{kj}, \tag{3.96}$$

and

$$x_{mn} = x_m - x_n \text{ and } y_{mn} = y_m - y_n. \tag{3.97}$$

From Eq. 3.95, it is obvious that both u and v vary linearly along each edge of the element, and they depend only on the displacements of the two nodes on a particular edge. Therefore, the compatibility of displacements on two adjacent elements with common boundary is satisfied.

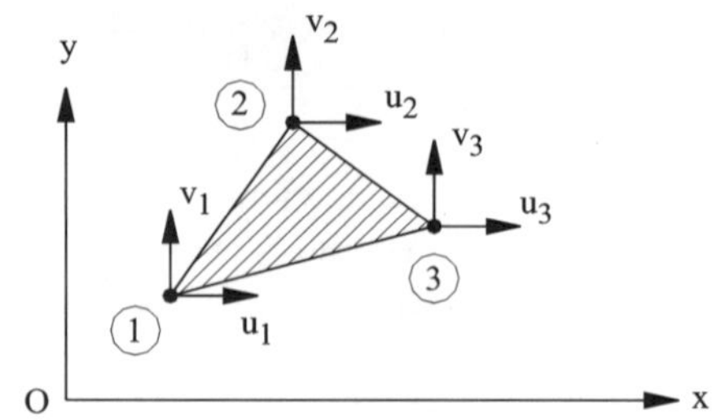

Fig. 3.17 A triangular element

From the theory of elasticity, the nodal displacements $\mathbf{u}_m^t = \{u_1, u_2, \ldots, u_6\}$ are related to total strains $e^t = \{e_{xx}, e_{yy}, e_{zz}\}$ by the following:

$$\mathbf{e} = \begin{bmatrix} e_{xx} \\ e_{yy} \\ e_{xy} \end{bmatrix} = \begin{bmatrix} \dfrac{\partial u}{\partial x} \\ \dfrac{\partial v}{\partial y} \\ \dfrac{\partial u}{\partial y} + \dfrac{\partial v}{\partial x} \end{bmatrix}$$

$$= \frac{1}{2A} \begin{bmatrix} y_{kj} & 0 & -y_{ki} & 0 & y_{ji} & 0 \\ 0 & -x_{kj} & 0 & x_{ki} & 0 & -x_{ji} \\ -x_{kj} & y_{kj} & x_{ki} & -y_{ki} & -x_{ji} & y_{ji} \end{bmatrix} \begin{bmatrix} u_1 \\ u_2 \\ u_3 \\ u_4 \\ u_5 \\ u_6 \end{bmatrix}. \tag{3.98}$$

This relationship can be written in matrix notation as,

$$\mathbf{e} = \frown{\mathbf{b}}\,\mathbf{u}, \tag{3.99}$$

where:

$$\frown{\mathbf{b}} = \frac{1}{2A} \begin{bmatrix} y_{kj} & 0 & -y_{ki} & 0 & y_{ji} & 0 \\ 0 & -x_{kj} & 0 & x_{ki} & 0 & -x_{ji} \\ -x_{kj} & y_{kj} & x_{ki} & -y_{ki} & -x_{ji} & y_{ji} \end{bmatrix}. \tag{3.100}$$

The above equation, indicates that for linearly varying displacement field, the strains are constant, and by Hooke's law it also leads to constant stresses. Substituting the total strain **e** in Eq. 3.96 gives the stress-displacement relationship,

$$\begin{bmatrix} \sigma_{xx} \\ \sigma_{yy} \\ \sigma_{xy} \end{bmatrix} = \frac{E}{2A(1-v^2)} \begin{bmatrix} y_{kj} & -vy_{kj} & -y_{ki} & vx_{ki} & y_{ji} & -vy_{ji} \\ vy_{kj} & -x_{kj} & -vy_{ki} & x_{ki} & vy_{ji} & -x_{ji} \\ -\Psi x_{kj} & \Psi y_{kj} & \Psi x_{ki} & -\Psi y_{ki} & -\Psi x_{ji} & \Psi y_{ji} \end{bmatrix} \begin{bmatrix} u_1 \\ u_2 \\ u_3 \\ u_4 \\ u_5 \\ u_6 \end{bmatrix}, \tag{3.101}$$

where v is the Poisson ratio and

$$\Psi = \frac{1-v}{2}.$$

The stiffness matrix is then calculated using Eq. 3.26, and for convenience is presented in two separate parts as:

$$\mathbf{k} = \mathbf{k}_n + \mathbf{k}_s, \tag{3.102}$$

where $\mathbf{k}_n$ represents the stiffness due to normal stresses and $\mathbf{k}_s$ represents the stiffnesses due to shearing stresses. Thus:

$$\mathbf{k}_n = \frac{Et}{4A(1-v^2)} \begin{bmatrix} y_{32}^2 & & & & \text{sym.} & \\ -vy_{32}x_{32} & x_{32}^2 & & & & \\ -y_{32}y_{31} & vx_{32}y_{31} & y_{31}^2 & & & \\ vy_{32}x_{31} & -x_{32}x_{31} & -vy_{31}x_{31} & x_{31}^2 & & \\ y_{32}y_{21} & -vx_{32}y_{21} & -y_{31}y_{21} & vx_{31}y_{21} & y_{21}^2 & \\ -vy_{32}x_{21} & x_{32}x_{21} & vy_{31}x_{21} & -x_{31}x_{21} & -vy_{21}x_{21} & x_{21}^2 \end{bmatrix},$$

and

$$\mathbf{k}_s = \frac{Et}{4A(1+v)} \begin{bmatrix} x_{32}^2 & & & & \text{sym.} & \\ -x_{32}y_{32} & y_{32}^2 & & & & \\ -x_{32}x_{31} & y_{32}x_{31} & x_{31}^2 & & & \\ x_{32}y_{31} & -y_{32}x_{21} & -x_{31}y_{31} & y_{31}^2 & & \\ x_{32}x_{21} & -y_{32}x_{21} & -x_{31}x_{21} & y_{31}x_{21} & x_{21}^2 & \\ -x_{32}y_{21} & y_{32}y_{21} & x_{31}y_{21} & -y_{31}y_{21} & -x_{21}y_{21} & y_{21}^2 \end{bmatrix}. \tag{3.103}$$

Using the same method, the stiffness matrices for other elements can be derived. Since there are many excellent books on finite element methods, no further studies are made here, and the interested reader may refer to McGuire and Gallagher [3], Przemieniecki [4], and Zienkiewicz [5], among many others. For the formation of well-conditioned stiffness matrices the reader may refer to Kaveh [6, 7].

3.6 Computational Aspects of the Matrix Displacement Method

The main advantage of the displacement method is its simplicity for computer programming. This is due to the existence of a simple kinematical basis formed on a special cutset basis known as cocycle basis of the graph model S of the structure. Such a basis does not correspond to the most sparse stiffness matrix; however, the sparsity is generally so good that there is usually no need to look further. However, if an optimal cutset basis of S is needed, the displacement method encounters all the problems met by the force method, described in Chap. 3. The algorithm for the displacement method is summarized below.

Algorithm

Step 1: Select a global coordinate system and number the nodes and members of the structure. An appropriate nodal ordering algorithm will be discussed in Chap. 5.

Step 2: After initialization of all the vectors and matrices, read or generate the data for the structure and its members.

Step 3: For each member of the structure:

(a) Compute L, L*, $\sin\alpha$, $\sin\beta$, $\sin\gamma$, $\cos\alpha$, $\cos\beta$, $\cos\gamma$;
(b) Compute the rotation matrix $\mathbf{T}$;
(c) Form the member stiffness matrix $\bar{\mathbf{k}}$ in its local coordinate system;
(d) Form the member stiffness matrix $\mathbf{k}$ in the selected global coordinate system;
(e) Plant $\mathbf{k}$ in the overall stiffness matrix $\mathbf{K}$ of the structure.

Step 4: For each loaded member:

(a) Read the fixed end actions;
(b) Transform the fixed end actions to the global coordinate system and reverse it to apply at joints;
(c) Store these joint loads in the specified overall joint load vector.

Step 5: For each loaded joint:

(a) Read the joint number and the applied joint loads;
(b) Store it in the overall joint load vector.

Step 6: Apply boundary conditions to the structural stiffness matrix $\mathbf{K}$, to obtain the reduced stiffness matrix $\mathbf{K}_{\mathrm{ff}}$. Repeat the same for the overall joint load vector.

Step 7: Solve the corresponding equations to obtain the joint displacements.

Step 8: For each member:

(a) Extract the member distortions from the joint displacements;
(b) Rotate the member distortions to the local coordinate system;

(c) Compute the member stiffness matrix;
(d) Compute the member forces and fixed end actions.

Step 9: Compute the joint displacements and the member forces.

The application of the above procedure is now illustrated by a simple example, so that the reader can use it to fully understand the computational steps.

Example. Consider a planar truss, as shown in Fig. 3.18. Member 1 has a uniform load of intensity 0.6 kN/m and at joint 2 a concentrated load of magnitude 1.05 kN is applied. The cross-section areas for members are 2A and 1.8A, respectively.

The selected global coordinate system and the equivalent nodal forces are illustrated in Fig. 3.19. The stiffness matrices are formed as:

for member 1:

$$\mathbf{k}_1 = \frac{2}{5}EA\begin{bmatrix} 0.64 & 0.48 & -0.64 & -0.48 \\ 0.48 & 0.36 & -0.48 & -0.36 \\ -0.64 & -0.48 & 0.64 & 0.48 \\ -0.48 & -0.36 & 0.48 & 0.36 \end{bmatrix}.$$

and for member 2:

$$\mathbf{k}_2 = \frac{1.8}{3}EA\begin{bmatrix} 0 & 0 & 0 & 0 \\ 0 & +1 & 0 & -1 \\ 0 & 0 & 0 & 0 \\ 0 & -1 & 0 & +1 \end{bmatrix}.$$

The overall stiffness matrix is then obtained as:

$$\mathbf{K} = EA\begin{bmatrix} 0.256 & 0.192 & -0.256 & -0.192 & 0 & 0 \\ 0.192 & 0.144 & -0.192 & -0.144 & 0 & 0 \\ -0.256 & -0.192 & 0.256 & 0.192 & 0 & 0 \\ -0.192 & -0.144 & 0.192 & 0.744 & 0 & -0.6 \\ 0 & 0 & 0 & 0 & 0 & 0 \\ 0 & 0 & 0 & -0.6 & 0 & 0.6 \end{bmatrix}.$$

The fixed end actions are shown in Fig. 3.19b, and calculated for member 1 as:

$$\mathbf{FEA}_1 = \begin{bmatrix} 0 \\ 1.5 \\ 0 \\ 1.5 \end{bmatrix}.$$

These forces are reversed and transformed into the global coordinate system as:

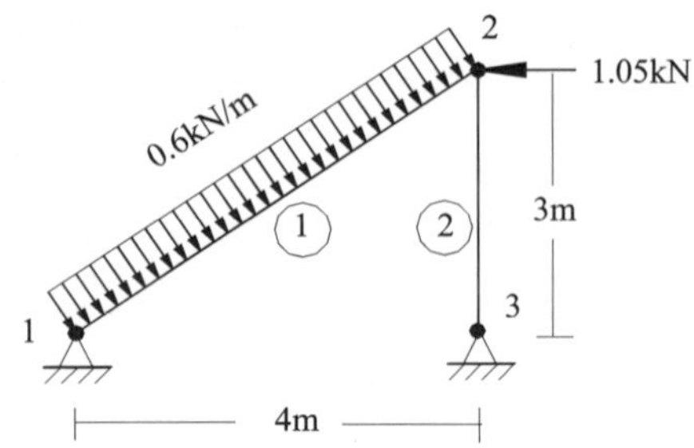

Fig. 3.18 A planar truss with general loading

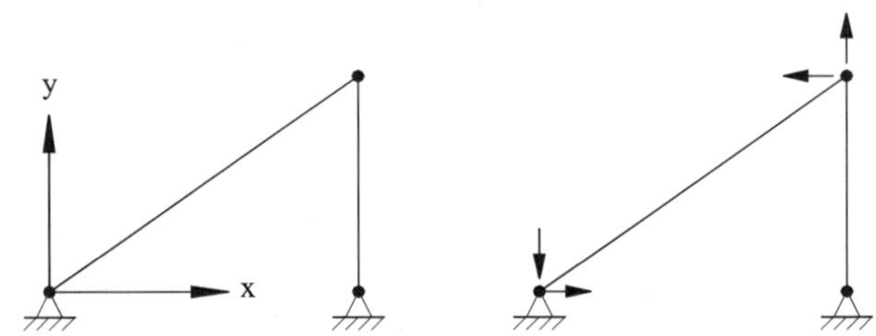

Fig. 3.19 The selected coordinate system and the equivalent nodal loads

$$\mathbf{T}_1^t(-\mathbf{FEA}_1) = \begin{bmatrix} 0.8 & -0.6 & 0 & 0 \\ 0.6 & 0.8 & 0 & 0 \\ 0 & 0 & 0.8 & -0.6 \\ 0 & 0 & 0.6 & 0.8 \end{bmatrix} \begin{bmatrix} 0 \\ -1.5 \\ 0 \\ -1.5 \end{bmatrix} = \begin{bmatrix} 0.9 \\ -1.2 \\ 0.9 \\ -1.2 \end{bmatrix}.$$

Superimposing the concentrated force at node 2 yields the final vector of external forces as:

$$\mathbf{p} = \{0.9 \quad -1.2 \quad -0.15 \quad -1.2 \quad 0 \quad 0\}^t.$$

Substituting a large number such as 1.E + 30 for the diagonal entries corresponding to the zero displacement boundary conditions,

$$\begin{bmatrix} 0 \\ 0 \\ -0.15 \\ -1.2 \\ 0 \\ 0 \end{bmatrix} = EA \begin{bmatrix} 1.E+30 & 0.192 & -0.256 & -0.192 & 0 & 0 \\ 0.192 & 1.E+30 & -0.192 & -0.256 & 0 & 0 \\ -0.256 & -0.192 & 0.256 & 0.192 & 0 & 0 \\ -0.192 & -0.256 & 0.192 & 0.714 & 0 & -0.6 \\ 0 & 0 & 0 & 0 & 1.E+30 & 0 \\ 0 & 0 & 0 & -0.6 & 0 & 1.E+30 \end{bmatrix} [\mathbf{v}].$$

Solving these equations results in:

$$\mathbf{v} = \frac{1}{\mathrm{EA}}\{0 \quad 0 \quad 0.845 \quad -1.907 \quad 0 \quad 0\}^{t}.$$

The member forces are now computed as:

$$\mathbf{r}_1 = \frac{2}{5}\begin{bmatrix} 1 & 0 & -1 & 0 \\ 0 & 0 & 0 & 0 \\ -1 & 0 & 1 & 0 \\ 0 & 0 & 0 & 0 \end{bmatrix}\begin{bmatrix} 0.8 & 0.6 & 0 & 0 \\ -0.6 & 0.8 & 0 & 0 \\ 0 & 0 & 0.8 & 0.6 \\ 0 & 0 & -0.6 & 0.8 \end{bmatrix}\begin{bmatrix} 0 \\ 0 \\ 0.845 \\ -1.907 \end{bmatrix} + \begin{bmatrix} 0 \\ 1.5 \\ 0 \\ 1.5 \end{bmatrix}$$

$$= \begin{bmatrix} 0.179 \\ 1.5 \\ -0.179 \\ 1.5 \end{bmatrix},$$

and

$$\mathbf{r}_2 = \frac{3}{5}\begin{bmatrix} 1 & 0 & -1 & 0 \\ 0 & 0 & 0 & 0 \\ -1 & 0 & 1 & 0 \\ 0 & 0 & 0 & 0 \end{bmatrix}\begin{bmatrix} 0 & -1 & 0 & 0 \\ 1 & 0 & 0 & 0 \\ 0 & 0 & 0 & -1 \\ 0 & 0 & 1 & 0 \end{bmatrix}\begin{bmatrix} 0.845 \\ -1.907 \\ 0 \\ 0 \end{bmatrix} + \begin{bmatrix} 0 \\ 0 \\ 0 \\ 0 \end{bmatrix}$$

$$= \begin{bmatrix} 1.091 \\ 0 \\ -1.091 \\ 0 \end{bmatrix}.$$

References

1. Argyris et al (1979) Finite element method: the natural approach. Comput Methods Appl Mech Eng 7:1–106
2. Livesley RK (1975) Matrix methods of structural analysis, 2nd edn. Pergamon Press, New York
3. McGuire W, Gallagher RH (1979) Matrix structural analysis. Wiley, New York
4. Przemieniecki JS (1963) Matrix structural analysis of substructures. AIAA J 1:138–147
5. Zienkiewicz OC (1977) The finite element method in engineering, 3rd edn. McGraw-Hill, Maidenhead
6. Kaveh A (2004) Structural mechanics: graph and matrix methods, 3rd edn. Research Studies Press, Somerset
7. Kaveh A (1977) Optimal structural analysis, 1st edn. Research Studies Press, Chichester

Chapter 4
Ordering for Optimal Patterns of Structural Matrices: Graph Theory Methods

4.1 Introduction

In this chapter, methods are presented for ordering to form special patterns for sparse structural matrices. Such transformation reduces the storage and the number of operations required for the solution, and leads to more accurate results. Graph theory methods are presented for different approaches to reordering equations to preserve their sparsity, leading to predefined patterns. Alternative, objective functions are considered and heuristic algorithms are presented to achieve these objectives. Three main methods for the solution of structural equations require the optimisation of bandwidth, profile and frontwidth, especially for those encountered in finite element analysis. Methods are presented for reducing the bandwidth of the flexibility matrices. Bandwidth optimisation of rectangular matrices is presented for its use in the formation of sparse flexibility matrices.

In this chapter entries of the stiffness and flexibility matrices are provided with the most appropriate specified patterns for solution of the corresponding equations. Realization of these patterns (or not) affects the formulation of the mathematical models and efficiency of solution. Many patterns can be designed depending on the solution scheme being used. Figure 4.1 shows some of the popular ones encountered in practice.

Pattern equivalence of the stiffness matrix of a structure and cutset basis adjacency matrix $\mathbf{C}^*\mathbf{C}^{*t}$ of its graph model, and pattern equivalence of the flexibility matrix of a structure with that of a generalized cycle basis adjacency matrix $\mathbf{C}\mathbf{C}^{t}$ of its graph model, reduce the size of the problem β-fold, β being the degrees of freedom of the nodes of the model for the displacement method, and $\beta = 1$ to 6 depending on the type of structure being studied by the force method.

A. Kaveh, *Computational Structural Analysis and Finite Element Methods*,
DOI 10.1007/978-3-319-02964-1_4,

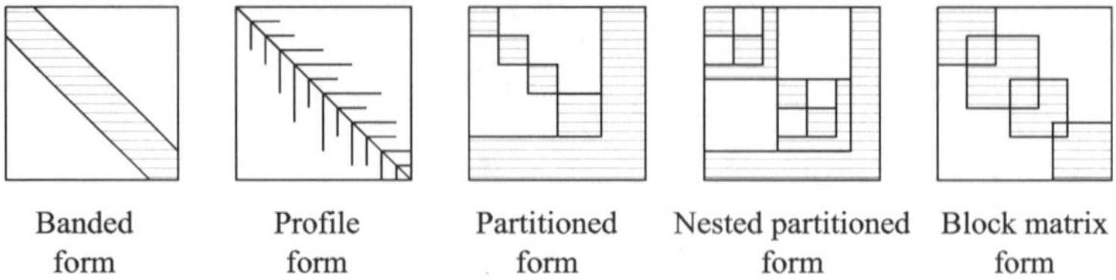

Fig. 4.1 Different matrix forms

4.2 Bandwidth Optimisation

The analysis of many problems in structural engineering involves the solution of a set of linear equations of the form,

$$\mathbf{Ax} = \mathbf{b}, \tag{4.1}$$

where **A** is a symmetric, positive definite and usually very sparse matrix. For large structures encountered in practice, 30–50 % of the computer execution time may be devoted to solving these equations. This figure may rise to about 80 % in non-linear, dynamic or structural optimisation problems.

Different methods can be used for the solution of the system of equations, of which the Gaussian elimination is the most popular among structural analysts, since it is simple, accurate and practical, producing some very satisfactory error bounds.

In the forward course of elimination, new non-zero entries may be created, but the back substitution does not lead to any new non-zero elements. It is beneficial to minimize the total number of such non-zero elements created during the forward course of the Gaussian elimination in order to reduce the round off errors and the computer storage. Matrix **A** of Eq. 4.1 can be transformed by means of row and column operations to a form which leads to the creation of a minimum number of non-zero entries during the forward course of the elimination. This is equivalent to the "a priori" determination of permutation matrices **P** and **Q**, such that:

$$\mathbf{PAQ} = \mathbf{G}. \tag{4.2}$$

When **A** is symmetric and positive definite, it is advantageous to have **G** also symmetric so that only the non-zero elements on and above the diagonal of **G** need to be stored, and only about half as many arithmetic operations are needed in the elimination. The diagonal elements of **A** and **G** are the same, only in different positions. In order to preserve symmetry, **P** is taken as $\mathbf{Q}^t$ so that Eq. 4.2 becomes:

$$\mathbf{Q}^t\mathbf{AQ} = \mathbf{G}. \tag{4.3}$$

For transforming a symmetric matrix **A** into the forms depicted in Fig. 4.1, various methods are available, some of which will be described in this chapter. However, due to the simplicity of the banded form, most of the material presented will be confined to optimising the bandwidth of the structural matrices, and other forms will only be introduced briefly.

In the Gaussian elimination method, the time required to solve the resulting equations by the banded matrix technique is directly proportional to the square of the bandwidth of **A**. As mentioned before, the solution of these equations forms a large percentage of the total computational effort needed for the structural analysis. Therefore it is not surprising that a lot of attention is being paid to the optimisation of the bandwidth of these sparse matrices. A suitable ordering of the elements of a kinematical basis for a structure reduces the bandwidth of **A**, hence decreasing the solution time, storage and round-off errors. Similarly, ordering the elements of a statical basis results in the reduction of the bandwidth of the corresponding flexibility matrix of the structure.

Iterative methods using different criteria for the control of the process of interchanging rows and columns of **A** are described by many authors, e.g. see Rosen [1] and Grooms [2]. For these methods, in general, the required storage and CPU time can be high, making them uneconomical.

The first direct method for bandwidth reduction was recognized by Harary [3] in 1967, who posed the following question:

> For a graph S with N(S) nodes, how can labels 1, 2, . . ., N(S) be assigned to nodes in order to minimize the maximum absolute value of the difference between the labels of all pairs of adjacent nodes?

For a graph labelled in such an optimum manner, the corresponding adjacency matrix will have unit entries concentrated as closely as possible to its main diagonal.

In structural engineering, Cuthill and McKee [4] developed the first graph-theoretical approach for reducing the bandwidth of stiffness matrices. In their work, a level structure was used which was called a "spanning tree" of a structure. The author's interest in bandwidth reduction was initially motivated by an interest in generating and ordering the elements of cycle bases and generalized cycle bases of a graph, as defined in Chap. 2, in order to reduce the bandwidth of the flexibility matrices, Refs. [5, 6]. For this purpose a *shortest route tree* (SRT) has been used. The application of this approach has been extended to the elements of a kinematical basis (cutset basis) in order to reduce the bandwidth of stiffness matrices. Subsequently it has been noticed that there is a close relation between Cuthill-McKee's *level structure* and the author's SRT. However, there is a difference between these trees in that an SRT contains additional information about the connectivity properties of the corresponding structure.

Further improvements have been achieved by employing special types of SRTs such as the longest and narrowest ones, Ref. [7]. Generation of a suitable SRT depends on an appropriate choice of starting node. Kaveh [5] used an end node of an arbitrary SRT, which was chosen from its last contour (level) having the least valency. Gibbs et al. [8] employed a similar node and called it a pseudo-peripheral node. Cheng [9] used an algebraic approach to select a single node or a set of nodes as the root of an SRT. Kaveh employed two simultaneous SRTs for selecting a pseudo-peripheral node. A comparison of six different algorithms was made in Ref. [10]. Algebraic graph theory has also been used for finding a starting node, Kaveh [11] and Grimes et al. [12]. Paulino et al. [13] used another type of algebraic graph-theoretical approach employing the Laplacian matrix of a graph for nodal ordering.

4.3 Preliminaries

A matrix **A** is called *banded*, when all its non-zero entries are confined within a band, formed by diagonals parallel to the main diagonal. Therefore, $A_{ij} = 0$ when $|i-j| > b$, and $A_{k,\ k-b} \neq 0$ or $A_{k,\ k+b} \neq 0$ for at least one value of k. b is the half-bandwidth and 2b + 1 is known as the *bandwidth* of **A**. As an example, for

$$\mathbf{A} = \begin{bmatrix} 1 & 6 & . & . & . \\ 6 & 2 & 7 & 9 & . \\ . & 7 & 3 & 8 & . \\ . & 8 & 9 & 4 & . \\ . & . & . & . & 5 \end{bmatrix}, \tag{4.4}$$

the bandwidth of **A** is 2b + 1 = 2×2 + 1 = 4.

A banded matrix can be stored in different ways. The *diagonal storage* of a symmetric banded n×n matrix **A** is an n×(b + 1) matrix **AN**. The main diagonals are stored in the last column, and lower co-diagonals are stored down-justified in the remaining columns. As an example, **AN** for the above matrix is:

$$\mathbf{AN} = \begin{bmatrix} \cdot & \cdot & 1 \\ \cdot & 6 & 2 \\ 0 & 7 & 3 \\ 9 & 8 & 4 \\ 0 & 0 & 5 \end{bmatrix}. \tag{4.5}$$

When **A** is a sparse matrix, this storage scheme is very convenient, since it provides direct access, in the sense that there is a simple one-to-one correspondence between the position of an entry in the matrix **A**(i, j) and its position in **AN**(i, j − i + b + 1).

Obviously, the bandwidth depends on the order in which the rows and columns of **A** are arranged. This is why iterative techniques seek a permutation of the rows and a permutation of columns to make the resulting bandwidth small. For symmetric matrices, identical permutations are needed for both the rows and the columns. When a system of linear equations has a banded matrix of coefficients and the system is solved by Gaussian elimination, with pivots being taken from the diagonals, all the operations are confined to the band and no new non-zero entries are generated outside the band. Therefore, the Gaussian elimination can be carried out in place, since a memory location is already reserved for any new non-zeros that might be introduced within the band.

For each row i of a symmetric matrix **A** define,

$$b_i = i - j_{min}(i), \tag{4.6}$$

where $j_{min}(i)$ is the minimum column index in row i for which $A_{ij} \neq 0$. Therefore, the first non-zero of row i lies b_i positions to the left of the diagonal, and b is defined as:

$$b = \max(b_i). \tag{4.7}$$

In Chap. 4, it has been shown that the stiffness matrix **K** of a structure is pattern equivalent to the cutset basis adjacency matrix $\mathbf{C^*C^{*t}}$, where $\mathbf{C^*}$ is the cutset basis-member incidence matrix of the structural model S. Similarly, the flexibility matrix **G** is pattern equivalent to the cycle basis adjacency matrix $\mathbf{CC^t}$, where **C** is the cycle basis-member incidence matrix of S.

Reducing the bandwidths of $\mathbf{C^*C^{*t}}$ and $\mathbf{CC^t}$ directly influences those of **K** and **G**, respectively. Notice that the dimensions of $\mathbf{C^*C^{*t}}$ and $\mathbf{CC^t}$, for general space structures, are sixfold smaller than those of **K** and **G**, and therefore simpler to optimise.

For the displacement method of analysis, there exists a special cutset basis whose elements correspond to stars of its nodes except for the ground node (cocycle basis). The adjacency matrix of such a basis naturally is the same as that of the node adjacency matrix of S, with the row and column corresponding the datum node being omitted. In this chapter, such a special cutset basis will be considered, and the nodes of S will be ordered such that the bandwidth of its node adjacency matrix is reduced to the smallest possible amount.

Let **A** be the adjacency matrix of a graph S. Let i and j be the nodal numbers of member k, and let $\alpha_k = |i - j|$. Then the bandwidth of **A** can be defined as:

$$b(\mathbf{A}) = 2\mathrm{Max}\{\alpha_k : k = 1, 2, \ldots, M(S)\} + 1, \tag{4.8}$$

where M(S) is the number of members of S. In order to minimize the bandwidth of **A**, the value of b(**A**) should be minimized. The bandwidth of the stiffness matrix **K** of a structure is related to that of **A** by:

$$b(\mathbf{k}) = \beta b(\mathbf{A}), \tag{4.9}$$

where β is the number of degrees of freedom of a typical node of the structure.

Papademetrious [14] has shown that the bandwidth minimization problem is an NP-complete problem. Therefore any approach to it is of interest primarily because of its heuristic value.

4.4 A Shortest Route Tree and Its Properties

The main tool for most of the ordering algorithms using graph-theoretical approaches is the shortest route tree of its model or its associate model. A shortest route tree rooted at a node O, called the *starting node* (root) of the tree, is denoted by SRT_O and has the following properties:

The path from any node to the root through the tree is a shortest path. An algorithm for generating an SRT is given in Sect. 1.4.7 and therefore only its properties relevant to nodal number are discussed here.

An SRT decomposes (partitions) the node set of S into subsets according to their distance from the root. Each subset is called a *contour* (level) of the SRT, denoted by C_i. The contours of an SRT have the following properties:

$$\begin{aligned} &\mathrm{Adj}\,(C_i) \subseteq C_{i-1} \cup C_{i+1}, \quad 1 < i < m \\ &\mathrm{Adj}\,(C_1) \subseteq C_2, \\ &\mathrm{Adj}\,(C_m) \subseteq C_{m-1}. \end{aligned} \tag{4.10}$$

The number of nodes in each contour is called the *width* of that contour, and the largest width of the contours of an SRT is called the *width of the SRT* rooted at the starting node O, denoted by $w(SRT_O)$. This number is known as the *width number* of O. The number of contours of an SRT (except the starting node contour) is the *height* of the tree denoted by $h(SRT_O)$. The *longest* SRT is the one with maximal height and the *narrowest* SRT is the one with minimal width.

As an example, an SRT of S as shown in Fig. 4.2a, rooted at O, denoted by SRT_O, has the following identities:

$$\begin{aligned} &w(C_1) = 1, w(C_2) = 2, w(C_3) = 3, w(C_4) = 4, w(C_5) = 5, w(C_6) = 5, \\ &w(C_7) = 4, w(C_8) = 3, w(C_9) = 2 \text{ and } w(C_{10}) = 4. \end{aligned}$$

Hence $w(SRT_O) = 5$ and $h(SRT_O) = 9$.

For the same graph model, an SRT rooted at O', as shown in Fig. 4.2b, leads to $w(SRT_{O'}) = 9$ and $h(SRT_{O'}) = 4$.

This simple example shows the importance of selecting an appropriate starting node. This will be discussed in some detail in subsequent sections.

4.5 Nodal Ordering for Bandwidth Reduction

The following four-step algorithm is employed for nodal ordering of graphs leading to banded node adjacency matrices. This method can directly be used for nodal ordering of skeletal structures resulting in banded stiffness matrices.

1. Finding a suitable starting node;
2. Decomposing the node set of S into ordered subsets (contours);

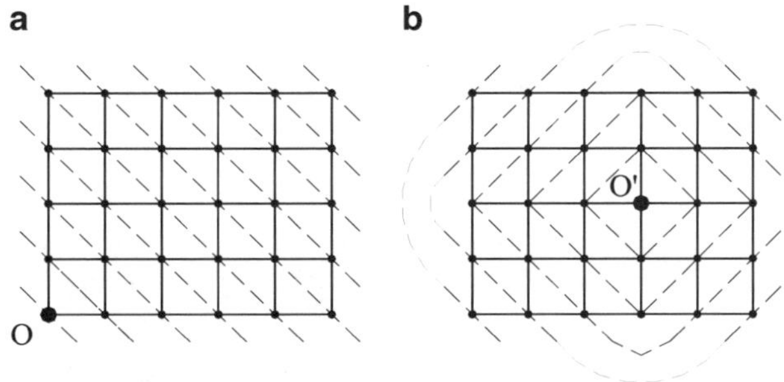

Fig. 4.2 A graph S and two of its SRTs. (**a**) An SRT rooted at O. (**b**) An SRT rooted at O′

3. Selecting a connected path (transversal) containing one representative node from each contour;
4. Ordering the nodes within each contour, to obtain the final nodal numbering of S.

All the above steps require the use of an SRT algorithm of Sect. 1.4.7, known as breadth-first-search algorithm. Therefore, a nodal ordering process may be considered as a multiple application of the SRT algorithm.

The node set of S can be decomposed into ordered subsets by means of a breadth-first-search algorithm. The quality of the results depends upon the choice of an appropriate starting node, as the root of this tree. The results corresponding to the ordering within each contour, however, also depend upon the use of a suitable transversal containing one representative node from each contour.

Methods for finding suitable starting nodes have been developed by Cheng [15], Kaveh [16, 17], Gibbs et al. [8], and Grimes et al. [12]. In the following, various graph-theoretical methods are presented for finding good starting nodes and selecting suitable transversals.

4.5.1 A Good Starting Node

The *distance* $d(n_i, n_j)$ between two nodes n_i and n_j is defined to be the length of the shortest path between these nodes. The *eccentricity* of a node n_i is defined as:

$$e(n_i) = \mathrm{Max}\ d\left(n_i, n_j\right) \text{ for } j = 1, \ldots, N(S). \tag{4.11}$$

The *diameter* of S is defined as:

$$\delta(S) = \mathrm{Max}\ e(n_i) \text{ for } i = 1, \ldots, N(S). \tag{4.12}$$

As an example, the eccentricity of n_2 in Fig. 4.3 is $e(n_2) = 3$, and the diameter of S is $\delta(S) = 4$.

Fig. 4.3 A graph S

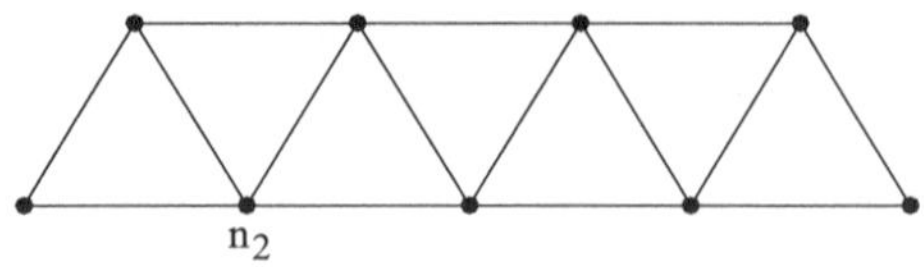

A node n_i of S is called *peripheral* if its eccentricity is the same as the diameter of S, i.e. $\delta(S) = e(n_i)$. If the eccentricity is close to the diameter, then n_i is called a *pseudo-peripheral* node or a *good starting node*.

In this section, three algorithms are described for selection of good starting node or nodes for nodal numbering. Other algorithms have been developed, the detail of which may be found in Kaveh [11].

Algorithm A

Step 1: Start from an arbitrary node of S. Construct an SRT on this node and take a node of least valency from its last contour.

Step 2: Form a new SRT from the selected node, and record all the nodes of the last contour of the selected SRT.

Step 3: Form SRTs rooted at each of the recorded nodes and choose the one that corresponds to the narrowest SRT. The process of constructing an SRT is terminated as soon as the width of one of its contours exceeds the width of the previously selected SRT.

This algorithm is similar to Gibbs et al. [8] algorithm, where the starting node O and another node of minimum valency from its last contour are selected as *pseudo-peripheral* or *diameteral* nodes

Algorithm B

Step 1: Start with an arbitrary node, form an SRT on this node and take a node n_i of least valency from its last contour.

Step 2: Generate an SRT on n_i and find all nodes contained in its even, first and last contours.

Step 3: Generate an SRT on each node of these contours, and find the narrowest one. The process of formation of an SRT is terminated as soon as the width of one of its contours exceeds the width of the previously selected SRT. Denote the selected node by n_j.

Step 4: Check adjacent nodes to n_j for possible reduction in width, to decide the final starting node.

Algorithm C

Step 1: From an arbitrary node generate an SRT, and from its last contour select a node X_1 of minimal valency. Observe the width of the selected SRT.

Step 2: Generate an SRT from X_1 and select X_2 of the least valency from its last contour, and observe the width.

Step 3: Generate two SRTs simultaneously rooted at X_1 and X_2 and find the node X_3 which is the last node of S included in one of the SR subtrees. Once X_3 is found, terminate the process of forming SRTs. Generate an SRT from X_3 and observe its width. X_1 and X_2 are called the *generators* of X_3.

Step 4: Repeat the process of Step 3, using the pairs (X_1, X_3) and (X_2, X_3) as the generators to find X_4 and X_5, respectively. Construct the corresponding SRTs and observe their widths.

Step 5: Repeat the process of Step 3 for X_i ($i = 3, 4, \ldots$) together with the corresponding generator, until no further improvement in width is observed. The narrowest SRT should be selected for nodal decomposition of S.

An example of the application of this algorithm is depicted in Fig. 4.4, where a cross-shaped grid S is considered. Starting from an arbitrary node "O", an SRT is generated and X_1 is obtained from its last contour. Generating a new SRT from X_1, node X_2 is chosen from its last contour. X_3 is the result of generating two simultaneous SRTs from X_1 and X_2. Using (X_1, X_3) and (X_2, X_3), nodes X_4 and X_5 are obtained, respectively. The widths of the selected SRTs rooted at X_1, X_2, X_3, X_4 and X_5 are 8, 8, 8, 11, and 10, respectively. Therefore the process is terminated and X_3 is taken as a good starting node of S.

Algorithms A and B may search for a good starting node in a single direction of a graph and do not meet nodes laying in other directions. Algorithm C has the feature of overcoming such problem. In this method, the control of overall connectivity properties of the graph becomes feasible. The following example will more clearly illustrate this point.

4.5.2 Primary Nodal Decomposition

Once a good starting node is selected, an SRT is constructed and its contours {C_1, C_2, …, C_m} are obtained. These subsets are now ordered according to their distances from the selected starting node. Obviously, many SRTs can be constructed on a node. Although all lead to the same nodal decompositions, different transversals will be obtained for different SRTs. Thus, in the generation process, the nodes of each contour C_i are considered in ascending order of their valencies for selecting the nodes in C_{i+1}, in order to provide the conditions for the possibility of generating a minimal (or optimal) transversal as defined in the next section. Finding an optimal transversal before an SRT is fixed, seems to be a time-consuming problem. However, for most of the models encountered in practice, an optimal transversal lies between the minimal ones. In the following, an algorithm is given for selecting a suboptimal transversal of an SRT.

Fig. 4.4 A cross-shaped grid and the selected X_i ($i = 1, \ldots, 5$) by Algorithm C

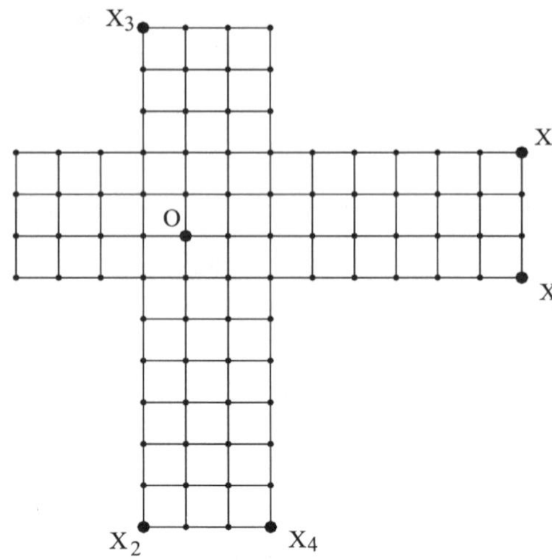

4.5.3 *Transversal P of an SRT*

A *transversal* of an SRT is defined as a connected path P containing one distinct node N_i from each contour C_i of an SRT. A *minimal transversal* is the one for which $\sum_{i=1}^{m} \deg(N_i)$ is minimum. An *optimal transversal* is the one leading to the best nodal numbering, i.e. a numbering corresponding to smallest bandwidth for the selected decomposition. The weight of a node is defined as its degree.

Algorithm

Step 1: Take a node N_m of minimal weight from the last contour C_m of the selected SRT.

Step 2: Find N_{m-1} from C_{m-1} which is connected to N_m by a branch of the SRT.

Step 3: Repeat the process of Step 2 selecting nodes $N_{m-2}, N_{m-3}, \ldots, N_1$, as the representative nodes of the contours $C_{m-2}, C_{m-3}, \ldots, C_1$, respectively.

The above algorithm is a backtracking process from a node of minimal weight in the last contour C_m, that selects a transversal $P = \{N_1, N_2, \ldots, N_m\}$ which can now be used for ordering the nodes of the contours of the corresponding SRT.

4.5.4 *Nodal Ordering*

Step 1: Number N_1 as "1".

Step 2: N_2 is given number "2" and an SR subtree is generated from N_2, numbering the nodes of C_2 in the order of their occurrence in this SR subtree.

Step 3: The process of Step 2 is repeated for numbering the nodes of $C_3, C_4, \ldots, C_m$, sequentially using $N_3, N_4, \ldots, N_m$ as the starting nodes of SR subtrees, until all the nodes of S are numbered.

Now the numbering can be reversed, in a way similar to that of the Reverse Cuthill-McKee algorithm, for possible reduction of fill-ins in the process of Gaussian elimination, which will be discussed in Sect. 4.7.

4.5.5 Example

The following simple example is chosen to illustrate the steps of the presented approaches, but the applications are by no means limited to such simple cases.

Let S be the graph model of a truss structure, as shown in Fig. 4.5a. Using one of the algorithms of Sect. 4.5.1, a good starting node A is found, and the corresponding SRTs are depicted in Fig. 4.5b. A transversal is selected as shown in bold lines, Fig. 4.5c. Then nodes are numbered contour by contour, employing the representative nodes as the starting nodes of SR subtrees, Fig. 4.5d.

In order to cast the concepts developed for nodal ordering in a mathematical form, a connectivity coordinate system is defined for nodal numbering of S. Separate study of planar and space graphs results in clarification of further interesting points about nodal numbering of space structures, as described in Kaveh [18].

4.6 Finite Element Nodal Ordering for Bandwidth Optimisation

Extensions and applications of the nodal numbering algorithms to element ordering are due to Kaveh [10], Everstine [19], Razzaque [20], Sloan [21], Burgess and Lai [22].

For finite element nodal ordering, different methods are developed. The application of a natural associate graph in a two-step approach, has been suggested by Kaveh [23], and later by Fenves and Law [24]. A corner-node method is developed by Kaveh [5], and Kaveh and Ramachandran [25]. The application of an element clique graph is due to Sloan [9], and Livesley and Sabin [26]. Additional graphs for transforming the information concerning the connectivity of the finite element model (FEM) to those of different simple graphs, are introduced and employed in efficient finite element nodal numbering by Kaveh and Roosta [27].

In this section, the connectivity properties of FE models are embedded in the topological properties of nine different graphs. A nodal ordering is then performed on these graphs, leading to the element ordering of the corresponding FEMs,

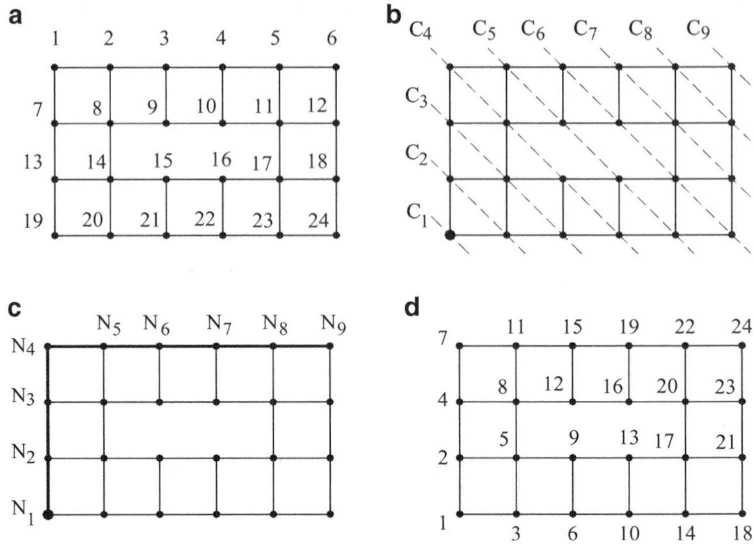

Fig. 4.5 Graph model S and its nodal numbering. (**a**) Initial numbering of S. (**b**) The selected SRT. (**c**) The selected transversal P. (**d**) Final nodal numbering of S

followed by their final nodal ordering. This process is summarised in the flow chart of the following page.

For the sake of clarity, in this section, the nodes of the constructed graphs are referred to as *vertices*.

The complexities of the presented methods are given for a logical comparison of their efficiency. The efficiency of the methods are also tested by some 2-dimensional and 3-dimensional FE models. For these models, the computational time and the bandwidth obtained are presented for comparison.

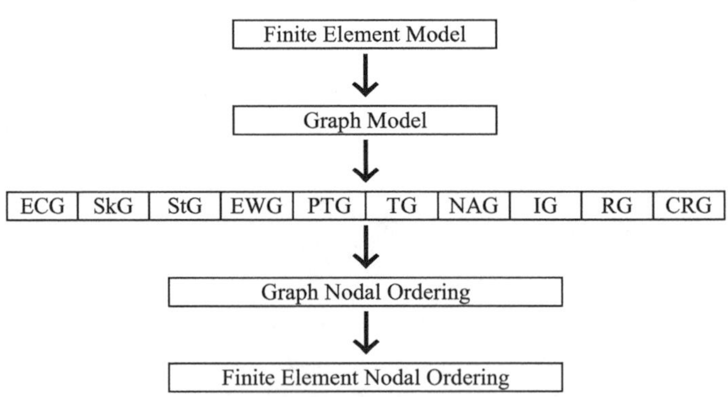

Notations: Element Clique Graph (ECG); Skeleton Graph (SkG); Element Star Graph (EStG); Element Wheel Graph (EWG); Partially Triangulated Graph (PTG); Triangulated Graph (TG); Natural Associate Graph (NAG); Incidence Graph (IG); Representative Graph (RG); Complete Representative Graph (CRG).

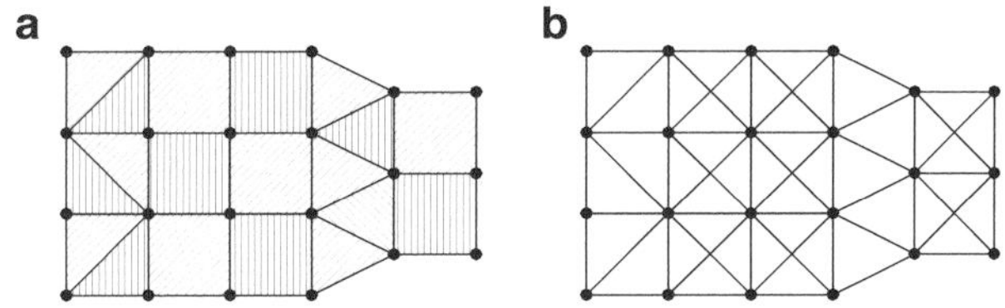

Fig. 4.6 An FEM and its element clique graph. (**a**) An FEM. (**b**) The element clique graph of the FEM

4.6.1 Element Clique Graph Method (ECGM)

Definition. The *element clique graph* S of an FEM is a graph whose vertices are the same as those of the FEM, and two vertices n_i and n_j of S are connected with a member if n_i and n_j belong to the same element in the FEM. The element clique graph of the FEM shown in Fig. 4.6a is illustrated in Fig. 4.6b.

This graph is particularly suitable for bandwidth optimisation, since in this graph each vertex corresponds to a node of the FEM, and a single step is needed for direct nodal numbering of the considered FEM.

Algorithm

Step 1: Construct the element clique graph S of the considered FEM.

Step 2: Use a nodal numbering algorithm available (e.g. the algorithm presented in Sect. 4.5.4).

In this method, all the nodes of an element will be contained in at most two adjacent contours of an SRT, hence the bandwidth becomes dependent on the width of the SRT.

4.6.2 Skeleton Graph Method (SkGM)

Definition. The *1-skeleton graph* S of an FEM is a graph whose vertices are the same as the nodes of the FEM, and its members are the edges of the FEM. Figure 4.7 illustrates the skeleton graph of the FEM shown in Fig. 4.6a.

Simultaneous application of the ECG and the skeleton graph provides very efficient tools. As an example, consider the small FE as shown in Fig. 4.8. Suppose an SRT is rooted from vertex 1 in the ECG of the FEM to find a good starting node with minimum degree from its last contour. Vertices 17 and 19 are found. They are the farthest from vertex 1 and have the same degree as 3 (in the ECG). However, vertex 19 is better than vertex 17, since $W(SRT_{19}) < W(SRT_{17})$. Instead of generating two SRTs from vertices 17 and 19, one can choose 19 by generating SRT_1 in the skeleton graph, because $d_{SG}(1, 19) > d_{SG}(1, 17)$, where $d_{SG}(i, j)$ denotes the distance between vertices i and j in the skeleton graph.

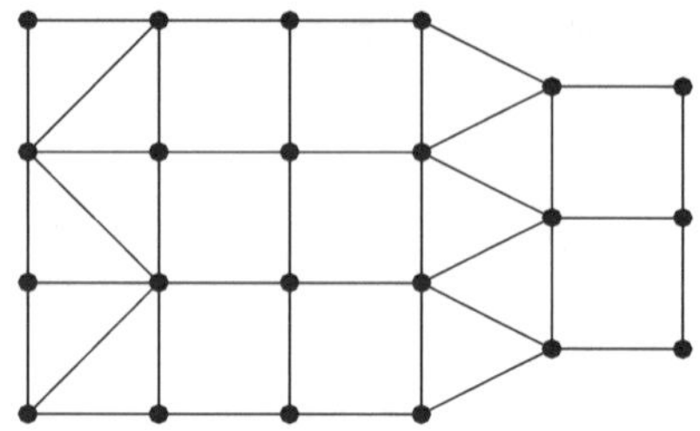

Fig. 4.7 The skeleton graph of the FEM of Fig. 4.6a

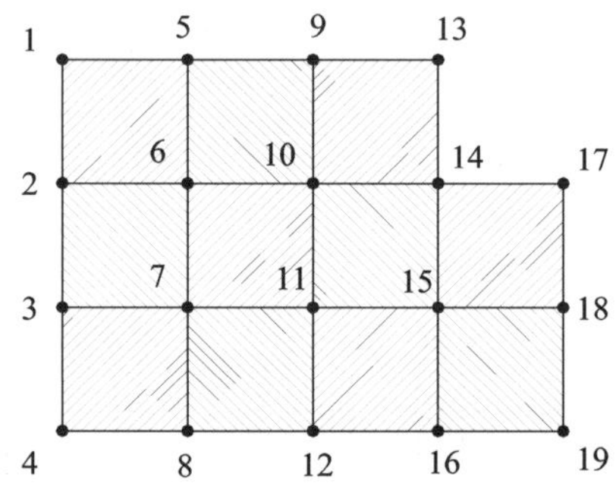

Fig. 4.8 A small finite element model

Algorithm

Step 1: Construct the skeleton graph S of the considered FEM. For each element i of the FEM connect two end nodes of each edge of element by a member. Such nodes should be connected only once.

Step 2: Order the vertices of S using any nodal numbering algorithm available (e.g. the algorithm presented in Sect. 4.5.4) thus obtaining a nodal ordering of S.

In order to generate the skeleton graph of An FEM, it is necessary to list the nodes of each element in a suitable order. In this method the number of members of S is less than that of the ECGM; however, in FEMs with triangular elements, these members are the same. Therefore, this method takes less computer storage for keeping the connectivity of S. Generating an SRT in a skeleton graph may lead to allocation of the nodes of an element in three or more adjacent contours. Therefore the width of the SRT being used, together with the number of contours containing the nodes of an element of the FEM, specify the bandwidth.

4.6.3 Element Star Graph Method (EStGM)

Definition. The *element star graph* S of An FEM has two sets of vertices, namely the main set containing the same nodes as those of the FEM and a virtual set consisting of the virtual vertices associated in a one-to-one correspondence with the elements of the FEM. The member set of S is constructed by connecting the virtual

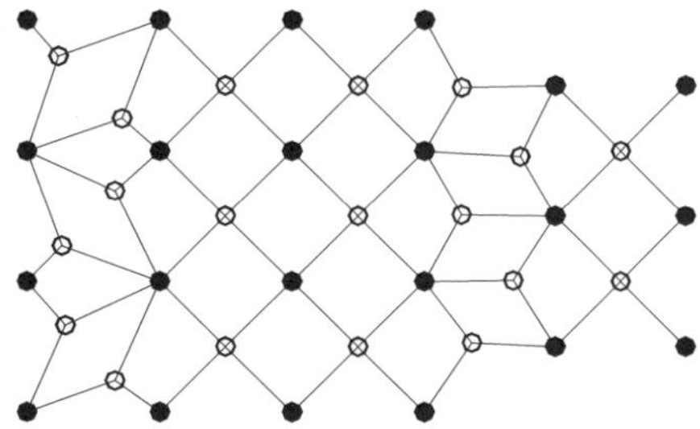

Fig. 4.9 The element star graph of the FEM of Fig. 4.6a

vertex of each element i to all the nodes of the element i. The element star graph of the FEM shown in Fig. 4.6a is illustrated in Fig. 4.9. The virtual vertices are shown by larger sized dots.

In the element star graph of An FEM, the distance between each pair of vertices corresponding to two nodes of the FEM which share an element is equal to 2, while in the ECG it is equal to 1. This difference does not cause the element star graph to lose the previously discussed property of the ECG, which is the existence of more than one pair of peripheral nodes in most of FEMs. Hence this graph model is efficient for algorithms in which multiple roots are needed to be found. In this graph the degree of each vertex corresponding to a node i of the FEM is the same as that of the number of elements of the FEM incident to node i.

Algorithm

Step 1: Construct the element star graph S of the considered FEM. For each element i, generate a virtual vertex labelling with i+α, and connect the nodes of i to the vertex i+α, where α is the total number of nodes of the FEM.

Step 2: Order the main vertices of S using a nodal numbering algorithm available, e.g. the method presented in Sect. 4.5.4. This step is similar to the previous methods but virtual vertices need not be labelled in the process of numbering of the nodes. The virtual vertices can easily be identified by their labels being above α.

4.6.4 Element Wheel Graph Method (EWGM)

Definition. The *element Wheel Graph* S of An FEM is the union of the element star graph and the skeleton graph of the FEM. The element wheel graph of the FEM shown in Fig. 4.6a is illustrated in Fig. 4.10. The virtual vertices are shown by larger sized dots.

Algorithm

Step 1: Construct the element wheel graph S of the considered FEM. This can be done by generating the union of the element star graph and skeleton graph.

Step 2: Order the main vertices of S using a nodal numbering algorithm available, e.g. the method presented in Sect. 4.5.4. This step should be carried out similarly to that of Step 2 in ESGM.

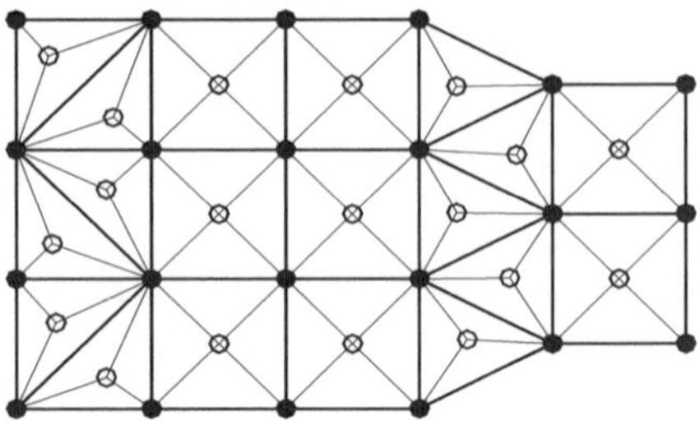

Fig. 4.10 The element wheel graph of the FEM of Fig. 4.6a

In order to generate the element wheel graph of an FEM, it is necessary to list the nodes of each element in a suitable order. In this method M(S) is higher than that of ESGM, and therefore it needs more computer storage than ESGM. The nodes of an element of FEM are at most contained in three contours of the generalized SRT of the element wheel graph.

4.6.5 Partially Triangulated Graph Method (PTGM)

Definition. The *partially triangulated graph* S of an FEM is a graph whose vertices are the same as the nodes of the FEM and an artificial vertex assigned to each element i is connected to all the original nodes of i. The selected nodes of the elements are found by generating all SR subtrees from a good starting vertex in the skeleton graph of the FEM and taking the first node of an element included in the SRT during the process of the generation. As an example, for the FEM shown in Fig. 4.6a, an SR subtree is routed from n_0 and shown in Fig. 4.11a, and the selected nodes of the elements are shown by larger sized dots. The partially triangulated graph of the FEM is shown in Fig. 4.11b.

In order to generate the partially triangulated graph of an FEM, the following steps can be executed:

1. Generate the SG of the FEM;
2. Form an SRT rooted from an arbitrary node n_0 and select a node n_1 from the last contour of SRT_{n0} with the minimum degree;
3. Form an SRT rooted from n_1 and select a node n_2 in the last contour of SRT_{n1}, with minimum degree;
4. Form an SRT rooted from n_2 and take n_s from n_0, n_1 and n_2 whose corresponding SRT has the least width;
5. Calculate the distance between each vertex of the SG and n_s;
6. For each element i select a vertex which is the nearest node to n_s;
7. Form the partially triangulated graph by connecting the vertex corresponding to the selected node of each element i to the vertices corresponding to other nodes of i; previously connected nodes should not be connected again.

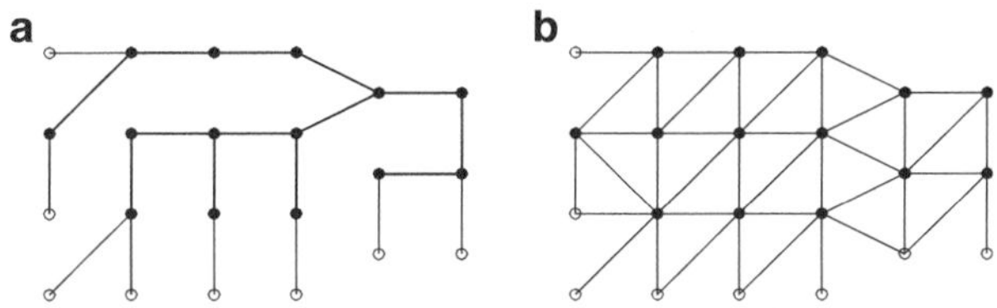

Fig. 4.11 The skeleton, an SR subtree and the partially triangulated graph of the FEM of Fig. 4.6a. (**a**) The skeleton graph and an SR subtree of the FEM. (**b**) The partially triangulated graph of the FEM

Algorithm
Step 1: Construct the partially triangulated graph S of the considered FEM.
Step 2: Order the vertices of S using a nodal numbering algorithm available, e.g. the algorithm of Sect. 4.5.4.

For generating the partially triangulated graph of an FEM, it is necessary to list the nodes of each element in a suitable order. In this method M(S) may or may not be higher than that of SKGM. In the process of forming an SRT in a partially triangulated graph, the nodes of an element may lie in one, two or three adjacent contours.

4.6.6 Triangulated Graph Method (TGM)

Definition. The *triangulated graph* S of an FEM is the union of the partially triangulated graph and the skeleton graph of the FEM. The triangulated graph of the FEM shown in Fig. 4.6a is illustrated in Fig. 4.12. The selected vertices of the elements are the same as those of Fig. 4.6a.

Algorithm
Step 1: Construct the triangulated graph S of the considered FEM. This step can be carried out by generating the partially triangulated graph and the skeleton graph.
Step 2: Order the vertices of S using a nodal numbering algorithm.

In this method the number of members is higher than that of the PTGM. For an SRT in a triangulated graph, the nodes of an element of an FEM are contained in at most three adjacent contours.

4.6.7 Natural Associate Graph Method (NAGM)

Definition. The *natural associate graph* S of an FEM has its vertices in a one-to-one correspondence with the elements of the FEM, and two vertices of S are

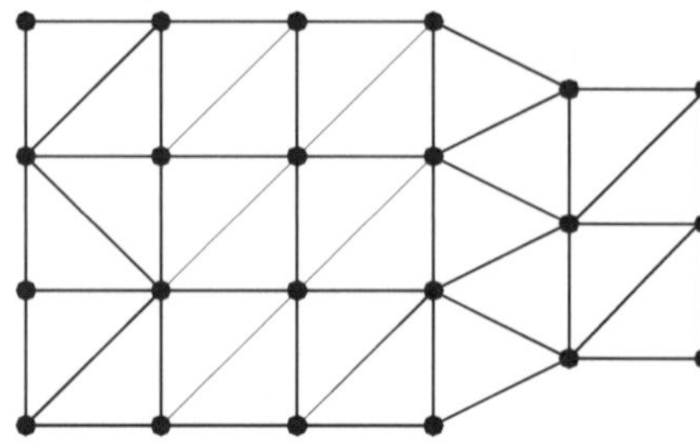

Fig. 4.12 The triangulated graph of the FEM of Fig. 4.6a

connected by a member if the corresponding elements have a common boundary. The natural associate graph of the FEM shown in Fig. 4.6a is illustrated in Fig. 4.13. In order to generate the natural associate graph of an FEM, one of the following two methods can be employed. The first is a direct scheme which takes high computer time but low computer storage. In this case only the element-node list should be provided. The second takes low computational time but uses larger memory. In this case, the node-element list, together with the element-node list are provided as input data.

Method 1. Check each pair of elements i and j of the FEM for a common boundary. If i and j have such a boundary, then the vertices corresponding to i and j should be connected by a member in the natural associate graph.

Method 2. Step 1: Generate the node-element list of the considered FEM.

Step 2: Take each pair of elements incident at a node, and note whether they have more than one corner node in common.

Step 3: When two elements of equal or different dimensions have common corner nodes equal to or more than the smallest dimension of the elements, then the corresponding vertices in the natural associate graph are connected by a member.

Algorithm

Step 1: Construct the natural associate graph S of the considered FEM.

Step 2: Order the vertices of S using a nodal numbering algorithm, to obtain an ordering for the elements of the FEM.

Step 3: Order the nodes of the FEM, element by element, in the same sequence as decided in Step 2. Within each element, priority is given to mid nodes, passive and active nodes, respectively. A node is called *passive* if it has no incident new element, otherwise it is *active*.

Step 3 of this method can be carried out using the following process:

I. Generate a matrix **NE** with a rows and ε columns, in which its *i*th row contains the labels of the elements containing node i, where ε is the same as the maximum number of elements incident to a specified node.

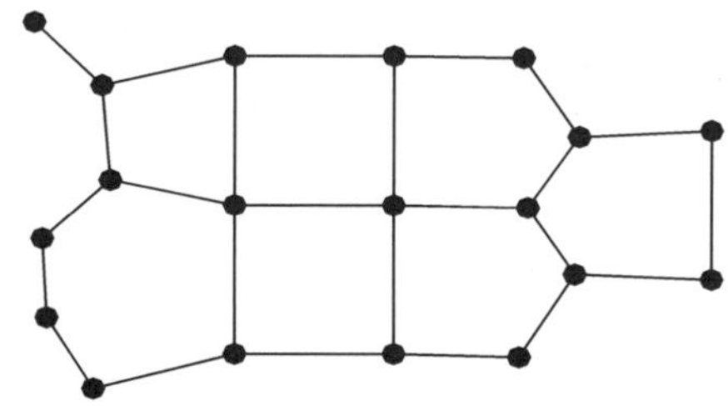

Fig. 4.13 The natural associate graph of the FEM of Fig. 4.6a

II. For each element j (j = 1, . . ., l) execute the following steps, in turn:

(a) If j has a mid node, label it first;
(b) Detect the active and passive nodes j using the matrix **NE**. It should be noted that using **NE** makes the process fast; however, one can instead check a node of j for incidence with a new element;
(c) Form a multiple root SR subtree from the active node of j;
(d) Label the passive nodes of j when they are selected in the multiple root SR subtree;
(e) Label the active nodes of j which are adjacent to the previously labelled nodes;
(f) Repeat Step (e) until all the active nodes of j are labelled.

In order to generate the natural associate graph of an FEM, it is necessary to list the nodes of each element in a suitable order. In this algorithm M(S) has the least value among the presented methods thus far; therefore it take less computer storage for keeping the connectivity data of S.

4.6.8 Incidence Graph Method (IGM)

Definition. The *incidence graph* S of an FEM has its vertices in a one-to-one correspondence with the elements of the FEM, and two vertices of S are connected by a member, if the corresponding elements have a common node. Figure 4.14 shows the incidence graph of the FEM shown in Fig. 4.6a.

In order to generate the incidence graph of an FEM, one of the following two methods can be employed. The first is a direct approach, which takes high computational time but low words of memory; for which only the element-node list should be provided. The second scheme takes short computational time but high computer storage; the node-element list together with the element-node list should be provided.

Method 1. Check each pair of elements i and j of the FEM for a common node, and if they have such a node, connect with a member to the corresponding vertices i and j in the incidence graph.

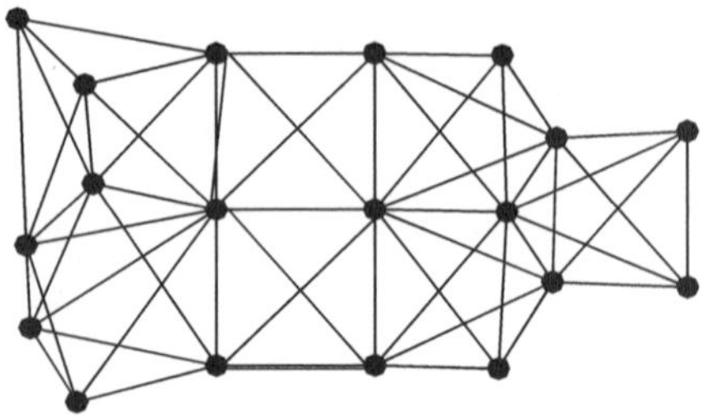

Fig. 4.14 The incidence graph of the FEM of Fig. 4.6a

Method 2. Step 1: Generate the node-element list of the considered FEM.
Step 2: Connect the representative vertices of each pair of elements, which contain a common node.

Algorithm
Step 1: Construct the incidence graph S of the considered FEM.
Step 2: Order the vertices of S using a nodal numbering algorithm, to obtain an ordering for the elements of the FEM.
Step 3: Order the nodes of the FEM, element by element, in the same sequence as decided in Step 2. Within each element priority is given to mid nodes, passive and active nodes, respectively.

4.6.9 Representative Graph Method (RGM)

Definition. Consider the skeleton graph of an FEM, and select an appropriate starting vertex, using any available algorithm. The nearest corner node of each element of the FEM is taken as the representative node of that element. The SR subtree of the skeleton graph of the FEM containing all representative nodes of the elements is called a *representative graph* S of the FEM. The representative graph of the FEM shown in Fig. 4.6a is illustrated in Fig. 4.15.

In order to generate the representative graph of an FEM the following steps should be executed:

Step 1: Execute Steps 1–4 of the algorithm for the formation of the PTG.
Step 2: Form a SR subtree step by step from n_s until each element of the FEM has a node whose corresponding vertex in SG is contained in the SR subtree. The first selected vertex (in the SR subtree) corresponding to the nodes of each element i should be taken as the representative node of i.

Algorithm
Step 1: Construct the representative graph of the FEM, and number its vertices, resulting in the ordering of the elements of the considered FEM.
Step 2: Use Step 3 of the NAGM to number the nodes of the FEM.

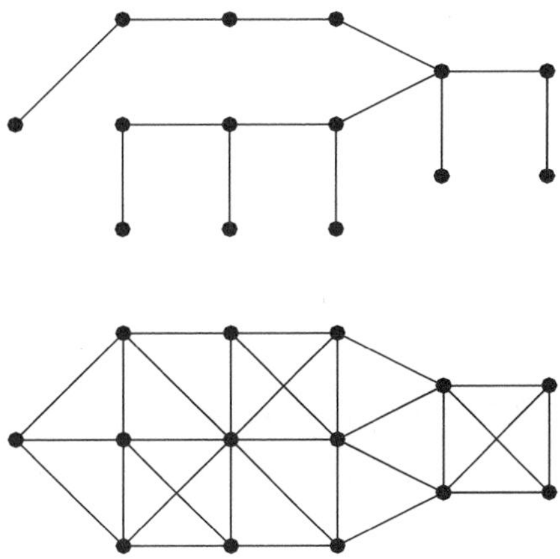

Fig. 4.15 The representative graph of the FEM in Fig. 4.6a

This method is the most efficient approach from the computational time and storage points of view for most of the practical models.

4.6.10 Computational Results

The presented algorithms are implemented on a PC and many examples are examined, some of which are included in this section. The bandwidth of **D** and the relative computational time for the algorithms are provided.

Example 1. A planar FEM with three types of elements consisting of 4-node, 8-node and 12-node elements is considered as shown in Fig. 4.16. This model contains 1,959 nodes and 2,250 elements. The combination of elements of this model may not be practical; however, it is purposely chosen to illustrate the generality of the methods in dealing with the presence of different elements of a model. The results are presented in Table 4.1.

Example 2. A three-dimensional finite element model consisting of 480 (5812) 20-node cubic elements (each edge of elements has a mid side node) is considered, having the total of 2,559 nodes. The results are depicted in Table 4.2.

Example 3. A planar FEM with two holes is considered as shown in Fig. 4.17. Six FEMs with 1,000 elements are studied with elements having 4 nodes, 4 nodes and a mid-node, 8 nodes, 8 nodes and a mid-node, 12 nodes, and 12 nodes and a mid-node. These models contain 1,134, 2,134, 3,269, 4,204 and 6,404 nodes, respectively. The results are depicted in Table 4.3.

Example 4. The finite element model of a buttress dam is considered, the section of which is illustrated in Fig. 4.18, consisting of 480 nodes and 603 elements. This model contains three layers of prismatic members and each element contains six nodes. The results are depicted in Table 4.4. The patterns of the nodes adjacency matrices are presented in Fig. 4.18.

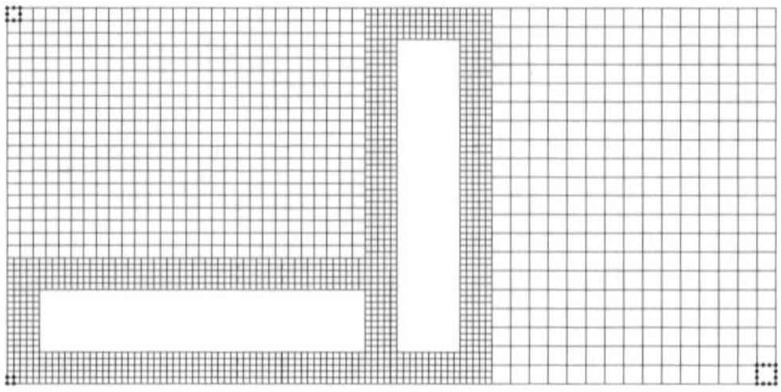

Fig. 4.16 A planar FEM

Table 4.1 Results of Example 1

Method	ECGM	SKGM	ESGM	EWGM	PTGM	TRGM	NAGM	INGM	REGM
b(**D**)	313	497	313	457	513	515	447	451	491
Time	29.77	27.02	21.92	36.09	20.65	22.03	18.29	14.71	9.72

Table 4.2 Results of Example 2

Method	ECGM	SKGM	ESGM	EWGM	PTGM	TRGM	NAGM	INGM	REGM
b(**D**)	843	1,173	843	787	1,103	1,103	1,185	845	1,195
Time	18.62	7.08	4.93	8.12	7.47	7.85	38.67	7.75	4.93

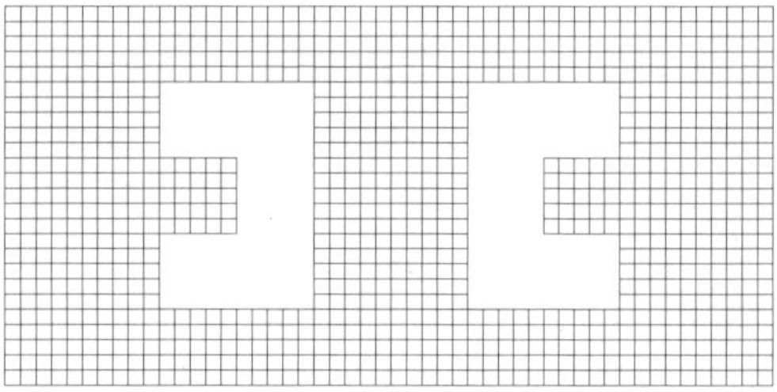

Fig. 4.17 A planar FEM with two holes

4.6.11 Discussions

The algorithms presented in this section transform the connectivity of FEMs into the topological properties of different graphs. Then a nodal ordering algorithm undertakes numbering the nodes of the graphs, leading to nodal numbering of the FEMs. All the methods presented are low order polynomial time algorithms. Analyses are considered for worst cases and compared. Such analysis is the most logical way of comparing the algorithms, since most of the combinatorial optimisation algorithms are configuration dependent. Each algorithm presented has advantages and disadvantages which become manifest when the algorithm is employed for models with different element types and connectivity properties. It should be noted that the relative performance of the algorithms depends also on the starting node selection algorithm and the nodal ordering algorithm being employed.

Table 4.3 Results of Example 3

Element		4 nodes	4 nodes + mid-node	8 nodes	8 nodes + mid-node	12 nodes	12 nodes + mid-node
ECGM	b(**D**)	111	217	333	437	553	657
	Time	4.12	7.25	14.32	19.55	30.49	37.90
SKGM	b(**D**)	95	179	269	347	439	519
	Time	4.29	7.14	14.93	20.32	33.23	39.71
EWGM	b(**D**)	97	185	313	417	541	639
	Time	7.31	10.60	17.13	20.82	26.04	29.94
PTGM	b(**D**)	159	309	477	633	807	963
	Time	4.29	6.59	10.60	12.97	16.70	19.45
TRGM	b(**D**)	167	327	479	619	791	945
	Time	4.17	6.98	11.15	13.90	17.74	21.09
NAGM	b(**D**)	95	177	271	353	447	529
	Time	4.22	6.87	10.28	12.31	16.03	18.84
INGM	b(**D**)	113	225	341	455	569	687
	Time	4.39	4.77	7.91	9.23	11.10	12.96
REGM	b(**D**)	95	177	271	353	447	529
	Time	2.70	4.18	6.48	8.07	10.05	12.24

Fig. 4.18 A planar FEM

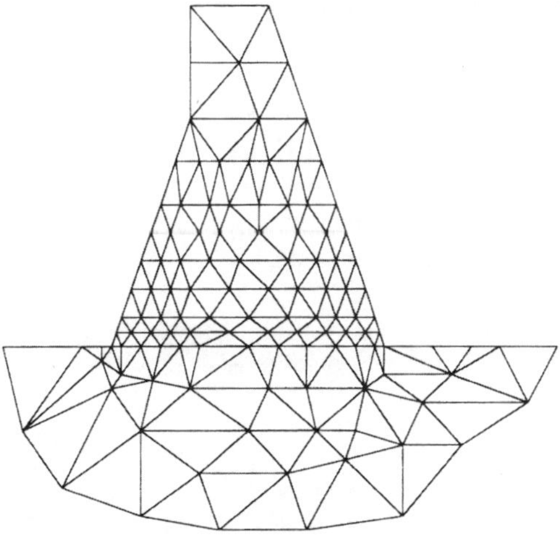

Table 4.4 Results of Example 4

Method	ECGM	SKGM	ESGM	EWGM	PTGM	TRGM	NAGM	INGM	REGM
b(**D**)	125	221	125	221	229	213	175	125	187
Time	1.70	1.32	1.70	2.42	1.76	1.76	6.43	4.45	1.54

Finally, it should be mentioned that the simultaneous use of two graphs out of the nine graphs presented in this section for nodal ordering may lead to a combined model more informative than individual models, Kaveh and Roosta [28].

4.7 Finite Element Nodal Ordering for Profile Optimisation

4.7.1 Introduction

When a banded matrix of high order has a wide band and a large number of zeros inside it, the diagonal storage scheme may become wasteful. Then a *profile* (variable band) scheme of Jennings [29], the so-called *skyline scheme* (Felippa [30]), may be used.

Nodal numbering algorithms can also be applied to profile reduction. As mentioned before, after nodal numbering for bandwidth reduction, by reversing the ordering, a numbering corresponding to a much smaller profile can be found. This has been found by George [31] and proved by Liu and Sherman [32]. The method is known as the *Reverse Cuthill-McKee algorithm*. For the Cuthill-McKee type of ordering the bandwidth remains unchanged when the order is reversed; however, the profile can never increase.

As an example, consider a nodal numbering for a graph as shown in Fig. 4.19a with corresponding adjacency matrix **A** in Fig. 4.19b. Reversing the nodal numbers as in Fig. 4.19c, leads to a matrix **A**′ as depicted in Fig. 4.19d, with a reduction of the profile from 15 to 13.

There are many algorithms for profile and frontwidth reduction, which can be categorized in different ways. In this section the general algorithm of Souza and Murray [33] is adopted for nodal ordering of all the graph models presented in the previous section, to reduce the profile of sparse matrices with symmetric structures. This algorithm incorporates the algorithm for selection of peripheral nodes, the re-sequencing scheme of Sloan [9], and the algorithm of Gibbs-King [26].

In order to proceed with main algorithms for profile reduction, some definitions will now be stated in the following:

The profile of an n×n matrix **A** is defined as,

$$P = \sum_{i=1}^{N} b_i, \tag{4.13}$$

where the row bandwidth, b_i, for row i is defined as the number of inclusive entries from the first non-zero element in the row to the (i + 1)th entry. The efficiency of any given ordering for the profile solution scheme is related to the number of active equations during each step of the factorisation process. Formally, row j is defined to be active during the elimination of column i if $j \geq i$ and there exists $a_{ik} = 0$ with $k \leq i$. Hence, at the *i*th stage of the factorisation, the number of active equations is the number of rows of the profile, which intersect column i, ignoring those rows already eliminated. Letting f_i denote the number of equations which are active during the elimination of the variable x_i, it follows from the symmetric structures of **A** that:

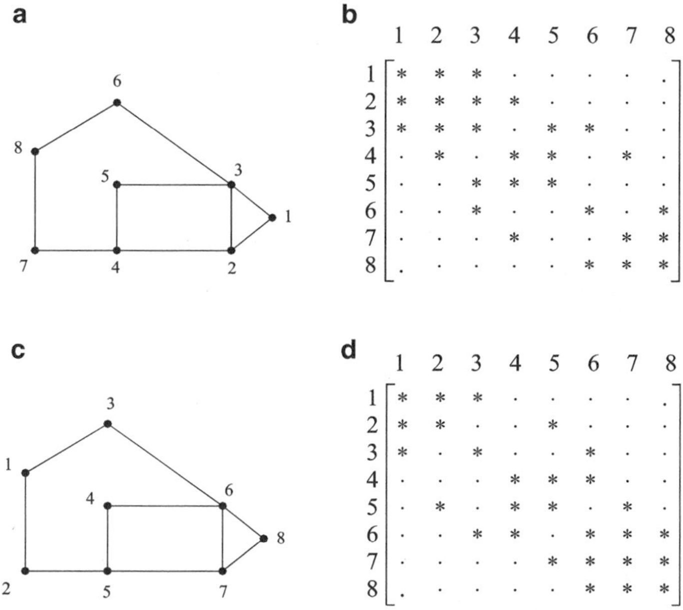

Fig. 4.19 A Reverse Cuthill-McKee for nodal numbering. (**a**) A nodal numbering. (**b**) Matrix A. (**c**) Reverse of the nodal numbering of (a). (**d**) Matrix A′

$$P = \sum_{i=1}^{N} f_i = \sum_{i-1}^{N} b_i, \tag{4.14}$$

where f_i is commonly known as the *wavefront* or *frontwidth*. Assuming that N and the average value of f_i are reasonably large, it can be shown that a complete profile or front factorisation requires approximately $O(Nf^2)$ operations, where F is the root-mean-square wavefront, which is defined as:

$$F = \left(\frac{1}{N} \sum_{i=1}^{N} f_i^2 \right)^{0.5}. \tag{4.15}$$

Everstine [19] has shown that $P/N \leq F \leq W_{max} \leq B$, where W_{max} is the maximum wavefront. Hence in order to minimize the storage requirement and solution time, it is imperative to reduce the profile and root-mean-square wavefront, respectively. As both P and F are related, any algorithm that seeks to minimize either will inevitably tend to reduce the other as well. We will call an algorithm efficient if in a reasonable computer time it results in significant profile reduction.

In the storage scheme due to Jennings, all elements which belong to the envelope are stored row by row including zeros, in a one-dimensional array, say **AN**. Diagonal elements are stored at the end of each row. The length of **AN** is equal to Profile (**A**) + n. An array of pointers **IN**, the entries of which are pointers to the locations of the diagonal elements in **AN**, is also necessary. Thus, the elements of row i, when i > 1, are in positions **IN**(i − 1) + 1 to **IN**(i). The only element of row 1 is A_{11}, stored in **AN**(1). The elements have consecutive, easily calculable column indices.

For example, the matrix of Eq. 4.4, has a profile equal to 4, and its envelope storage is

$$\begin{array}{lllllllllll} \text{Postion} & = & 1 & 2 & 3 & 4 & 5 & 6 & 7 & 8 & 9 \\ \text{AN} & = & [1 & 6 & 2 & 7 & 3 & 9 & 8 & 4 & 5] \\ \text{IN} & = & [1 & 3 & 5 & 8 & 9] & & & & \end{array}$$

A variant of Jennings' scheme is obtained when the transpose of the lower envelope is stored. In this case elements are stored column-wise, and since the columns of the matrix retain their length, the scheme is often termed *skyline storage*. The profile of a matrix also changes if the rows and columns are permuted.

4.7.2 Graph Nodal Numbering for Profile Reduction

Graph models defined in the previous section are incorporated in a general algorithm of Souza and Murray [27] to obtain ten approaches for profile reduction.

This algorithm is based on Sloan's algorithm, using priorities to control the selection of nodes from a priority queue. Some of its features are adapted in the following algorithms.

The numbering and control of nodes in the priority queue are carried out through the assignment of status, based on the numbering strategy of King [28], which operates as follows:

Take a node of minimum valency and number it "1". The set of nodes is now divided into three subsets, A, B and C. The subset A consists of nodes already numbered. The subset B is defined as B = Adj (A), i.e. consists of all nodes adjacent to any node of A. The subset C contains the remaining nodes. Then, at each step, number the node of subset B which causes the smallest number of nodes of subset C to be transferred to subset B, and redefine A, B and C, accordingly.

As an example, consider a graph S with original nodal numbering as in Fig. 4.20a.

Take node "5" as a starting node and number it as "1". Then:

$$A = \{5\}, B = \{1, 8\} \text{ and } C = \{\text{the remaining nodes}\}.$$

At this stage 1, 8 are the next candidates. If 1 is taken to A, then 2 will come to B; and for 8, node 7 will join B. Therefore, arbitrarily, 1 is taken to A and numbered as "2". Now we have:

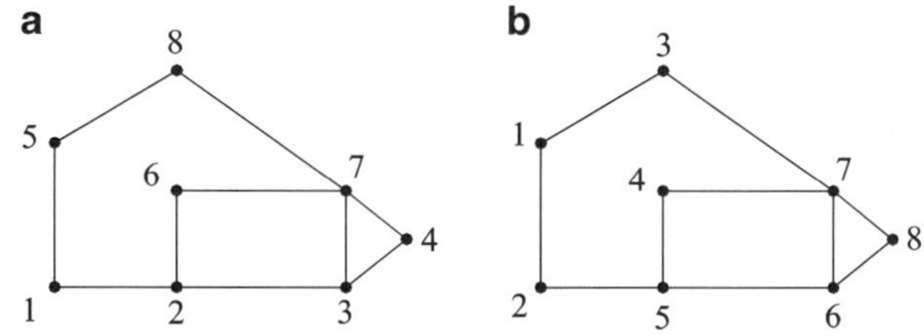

Fig. 4.20 An example of numbering by King's algorithm

$$A = \{5, 1\}, B = \{8, 2\} \text{ and } C = \{\text{the remaining nodes}\}.$$

From new candidates 8 and 2, naturally 8 will be selected because it brings only 7 to B, while 2 brings 3 and 6. Therefore 8 is numbered as 3. This process is continued until the nodal numbering of Fig. 4.20b is obtained, which corresponds to a profile equal to 14.

The nodes in the above strategy can be categorized more formally as follows:

Prior to the numbering all the nodes of a graph model G of the considered FEM are assigned *inactive status*. When a node of G is inserted in the priority queue, it is assigned *preactive status*. After a node is numbered, it is assigned *postactive status*. Nodes which are adjacent to a postactive node and do not have postactive status are defined as having *active status*, Fig. 4.21. King's algorithm is generalized by Sloan [21] through introducing a priority queue to control the order to be followed in the numbering of the nodes. This algorithm consists of the following two phases:

Phase 1: Selection of pseudo-peripheral nodes

The pair of starting nodes is determined according to the following steps:

Step 1: Choose an arbitrary node v of minimum degree.

Step 2: Generate an $SRT_v = \{C_1^v, C_2^v, \ldots, C_d^v\}$ rooted from v. Let S be the list of the nodes of C_d^v which is stored in order of increasing degree.

Step 3: Decompose S into subsets S_j of cardinality $|S_j|$, $j = 1, 2, \ldots, \Delta$, where Δ is the maximum degree of any node of S, such that all nodes in S_j have degree j. Generate an SRT from each node y in S, for the first $1 \leq m_j \leq \Delta$. If $d(SRT_y) > d(SRT_v)$, then set $v = y$ and go to Step 2.

Step 4: Let u be the root of the longest SRT which has the smallest width. When the algorithm terminates, v and u are end points of a pseudodiameter.

Phase 2: Numbering

The general algorithm for nodal numbering of an arbitrary graph associated with an FEM consists of the following steps:

Step 1: The priority queue denoted by Q is initialised with a starting node s, i.e. $Q_1 = s$. Set $n = 1$, where n is the length of the queue. The node s is assigned preactive status. Let k be the node count, which is initially set equal to zero or equal to the last number being used, in the case of disconnected graph models.

Step 2: Assign initial status and priorities to all the nodes.

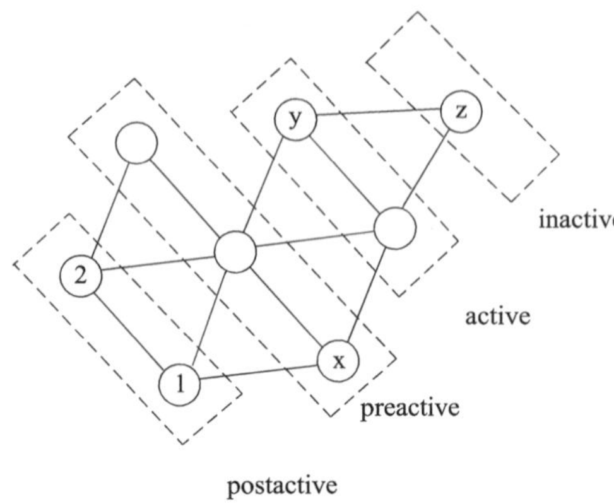

Fig. 4.21 Nodes in different status

Step 3: Select the node $u \in Q$ which has the maximum priority. Let i be the index of node u in the queue such that $Q_i = u$.

Step 4: Update queue, priority and status. Delete u from Q by setting $Q_i = Q_n$ and $n \leftarrow n - 1$. Insert nodes in queue: for each node x adjacent to u, whose status is inactive, set $n \leftarrow n + 1$ and $Q_n = x$. Assign node x preactive status and update priorities.

Step 5: Increment the node count by setting $k \leftarrow k + 1$ and label node u by *label* (u) $\leftarrow$ k, where *label*(.) contains the new labels of the nodes of the graph model. The node u is assigned postactive status.

Step 6: If $n > 0$, i.e. there are still nodes in the queue, then update priorities and status and go back to Step 3.

Step 7: Exit; i.e. the new ordering is now completed and the number of each node u is obtained as *label*(u).

4.7.3 Nodal Ordering with Element Clique Graph (NOECG)

In this method Sloan's criteria and definition for profile reduction are adapted and the general algorithm of the previous section together with the element clique graph of the considered FEM are employed for ordering. In Sloan's algorithm a quantity is defined and used as the current degree. The initial priority for each node is set to:

$$P_v = W_1 \times d(e, v) - W_2 \times cd(v), \tag{4.16}$$

where W_1 and W_2 are integers (set to $W_1 = 1$ and $W_2 = 2$ in the original algorithm of Sloan [21]), d(e, v) is the distance of node v from the end node e, and cd(v) is the current degree of v.

In Step 4 of the general algorithm, if u has preactive status, then each node x which is adjacent to it has its priority incremented according to $p_x \leftarrow p_x + W_2$. This is equivalent to decreasing the current degree of node x by unity.

In Step 6, each node x which is adjacent to the node u has its priority and status updated if it is preactive. Then it is assigned an active status and its priority is increased by setting $p_x \leftarrow p_x + W_2$. Each node y which is adjacent to x is examined next, according to:

(i) If y is not postactive, its priority is incremented by setting $p_y \leftarrow p_x + W_2$,
(ii) Else, if y is inactive then it is assigned preactive status and increased in the priority queue by setting $n \leftarrow n + 1$ and $Q_n = y$. The time complexity of this method is $O(\alpha^2)$ for the worst case.

4.7.4 *Nodal Ordering with Skeleton Graph (NOSG)*

The method for ordering the nodes of the skeleton graph of an FEM, to reduce the profile differs in two ways from the method of NOECG (i.e. Sloan's method):

1. The distance between each node of SG and s (not e) is considered.
2. The initial priorities of nodes are calculated in a different manner.

The steps of the algorithm are outlined in the following:

Step 1: Form an SRT from S and compute the distance d(s, v) between each node v of the SG and the starting node s.
Step 2: Assign each node in the graph an inactive status and compute its initial priority, p_v, according to

$$P_v = -d(s, v) - 3 \times \deg(v), \tag{4.17}$$

where deg(v) is the degree of node v.
Step 3: Initialise the priority queue Q with the starting node s, i.e. $Q_1 = s$. Set $n = 1$, where n is the length of the queue. The node s is assigned preactive status. Let k be the node count.
Step 4: While the priority queue is not empty, which is signified by $n > 0$, execute Steps 5–8.
Step 5: Select node $u \in Q$ which has the maximum priority. Let i be the index of the node u in the queue such that $Q_i = u$.
Step 6: Delete node u from the priority queue by setting $Q_i = Q_n$ and decreasing according to $n \leftarrow n - 1$. If node u is not pre-active go to Step 7. Otherwise, examine each node w which is adjacent to node u and increment its priority according to $p(w) = p(w) + 2$. If node w is inactive, then insert it in the priority queue with a pre-active status by setting $n \leftarrow n + 1$ and $Q_i = w$.
Step 7: Label node u with its new number by incrementing the node count according to $k \leftarrow k + 1$ and setting *label*(u) $\leftarrow k$. Assign node u a postactive status.
Step 8: Examine each node w which is adjacent to node u. If node w is pre-active, assign node w an active status, set $p(w) = p(w) + 2$ and examine each node x which is adjacent to node w. If node x is not postactive, increment its priority to p

(x) = p(x) + 2. If node x is inactive, insert it in the priority queue with a pre-active status by setting $n \leftarrow n + 1$ and $Q_n = x$.

Once the above steps are carried out, the new label of each node v will be *label* (v). The time complexity of this method is the same as that of the NOECG method, for worst case. However, it is interesting to note that the NOSG must be executed faster than the NOECG in average cases, since the value of n in the process of the NOSG is mostly less than that of the process in the NOECG. This is because, for FEMs containing elements with four or more nodes, the degree of nodes of the SG is less than those of the ECG. These two methods need the same lists for nodal ordering of the considered graph; however, one should note that the compact adjacency list of the SG occupies usually less memory than that of the ECG.

4.7.5 Nodal Ordering with Element Star Graph (NOESG)

The profile reduction algorithm which employs the element star graph of an FEM is the same as the method of NOECG with the following modifications being imposed:

If a virtual node u (a node whose old label is more than λ) is selected for being labelled, it should be labelled with its new number by λ plus another node count without incrementing according to $k' \leftarrow k' + 1$ and setting *label*(u) = $\lambda + k'$.

This modification enables the numbering of the elements of the main set to be varied continuously from 1 to α.

4.7.6 Nodal Ordering with Element Wheel Graph (NOEWG)

The same method as that of Sect. 4.7.5 is employed for ordering the nodes of the EWG of an FEM. The time complexity of the NOEWG is the same as that of the NOESG for worst case. However, the NOESG will be executed faster than the NOEWG in average cases, since the value of n in the process of the NOESG is, in general, less than in the process of the NOEWG, since for all FEMs the degrees of the nodes of the ESG are, in general, less than those of the EWG. These two methods require the same lists to be provided for nodal ordering of the considered graph model. However, the compact adjacency list of the ESG uses fewer words of memory than that of the EWG.

4.7.7 Nodal Ordering with Partially Triangulated Graph (NOPTG)

Ordering the nodes of the partially triangulated graph of an FEM for profile reduction does not require the selection of a pair of pseudo-peripheral nodes. The same good starting node used for the formation of the PTG (the node found in the SG for the formation of the PTG) can be used again in the NOPTG as the starting nodes s. The following two steps together with Steps 3–8 of the NOSG presented in Sect. 4.7.4 complete the process of the NOPTG.

Step 1: Form an SRT from the good starting node s used for the formation of the PTG and compute the distance d(s, v) between each node v of the PTG and the starting node s.
Step 2: Assign each node in the graph an inactive status and compute its initial priority, p_v according to

$$P_v = -d(s, v) - 2 \times (\deg(v) + 1). \tag{4.18}$$

The time complexity of this method is clearly the same as that of the NOECG and NOSG methods for worst case. In the method NOPTG the same lists needed for the previous four methods should be provided. However, some of these lists such as the compact adjacency list do not take the same number of words of memory in different graph models.

4.7.8 Nodal Ordering with Triangulated Graph (NOTG)

In order to number the nodes of the triangulated graph of an FEM for profile reduction, it is not necessary to find a pair of pseudoperipheral nodes. The same good starting node used for the formation of the TG is employed again in the NOTG as the starting node s. The following steps together with Steps 2–8 of the NOSG method complete the process of NOTG.

Step 1: Form an SRT from the good starting node s used in the formation of the TG and compute the distance d(s, v) between each node v of the TG and the starting node s.

The time complexity of this method is the same as methods NOECG, NOSG and NOPTG for the worst case.

The value of n in the process of the NOTG is mostly greater than those in the process of the NOSG and NOPTG, since the degrees of the nodes of the TG are mostly more than those of the PTG and the SG. Thus NOTG is executed more slowly than NOSG and NOPTG in average cases. An advantage of NOTG, similar to NOPTG, is that no pseudo-peripheral nodes are needed.

4.7.9 Nodal Ordering with Natural Associate Graph (NONAG)

The profile reduction algorithm which employs the natural associate graph of the FEM consists of two phases. In the first phase, which is the same as NOECG, the nodes of the NAG are ordered. In the second phase the nodes of the considered FEM are ordered based on the new labels of the nodes of the NAG. This step contains the following steps:

Step 1: For each node i of the graph model set *n(label*(i)) = i.
Step 2: For each element e corresponding to the node u, u = n(j), j = 1, 2, ..., a, label the unlabelled nodes of e, in turn.

This algorithm needs the same lists as the previous methods; however, the number of nodes of the graph model is equal to λ. Therefore it is very efficient for FEMs containing higher order elements. In the second phase of this method, an additional list with λ integer words of memory is needed which is denoted by n(.) in the steps of the process. However, this list can be created when most of the lists needed for the first phase are not required, and can be erased from the working memory.

4.7.10 Nodal Ordering with Incidence Graph (NOIG)

The profile heuristic which employs the IG of an FEM contains two parts as the NONAG method. These phases are the same as those of NONAG, with IG being employed in place of NAG.

Time and memory complexities of the NOIG are the same as those of NONAG. However, the value of n is higher than that of the NONAG, since degrees of the IG are more than those of the NAG. Therefore the NOIG should have slower execution than NONAG in average cases.

4.7.11 Nodal Ordering with Representative Graph (NORG)

This method consists of two parts. The first part orders the nodes of the RG, i.e. the representative nodes of the elements of the considered FEM. The second phase orders the nodes of the considered FEM based on the new labels of the representative nodes of the elements of the FEM.

The first part contains the following steps:

Step 1: Form an SRT from a good starting node s used for the formation of the RG and compute the distance d(s, v) between each node v of the RG and the starting node s.

Step 2: Assign each node in the graph an inactive status and compute its initial priority p_v, according to,

$$P_v = -3 \times d(s, v) - \varepsilon(v), \tag{4.19}$$

where $\varepsilon(v)$ denotes the number of elements incident to node v.

Step 3: Initialise the priority queue Q with the starting node s used for the formation of the RG, i.e. $Q_1 = s$. Set $n = 1$, where n is the length of the queue. The node s is assigned preactive status. Let k be the node count.

Step 4: While the priority queue is not empty, signified by $n > 0$, execute Steps 5–8.

Step 5: Select node $u \in Q$ which has the maximum priority. Let i be the index of the node u in the queue such that $Q_i = u$.

Step 6: Delete node u from the priority queue by setting $Q_i = Q_n$ and decrementing n according to $n \leftarrow n - 1$. If node i is not pre-active, go to Step 7, otherwise examine each node w which is adjacent to node u and increment its priority according to $p(w) = p(w) + 1$. If node w is inactive, then insert it in the priority queue with a pre-active status by setting $n \leftarrow n + 1$ and $Q_n = w$.

Step 7: Label node u with its new number by incrementing the node count according to $k \leftarrow k + 1$, and setting *label*(u) $\leftarrow$ k. Assign node u a postactive status.

Step 8: Examine each node w which is adjacent to node u. If node w is pre-active, assign node w an active status, set $p(w) = p(w) + 1$ and examine each node x which is adjacent to node w. If node x is not postactive, increment its priority to p (x) = p(x) + 1. If node x is inactive, insert it in the priority queue with a pre-active status by setting $n \leftarrow n + 1$ and $Q_n = x$.

When the above steps are completely performed, the new label of each node v is label(v). In this method there is no need to find any pseudo-peripheral, and the same good starting node used for generating the RG is employed again in the process of numbering.

The second phase of the algorithm contains the following steps:

Step 1: For each node i of the graph model set n(label(i)) = i.

Step 2: Set $k = 0$. Check each element e containing node u, u = *n*(j), j = 1, 2, . . ., α, in turn, if e does not contain a node v corresponding to *n*(l) and $i < j$, then set $k \leftarrow k + 1$ and *m*(k) = e.

Step 3: Set *label*(i) = 0, where i = 1, 2, . . ., α.

Step 4: Set l = 0. Check each node w of element e, e = *m*(j), j = 1, 2, . . ., α, in turn, if *label*(w) = 0 then set i = 0 +1 and *label*(w) = 1.

The time complexity of the first part of this algorithm is $O(\alpha^2)$ and the second part uses $O(\lambda\theta^2)$ operations. One can reduce the time complexity of the second phase by using an additional list in Step 2 to show whether element e has been previously detected. This procedure uses O(λθ) operations. The first phase of the algorithm NORG needs the same lists as the previous methods of profile reduction.

4.7.11.1 Complete Representative Graph (CREG)

This graph is the same as the REG with additional members connecting each pair of nodes in the CREG if their corresponding nodes in the FEM are contained in the same element.

4.7.12 Nodal Ordering with Element Clique Representative Graph (NOECRG)

The profile reduction of this method consists of two steps as in NONAG and NOIG and NORG methods. The first process is the same as that of Sloan's algorithm (NOECG) and the second step is similar to the second step of the NORG approach.

The time complexity and memory complexity of the NOECRG method are the same as those of NORG, but the magnitude of n in the process of NOECRG is, in general, far higher than that of the NORG, since the degrees of the nodes of ECRG are generally much greater than those of RG. Therefore NOECRG should be slower in execution than RG.

4.7.13 Computational Results

A program is developed to implement the algorithms, and many FEMs are studied. Six examples are presented here. For each problem illustrated in Figs. 4.22, 4.23, 4.24, and 4.25, the results of executing the program are provided in Tables 4.5, 4.6, 4.7, 4.8, respectively. The numbers of nodes α and elements λ for each FEM are provided in the caption of the corresponding figure.

4.7.14 Discussions

The algorithms presented for the profile reduction of sparse matrices with symmetric structures are analysed for the worst case to show their time and memory complexities.

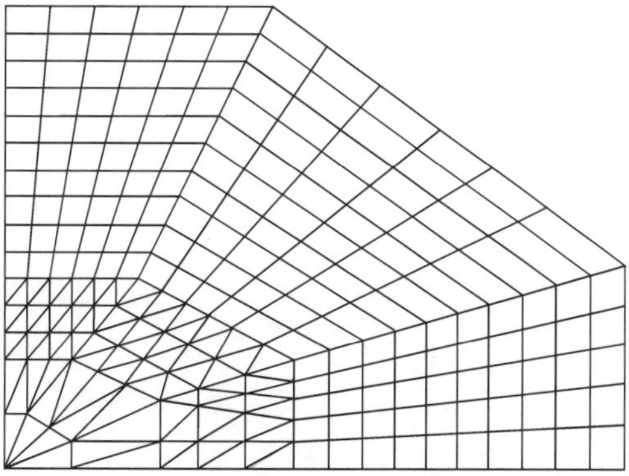

Fig. 4.22 $\alpha = 240$ and $\lambda = 499$

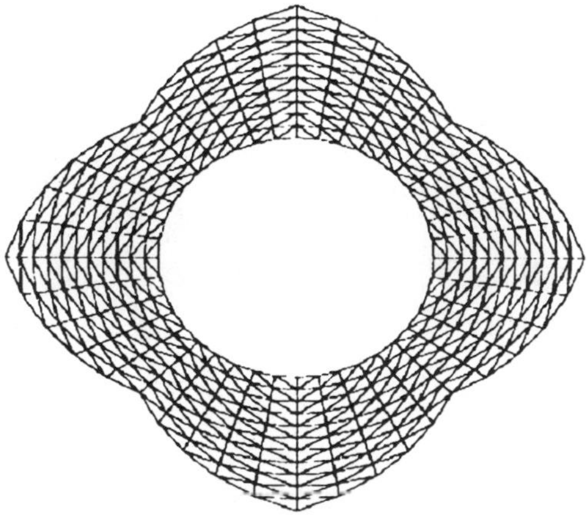

Fig. 4.23 $\alpha = 748$ and $\lambda = 1{,}236$

The programs developed for these algorithms have been tested on very many examples, and the following table is obtained which illustrates the average computational time (in seconds) of the methods (Table 4.9).

4.8 Element Ordering for Frontwidth Reduction

For the solution of sparse systems of simultaneous equations arising from the finite element method, the frontal methodology due to Irons [34] and the profile method described by George [31], as well as band-matrix techniques, are commonly used. These methods exploit the sparsity of the coefficient matrices generated by the finite element approximation. They differ, however, in one significant respect: the band and profile methods first construct the coefficient matrix explicitly, while the frontal method arranges for elimination of variables as it assembles the matrix.

The most suitable ordering of the equation is dependent on the type of equation solving scheme adopted (i.e. whether a band, profile or frontal solver is used).

Fig. 4.24 $\alpha = 936$ and $\lambda = 1{,}640$

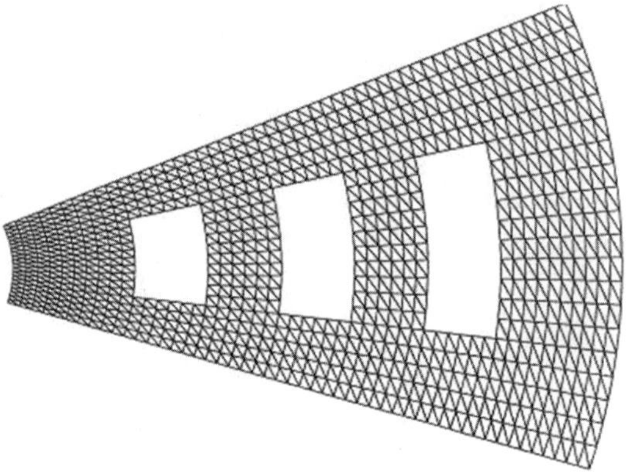

Fig. 4.25 $\alpha = 936$ and $\lambda = 1{,}640$

Table 4.5 Results of the finite element model of Fig. 4.22

Algorithm	Profile	Elapsed time
NOECG	3,207	0.22
NOSG	3,236	0.22
NOESG	3,367	0.44
NOEWG	3,465	0.66
NOPTG	3,194	0.28
NOTG	3,237	0.27
NONAG	3,365	0.71
NOIG	3,365	0.60
NORG	3,460	0.33
NOECRG	3,185	0.44

Table 4.6 Results of the finite element model of Fig. 4.23

Algorithm	Profile	Elapsed time
NOECG	7,444	0.39
NOSG	8,436	0.39
NOESG	8,336	0.87
NOEWG	8,256	1.27
NOPTG	8,527	0.65
NOTG	8,514	0.66
NONAG	7,320	0.93
NOIG	7,204	1.32
NORG	9,388	0.66
NOECRG	7,818	0.88

Table 4.7 Results of the finite element model of Fig. 4.24

Algorithm	Profile	Elapsed time
NOECG	12,248	0.72
NOSG	13,142	0.71
NOESG	13,016	1.37
NOEWG	13,049	2.03
NOPTG	13,282	1.16
NOTG	13,113	1.21
NONAG	12,631	1.54
NOIG	12,665	1.98
NORG	16,055	1.16
NOECRG	12,894	1.65

Table 4.8 Results of the finite element model of Fig. 4.25

Algorithm	Profile	Elapsed time
NOECG	15,223	0.88
NOSG	16,217	0.93
NOESG	16,008	1.87
NOEWG	15,852	2.63
NOPTG	15,391	1.48
NOTG	16,204	1.60
NONAG	15,482	2.15
NOIG	15,345	2.69
NORG	17,474	1.42
NOECRG	15,343	2.09

In finite element analysis, in the case of one degree of freedom per node, performing nodal ordering is equivalent to reordering the equations. In a more general problem with β degrees of freedom per node, there are β coupled equations produced by each node. In this case re-sequencing is usually performed on the nodal numbering to reduce the bandwidth, profile or frontwidth, because the size of this problem is β times less than that for degree of freedom numbering.

Table 4.9 Average computational time for different graphs

Algorithm	Average of the computational time
NOECG	0.99
NOSG	0.77
NOESG	1.39
NOEWG	1.96
NOPTG	1.25
NOTG	1.32
NONAG	1.33
NOIG	1.45
NORG	1.17
NOECRG	1.90

An efficient graph-theoretical approach for element renumbering of finite element meshes for frontwidth reduction of sparse matrices with symmetric structures can be found in the work of Kaveh [35].

4.9 Element Ordering for Bandwidth Optimisation of Flexibility Matrices

The elements of a generalized cycle basis (GCB), as defined in Chap. 3, must be ordered to obtain a banded flexibility matrix **G**. This is similar to ordering the elements of a cutset basis (nodal numbering) for reducing the bandwidth of the stiffness matrix **K**. This problem can be transferred to a nodal ordering algorithm by defining appropriate mathematical structures for the transformation of the connectivity properties, Kaveh [11]. Two approaches for this problem are developed in the following.

4.9.1 An Associate Graph

An *associate graph* A(B(S)) of a generalized cycle basis B(S) of S is a graph whose nodes are in a one to one correspondence with the elements of B(S), and two nodes are connected if two elements of B(S) have at least one common member. As an example the associate graph of the mesh basis in Fig. 4.26a is depicted in Fig. 4.26b.

A *weighted associate graph* can similarly be defined. For this graph, the nodes and members are assigned integer numbers. The *weight* of a node in A(B(S)) is taken as the number of members of the corresponding cycle in S, and the weight of a member $m_k = (n_i, n_j)$ in A(B(S)) is taken as the number of members of $C_i \cap C_j$, where C_i and C_j are the cycles of S corresponding to the nodes n_i and n_j of A(B(S)), respectively.

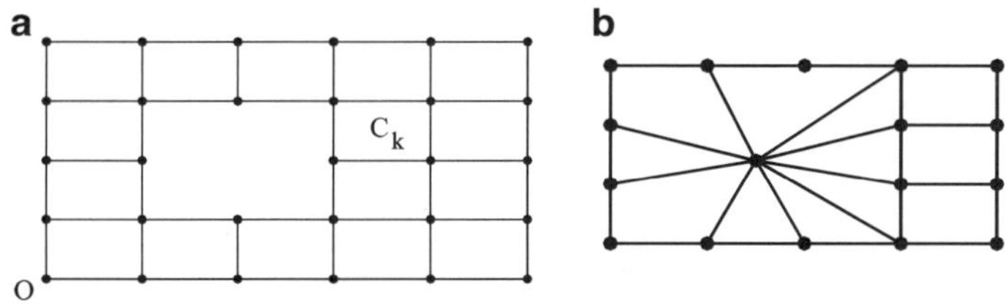

Fig. 4.26 A mesh basis and its associate graph. (**a**) A mesh basis B(S) of S. (**b**) The associate graph of B(S)

4.9.2 Distance Number of an Element

The *distance* d_i of a node n_i of S from a selected node O is the length of the shortest path connecting n_i to O. The *distance number* of a cycle or a γ-cycle or an element C_k from O is defined as one of the following:

(a) The distance of the nearest node of C_k from O, denoted by d_k^n.
(b) The distance of the furthest node of C_k from O, denoted by d_k^f.
(c) The mean value of d_k^n and d_k^f; i.e. $\left|\left(\frac{1}{2}\right)\left(d_k^n + d_k^f\right)\right|$, where |.| means the integer part of the number.
(d) The sum of $d_k^n + \left|\left(\frac{1}{2}\right)L(C_k)\right|$, where $L(C_k)$ is the length of C_k.
(e) The mean value of the distance of the nodes of C_k; i.e. $\sum_{i=1}^{L(C_k)} |d_i/L(C_k)|$.

As an example, the values defined above for a cycle C_k are shown in bold lines in Fig. 4.26a, and with respect to a reference node O are 5, 6, 5, 7 and 5, respectively. For simplicity only the integer parts of the divisions are considered.

Any of the definitions (a)–(e) can be used as the distance number of a cycle, a γ-cycle or an element of a finite element model (FEM).

4.9.3 Element Ordering Algorithms

In the following, two algorithms are presented for ordering the elements of a cycle basis, a GCB, an FEM or the substructures of a structure. However, for simplicity we will refer to a GCB only.

Algorithm A

Step 1: Order the nodes of S with a nodal numbering algorithm.

Step 2: Use the same starting node as in Step 1 to form an SRT and find the distance numbers of the elements of the GCB.

Step 3: Assign these distance numbers to the nearest (furthest or any other appropriate intermediate) nodes of the elements of the GCB. In this process a node may become the representative node of p elements. Then p independent distance numbers will be assigned to the representative nodes.
Step 4: Order these nodes in ascending order of distance number. A node representing p elements will receive p different (independent) numbers. For equi-distant nodes the same sequence as the nodal numbering of Step 1 should be used, to effect the connectivity properties of S.
Step 5: Order the elements of the GCB with the same numbers received by their representative nodes. This provides an efficient ordering for the elements of the GCB.

Algorithm B
Step 1: Construct the associate graph A(B(S)) of the GCB.
Step 2: Generate an SRT of S, starting from an appropriate node O, and find the distance numbers of the elements of the GCB.
Step 3: Assign these numbers to the nodes of A(B(S)), and order its nodes by a nodal numbering algorithm, with a starting node which corresponds to an element containing O .
Step 4: Reorder the nodes of A(B(S)) in ascending order of their distance numbers obtained in Step 2. For equi-distant nodes the same sequence as that obtained by the nodal numbering algorithm of Step 3 should be used.
Step 5: Number the elements in the same order as that obtained for their representative nodes in A(B(S)). This leads to an efficient numbering of the elements of the considered GCB.

Example. Let S be the model of a rigid-jointed planar frame. Suppose the selected cycle basis consists of the boundaries of the bounded regions of S (a mesh basis), Fig. 4.27a.

For Algorithm A, an SRT starting from O is generated, Fig. 4.27a, and the distance numbers of the cycles corresponding to definitions (a) and (e) of Sect. 4.7.2 are calculated and assigned to the representative nodes of the cycles. The nearest node of a cycle to O is taken as its representative node, Fig. 4.27b, c. These nodes are then ordered, leading to an ordered cycle basis. The bandwidth of the cycle adjacency matrices for these orderings are 15 and 13. The latter result can further be reduced to 11 by imposing additional restrictions in the process of ordering. Since the frame is planar, the bandwidths of the corresponding flexibility matrices will be 45 and 39, respectively.

Algorithm B is also applied to this example. The associate graph A(B(S)) of the mesh basis is formed, Fig. 4.27d, and using definition (e) for distance number of the elements, the order of the nodes of A(B(S)) is obtained. The numbering of the cycles is shown in Fig. 4.27d, which corresponds to a bandwidth of 13 for its cycle adjacency matrix, and 39 for its flexibility matrix.

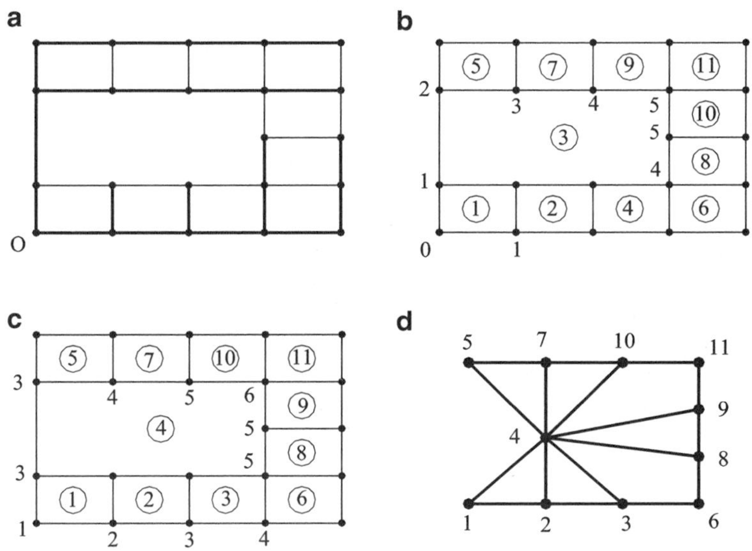

Fig. 4.27 S and ordering the elements of its cycle basis. (**a**) An SRT of S. (**b**) Cycle ordering by definition (a). (**c**) Cycle ordering by definition (e). (**d**) A(B(S)) and its nodal ordering

4.10 Bandwidth Reduction for Rectangular Matrices

In previous sections the bandwidth optimisation of square matrices has been discussed. In structural analysis, it may also be desirable to reduce the bandwidth of some sparse rectangular matrices. As an example, it may be beneficial to reduce the bandwidth of the equilibrium equations of a structure, Kaneko et al. [36]. This can be done by optimising the bandwidth of the corresponding cutset basis incidence matrix **L**. Similarly for compatibility equations, one can optimise the bandwidth of **C**.

In this section a K-total graph is defined and two algorithms are presented for the bandwidth reduction of rectangular matrices.

4.10.1 Definitions

Let **B** be a rectangular matrix with m rows and n columns, whose entries are denoted by b_{ij}. For each row like i (except the first and the last row, where $i_d = 1$ and $i_d = n$, respectively), the integer part of the real number i(n/m) is defined as i_d. Therefore, the entry of **B** at position (i, i_d) is considered as the *i*th diagonal entry. For square matrices m = n and i = i_d. The bandwidth of **B** is then defined as

$$b(\mathbf{B}) = m_r + m_l + 1, \tag{4.20}$$

Where

$$m_r = \max_{1 \leq i \leq n}\{k - i_d | b_{ik} \neq 0,\ \ k > i_d\},$$

and

$$m_l = \max_{1 \leq i \leq n}\{i_d - k | b_{ik} \neq 0,\ \ k < i_d\}. \tag{4.21}$$

If **B** is a symmetric square matrix, then $m_r = m_l$ and b(**B**) reduces to the conventional definition of square matrices. A rectangular matrix is called *banded* if b(**B**) is small compared to m.

Matrix **B** in block submatrix form has the same pattern as **L**, i.e. each non-zero entry of **L** corresponds to a $\eta \times \eta$ submatrix in **B**, where η is the degree of freedom of a node of the structure. Obviously, reduction of the bandwidth of **L** leads to a banded matrix **B**.

The terms "nodes" and "members" have been used for a graph S, and now we use "vertices" and "edges" for the elements of a *K-total graph* which is defined as follows:

Associate one vertex with each member and each element of the selected cutset basis or a cycle (γ-cycle) basis of S. Connect two vertices with an edge if

(a) The corresponding members are incident,
(b) The corresponding cutsets (cycles or γ-cycles) are adjacent,
(c) The corresponding member and cutset (cycle or γ-cycle) are incident.

When a cutset or cycle is changed to a node of S, then the K-total graph becomes a total graph as defined in graph theory (see Behzad [37]).

Examples of K-T(S) are shown in Figs. 4.28 and 4.29, when the cocycle basis and cycle basis are considered, respectively. In these figures small squares are used to represent members, and circles are employed to show the elements of the considered basis.

4.10.2 Algorithms

Algorithm A

Construct the K-total graph of S and order its vertices. The corresponding sequence leads to a favourable order of cutsets (nodes) and members of S, to reduce the bandwidth of **L**, which is pattern equivalent to the coefficient matrix of the

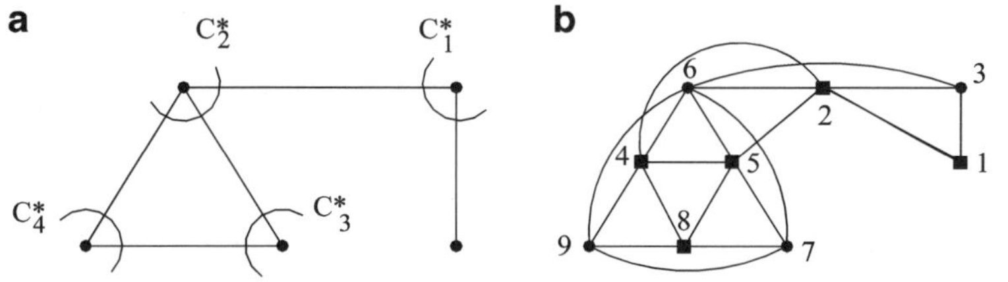

Fig. 4.28 Reduction of bandwidth for a cutset basis incidence matrix. (**a**) S and the considered cocycle basis. (**b**) K-T(S) and its nodal ordering

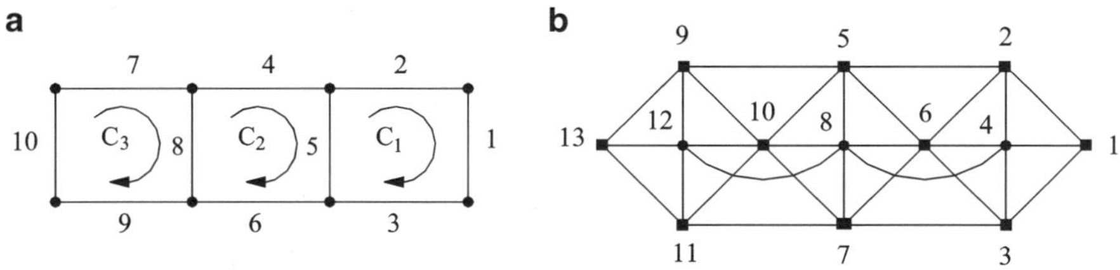

Fig. 4.29 Reduction of bandwidth for a cycle basis incidence matrix. (**a**) S and the considered cycle basis. (**b**) K-T(S) and its nodal ordering

equilibrium equations. A similar approach reduces the bandwidth of **C**, when cycles (γ-cycles) are considered in place of cutsets.

This algorithm will now be applied to the examples of Figs. 4.28 and 4.29, from which the corresponding orders for the elements of the bases and members of S are obtained.

Algorithm B

Order the nodes of S. Then order the unnumbered members of the stars of the nodes in the selected sequence, to obtain a reasonably banded **L** matrix.

In general, Algorithm A leads to a better result than Algorithm B, at the expense of additional computer time.

4.10.3 Examples

Consider a graph S as shown in Fig. 4.30 with the corresponding member and cutset orders.

The cutset basis incidence matrix of S can be written as,

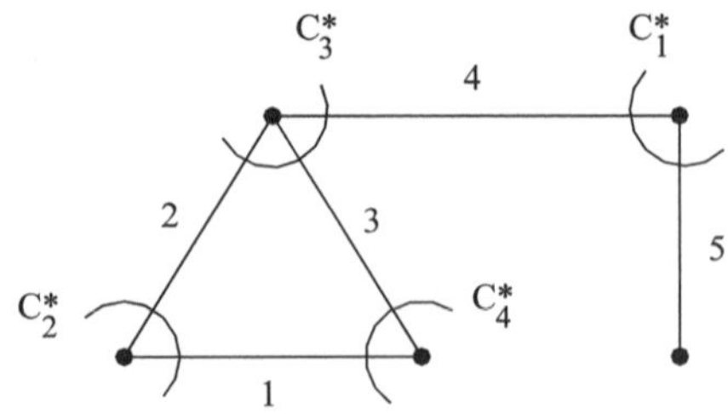

Fig. 4.30 S with an arbitrarily ordered members and cutsets

$$\mathbf{C}* = \begin{matrix} & m_1 & m_2 & m_3 & m_4 & m_5 \\ C_1^* & . & . & . & 1 & 1 \\ C_2^* & 1 & 1 & . & . & . \\ C_3^* & . & 1 & 1 & 1 & . \\ C_4^* & 1 & . & 1 & . & . \end{matrix} \qquad b(\mathbf{L}) = 4 + 4 + 1 = 9,$$

Using the ordering obtained by K-T(S), the cutset basis incidence matrix becomes,

$$\mathbf{C}* = \begin{matrix} & m_1 & m_2 & m_3 & m_4 & m_5 \\ C_1^* & 1 & 1 & . & . & . \\ C_2^* & . & 1 & 1 & 1 & . \\ C_3^* & . & . & . & 1 & 1 \\ C_4^* & . & . & 1 & . & 1 \end{matrix} \qquad b(\mathbf{L}) = 2 + 2 + 1 = 5,$$

in which the non-zero entries are clustered to the diagonal of the matrix.

As a second example, consider S as shown in Fig. 4.31, in which the regional cycles and members are arbitrarily numbered.

The cycle basis incidence matrix for S is given as:

$$\mathbf{C} = \begin{matrix} & m_1 & m_2 & m_3 & m_4 & m_5 & m_6 & m_7 & m_8 & m_9 & m_{10} \\ C_1 & 1 & 0 & 0 & 1 & 1 & 0 & 0 & 1 & 0 & 0 \\ C_2 & 0 & 0 & 1 & 0 & 0 & 1 & 1 & 0 & 0 & 1 \\ C_3 & 0 & 1 & 0 & 0 & 1 & 1 & 0 & 0 & 1 & 0 \end{matrix}.$$

For this matrix, $b(\mathbf{C}) = 7 + 8 + 1 = 16$. With ordering the cycles and members simultaneously, using Algorithm A, the following cycle basis incidence matrix is obtained.

$$\mathbf{C} = \begin{matrix} & m_1 & m_2 & m_3 & m_4 & m_5 & m_6 & m_7 & m_8 & m_9 & m_{10} \\ C_1 & 1 & 1 & 1 & 0 & 1 & 0 & 0 & 0 & 0 & 0 \\ C_2 & 0 & 0 & 0 & 1 & 1 & 1 & 0 & 1 & 0 & 0 \\ C_3 & 0 & 0 & 0 & 0 & 0 & 0 & 1 & 1 & 1 & 1 \end{matrix}.$$

The bandwidth fir this matrix is obtained as $b(\mathbf{C}) = 4 + 3 + 1 = 8$.

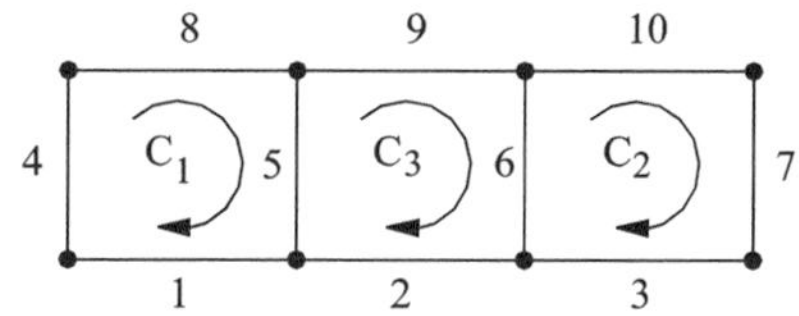

Fig. 4.31 S with arbitrarily numbered members and cycles

For the force method of frames, the coefficient matrix of the equilibrium equations can be made banded through reducing the bandwidth of its member-cycle incidence matrix. After an algebraic force method is employed, a repeated application of the developed method makes the null basis matrix a banded one for subsequent applications. Similarly, if a combinatorial approach is used, the bandwidth reduction algorithm makes the cycle basis incidence matrix banded, leading to a banded statical basis (null basis) matrix.

4.10.4 Bandwidth Reduction of Finite Element Models

The algorithms presented in the previous section can also be applied to finite element models, for their analysis by the algebraic force method, Kaveh and Mokhtar-zadeh [38]. For such models, the K-total graph of an FEM is defined as follows:

Associate one vertex with each side and each element of the FEM, and connect two vertices with an edge if any of the following conditions hold:

1. Sides are adjacent;
2. Elements are adjacent;
3. A side and an element are incident.

The Algorithm A can now be adapted to FEMs as follows:

Step 1: Generate the K-total graph of the finite element mesh S.
Step 2: Order the vertices of K-T(S) by any nodal ordering algorithm available.
Step 3: Assign numbers the members of K-T(S) and to the elements of the considered FEM, in the order of their occurrence in the sequence selected in Step 2.

Example. Four groups of examples are considered as shown in Fig. 4.32(a–d). In these figures, Ω is the aspect ratio of the element numbers in two perpendicular directions (x and y directions) which is taken as unity. The ratio of the length of the elements side in x direction to that of the y direction, is taken as 1.2. S is the refinement index of a group. In the group UT, Ω_1, Ω_2 are the aspect ratios of the element numbers in the two sides of the general configuration with respect to the central part of the model.

The sparsity of the self-stress and flexibility matrices of LQ and HQ groups is illustrated in Fig. 4.33(a–d).

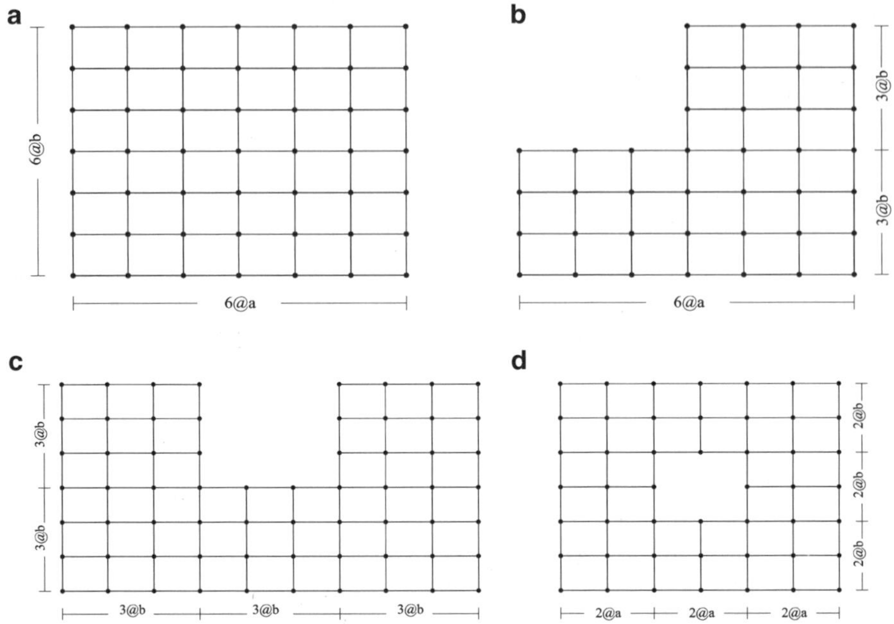

Fig. 4.32 Test group examples. (**a**) Group RQ-Ω-S (Ω=1, S = 6). (**b**) Group LQ-Ω-S (Ω=1, S = 3). (**c**) Group UQ-Ω_1-$\Omega_{2\text{-}S}$ (Ω_1=Ω_2 = 1, S = 3). (**d**) Group HQ-Ω-S (Ω=1, S = 2)

4.11 Graph-Theoretical Interpretation of Gaussian Elimination

In this section, a simple graph-theoretical interpretation of the Gaussian elimination is presented, in order to establish a closer link between the matrix algebra on one hand and graph-theoretical concepts on the other hand.

Let **A** be a symmetric sparse matrix of order N and let S be the corresponding graph. Suppose that Gaussian elimination by columns is performed on **A** until the factorization $\mathbf{A} = \mathbf{U}^t\mathbf{D}\mathbf{U}$ is obtained. At the beginning of the *k*th step all non-zeros in columns 1, 2, . . ., k−1 below the diagonal have been eliminated. Multiples of the *k*th row are then subtracted from all rows which have a non-zero in column k below the diagonal. On performing this operation, new non-zero entries may be introduced in row k + 1, . . ., N to the right of column k. Cancellations may also occur, producing new zeros, but this is rare in practice and will be neglected. Consider the active submatrix at the *k*th step (an active submatrix contains all elements $A_{ij}^{(k)}$ with i, j ≥ k). Let S^k be the graph associated with the active submatrix. S^k is called *an elimination graph*, Parter [39]. The nodes of this graph are N−k + 1 last numbered nodes of S. S^k contains all members connecting those nodes which were present in S, and additional members corresponding to fill-ins produced during the k−1

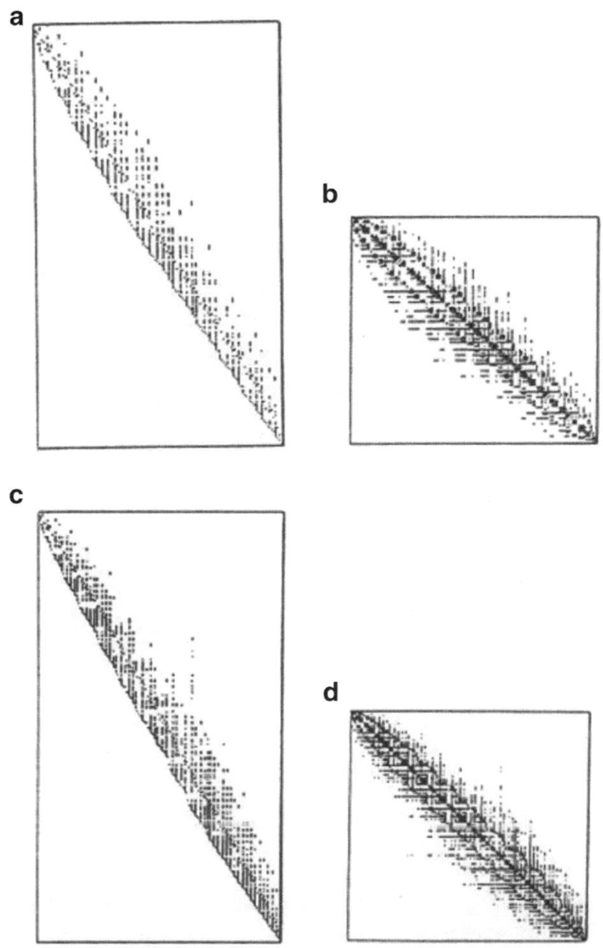

Fig. 4.33 Self-stress and flexibility matrices. (**a**) Self-stress matrix of LQ-1-4. (**b**) Flexibility matrix of LQ-1-4. (**c**) Self-stress matrix of HQ-1-4. (**d**) Flexibility matrix of HQ-1-4

initial elimination steps. The sequence $S = S^1, S^2, S^3, \ldots$ can be obtained using the following rule:

To obtain S^{k+1} from S^k, delete node k and add all possible members between nodes, which are adjacent to node k in S^k.

As an example, consider a graph S and the corresponding adjacency matrix, as shown in Fig. 4.34. Two steps of the Gaussian elimination and the corresponding elimination graphs are also illustrated.

Eliminating the rest of the nodes, and considering a clique (a complete graph) between the nodes adjacent to each eliminated node (when such members are not present), matrix **U** is obtained. The structure of $\mathbf{U} + \mathbf{U}^t$ and the corresponding filled graph are shown in Fig. 4.35.

There are algorithms which try to reduce the number of fill-ins caused by elimination. The minimum degree algorithm of Tinney [40] is perhaps the best method for such a reduction.

a

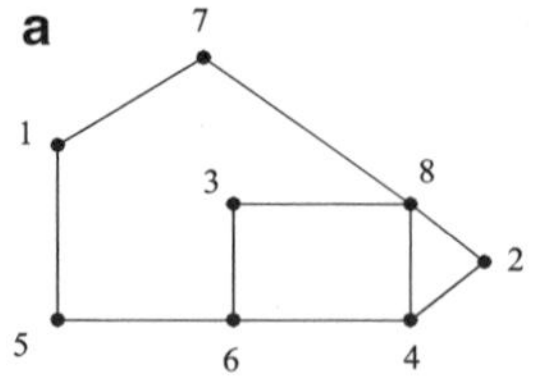

b

$$
\mathbf{A}^1 = \begin{array}{c|cccccccc|}
 & 1 & 2 & 3 & 4 & 5 & 6 & 7 & 8 \\
1 & \ast & \cdot & \cdot & \cdot & \ast & \cdot & \ast & \cdot \\
2 & \cdot & \ast & \cdot & \ast & \cdot & \cdot & \cdot & \ast \\
3 & \cdot & \cdot & \ast & \cdot & \cdot & \ast & \cdot & \ast \\
4 & \cdot & \ast & \cdot & \ast & \cdot & \ast & \cdot & \ast \\
5 & \ast & \cdot & \cdot & \cdot & \ast & \ast & \cdot & \cdot \\
6 & \cdot & \cdot & \ast & \ast & \ast & \ast & \cdot & \cdot \\
7 & \ast & \cdot & \cdot & \cdot & \cdot & \cdot & \ast & \ast \\
8 & \cdot & \ast & \ast & \ast & \cdot & \cdot & \ast & \ast
\end{array}
$$

c

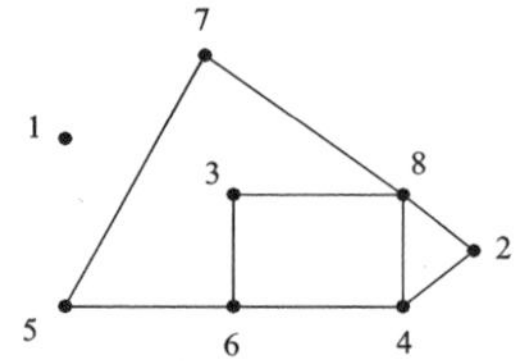

d

$$
\mathbf{A}^2 = \begin{array}{c|cccccccc|}
 & 1 & 2 & 3 & 4 & 5 & 6 & 7 & 8 \\
1 & \ast & \cdot & \cdot & \cdot & \ast & \cdot & \ast & \cdot \\
2 & \cdot & \ast & \cdot & \ast & \cdot & \cdot & \cdot & \ast \\
3 & \cdot & \cdot & \ast & \cdot & \cdot & \ast & \cdot & \ast \\
4 & \cdot & \ast & \cdot & \ast & \cdot & \ast & \cdot & \ast \\
5 & \cdot & \cdot & \cdot & \cdot & \ast & \ast & \otimes & \cdot \\
6 & \cdot & \cdot & \ast & \ast & \ast & \ast & \cdot & \cdot \\
7 & \cdot & \cdot & \cdot & \cdot & \otimes & \cdot & \ast & \ast \\
8 & \cdot & \ast & \ast & \ast & \cdot & \cdot & \ast & \ast
\end{array}
$$

e

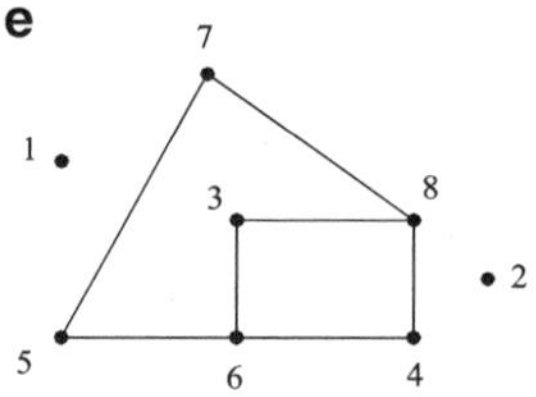

f

$$
\mathbf{A}^3 = \begin{array}{c|cccccccc|}
 & 1 & 2 & 3 & 4 & 5 & 6 & 7 & 8 \\
1 & \ast & \cdot & \cdot & \cdot & \ast & \cdot & \ast & \cdot \\
2 & \cdot & \ast & \cdot & \ast & \cdot & \cdot & \cdot & \ast \\
3 & \cdot & \cdot & \ast & \cdot & \cdot & \ast & \cdot & \ast \\
4 & \cdot & \cdot & \cdot & \ast & \cdot & \ast & \cdot & \ast \\
5 & \cdot & \cdot & \cdot & \cdot & \ast & \ast & \otimes & \cdot \\
6 & \cdot & \cdot & \ast & \ast & \ast & \ast & \cdot & \cdot \\
7 & \cdot & \cdot & \cdot & \cdot & \otimes & \cdot & \ast & \ast \\
8 & \cdot & \cdot & \ast & \ast & \cdot & \cdot & \ast & \ast
\end{array}
$$

Fig. 4.34 Illustration of two steps of the Gaussian elimination. (**a**) $S = S^1$. (**b**) Matrix $\mathbf{A}^1$. (**c**) S^2. (**d**) Matrix $\mathbf{A}^2$. (**e**) S^3. (**f**) Matrix $\mathbf{A}^3$

a

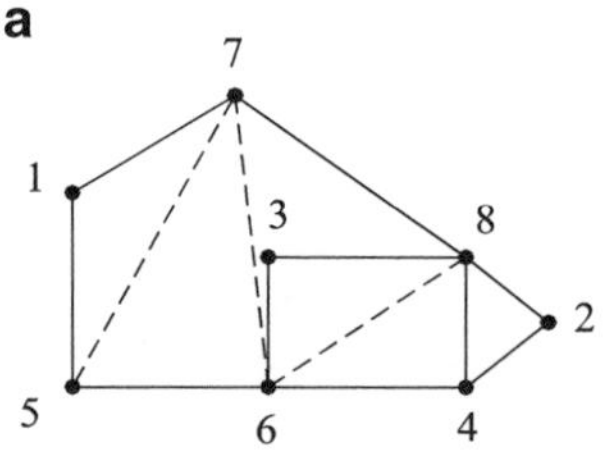

b

$$
\begin{array}{c|cccccccc|}
 & 1 & 2 & 3 & 4 & 5 & 6 & 7 & 8 \\
1 & \ast & \cdot & \cdot & \cdot & \ast & \cdot & \ast & \cdot \\
2 & \cdot & \ast & \cdot & \ast & \cdot & \cdot & \cdot & \ast \\
3 & \cdot & \cdot & \ast & \cdot & \cdot & \ast & \cdot & \ast \\
4 & \cdot & \ast & \cdot & \ast & \cdot & \ast & \cdot & \ast \\
5 & \ast & \cdot & \cdot & \cdot & \ast & \ast & \otimes & \cdot \\
6 & \cdot & \cdot & \ast & \ast & \ast & \ast & \otimes & \otimes \\
7 & \ast & \cdot & \cdot & \cdot & \otimes & \otimes & \ast & \ast \\
8 & \cdot & \ast & \ast & \ast & \cdot & \otimes & \ast & \ast
\end{array}
$$

Fig. 4.35 The structure of $\mathbf{U} + \mathbf{U}^t$ and the corresponding graph. (**a**) S^F. (**b**) Matrix $\mathbf{U} + \mathbf{U}^t$

References

1. Rosen R (1968) Matrix bandwidth minimization. In: Proceedings of the 23rd national conference of the ACM. Brandon System Press, pp 585–595
2. Grooms HR (1972) Algorithm for matrix bandwidth reduction. J Struct Div ASCE 98:203–214
3. Hall M Jr (1956) An algorithm for distinct representatives. Amer Math Mon 63:716–717
4. Cuthill E, McKee J (1969) Reducing the bandwidth of sparse symmetric matrices. In: Proceedings of the 24th national conference of the ACM. Brandon System Press, pp 157–172
5. Kaveh A (1974) Application of topology and matroid theory to the analysis of structures. Ph.D. thesis, Imperial College, London University
6. Kaveh A (1976) Improved cycle bases for the flexibility analysis of structures. Comput Methods Appl Mech Eng 9:267–272
7. Kaveh A (1977) Topological study of the bandwidth reduction of structural matrices. J Sci Tech 1:27–36
8. Gibbs NE, Poole WG, Stockmeyer PK (1976) An algorithm for reducing the bandwidth and profile of a sparse matrix. SIAM J Numer Anal 12:236–250
9. Sloan SW (1989) A Fortran program for profile and wavefront reduction. Int J Numer Methods Eng 28:2651–2679
10. Kaveh A (1986) Ordering for bandwidth reduction. Comput Struct 24:413–420
11. Kaveh A (2004) Structural mechanics: graph and matrix methods, 3rd edn. Research Studies Press, Baldock
12. Grimes RG, Pierce DJ, Simon HD (1990) A new algorithm for finding a pseudo-peripheral node in a graph. SIAM J Anal Appl 11:323–334
13. Paulino GH, Menezes IFM, Gattass M, Mukherjee S (1994) Node and element resequencing using the Laplacian of a finite element graph: part I-general concepts and algorithms. Int J Numer Methods Eng 37:1511–1530
14. Papademetrious CH (1976) The NP-completeness of bandwidth minimization problem. Comput J 16:177–192
15. Cheng KY (1973) Note on minimizing the bandwidth of sparse symmetric matrices. Comput J 11:27–30
16. Kaveh A (1992) Recent developments in the force method of structural analysis. Appl Mech Rev 45:401–418
17. Kaveh A, Rahimi Bondarabady HA (2001) Spectral trisection of finite element models. Int J Numer Methods Heat Fluid Flow 11:358–370
18. Kaveh A (1991) A connectivity coordinate system for node and element ordering. Comput Struct 41:1217–1223
19. Everstine GC (1979) A comparison of three resequencing algorithms for the reduction of profile and wavefront. Int J Numer Methods Eng 14:837–853
20. Razzaque A (1980) Automatic reduction of frontwidth for finite element analysis. Int J Numer Methods Eng 15:1315–1324
21. Sloan SW (1986) An algorithm for profile and wavefront reduction of sparse matrices. Int J Numer Methods Eng 23:1693–1704
22. Bykut A (1977) A note on an element ordering scheme. Int J Numer Methods Eng 11:194–198
23. Kaveh A (1984) A note on a two-step approach to finite element ordering. Int J Numer Methods Eng 19:1753–1754
24. Fenves SJ, Law KH (1983) A two-step approach to finite element ordering. Int J Numer Methods Eng 19:891–911
25. Kaveh A, Ramachandran K (1984) Graph theoretical approach for bandwidth and frontwidth reductions. In: Nooshin H (ed) Proceedings of the 3rd international conference on space structures .Surrey University, pp 244–249
26. Livesley RK, Sabin MA (1991) Algorithms for numbering the nodes of finite element meshes. Comput Syst Eng 2:103–114

27. Kaveh A, Roosta GR (1998) Comparative study of finite element nodal ordering methods. Eng Struct 20(1–2):86–96
28. Kaveh A, Roosta GR (1995) An efficient method for finite element nodal ordering. Asian J Struct Eng 1:229–242
29. Jennings A (1966) A compact storage scheme for the solution of symmetric linear simultaneous equations. Comput J 9:281–285
30. Felippa CA (1975) Solution of linear equations with skyline-stored symmetric matrix. Comput Struct 5:13–29
31. George A (1977) Solution of linear system of equations; direct methods for finite element problems. In: Dold A, Eckmann E (eds) Lecture notes in mathematics, vol 572. Springer-Verlag, pp 52–101
32. Liu WH, Sherman, AH (1975) Comparative analysis of the Cuthill-McKee and Reverse Cuthill-McKee ordering algorithms for sparse matrices. Department of Computer Science, Yale University, New Haven, Report 28
33. Souza LT, Murray DW (1995) A unified set of resequencing algorithms. Int J Numer Methods Eng 38:565–581
34. Irons BM (1970) A frontal solution program for finite element analysis. Int J Numer Methods Eng 2:5–32
35. Kaveh A (2013) Optimal structural analysis by concepts of symmetry and regularity. Springer Verlag, GmbH, Wien-New York
36. Kaneko I, Lawo M, Thierauf G (1982) On computational procedures for the force methods. Int J Numer Methods Eng 18:1469–1495
37. Behzad M (1970) A characterization of total graphs. Proc Amer Math Soc 26(1970):383–389
38. Kaveh A, Mokhtar-zadeh A (1993) A comparative study of the combinatorial and algebraic force methods. In: Proceedings of the Civil-Comp93, Edinburgh, pp 21–30
39. Parter SV (1961) The use of linear graphs in Gauss elimination. SIAM Rev 3:119–130
40. Tinney WF (1969) Comments on using sparsity technique for power systems problem. In: Willoughby RA (ed) Report No. RA1 (11707), IBM, pp 25–34

Chapter 5
Ordering for Optimal Patterns of Structural Matrices: Algebraic Graph Theory and Meta-heuristic Based Methods

5.1 Introduction

There are different matrices associated with a graph, such as incidence matrix, the adjacency matrix and the Laplacian matrix. One of the aims of algebraic graph theory is to determine how properties of graphs are reflected in algebraic properties of these matrices. The eigenvalues and eigenvectors of these matrices provide valuable tools for combinatorial optimisation and in particular for ordering of sparse symmetric matrices such as the stiffness and flexibility matrices of the structures.

In this chapter, algebraic graph-theoretical methods are discussed for nodal ordering for bandwidth reduction. Hybrid methods are also applied to nodal ordering, using graph theory and algebraic graph theory.

Though graph theoretical methods are highly efficient for ordering; however, bandwidth minimization is a NP-complete problem. To tackle this problem partially, meta-heuristic algorithms seem to good alternatives, though these do not lead to absolute minimum either. Here, the recently developed meta-heuristic, known as the charged system search, is applied to nodal ordering for bandwidth and profile reduction. Meta-heuristic algorithms are rapidly developing and these can provide powerful methods for ordering in the near future [1–3].

5.2 Adjacency Matrix of a Graph for Nodal Ordering

5.2.1 Basic Concepts and Definitions

There are several geographical papers dealing with the question of whether important places or well connected sets of towns in a traffic network can be identified by an inspection of certain eigenvalues and corresponding eigenvectors of the

DOI 10.1007/978-3-319-02964-1_5,

adjacency matrix A of the underlying graph model. Gould [4] appears to be the first important publication on this subject.

In structural analysis, Kaveh [5] used the first eigenvalue and eigenvector of $[\mathbf{A} + \mathbf{I}]$, for nodal ordering for bandwidth reduction. Grimes et al. [6] employed this concept for finding pseudo-peripheral nodes of a graph. This algebraic graph-theoretical method in studied in the following.

A node n_i of S is called *peripheral,* if its eccentricity is the same as the diameter of S, i.e. $\delta(S) = e(n_i)$. If the eccentricity is close to the diameter, then n_i is called a *pseudo-peripheral* node or a *good starting node.*

Reordering the nodes of the graph model of a structure does not change the properties of the stiffness matrix. This fact stays true for the properties of the graph itself. Therefore, a natural question is: what can the theory of matrices and in particular the eigenvalues of the matrices associated with graphs tell us about the structure of the graph itself. In the following, we shall endeavour to find out to what extent the eigenvalues of the adjacency matrix of a given graph, reflect the properties of that graph.

Let **A** be the adjacency matrix of the graph S, which is a real symmetric (0, 1) matrix, and the sum of entries of any row or column is equal to the valency of the corresponding node. Denote the characteristic polynomial of **A** by $\phi(S;x)$. Since $\phi(S;x)$ is uniquely determined by the graph S, it is referred to as the *characteristic polynomial of S* and expressed as:

$$\phi(S;x) = \det(x\mathbf{I} - \mathbf{A}) = \sum_{i=0}^{N} a_i x^{N-i}. \tag{5.1}$$

Since **A** is a real symmetric matrix, its eigenvalues (the roots of this polynomial) must be real, and can be ordered $\lambda_1 \geq \lambda_2 \geq \lambda_3 \geq \ldots \geq \lambda_N$. These eigenvalues are called the eigenvalues of S, and the sequence of N eigenvalues is called the *spectrum* of G.

The following important results are stated, however, the reader may refer to Schwenk and Wilson [7] for further details and proofs.

1. The sum of the eigenvalues of a graph is equal to the trace of **A**, and therefore zero.
2. If S is connected with N nodes, then $2\cos\left(\frac{\pi}{N+1}\right) \leq \lambda_1 \leq N - 1$. The lower bound occurs only when S is a path graph, and the upper bound occurs when S is a complete graph.
3. If S is a connected graph with m distinct eigenvalues and with diameter d, then $m > d$.

The spectrum by no means specifies its graph uniquely, however, it does provide a wealth of information about the graph and hence about the structure. Some applications of such information will be given in this chapter and Chap. 8.

Table 5.1 Simple examples

Graph	Adjacency matrix	Characteristic polynomial	Eigenvalues
K_2	$\begin{bmatrix} 0 & 1 \\ 1 & 0 \end{bmatrix}$	$x^2 - 1$	1, −1
P_3	$\begin{bmatrix} 0 & 0 & 1 \\ 0 & 0 & 1 \\ 1 & 1 & 0 \end{bmatrix}$	$x^3 - 2x$	$\sqrt{2}, -\sqrt{2}, 0$
C_4	$\begin{bmatrix} 0 & 1 & 0 & 1 \\ 1 & 0 & 1 & 0 \\ 0 & 1 & 0 & 1 \\ 1 & 0 & 1 & 0 \end{bmatrix}$	$x^4 - 4x^2$	2,−2, 0, 0

However, the writer strongly believes that in future many other applications in structural mechanics will be found.

Table 5.1 shows some simple examples to verify the results stated.

Perron-Frobenius Theorem. If S is a connected graph with at least two nodes, then:

(i) Its largest eigenvalue λ_1 is a simple root of $\phi(S;x)$;
(ii) Corresponding to the eigenvalue λ_1, there is an eigenvector $\mathbf{w}_1$ all of whose entries are positive;
(iii) If λ is any other eigenvalue of S, then $-\lambda_1 \leq \lambda \leq \lambda_1$;
(iv) The deletion of any member of S decreases the largest eigenvalue.

The largest eigenvalue λ_1 is often known as the *spectral radius* of S. Since the eigenvectors corresponding to any eigenvalue other than λ_1 must be orthogonal to $\mathbf{w}_1$, we observe that the multiples of $\mathbf{w}_1$ are the only eigenvectors all of whose entries are positive.

Consider the node adjacency matrix **A** of S. Let,

$$\mathbf{Q} = \mathbf{A} + \mathbf{I}, \tag{5.2}$$

where **I** is an N(S)×N(S) identity matrix. The eigenvalues of **Q** are one unit bigger than those of **A**, and the eigenvectors of **Q** are exactly the same as those of **A**. Matrix **Q** is real and symmetric, and it can easily be shown that all the entries of $\mathbf{Q}^k$ are positive; thus it is primitive and, according to the Perron-Frobenius theorem, λ_1 is real and positive and a simple root of the characteristic equation, $\lambda_1 > |\lambda|$ for any eigenvalue $\lambda \neq \lambda_1$, and λ_1 has a unique corresponding eigenvector $\mathbf{w}_1$ with all entries positive.

As $\mathbf{w}_i$ is the eigenvector corresponding to λ_i, therefore $\mathbf{Q}\mathbf{w}_i = \lambda_i\mathbf{w}_i$ for i = 1, . . ., N(S). Multiplying the two sides by **Q**, one obtains $\mathbf{Q}\mathbf{Q}\mathbf{w}_i = \lambda_i\mathbf{Q}\mathbf{w}_i = \lambda_i^2\mathbf{w}_i$ Repeating this process results in $\mathbf{Q}^k\mathbf{w}_i = \lambda_i^k\mathbf{w}_i$. Now consider any vector **x** not orthogonal to $\mathbf{w}_1$ as:

$$\mathbf{x} = \alpha_1 \mathbf{w}_1 + \alpha_2 \mathbf{w}_2 + \ldots + \alpha_{N(S)} \mathbf{w}_{N(S)} \qquad \alpha_1 \neq 0. \tag{5.3}$$

Multiplying the two sides with $\mathbf{Q}^k$, and using $\mathbf{Q}^k \mathbf{w}_i = \lambda_i^k \mathbf{w}_i$ for i = 1, ..., N(S), leads to,

$$\mathbf{Q}^k \mathbf{x} = \lambda_1^k \alpha_1 \mathbf{w}_1 + \lambda_2^k \alpha_2 \mathbf{w}_2 + \ldots + \lambda_{N(S)}^k \alpha_{N(S)} \mathbf{w}_{N(S)}, \tag{5.4}$$

and as $k \to \infty$, we have,

$$\mathbf{Q}^k \mathbf{x} / \lambda_1^k = \alpha_1 \mathbf{w}_1 + (\lambda_2/\lambda_1)^k \alpha_2 \mathbf{w}_2 + \ldots + \left(\lambda_{N(S)}/\lambda_1\right)^k \alpha_{N(S)} \mathbf{w}_{N(S)} \to \alpha_1 \mathbf{w}_1, \tag{5.5}$$

since λ_1 is the eigenvalue of strictly largest modulus and (λ_i/λ_1) is less than unity and approaches to zero when $k \to \infty$. In other words, the ratios of the components of $\mathbf{Q}^k\mathbf{x}$ approach the ratios of the components of $\mathbf{w}_1$ as k increases.

Let $\mathbf{v} = \{1,1, \ldots,1\}^t$, then the *i*th component of $\mathbf{Q}^k\mathbf{v}$, is the number of walks of length k beginning at an arbitrary node of S and ending at n_i. If n_i is a good starting node (peripheral node), this number will be smaller. Thus, for $k \to \infty$, one should obtain some average number, defined as the *accessibility index* by Gould [4]. This number indicates how many walks go on average through a node. With a suitable normalization, $\mathbf{Q}^k\mathbf{v}$ converges to the largest eigenvector $\mathbf{w}_1$ of $\mathbf{Q}$, Straffing [8].

5.2.2 A Good Starting Node

Algorithm A

Step 1: Calculate the dominant eigenvector $\mathbf{w}_1 = \{w_1, w_2, \ldots, w_{N(S)}\}^t$ of matrix $\mathbf{Q}$.

Step 2: Find Min w_i in $\mathbf{w}_1$. The node corresponding to this entry is taken as a good starting node of S.

For calculating the dominant eigenvector $\mathbf{w}_1$ of $\mathbf{Q}$, an iterative method is used, which starts with $\mathbf{v} = \{1, 1, \ldots, 1\}^t$ and calculates $\mathbf{Qv}$. This vector is then normalized and multiplied by $\mathbf{Q}$. This process is repeated until the difference between two consecutive eigenvalues, obtained from $\mathbf{Qv} = \lambda\mathbf{v}$, is reduced to a small value which, for example, can be taken as 10^{-3}.

5.2.3 Primary Nodal Decomposition

Once a good starting node is selected, an SRT is constructed and its contours $\{C_1, C_2, \ldots, C_m\}$ are obtained. These subsets are then ordered according to their distances from the selected starting node. Obviously many SRTs can be constructed on a node. Although all lead to the same nodal decompositions, different transversals will be obtained for different SRTs. Thus in the generation process, the nodes of each contour C_i are considered in ascending order of their entries in eigenvector W_1 for selecting the nodes in C_{i+1}, in order to provide the conditions for the possibility of generating a minimal (or optimal) transversal as defined in the next section.

5.2.4 *Transversal P of an SRT*

For selection of an optimal transversal, the weight of a node is defined as its value w_i in $\mathbf{w}_1$, when an algebraic graph-theoretical method is employed.

Algorithm B
Let C_1, C_2, ..., C_m be the selected contours of the SRT, and correspondingly put these subsets in $\mathbf{w}_1$ into a similar order, i.e.

$$\mathbf{w}_1 = \{W(C_1), W(C_2), \ldots, W(C_m)\}, \tag{5.6}$$

where $W(C_i)$ contains the entries of $\mathbf{w}_1$ corresponding to the nodes of C_i. Now the algorithm can be described as follows:

Step 1: Label the root as N_1 and assign w_i of this node as its new weight, denoted by $\overline{w}_1$.

Step 2: Calculate the new weight $\overline{w}_i$ of each node of C_2 by adding the w_i's from W (C_2) to $\overline{w}_1$.

Step 3: Repeat the process of Step 2, calculating $\overline{w}_i$ for each node of C_3, C_4, ..., C_m.

Step 4: Take a node N_m of minimal weight from the last contour C_m of the selected SRT.

Step 5: Find N_{m-1} from C_{m-1}, which is connected to N_m by a branch of the SRT.

Step 6: Repeat the process of Step 5, selecting N_{m-2}, N_{m-3}, ..., N_1 as the representative nodes of the contours C_{m-2}, C_{m-3}, ..., C_1.

The set $P = \{N_1, N_2, \ldots, N_m\}$, forms a suboptimal transversal of the selected SRT.

5.2.5 *Nodal Ordering*

Step 1: Number N_1 as "1".

Step 2: N_2 is given number "2" and an SR subtree is generated from N_2, numbering the nodes of C_2 in the order of their occurrence in this SR subtree.

Step 3: The process of Step 2 is repeated for numbering the nodes of $C_3, C_4, \ldots, C_m$, sequentially using $N_3, N_4, \ldots, N_m$ as the starting nodes of SR subtrees, until all the nodes of S are numbered.

Now the numbering can be reversed, in a way similar to that of the Reverse Cuthill-McKee algorithm, for possible reduction of fill-ins in the process of Gaussian elimination.

5.2.6 Example

S is the model of a grid with uniform valency distribution, as shown in Fig. 5.1a. Using algorithm A, the following dominant eigenvector is obtained for matrix Q of S, in which for simplicity only four digits are provided:

$$\mathbf{w}_1 = \left\{ \begin{array}{c} 0.3344, 0.5298, 0.6161, 0.5951, 0.4791, 0.3011, 0.1180, 0.3972, 0.7432 \\ 0.9540, 1.0000, 0.8786, 0.6183, 0.2875, 0.2875, 0.6183, 0.8786, 1.000, 0.9540 \\ 0.7432, 0.3972, 0.1180, 0.3011, 0.4791, 0.5951, 0.6160, 0.5298, 0.3344 \end{array} \right\}^t.$$

Thus node "7" is selected as a good starting node. An SRT is generated from this node and using Algorithm B, a transversal P = {7, 14, 21, 28, 27, 26, 25, 24, 23, 22} is selected, which is shown in bold lines in Fig. 5.1a. Final nodal numbering is illustrated in Fig. 5.1b.

5.3 Laplacian Matrix of a Graph for Nodal Ordering

5.3.1 Basic Concepts and Definitions

Another interesting matrix associated with a graph is the Laplacian matrix of S, denoted by **L**(S).

Consider a directed graph S with an arbitrary nodal numbering and member orientations. The adjacency matrix **A**(S), degree matrix **D**(S), node-member incidence matrix C(S), and Laplacian matrix **L**(S) are defined as follows:

The *adjacency matrix* $\mathbf{A}(S) = [a_{ij}]_{N \times N}$ of the labelled graph S is defined as:

$$a_{ij} = \begin{cases} 1 & \text{if node } n_i \text{ is adjacent to } n_j, \\ 0 & \text{otherwise.} \end{cases}$$

The *degree matrix* $\mathbf{D}(S) = [d_{ij}]_{N \times N}$ is the diagonal matrix of node degrees:

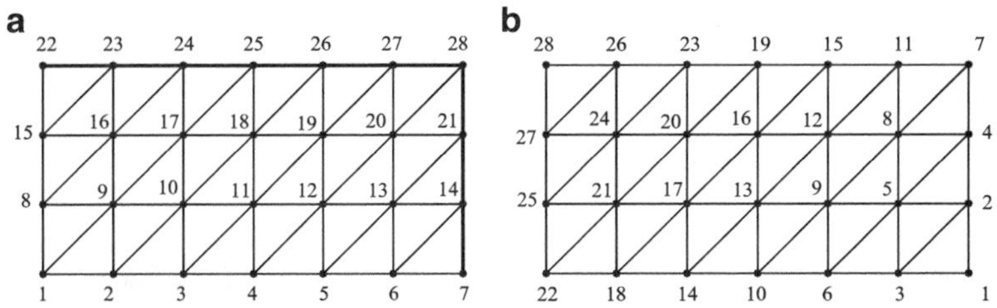

Fig. 5.1 The graph model S and its nodal numbering. (**a**) Initial numbering and the selected transversal. (**b**) Final numbering

$$d_{ij} = \begin{cases} \deg(n_i) & \text{if } i = j, \\ 0 & \text{otherwise.} \end{cases}$$

The *Laplacian matrix* $\mathbf{L}(S) = [l_{ij}]_{N \times N}$ is defined as,

$$\mathbf{L}(S) = \mathbf{D}(S) - \mathbf{A}(S); \tag{5.7}$$

therefore, the components of **L**(S) are given as:

$$l_{ij} = \begin{cases} -1 & \text{if } n_l \text{ is adjacent to } n_j, \\ \deg(n_j) & \text{if } i = j, \\ 0 & \text{otherwise.} \end{cases}$$

The node-member incidence matrix $\mathbf{C}(G) = [c_{ij}]_{N \times M}$ for the arbitrarily oriented graph is defined as:

$$c_{ij} = \begin{cases} +1 & \text{if } m_j \text{ points toward } n_i, \\ -1 & \text{if } m_j \text{ points away form } n_i, \\ 0 & \text{otherwise.} \end{cases}$$

Two distinct rows of **C**(S) have non-zero entries in the same column if and only if a member joins the corresponding nodes. These entries are 1 and −1. It can be shown that:

$$\mathbf{L} = \mathbf{C}\mathbf{C}^t \tag{5.8}$$

It can also be shown that **L** is independent of the orientation of the members of the graph.

Hall [9] considered the problem of finding the minimum of the weighted sum,

$$Z = \frac{1}{2}\sum\nolimits_{i,j} (x_i - x_j)^2 a_{ij}, \tag{5.9}$$

where a_{ij} are the elements of the adjacency matrix **A**. The sum over all pairs of squared distances between nodes which are connected, and so the solution should result in nodes with large numbers of inter-connection being clustered together.

The above equation can be rewritten as:

$$\begin{aligned}\frac{1}{2}\sum_{i,j}\left(x_i^2 - 2x_ix_j + x_j^2\right)a_{ij} &= \frac{1}{2}\sum_{i,j}x_i^2a_{ij} - \frac{1}{2}\sum_{i,j}2x_ix_ja_{ij} + \frac{1}{2}\sum_{i,j}x_j^2a_{ij} \\ &= \sum_{i,j}x_i^2a_{ij} + \sum_{i,j}x_ix_ja_{ij} = \mathbf{x}^t\mathbf{L}\mathbf{x}.\end{aligned} \tag{5.10}$$

where **L** is the Laplacian. Hall also supplied the condition that $\mathbf{x}^t\mathbf{x} = 1$, i.e. the distances are normalized. Using Lagrange multiplier, we have,

$$Z = \mathbf{x}^t\mathbf{L}\mathbf{x} - \lambda\mathbf{x}^t\mathbf{x}, \tag{5.11}$$

and to minimize this expression, the derivative with respect to **X** is taken as,

$$\mathbf{L}\mathbf{x} - \lambda\mathbf{x} = 0, \tag{5.12}$$

or

$$\mathbf{L}\mathbf{x} = \lambda\mathbf{x}, \tag{5.13}$$

which is the eigenvalue equation. The smallest eigenvalue of **L** is $\lambda_1 = 0$, and the corresponding eigenvector $\mathbf{y}_1$ has all its normalized components equal to 1. The second eigenvalue λ_2, and the associated eigenvector $\mathbf{y}_2$ have many interesting properties, which will be used for nodal numbering in this chapter, and for domain decomposition in Chap. 8.

In order to get a feeling of the magnitude of $\lambda_2 = \alpha(S)$, known also as the *algebraic connectivity* of a graph, some simple theorems are restated from the results of Fiedler [10] in the following:

1. For a complete graph K_N with N nodes, $\alpha(K_N) = N$.
2. If $S_1 \subseteq S_2$ (S_1 and S_2, have the same nodes), then $\alpha(S_1) \leq \alpha(S_2)$.
3. Let S be a graph, let S_1 arise from S by removing k nodes from S and all adjacent members, then

$$\alpha(S_1) \geq \alpha(S) - k. \tag{5.14}$$

4. For a non-complete graph S,

$$\alpha(S) \leq v(S) \leq e(S), \tag{5.15}$$

where v(S) and e(S) are the node connectivity and edge connectivity of S, respectively. The *node connectivity* of a graph S is the smallest number of nodes whose removal from S, along with members incident with at least one of the removed nodes, leaves either a disconnected graph or a graph with a single node. The *edge connectivity* of S is the smallest number of edges whose removal from S, leaves a disconnected graph or a graph with one node. As an example, the node and edge connectivity of a complete graph K_N is equal to N−1.

5. For a graph with N(S) nodes

$$\alpha(G) = \lambda_2 \leq \frac{N}{N-1}\min\{\deg(n); \quad n \in N(G)\}. \tag{5.16}$$

and the largest eigenvalue has the following bound:

$$\lambda_N \geq \frac{N}{N-1}\max\{\deg(n); \quad n \in N(S)\}. \tag{5.17}$$

6. Let U be the set of all real N-tuple $\mathbf{x}$ such that $\mathbf{x}^t\mathbf{x} = 1$ and $\mathbf{x}^t\mathbf{e}_N = 0$. From the theory of symmetric matrices, the following characterization for $\alpha(S)$ is obtained,

$$\alpha(S) = \min\{\mathbf{x}^t\mathbf{L}\mathbf{x} | \mathbf{x} \in U\}, \tag{5.18}$$

where

$$e_N = \{1, 1, \ldots, 1\}^t. \tag{5.19}$$

7. The following theorem is interesting since it relates the properties of the adjacency matrix **A** of a graph to those of its Laplacian matrix **L**. Such theorems may establish firm relationships between the application of the largest eigenvalue and eigenvector of **A** for ordering to the second smallest eigenvector and eigenvalue of the Laplacian matrix **L** of the graph for ordering and partitioning.

Theorem. Let S be a graph with adjacency matrix A and Laplacian matrix **L**. Let D and d be the maximum and minimum node degrees of S, respectively. The second largest eigenvalue μ_2 of **A** and the second smallest eigenvalue λ_2 of **L** are then related as:

$$\delta - \lambda_2 \leq \mu_2 \leq \Delta - \lambda_2. \tag{5.20}$$

Proof. μ_2 is the second largest eigenvalue of A, and $\delta - \lambda_2$ is the second largest eigenvalue of $\delta\mathbf{I} - \mathbf{L} \leq \mathbf{A} - (\text{diag}(\deg(v)) - \delta\mathbf{I})$, which differs from A only on the diagonal, where the non-negative values deg(v)−δ are subtracted. Consequently, $\delta - \lambda_2 \leq \mu_2$. In a similar way also the other inequality is obtained.

Lemma. If S is not a complete graph, then $\mu_2 \geq 0$ and $\lambda_2 \leq \Delta$.

5.3.2 Nodal Numbering Algorithm

Based on the concepts presented in the previous section, the method can be described as follows:

Step 1: Construct the Laplacian matrix **L**(S) of the given graph S.
Step 2: Compute the second eigenvalue λ_2 of L and its corresponding eigenvector $\mathbf{y}_2$. Different methods are available for such calculation. Paulino et al. [11] used a special version of the subspace iteration method. However, the algorithm of Lanczos described in the next chapter can also be efficiently applied; $\mathbf{y}_2$ is also known as Fiedler vector.
Step 3: Reorder the nodes of S in ascending order of the vector components in y_2.

Similar to the previous algebraic method, this algorithm has the advantage of using global information of the graph model. However, although it does not use the pseudo-peripheral nodes and SRT and its transversal, its efficiency is very sensitive to the initial ordering of the nodes of the model. Preconditioning by pre-ordering can be utilised for improving the running time of the method, resulting on some kind of dependency on graph-theoretical properties.

5.3.3 Example

Consider a graph with 12 nodes, as shown in Fig. 5.2a, with an arbitrary nodal numbering.

The Laplacian matrix **L**(S) is constructed and its eigenvalue λ_2 and eigenvector $\mathbf{y}_2$ are calculated as follows:

$$\lambda_2 = 1.1071,$$

$$\mathbf{y}_2 = \left\{ \begin{matrix} -0.0608, -0.2023, 1.0000, -0.5303, 0.0658, -0.4721, 0.3099, 0.3106, \\ -0.2399, 0.5829, -0.3125, -0.4514 \end{matrix} \right\}^t .$$

Using $\mathbf{y}_2$, the new labelling is obtained, as illustrated in Fig. 5.2b.

This method can also applied to finite element nodal numbering, using any of the ten graphs defined in Chap. 5.

5.4 A Hybrid Method for Ordering

In this method, the advantages of both graph and algebraic graph methods are incorporated into an algorithm for ordering. In the algebraic graph method, general approaches are used to calculate the eigenvalues and eigenvectors, and the information available from the connectivity of their graph models are ignored. This is

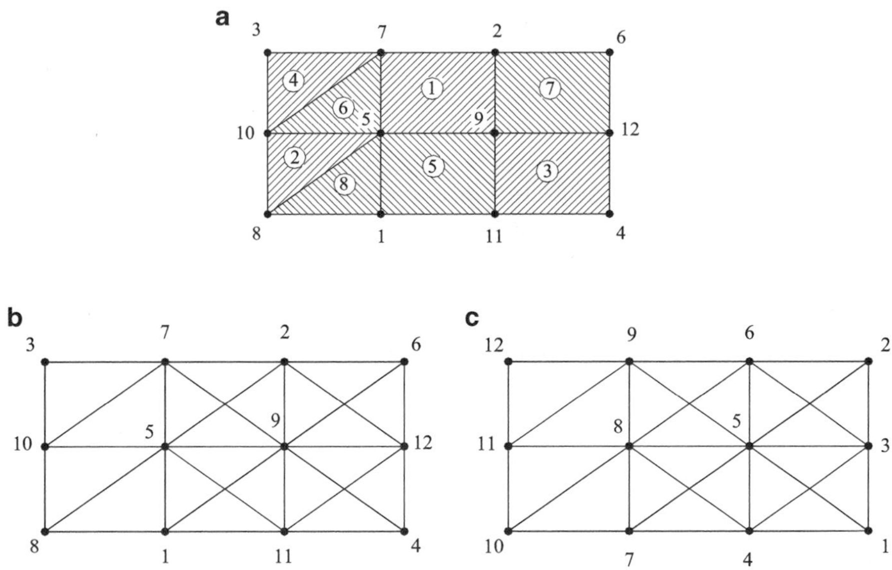

Fig. 5.2 A graph G. (**a**) A simple finite element model. (**b**) Numbering before ordering. (**c**) Numbering after ordering

why the computational time and complexity of these algorithms are not low enough to compete with pure graph theory methods. In this section, graph parameters are used to increase the efficiency of the algebraic graph theory approaches. Typical graph parameters can be taken as the degrees of the nodes, the 1-weighted degrees of the nodes, the distances of the nodes from two pseudo-peripheral nodes, and 2-weighted degrees of the nodes of the graph.

The algebraic graph theory method employed here is not the same as those employed in a general eigenproblem, but rather a specific method is used in which the valuable features of graph parameters are incorporated.

5.4.1 Development of the Method

Here, the graph parameters are considered as Ritz vectors, and the first eigenvector of the complementary Laplacian matrix $\mathbf{L}_c$ (Fiedler vector) is considered as a linear combination of Ritz vectors. The coefficients for these vectors are in fact the weights of the graph parameters, which are usually determined either by heuristic approaches or by experience.

Consider the following vector,

$$\overline{\phi} = \sum_{i=1}^{p} w_i \mathbf{v}_i, \tag{5.21}$$

where $\overline{\phi}$ is an approximation to the Fiedler vector, $\mathbf{v}_i$ (i = 1, ..., p) are the normalized Ritz vectors representing the graph parameters, and w_i (i = 1,..., p) are the coefficients of the Ritz vectors (Ritz coordinates) which are unknowns, and p is the number of parameters being employed. Equation 5.21 can be written as,

$$\overline{\phi} = \mathbf{vw}, \tag{5.22}$$

where $\mathbf{w}$ is a p×1 vector and $\mathbf{v}$ is an N×p matrix containing the Ritz vectors.

Consider the eigenproblem of the complementary Laplacian as:

$$\mathbf{L}_c\phi = \rho\phi. \tag{5.23}$$

Approximating ϕ by $\overline{\phi}$ and multiplying by $\mathbf{v}^t$, results in,

$$\mathbf{v}^t\mathbf{L}_c\mathbf{vw} = \rho\mathbf{v}^t\mathbf{vw}, \tag{5.24}$$

or

$$\mathbf{Aw} = \rho\mathbf{Bw}, \tag{5.25}$$

where $\mathbf{A} = \mathbf{v}^t\mathbf{L}_c\mathbf{v}$ and $\mathbf{B} = \mathbf{v}^t\mathbf{v}$. Both $\mathbf{A}$ and $\mathbf{B}$ are p×p matrices and therefore Eq. 5.21 has much smaller dimension compared to Eq. 5.23; ρ is the approximate eigenvalue of the original problem.

Solution of the reduced problem, with dimensions far less than the original one, results in the first eigenvector $\mathbf{w}_1$ and hence $\overline{\phi}$. Nodal ordering is then performed considering the relative entries of $\overline{\phi}$ in an ascending order.

The present methods lead not only to a set of suitable coefficients for graph parameters, but also provide efficient means for measuring the relative significance of each considered graph parameter. These coefficients may also be incorporated in the design of other specific graph-theoretical algorithms for ordering.

5.4.2 *Numerical Results*

Many examples are studied and the results for three models are presented in this section. In the tables presented, Column 2 contains the results of the Pure Algebraic Graph Method (PAGM) of Ref. [11].

For the first case, four vectors, representing Ritz vectors are considered. For these vectors, $\mathbf{v}_1$ contains the degrees of the nodes, $\mathbf{v}_2$ comprises of the 1-weighted degrees of the nodes, and $\mathbf{v}_3$ and $\mathbf{v}_4$ are distances of the nodes from two pseudo-peripheral nodes. These nodes can be obtained using different algorithms, Kaveh [5]. The results are provided in column 3 of the tables denoted by $\mathbf{v}^4$.

For the second case, five Ritz vectors are employed. The first four vectors are the same as those of the previous case, and the fifth vector $\mathbf{v}_5$ contains the 2-weighted degrees of the nodes of the graph. The results are provided in column 4 of the tables labelled as $\mathbf{v}^5$.

It should be noted that other vectors containing graph properties which influence the ordering may be considered additional to the above five vectors. However, the formation of such additional vectors may require some extra computational time, reducing the efficiency of the algorithm.

Example 1. An FE mesh with one opening comprising of 1,248 nodes and 1,152 rectangular elements is considered as shown in Fig. 5.3. The results for different methods and their computational time are illustrated in Table 5.2, for comparison of their efficiency.

Example 2. An H-shaped FE mesh comprising of 2,096 nodes and 3,900 triangular elements is considered as shown in Fig. 5.4. The results for different methods and their computational time are illustrated in Table 5.3, for comparison of their efficiency.

Example 3. A two dimensional FEM of a tunnel comprising of 6,888 nodes and 6,720 rectangular elements is considered as shown in Fig. 5.5. The results of using different methods and their computational time are presented in Table 5.4.

Example 4. An FE mesh with four openings comprising of 748 nodes and 1,236 triangular elements is considered as shown in Fig. 5.6. The results for different methods and their computational time are illustrated in Table 5.5, for comparison of their efficiency.

Example 5. A three-dimensional finite element model of a nuzzle is considered, as shown in Fig. 5.7. This model contains 4,000 rectangular shell elements. The results for different methods and their computational time are illustrated in Table 5.6 in order to compare their efficiency.

5.4.3 Discussions

The performance of the hybrid method, compares well with a pure algebraic graph method, with a substantial reduction in the computational time. Naturally, addition of extra graph parameters will increase the computational time required. Relative values of the coefficients of the Ritz vectors show the importance of the corresponding parameters in the ordering algorithm. For the examples presented in the previous section, the coefficient corresponding to $\mathbf{v}_3$ and $\mathbf{v}_4$ (the distances from the pseudo-peripheral nodes) seem to be more important, since most of the examples have a more or less uniform distribution of nodal degrees. Naturally for

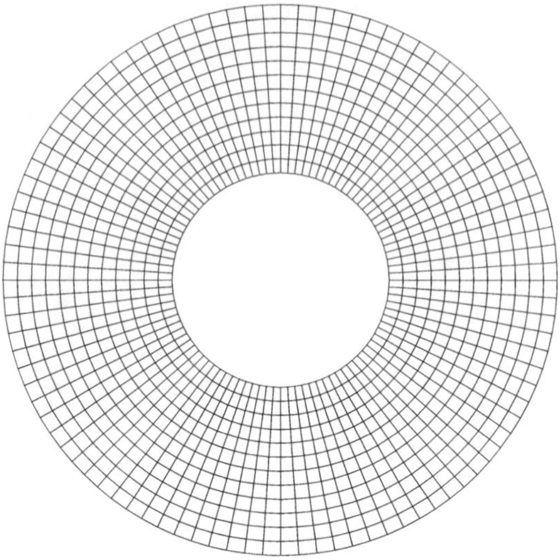

Fig. 5.3 A FE mesh with one opening

Table 5.2 Results of Example 1

	PAGM	v^4	v^5
B	46	43	45
P	34,848	36,243	36,189
$\widetilde{F}$	28.07	29.44	29.25
F_{max}	35	39	39
Time (s)	1,400.3	2.8	2.9

Fig. 5.4 An H-shaped FE mesh

Table 5.3 Results of Example 2

	PAGM	v^4	v^5
B	74	77	77
P	47,741	49,400	48,936
$\tilde{F}$	23.97	25.63	25.32
F_{max}	37	42	42
Time (s)	Large	2.63	2.89

Fig. 5.5 A two dimensional FEM of a tunnel

Table 5.4 Results of Example 3

	PAGM	v^4	v^5
B	455	331	332
P	731,694	733,738	733,738
$\tilde{F}$	112.99	112.93	112.93
F_{max}	164	175	175
Time (s)	10.6	27.6	28.9

models with non-uniform degree distributions, the significance of the other graph parameters will also become apparent.

Though only nodal ordering is addressed in here, however, the application of the present method can easily be extended to the element ordering. For this purpose the natural associate graph or the incidence graph of a FE mesh, should be used in place of the element clique graph.

Fig. 5.6 A FEM with four openings

Table 5.5 Results of Example 4

	PAGM	v^4	v^5
B	39	49	47
P	13,118	13,162	13,126
$\tilde{F}$	18.42	18.61	18.56
F_{max}	29	29	29
Time (s)	1,677	1.2	1.3

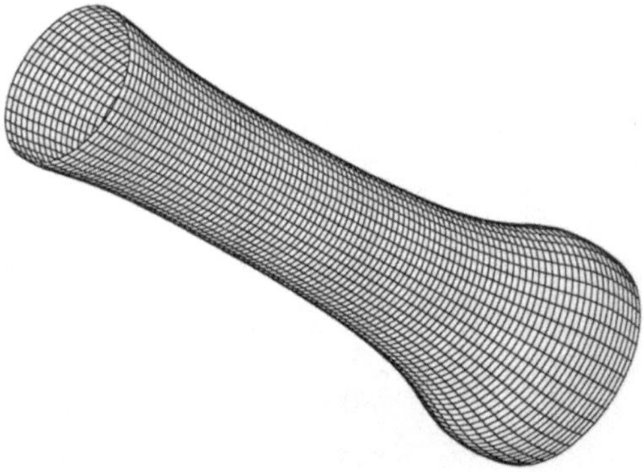

Fig. 5.7 A three dimensional FEM of a nuzzle

Table 5.6 Results of Example 5

	PAGM	v^4	v^5
B	39	49	47
P	13,118	13,162	13,126
$\tilde{F}$	18.42	18.61	18.56
F_{max}	29	29	29
Time (s)	1,677	1.2	1.3

5.5 Ordering via Charged System Search Algorithm

Meta-heuristic algorithms can be used for nodal ordering to structure the stiffness and flexibility matrices. Ant colony optimization is utilized for this purpose by Kaveh and Sharafi [1, 2]. In this section the recently developed meta-heuristic optimization method, known as charged system search (CSS) [12], is employed for optimum nodal ordering to minimize bandwidth and profile of sparse matrices that is based on [3]. This algorithm is presented in Appendix C. The bandwidth and profile of some graph matrices, pattern equivalent to structural matrices, are minimized using this approach.

5.5.1 Charged System Search

In this section, Charged system search (CSS) developed by Kaveh and Talatahari [12] is brifiely described. This is a powerful meta-heuristic algorithm developed for optimization and have found many applications is structural optimisation. Some applications of this algorithm in structural mechanics are briefly discussed in Chap. 10 of this book.

5.5.1.1 Background Definitions

In physics, the space surrounding an electric charge has a property known as the electric field. This field exerts a force on other electrically charged objects. The electric field surrounding a point charge is given by Coulomb's law. Coulomb has confirmed that the electric force between two small charged spheres is proportional to the inverse square of their separation distance r_{ij}. Therefore, this law provides the magnitude of the electric force (Coulomb force) between the two point charges. This force on a charge, q_j at position r_i, experiencing a field due to the presence of another charge, q_j at position r_i, can be expressed as

$$F_{ij} = k_e \frac{q_i q_j}{{r_{ij}}^2} \frac{r_i - r_j}{\left\| r_i - r_j \right\|} \tag{5.26}$$

where k_e is a constant known as the Coulomb constant; r_{ij} is the separation of the two charges (Halliday et al. [13]).

Consider an insulating solid sphere of radius "a" which has a uniform volume charge density and carries a total charge of magnitude q_i. The magnitude of the electric force at a point outside the sphere is defined as Eq. 5.26, while this force can be obtained using Gauss's law at a point inside the sphere as

$$F_{ij} = k_e \frac{q_i q_j}{a^3} r_{ij} \frac{r_i - r_j}{\|r_i - r_j\|} \tag{5.27}$$

In order to calculate the electric force on a charge (q_j) at a point (r_j) due to a group of point charges, the principle of superposition is applied to electric forces as

$$F_j = \sum_{i=1, i \neq j}^{N} F_{ij} \tag{5.28}$$

where N is the total number of charged particles and F_{ij} is equal to

$$F_{ij} = \begin{cases} k_e \dfrac{q_i}{a^3} r_{ij} \dfrac{r_i - r_j}{\|r_i - r_j\|} & \text{if } r_{ij} < a \\ k_e \dfrac{q_i}{r_{ij}^2} \dfrac{r_i - r_j}{\|r_i - r_j\|} & \text{if } r_{ij} \geq a \end{cases} \tag{5.29}$$

Therefore, the resulted electric force can be obtained as [12]

$$F_j = k_e q_j \sum \left(\frac{q_i}{a^3} r_{ij}.i_1 + \frac{q_i}{r_{ij}^2}.i_2 \right) \frac{r_i - r_j}{\|r_i - r_j\|} \quad \left\{ \begin{array}{l} i_1 = 1, i_2 = 0 \Leftrightarrow r_{ij} < a \\ i_1 = 0, i_2 = 1 \Leftrightarrow r_{ij} \geq a \end{array} \right\} \tag{5.30}$$

5.5.1.2 Newtonian Mechanics Laws

Newtonian mechanics studies the motion of objects. In the study of motion, the moving object is described as a particle regardless of its size. In general, a particle is a point-like mass having infinitesimal size. The motion of a particle is completely known if the particle's position in space is known at all times. The displacement of a particle is defined as its change in position. As it moves from an initial position r_{old} to a final position r_{new}, its displacement is given by

$$\Delta r = r_{new} - r_{old} \tag{5.31}$$

The slope of tangent line of the particle position represents the velocity of this particle as

$$v = \frac{r_{new} - r_{old}}{t_{new} - t_{old}} = \frac{r_{new} - r_{old}}{\Delta t} \tag{5.32}$$

The acceleration of the particle is defined as the change in the velocity divided by the time interval Δt during which that change has occurred:

$$\alpha = \frac{v_{new} - v_{old}}{\Delta t} \tag{5.33}$$

The displacement of any object can now be obtained as a function of time as

$$r_{new} = \frac{1}{2}\alpha.\Delta t^2 + v_{old}.\Delta t + r_{old} \tag{5.34}$$

Also according to Newton's second law, we have

$$F = m.\alpha \tag{5.35}$$

where *m* is the mass of the objective. Substituting Eq. 5.35 in Eq. 5.34, we have

$$r_{new} = \frac{1}{2}\frac{F}{m}.\Delta t^2 + v_{old}.\Delta t + r_{old} \tag{5.36}$$

5.5.1.3 The Rules of the Charged System Search

In this section, the recently developed optimization algorithm in [12] is briefly presented utilizing the aforementioned physics laws, which is called Charged System Search. In the CSS, each solution candidate X_i containing a number of decision variables (i.e. $X_i = \{x_{i,j}\}$) is considered as a charged particle. The charged particle is affected by the electrical fields of the other agents. The quantity of the resultant force is determined by using the electrostatics laws, and the quality of the movement is determined using the Newtonian mechanics laws. Thus an agent with good results must exert a stronger force than the bad ones, so the amount of the charge will be defined considering the objective function value, fit(i). In order to introduce CSS, the following rules are introduced:

Rule 1: In CSS each CP has a magnitude of charge (q_i) and as a result creates an electrical field around its space. The magnitude of the charge is defined considering the quality of its solution, as follows:

$$q_i = \frac{\text{fit}(i) - \text{fit}_{best}}{\text{fit}_{best} - \text{fit}_{worst}} \quad i = 1, 2, \ldots, N \tag{5.37}$$

where fit_{best} and fit_{worst} are the so far best and the worst fitness of all particles; fit(i) represents the objective function value or the fitness of the agent *i*; and *N* is the total number of CPs. The separation distance r_{ij} between two charged particles is defined as follows:

$$r_{ij} = \frac{\left\| X_i - X_j \right\|}{\left\| (X_i + X_j)/2 - X_{best} \right\| + \varepsilon} \tag{5.38}$$

where X_i and X_j are the positions of the *i*th and *j*th CPs, X_{best} is the position of the best current CP, and ε is a small positive number to avoid singularities.

Rule 2: The initial positions of CPs are determined randomly in the search space and the initial velocities of charged particles are assumed to be zero.

Rule 3: Electric forces between any two charged particles are attractive. Utilizing this rule increases the exploitation ability of the algorithm. Though it is possible to define repelling force between CPs as well, however, for our problems this seems to be unnecessary. When a search space is a noisy domain, having a complete search before converging to a result is necessary; in such conditions the addition of the ability of repelling forces to the algorithm may improve its performance.

Rule 4: Good CPs can attract the other agents and bad CPs repel the others, proportional to their rank, that is

$$c_{ij} \propto \text{rank}(CP_j) \wedge \begin{cases} 0 < c_{ij} \le +1 & \text{if the CP is above average} \\ -1 \le c_{ij} < 0 & \text{if the CP is below average} \end{cases} \tag{5.39}$$

where c_{ij} is a coefficient determining the type and the degree of influence of each CP on the other agents, considering their fitness and apart from their charges. This means that good agents are awarded the capability of attraction and bad ones are given the repelling feature, which will improve the exploration and exploitation abilities of the algorithm. On the one hand, when a good agent attracts a bad one, the exploitation ability for the algorithm is provided. On the other hand, when a bad agent repels a good CP, the exploration is provided.

Rule 5: The value of the resultant electrical force affecting a CP is determined using Eq. 5.30, as

$$F_j = q_j \sum_{i, i \ne j} \left(\frac{q_i}{a^3} r_{ij}.i_1 + \frac{q_i}{r_{ij}^2}.i_2 \right).c_{ij}.(X_i - X_j) \quad \left\langle \begin{array}{l} j = 1, 2, \ldots, N \\ i_1 = 1, i_2 = 0 \Leftrightarrow r_{ij} < a \\ i_1 = 0, i_2 = 1 \Leftrightarrow r_{ij} \ge a \end{array} \right. \tag{5.40}$$

where F_j is the resultant force acting on the *j*th CP, as illustrated in Fig. 5.8. In this algorithm, each CP is considered to be a charged sphere with radius a having a uniform volume charge density. Here a is set to unity.

Rule 6: The new position and velocity of each CP is determined considering Eqs. 5.32 and 5.36, as follows:

$$x_{j,new} = \text{Fix}\left(rand_{j1}.k_a.\frac{F_j}{m_j}.\Delta t^2 + rand_{j2}.k_v.v_{j,old}.\Delta t + x_{j,old} \right) \tag{5.41}$$

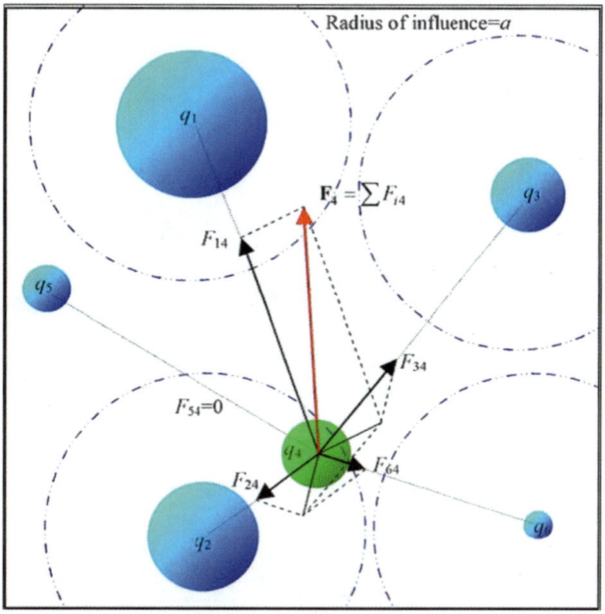

Fig. 5.8 The resultant electrical force acting on a CP [12]

$$v_{j,new} = \frac{x_{j,new} - x_{j,old}}{\Delta t} \tag{5.42}$$

where Fix(X) is a function which rounds each elements of X to the nearest permissible discrete value; k_a is the acceleration coefficient; k_v is the velocity coefficient to control the influence of the previous velocity; and $rand_{j1}$ and $rand_{j2}$ are two random numbers uniformly distributed in the range of (0,1). m_j is the mass of the CPs which is equal to q_j in this chapter. Δt is the time step and is set to one. Figure 5.9 illustrates the movement of a CP to its new position using this rule.

The effect of the pervious velocity and the resultant force acting on a CP can be decreased or increased based on the values of the k_v and k_a, respectively. Excessive search in the early iterations may improve the exploration ability; however, it must be deceased gradually, as described before. Since k_a is the parameter related to the attracting forces, selecting a large value for this parameter may cause a fast convergence, and a small value can increase the computational time. In fact k_a is a control parameter of the exploitation. Therefore, choosing an incremental function can improve the performance of the algorithm. Also, the direction of the pervious velocity of a CP is not necessarily the same as the resultant force. Thus, it can be concluded that the velocity coefficient k_v controls the exploration process and therefore a decreasing function can be selected. Thus, k_v and k_a are defined as

$$k_v = 0.5(1 - iter/iter_{max}), \quad k_a = 0.5(1 + iter/iter_{max}) \tag{5.43}$$

Rule 7: Charged memory (CM) is utilized to save a number of the best so far solutions. Here, the size of the CM is taken as *N/4*. The vectors stored in the CM

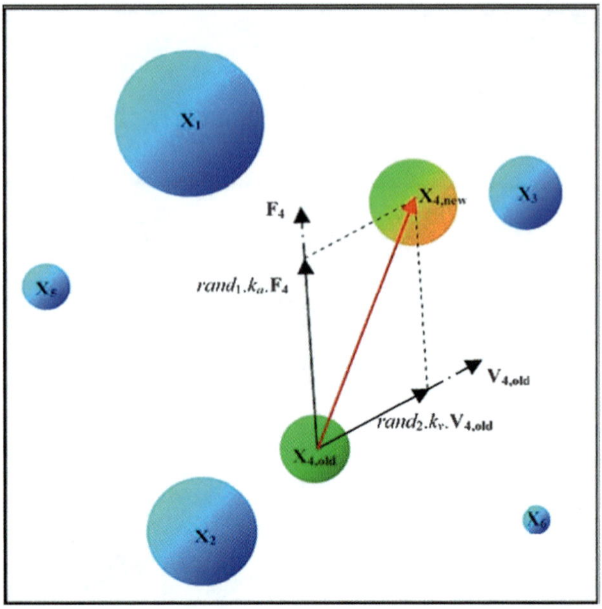

Fig. 5.9 Movement of a CP to its new position [12]

can have influence on the CPs. This may increase the computational cost, and therefore it is assumed that the same number of the worst particles cannot attract others.

Rule 8: The agents violating the limits of the variables are regenerated using the harmony search-based handling approach as described in Ref. [12].

Rule 9: Maximum number of iterations is considered as the terminating criterion.

The general flowchart of the CSS algorithm is illustrated in Fig. 5.10.

5.5.2 The CSS Algorithm for Nodal Ordering

This algorithm attempts to find an optimal assignment for nodal ordering of a graph to reduce the bandwidth or profile of the associated matrix employing a charged system search algorithm. The basis of the algorithms for both bandwidth and profile reduction are identical and it is based on reordering or assigning new labels to the graph nodes to achieve the optimal bandwidth or profile. The only difference is in defining the objective functions. That is, the main procedure of the CSS algorithm for reordering is the same but the objective function for bandwidth reduction and profile reduction are different.

For an $n \times n$ sparse matrix associated to graph G, each permutation of rows and columns leads to a new reordering called the assigned set. If the initial numbering of the graph is $\{1, 2, 3, \ldots, n\}$, each permutation of this set will be a new assigning list.

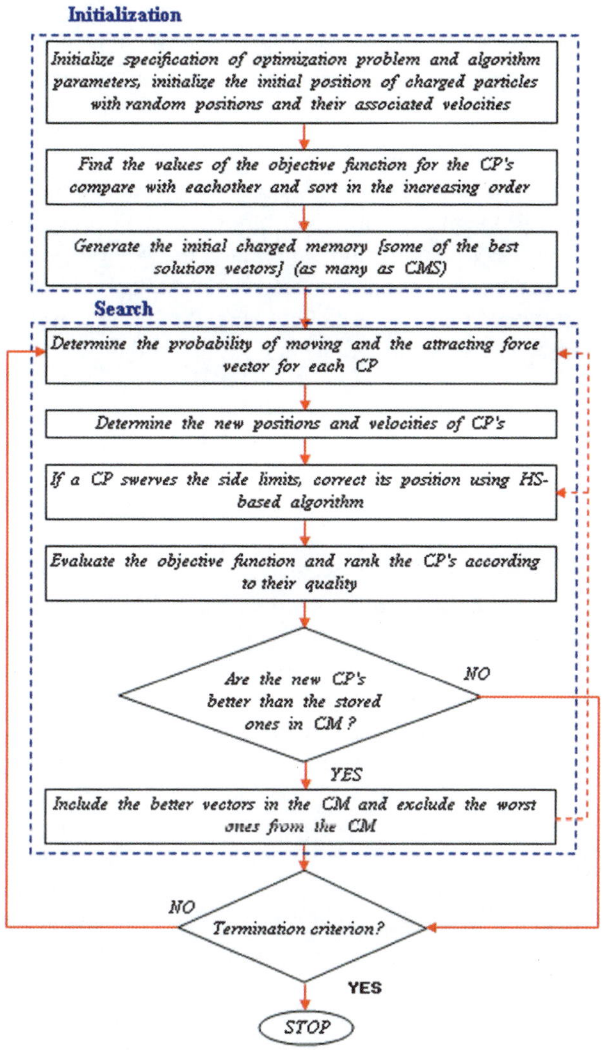

Fig. 5.10 The general flowchart of the CSS algorithm [12]

The aim is to find the optimal assigning list in order to reach the best bandwidth or profile.

Each probable permutation of set {1, 2, 3, . . ., n} is considered to be a potential solution which is called an agent. In CSS these agents are regarded as CPs. In fact, each solution candidate X_i containing a number of decision variables $x_{i,j}$, is considered to be a charged particle and each $x_{i,j}$ presents the number assigned to the node j in the original graph. Thus a solution candidate X_i which represents the position of CP_i, contains n arrays $x_{ij}(j = 1, 2, \ldots, n)$ which stand for the assigned numbers.

The algorithm for nodal ordering follows the above mentioned nine general rules of the CSS algorithms. As stated before, due to the nature of nodal ordering problem, the discrete version of CSS, consisting of nine steps is utilised.

Step 1: The number of CPs, i.e. candidate agents, is determined. For nodal ordering of sparse matrices this number is set to N = [fix(n/100) + 5] which means that for each 100 nodes one additional CP is added and at least 5 CPs are needed for any problems. Using larger number of CPs may result in more accurate results, however it significantly increases the computational time. On the other hand, using smaller number of CPs may leads to undesirable results. The considered number of CPs is capable to keep the balance as a moderate level.

Step 2: The CPs are defined and settled in their initial positions. For this purpose a random permutation of set {1, 2, 3, ..., n} is assigned to each agent as initial candidate solutions. That is, the initial candidate solutions X_i and as a result, their positions $\{x^{(0)}{}_{i,j}\}$ are randomly nominated. In other words, in this phase, N candidate solution $X_i(i = 1, 2, \ldots, N)$ which are located in their positions presented by $x^{(0)}{}_{i,j}$ are defined. (j = 1, 2, ..., n). The initial velocity for all CPs are considered to be zero. $((v^{(0)}{}_{i,j} = 0\ \forall\ ij)$

Step 3: The magnitude of charge for each CP is calculated using Rule 1. For this purpose the objective functions for each agent must be calculated. As mentioned before, this phase is the only distinction between bandwidth and profile reduction algorithm. The objective function for bandwidth reduction is obtained from Eq. 4.8 while for profile reduction it is calculated from Eq. 4.12. In this step when objective functions are calculated, they should be put in order and the best and the worst ones and the best and the worst N/5 agents are saved. This will help the algorithm to judge better in next steps. Then the magnitude of charge for each CP, i.e. q_i, is obtained through the Eq. 5.37.

Step 4: The separation distance between CPs are calculated. In the previous step, the position of each CP is defined by a coordinate of n arrays. Having the X_i for all CPs, the separation distance between them are calculated using the Eq. 5.38. It should be mentioned that in such discrete problems that X_i is an n-dimensional array, the intention of calculating distance between every two CPs is to find how far the two assumed nodes are in the n-dimensional space. In fact, the calculations of distance, velocity and acceleration are all made in a multidimensional space.

Step 5: The type and the degree of influence of each CP on the other agents are determined. For this reason, using the rank of the CPs obtained in step 3, a number between +1 and −1, is assigned to each agent proportional to its rank. That is, the number +1 is assigned to the best agent and −1 to the worst one and so on. Such an assignment leads to improvement of the abilities of exploration and exploitation simultaneously.

Step 6: The value of the resultant electrical force affecting a CP is determined using the Eq. 5.40. Each F_j is an n-dimensional array and shows the tendency of agent j toward other CPs.

Step 7: New position and velocity of each CP is determined considering Eq. 5.41 and 5.42, respectively. In Eq. 5.41 the function Fix(**X**) shifts each $x_{i,j}$ to its nearest position. That is the nearest permissible digit assigns to each $x_{i,j}$. As mentioned before, each new position determined by an n-dimensional array shows the new renumbering of CPs i.e. the new numbers assigned to nodes.

Step 8: The agents violating the limits of the variables are regenerated using the harmony search-based handling approach. Then the best so far solutions are saved.

Step 9: Maximum number of iterations is considered as the terminating criterion

5.5.3 Numerical Examples

In this section, three examples are presented and the results are compared to those of the other algorithm in Table 5.7. Then a comparison is made for convergence rate of different algorithms for each example. For profile reduction, results are compared to those of Sloan [14], King [15], and Kaveh and Sharafi [1–3]; and for bandwidth minimization, the 4-step algorithm of the previous chapter, and an ACO algorithm [1, 2] are used to perform the comparison.

The topological properties of the finite element models are transferred to the connectivity properties of graphs, by the clique graphs [5]. This graph has the same nodes as those of the corresponding finite element model, and the nodes of each element are cliqued, avoiding the multiple edges for the entire graph.

All computations are performed on P9700 @2.40 GHz computer running MATLAB R2009b. In order to ensure that the obtained solution from ACO is global or near global optimum, many runs are performed in parallel. Since each run is fully independent of the others, the program could be run in parallel so that the total execution time practically became the same as required for a single run.

Example 1. Consider a finite element mesh (FEM) of a fan. The element clique graph of this model contain 1,575 nodes as shown in Fig. 5.11. The performance of the CSS algorithm and some other algorithms are tested on this model, and the results are presented in Table 5.7.

Example 2. The FEM of a shear wall with 760 nodes and four openings is shown in Fig. 5.12. Similar to the previous example, the performance of the CSS algorithm and some other algorithms are tested on this model and the results are presented in Table 5.7.

Example 3. An H-shape finite element mesh (FEM) with 4,949 nodes is considered, as shown in Fig. 5.13. The element clique graph of this model contain 4,949 nodes and 9,688 beam element (edges). The performance of the CSS algorithm and some other algorithms are tested on this model, and the results are presented in Table 5.7.

Table 5.7 Comparison of the results

Example	Type of ordering	Initial value	Algorithm	Optimized value	CPU time (s)
1	Profile minimization	144,351	Sloan	31,002	32.7
			King	31,982	24.1
			ACO	29,665	55.9
			CSS	28,770	17.3
	Bandwidth minimization	461	4-step	23	4.9
			ACO	23	10.7
			CSS	21	4.4
2	Profile minimization	37,584	Sloan	19,110	11.1
			King	19,613	9.8
			ACO	19,007	8.3
			CSS	19,232	8.2
	Bandwidth minimization	382	4-step	46	1.8
			ACO	42	4.4
			CSS	41	2.0
3	Profile minimization	345,437	Sloan	210,845	117.8
			King	211,731	98.2
			ACO	208,945	296.6
			CSS	206,649	98.7
	Bandwidth minimization	407	4-step	66	17.7
			ACO	60	29.5
			CSS	58	13,3

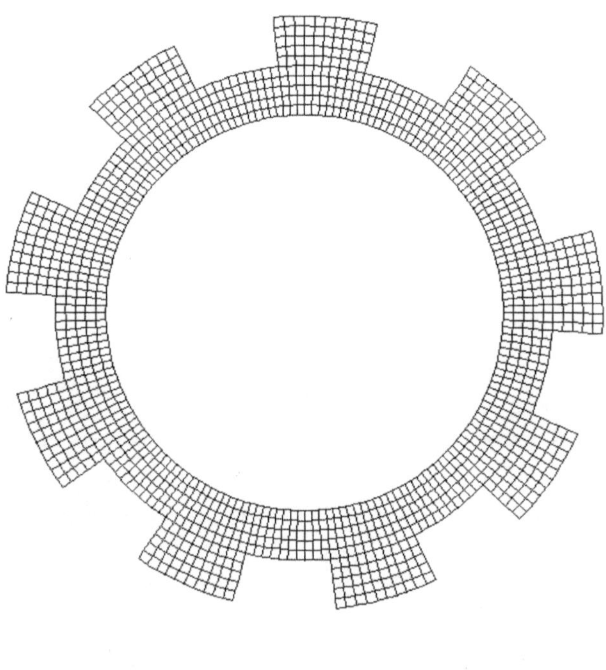

Fig. 5.11 Finite element mesh of a fan represented by its clique graph

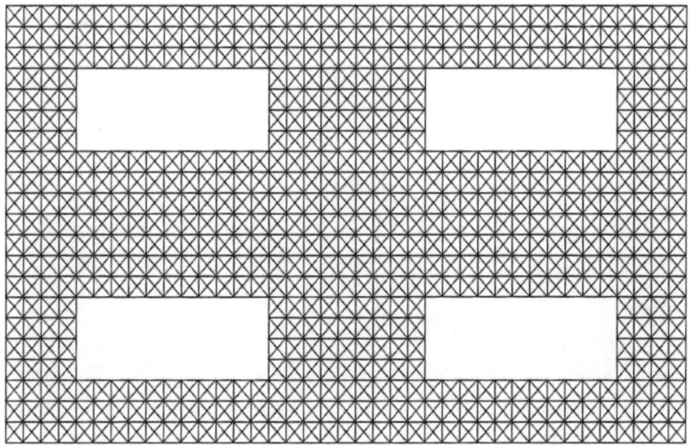

Fig. 5.12 The element clique graph of a rectangular FEM with four opening

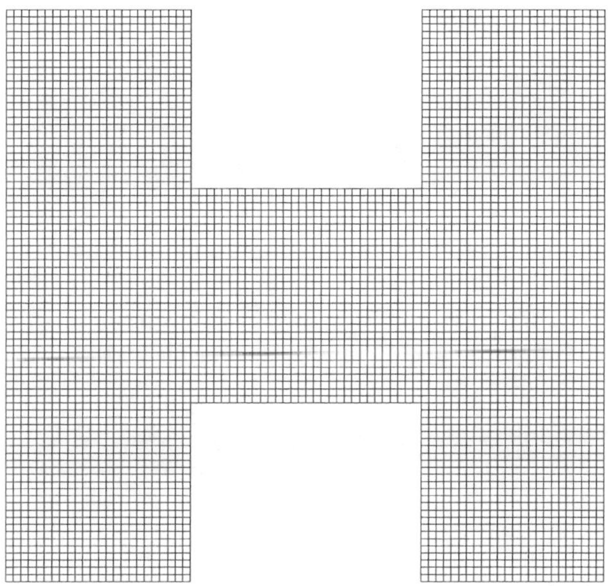

Fig. 5.13 An H-shaped finite element grid

References

1. Kaveh A, Sharafi P (2008) Optimal priority functions for profile reduction using ant colony optimization. Finite Elem Anal Des 44(3):131–143
2. Kaveh A, Sharafi P (2008) Nodal ordering for bandwidth reduction using ant system algorithm. Eng Comput 26(3):313–337
3. Kaveh A, Sharafi P (2012) Ordering for bandwidth and profile minimization problems via charged system search method. Iran J Sci Technol 36:39–52
4. Gould P (1967) The geographical interpretation of eigenvalues. Trans Inst Br Geogr 42:53–58
5. Kaveh A (2004) Structural mechanics: graph and matrix methods, 3rd edn. Research Studies Press, Somerset

6. Grimes RG, Pierce DJ, Simon HD (1990) A new algorithm for finding a pseudo-peripheral node in a graph. SIAM J Anal Appl 11:323–334
7. Schwenk AJ, Wilson RJ (1978) On the eigenvalues of a graph. In: Beineke LW, Wilson RJ (eds) Selected topics in graph theory. Academic Press, New York
8. Straffing PD (1980) Linear algebra in geography; eigenvectors of networks. Math Mag 53:269–276
9. Hall K (1970) R-dimensional quadratic placement algorithm. Manag Sci 17:219–229
10. Fiedler M (1973) Algebraic connectivity of graphs. Czech Math J 23:298–305
11. Paulino GH, Menezes IFM, Gattass M, Mukherjee S (1994) Node and element resequencing using the Laplacian of a finite element graph: part I—general concepts and algorithms. Int J Numer Methods Eng 37:1511–1530
12. Kaveh A, Talatahari S (2010) A novel heuristic optimization method: charged system search. Acta Mech 213(3–4):267–286
13. Halliday D, Resnick R, Walker J (2008) Fundamentals of physics, 8th edn. Wiley, Hoboken
14. Sloan SW (1986) An algorithm for profile and wavefront reduction of sparse matrices. Int J Numer Methods Eng 23:1693–1704
15. King IP (1970) An automatic reordering scheme for simultaneous equations derived from network systems. Int J Numer Methods Eng 2:523–533

Chapter 6
Optimal Force Method for FEMs: Low Order Elements

6.1 Introduction

In this chapter force method finite element models comprising of low order elements are presented.

In the first part an efficient method is developed for the formation of null bases of triangular and rectangular plane stress and plane strain finite element models, corresponding to highly sparse and banded flexibility matrices [1–3]. This is achieved by associating a special graph to the finite element model, and selecting subgraphs (γ-cycles as defined in Chap. 2) for the formation of localized self stress systems (null basis) [4].

In the second part, a graph theoretical method is presented for the formation of sparse, banded and accurate null basis matrices for finite element models with triangular and rectangular plate bending elements [5, 6].

In the third part, a similar approach is extended to 3D models with tetrahedron elements [7].

In the fourth part, an efficient method is presented for the analysis of FEMs composed of brick elements [8]. In this method, special graphs are associated with the considered FEM and minimal subgraphs are selected using these graph models. Localized self-equilibrating systems are constructed on these subgraphs, forming a suitable statical basis of the FEM.

6.2 Force Method for Finite Element Models: Rectangular and Triangular Plane Stress and Plane Strain Elements

In this section an efficient algorithm is presented for the formation of null bases for the models consisting of rectangular and triangular plane stress and plane strain finite element models, corresponding to highly sparse and banded flexibility matrices. The bases obtained by this algorithm require low computational effort leading

A. Kaveh, *Computational Structural Analysis and Finite Element Methods*, 215
DOI 10.1007/978-3-319-02964-1_6,

to highly sparse flexibility matrices with very small bandwidth. Thus using this algorithm, optimal flexibility analysis of such FEMs becomes feasible.

Here, first an interface graph is defined for 2D finite element models and then subgraphs corresponding to self stress systems are generated. By applying unit bi-actions and solving the corresponding statically determinate substructures, some null vectors are obtained. This is repeated for all the selected subgraphs to obtain the null basis. The efficiency of the present method and is illustrated through simple examples.

6.2.1 Member Flexibility Matrices

In the force method of analysis, the determination of the member flexibility matrix is an important step. Methods for the formation of an element with n nodes is already discussed in Sect. 3.2.2 and applied to bar and beam elements.

In this section, the force-displacement relationship is established for plane stress and plain strain problems. Triangular and rectangular elements are considered with constant and linearly varying stress fields, respectively.

Constant Stress Triangular Element. For this element, the nodal forces in global coordinate system have 6 components, as shown in Fig. 6.1a. The element forces are taken as natural forces acting along the sides of the triangle, as shown in Fig. 6.1b.

The nodal forces and element forces are related by projection as,

$$\begin{bmatrix} r_{1x} \\ r_{1y} \\ r_{2x} \\ r_{2y} \\ r_{3x} \\ r_{3y} \end{bmatrix} = \begin{bmatrix} -l_{12} & 0 & l_{31} \\ -m_{12} & 0 & m_{31} \\ l_{12} & -l_{23} & 0 \\ m_{12} & -m_{23} & 0 \\ 0 & l_{23} & -l_{31} \\ 0 & m_{23} & -m_{31} \end{bmatrix} \begin{bmatrix} F_1 \\ F_2 \\ F_3 \end{bmatrix}, \tag{6.1}$$

where l_{ij} and m_{ij} are the direction cosines of the side ij of the triangle.

The element forces are now related to stress resultants, Fig. 6.2. First F_1 is considered as the only natural force acting on the element, and the internal stresses are calculated as:

$$y_{23}\sigma_x + x_{32}\sigma_{xy} = \frac{2l_{12}}{t}F_1 \tag{6.2a}$$

$$-y_{31}\sigma_x + x_{31}\sigma_{xy} = \frac{2l_{12}}{t}F_1 \tag{6.2b}$$

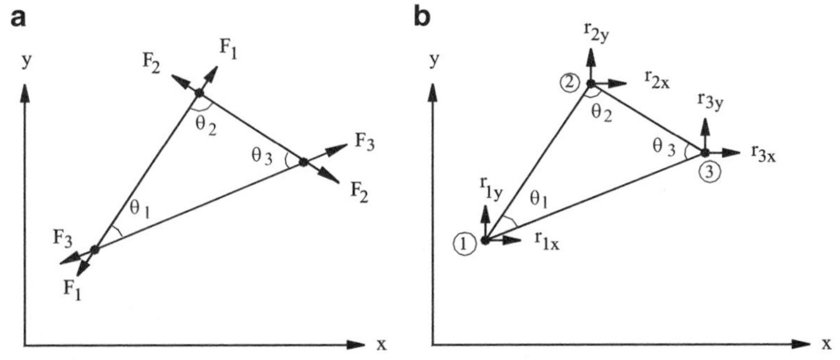

Fig. 6.1 A triangular element. (**a**) Element forces. (**b**) Nodal forces

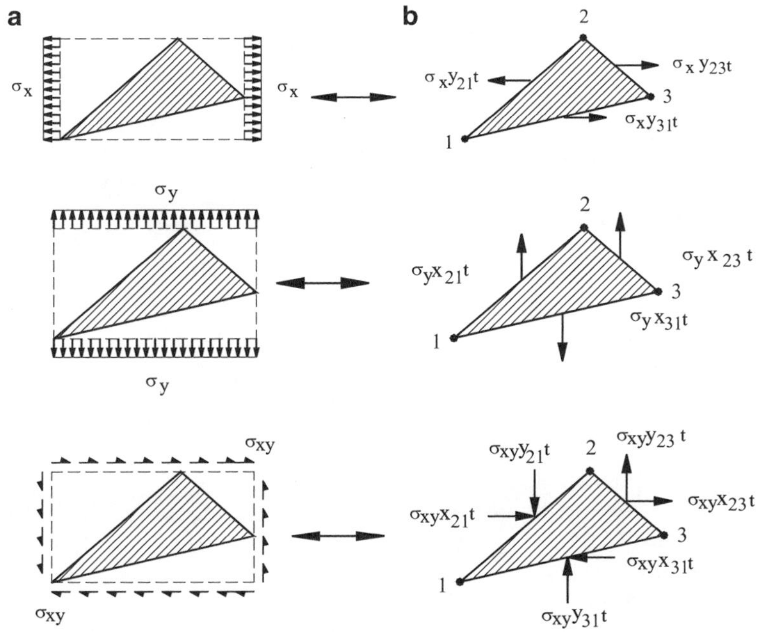

Fig. 6.2 The stress fields and their equivalent nodal forces. (**a**) Stress fields (**b**) Equivalent nodal forces

$$y_{31}\sigma_y + x_{31}\sigma_{xy} = \frac{2m_{12}}{t}F_1 \qquad (6.2c)$$

Solution of the Eqs. (6.2a, b , c) is obtained as follows,

$$\sigma_x = \frac{2l_{12}^2}{th_3}F_1, \sigma_y = \frac{2m_{12}^2}{th_3}F_1, \quad \text{and} \quad \sigma_{xy} = \frac{2l_{12}m_{12}}{th_3}F_1, \tag{6.3}$$

where:

$$\begin{cases} x_{ij} = x_i - x_j \\ y_{ij} = y_i - y_j \end{cases} \quad \text{for} \quad i, j = 1, 2, 3.$$

In the above relations, h_3 is the height of the triangle corresponding to corner 3. Permutation of the indices results in the stresses produced by $\mathbf{F}_2$ and $\mathbf{F}_3$ and in matrix form these equations can be collectively written as,

$$\begin{bmatrix} \sigma_x \\ \sigma_y \\ \sigma_{xy} \end{bmatrix} = \frac{2}{t} \begin{bmatrix} \frac{l_{12}^2}{h_3} & \frac{l_{23}^2}{h_1} & \frac{l_{31}^2}{h_2} \\ \frac{m_{12}^2}{h_3} & \frac{m_{23}^2}{h_1} & \frac{m_{31}^2}{h_2} \\ \frac{m_{12}l_{12}}{h_3} & \frac{m_{23}l_{23}}{h_1} & \frac{m_{31}l_{31}}{h_2} \end{bmatrix} \begin{bmatrix} F_1 \\ F_2 \\ F_3 \end{bmatrix}, \tag{6.4}$$

or

$$\boldsymbol{\sigma} = \bar{\mathbf{c}}\mathbf{F}. \tag{6.5}$$

The matrix $\bar{\mathbf{c}}$ represents statically equivalent stress system due to unit force $\mathbf{F}$. The flexibility matrix of the element can be written as:

$$\mathbf{f}_m = \int_V \bar{\mathbf{c}}^t \boldsymbol{\varphi} \bar{\mathbf{c}} dV. \tag{6.6}$$

The integration is taken over the volume of the element, where,

$$\varphi = \frac{1}{E} \begin{bmatrix} 1 & -\nu & 0 \\ -\nu & 1 & 0 \\ 0 & 0 & 2(1+\nu) \end{bmatrix}, \tag{6.7}$$

is the matrix relating the stresses to strains, $\boldsymbol{\varepsilon} = \boldsymbol{\varphi\sigma}$, in plane stress problems, and E and ν are the Young's modulus and Poisson's ratio, respectively. The force-displacement relationship for a triangular element becomes,

$$\mathbf{u}_m = \mathbf{f}_m \mathbf{r}_m, \tag{6.8}$$

where $\mathbf{u}_m$ and $\mathbf{r}_m$ are the element displacements and element forces, respectively. The flexibility matrix of the element can now be written as,

$$\mathbf{f}_m = \frac{2}{Et}\begin{bmatrix} A(\theta_3,\theta_2,\theta_1) & B(\theta_2) & B(\theta_1) \\ B(\theta_2) & A(\theta_1,\theta_2,\theta_3) & B(\theta_3) \\ B(\theta_1) & B(\theta_3) & A(\theta_2,\theta_1,\theta_3) \end{bmatrix}, \tag{6.9}$$

where t is the thickness of the element, and A and B are functions defined as follows,

$$A(\theta_i,\theta_j,\theta_k) = \frac{\sin\theta_i}{\sin\theta_j \sin\theta_k}, (i,j,k = \text{permutation of } 1,2,3), \tag{6.10a}$$

$$B(\theta_i) = \cos\theta_i \cot\theta_i - \nu\sin\theta_i, (i = 1,2,3), \tag{6.10b}$$

where θ_i, θ_j, and θ_k are the angles of the triangle.

Linear Stress Rectangular Element. For this element, the nodal forces in global co-ordinate system have 8 components, as shown in Fig. 6.3a. The element forces are taken as natural forces along the sides and one diagonal, as shown in Fig. 6.3b. The nodal forces and element forces are related similar to triangular element as,

$$\begin{bmatrix} r_{1x} \\ r_{1y} \\ r_{2x} \\ r_{2y} \\ r_{3x} \\ r_{3y} \\ r_{4x} \\ r_{4y} \end{bmatrix} = \begin{bmatrix} -1 & -\Omega & 0 & 0 & 0 \\ 0 & -\beta\Omega & -1 & 0 & 0 \\ 1 & 0 & 0 & 0 & 0 \\ 0 & 0 & 0 & 0 & -1 \\ 0 & 0 & 0 & -1 & 0 \\ 0 & 0 & 1 & 0 & 0 \\ 0 & \Omega & 0 & 1 & 0 \\ 0 & -\beta\Omega & 0 & 0 & 1 \end{bmatrix} \begin{bmatrix} F_1 \\ F_2 \\ F_3 \\ F_4 \\ F_5 \end{bmatrix}, \tag{6.11}$$

where $\beta = \frac{b}{a}$ and $\Omega = \frac{1}{\sqrt{1+\beta^2}}$

For this element the plane stresses are written as,

$$\begin{cases} \sigma_x = c_1 + c_2\eta \\ \sigma_y = c_3 + c_4\xi \\ \sigma_{xy} = c_5 \end{cases} \tag{6.12}$$

where c_1, c_2, ...,c_5 are constants and,

$$\xi = \frac{x}{a} \quad \text{and} \quad \eta = \frac{y}{b}$$

a and b being the length and width of the element, respectively.

The stress fields and the corresponding nodal forces are shown in Fig. 6.3c.

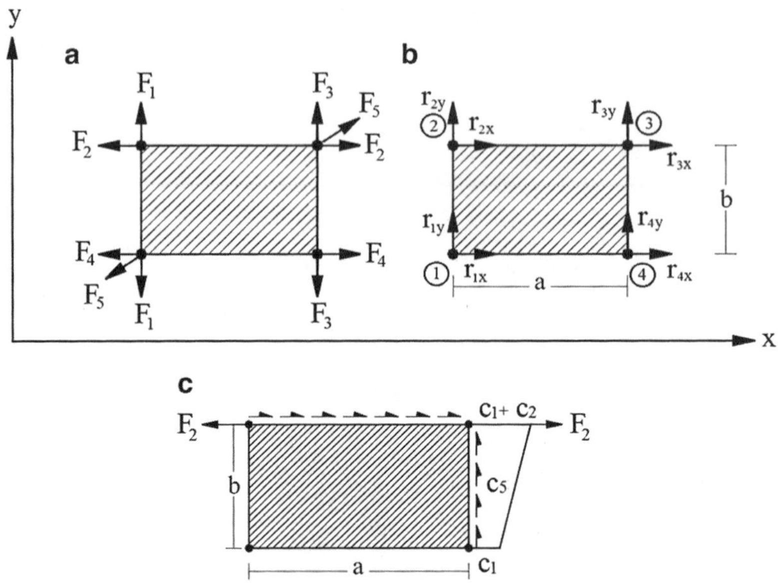

Fig. 6.3 A rectangular element. (**a**) Element forces. (**b**) Nodal forces. (**c**) The stress fields and their equivalent nodal forces

$$\mathbf{f}_{\mathrm{m}} = \frac{1}{\mathrm{Et}} \begin{bmatrix} 4\beta & & & & \\ -\nu & 4/\beta & & \text{sym.} & \\ \dfrac{\beta^2-\nu}{\sqrt{1+\beta^2}} & \dfrac{1-\nu\beta^2}{\sqrt{1+\beta^2}} & \dfrac{1+\beta^2}{\beta} & & \\ -2\beta & -\nu & \dfrac{\beta^2-\nu}{\sqrt{1+\beta^2}} & 4\beta & \\ -\nu & -2/\beta & \dfrac{1-\nu\beta^2}{\beta\sqrt{1+\beta^2}} & -\nu & 4/\beta \end{bmatrix}. \tag{6.13}$$

The unassembled flexibility matrix of the structure can now be formed, by using the above matrix for each element as block diagonal entries. This matrix is incorporated in algebraic force method of the next section.

6.2.2 Graphs Associated with FEMs

In order to transfer the topological property of a finite element model to the connectivity of a graph, ten different graphs are introduced in Ref. [9]. Here, the natural associate graph is used and an additional new graph is defined.

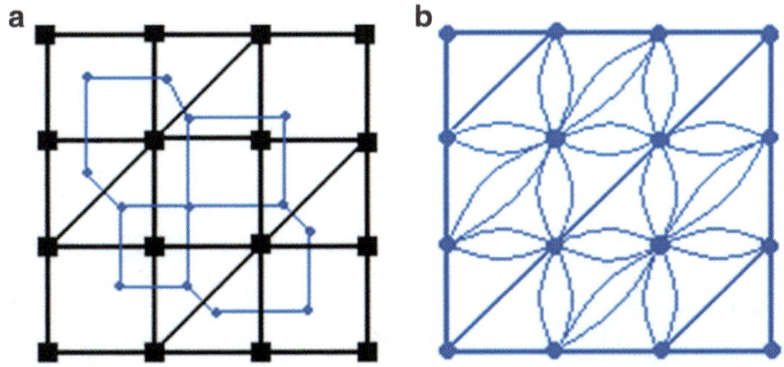

Fig. 6.4 A finite element model, the corresponding natural associate graph, and the interface graph. (**a**) A FEM and its natural associate graph (**b**) Interface graph

Natural associate graph. The natural associate graph of a FEM is constructed as follows:

1. Associate one node of the associate graph with an element of the FEM.
2. Two nodes of the associate graph are connected by a member if the corresponding elements in the FEM have a common edge.

A FEM and the corresponding associate graph are shown in Fig. 6.4a.

Interface graph. The Interface graph of a FEM is constructed by the following rules:

1. This graph contains all the nodes of the FEM.
2. For each edge of a finite element of the model, associate one independent member. Thus, a typical overlap of two elements in FEM is represented by double members in interface graph.
3. One diagonal member is associated with each rectangular element of the model.

Figure 6.4b shows the corresponding interface graph.

The member of the interface graph should be numbered according to the numbering of the FEM. A typical numbering is shown in Fig. 6.5. For each rectangular element like R, five members of the interface graph and for each triangular element like T, three members of interface graph should be numbered consequently. The numbering is performed according to the direction of the independent element forces (Fig. 6.6a, b).

6.2.3 *Pattern Corresponding to the Self Stress Systems*

The nodal forces and independent element forces of a rectangular and triangular element are defined as shown in Fig. 6.3. This is the same convention used by Przemieniecki [10].

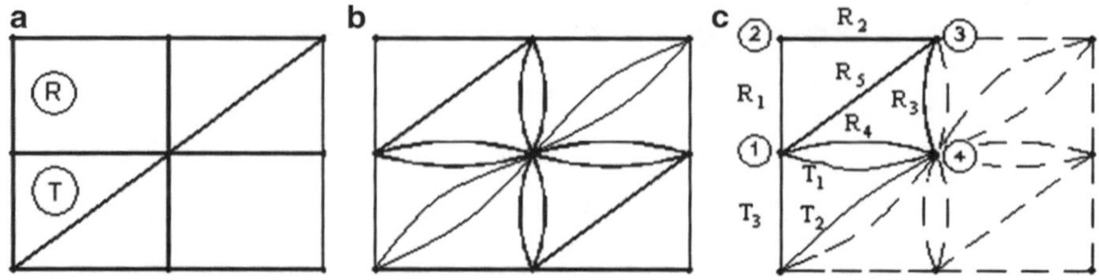

Fig. 6.5 (**a**) A 2D finite element model; (**b**) The interface graph; (**c**) Numbering for a typical rectangular element R and a triangular element T

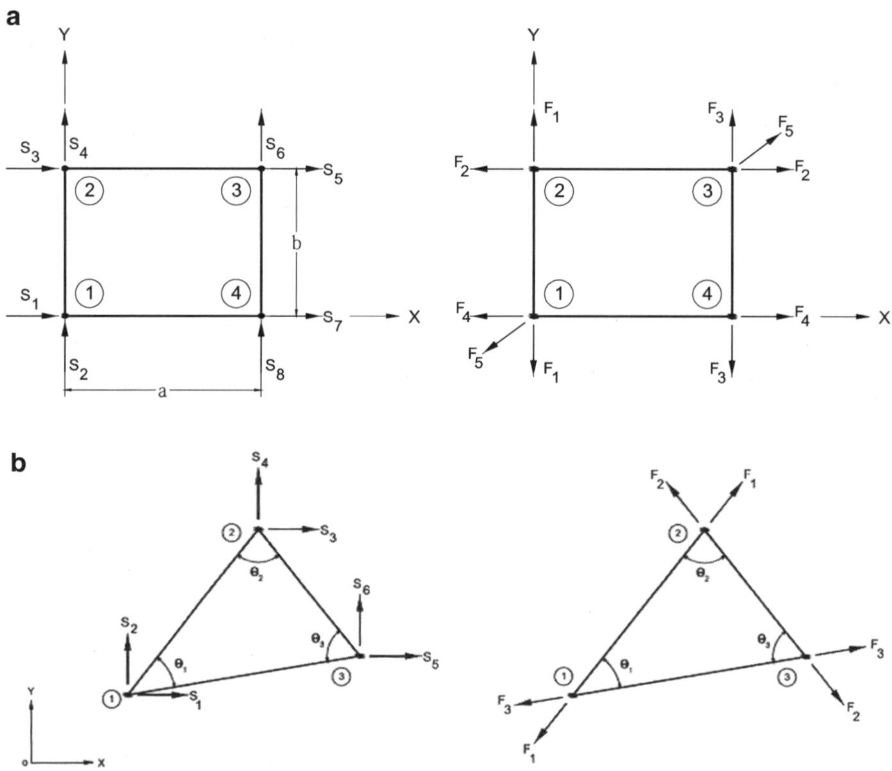

Fig. 6.6 The nodal and element forces for a rectangular element and a triangular element (**a**) The nodal and element forces for a rectangular element; (**b**) The nodal and element forces for a triangular element

Considering Fig. 6.6a, in order to find the patterns corresponding to the self stress systems, a rectangular is simulated as a planar truss formed as the 1-skeleton of the rectangular element together with a diagonal member. This is possible since the independent element forces F_1 to F_5 are applied to the nodes and are along the edges of the rectangular. Also, a FEM with plane strain and plane stress triangular elements can be viewed as a planar truss (Fig. 6.6b). The statical indeterminacy of planar truss with m members and n nodes is given as $\gamma(S) = m - 2n + 3$.

The patterns of the underlying subgraphs of self stress systems (null vectors) are identified as follows:

6.2.3.1 Type I Self Stress System

Each double member of the interface graph is the underlying subgraph of a self stress system. In other words, a double member consisting of members numbered as i and j with ($i < j$), have two non-zero entries in the null basis matrix in the corresponding rows i and j in which the entry in row i is -1 and the entry in row j is $+1$. These double members are called type-1 γ-cycles. The number of these double members is equal to the number of members of the natural associate graph, (see Kaveh [11] for the definition of a γ-cycle).

Using these double members nearly 80 % of the columns of a null basis matrix can easily be generated. For finding these double members one can use the adjacency matrix or the node-member incident matrix of the interface graph.

6.2.3.2 Type II Self Stress System

There are other types of self stress systems in the FEM which are topologically identical to the minimal self stress systems of corresponding planar trusses. The underlying subgraphs of these systems are known as type II γ-cycles, corresponding to the regional cycles of the natural associate graph bounding a single node of the FEM. In other words, if each multiple member from the interface graph is substituted by a member or generators of the Type I self stress systems are removed from S_I, and then the remaining subgraph is a graph, denoted by S.

In general the self stress systems built on S are called Type II self stress systems. In fact these systems are γ-cycles which correspond to minimal cycles of associate graph of finite element model (see Refs. [11] for definition). A finite element model with six elements is shown in Fig. 6.7a, and its associate graph is depicted in Fig. 6.7b.

6.2.3.3 Type III Self Stress System

Each regional cycle bounding a cut-out in the FEM corresponds to a regional cycle of the natural associate graph with 3° of statical indeterminacy, correspond to 3 self stress systems.

For a FEM with n_c cut-outs, apart from the self stress systems corresponding to double members of the interface graph, b_1 (natural associate graph) $- n_c + 3n_c$ additional self stress systems should be generated. This is obvious since for each γ-cycle of S corresponding to non-cut out cycle of natural associate graph one self

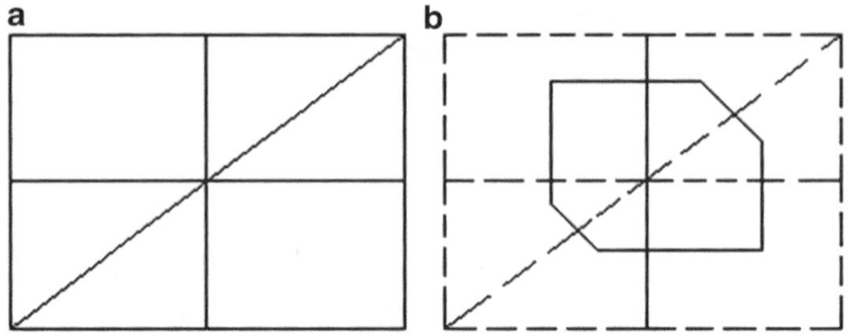

Fig. 6.7 A finite element model with six elements; the corresponding associate graph

stress system, and for each general subgraph corresponding to a cut out cycle of natural associate graph, three independent self stress systems can be generated. Such a general subgraph consists of three independent γ-cycles. In the above relations, b_1 (natural associate graph) is the first Betti number of the natural associate graph of the FEM.

6.2.4 *Selection of Optimal γ-Cycles Corresponding to Type II Self Stress Systems*

Thus far, it is found out that each γ-cycle corresponds to a cycle of the associate graph. Also each cycle with n nodes from A(S) such as c passes through n elements. The subgraph $s_I^c(s_I^c \subset S_I)$ which is relevant to these n elements and cycle c, is a base for the selection of an optimal γ-cycle. Such a subgraph may contain simple and multiple members, where each multiple member with k members corresponds to the overlap of k elements, and each simple member corresponds to the edge of a boundary element. By imposing a special condition on such subgraphs s_I^c, the lists corresponding to optimal γ-cycles can be obtained.

A finite element model with six elements and its associate graph are shown in Fig. 6.8a. The corresponding s_I^c which contains multiple and simple members is illustrated in Fig. 6.8b, and the corresponding γ-cycle is depicted in Fig. 6.8c.

In general, from each s_I^c many γ-cycles (self stress systems) can be extracted, since each simple member of a multiple member can be present in the final graph, while the presence of a simple member in the final graph is obvious.

Thus for obtaining an optimal self stress system, on each s_I^c, two basic selections should be performed which are as follows:

1. Selection of the generator or the last member of a self stress system, which is required for independency of null vectors.
2. Selection of a list of members from the subgraph s_I^c with maximum possible number for the first member. This selection reduces the bandwidth of the null basis matrix considerably.

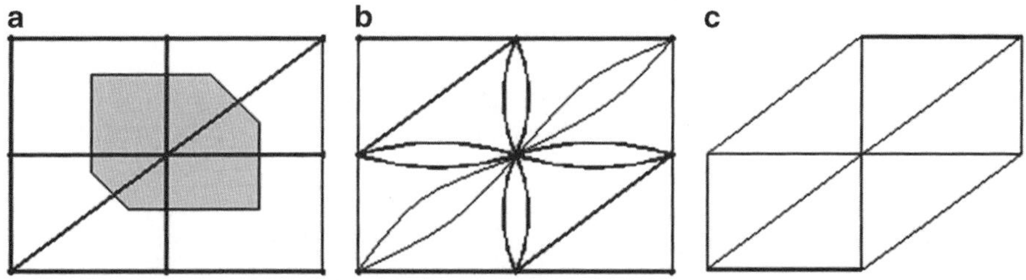

Fig. 6.8 (**a**) A finite element model with its associate graph; (**b**) The corresponding s_I^c; (**c**) The corresponding γ-cycle

The mathematical representation of this selection can be expressed as

$$\text{Minimize}\,(j - i) \tag{6.14}$$

where j is the generator's member number and i is the least member number of current γ-cycle. In the following a simple and fast method is presented for these selections.

6.2.5 Selection of Optimal Lists

First for each subgraph s_I^c we have to delete all the nodes with degree 2 and the members connected to such nodes (Fig. 6.8c) since the numeric values of these members in the corresponding null vector are zero, and therefore they play no role in the formation of a γ-cycle. It should be mentioned that if these members are not omitted then the process of finding a generator will be disturbed. Since each double member corresponds a self stress system which has already been selected, then for independency of a new null vector, each member of a double member which has greater member number cannot be selected as the last member. Then between the lower numbers of double members, the maximum number should be chosen as the last member of the current γ-cycle.

For finding a typical optimal list of members of a γ-cycle two row matrices T_1 and T_2 are considered as follows:

$$T_1 = \{d_1^1, d_2^1, \ldots, s_1, s_2, \ldots, d_i^1, s_j, \ldots, d_k^1\}$$
$$T_2 = \{d_1^2, d_2^2, \ldots, d_i^2, \ldots, d_k^2\}$$

In the above matrices, d_i^2, d_i^1(i = 1, ..., k) are lower and higher member numbers of the double members, in which $d_i^1 < d_i^2$(i = 1, ..., k) and s_j(j = 1, ..., t) are simple members of the selected subgraph. The entries of row matrix, T_2, (d_i^2) show the last members of previous self stress systems. Obviously, the maximum number of the entries of matrix, T_1, $\{d_i^1, s_j\}(i = 1, \ldots, k; j = 1, \ldots, t)$, must be selected as the last member (d_L^1) of this self stress system.

In order to increase the first member number and minimizing the difference between the first member number and the generator of the current self stress system, the following condition can be used.

$$\text{if}\left(\left(d_i^2 > d_i^1\right) \quad \text{and} \quad \left(d_i^1 < d_L^1\right)\right) \quad \text{then} \quad d_i^1 = d_i^2\,(i = 1, \ldots, k) \tag{6.15}$$

After using the above condition, the desired optimal list $\{d_i^1, s_j\}(i = 1, \ldots, k; j = 1, \ldots, t)$ will be obtained.

After finding optimal lists corresponding to type II γ-cycles, using relevant equilibrium submatrix, numerical values for each null vector are calculated.

The list corresponding to the remaining subgraph will have DSI equal to 3. Three null vectors corresponding to such cycles will be obtained directly from the equilibrium submatrix which leads to suboptimal basis.

For type III γ-cycles, finding an optimal list is a time consuming process and considering the fact that the number of cut outs is low in the real structures, the use of this process is not economical for improvement of the final null basis. Thus for each cycle of this type, graph s_I^c is decomposed and all members corresponding to Type I and Type II self stress systems and all the nodes of degree 2 are removed.

Algorithm

Step 1: Generate the associate graph of finite element model and use an efficient method for its node numbering. It is obvious that a good numbering of this graph corresponds to a good numbering of the elements of a finite element model. This numbering leads to a banded adjacency matrix of the graph and correspondingly to a banded flexibility matrix. Since numbering the members of the interface graphs correspond to the element numbering of the finite elements, therefore such a numbering is the only parameter for controlling the bandwidth of the flexibility matrix.

Step 2: Setup the equilibrium matrix of the finite elements model.

Step 3: Generate the interface graph and perform its numbering. The numbering of this graph should be performed according to the element numbering of the considered finite elements model. After this numbering, the interface graph can easily be formed and its members can be numbered.

Step 4: Find the Type I self stress systems. All multiple members of interface graph are identified and the values −1 and 1 are assigned to appropriate rows (corresponding to the member numbers) and the corresponding null vectors are created.

Step 5: Find the Type II self stress systems. Using the Type I and Type II minimal cycles of the associate graph, the subgraphs s_I^c relevant subgraphs are identified and their corresponding optimal lists are found.

Step 6: Calculate numerical values of the optimal lists. Using optimal lists selected in Step 5, null vectors corresponding to the Type I and Type II cycles are calculated from the relevant equilibrium submatrix. For each generator, unit load is applied at a cut in the generator and the internal forces are calculated to form a null vector.

Step 7: Order the null vectors. At this step the constructed null vectors should be ordered such that their generators form a list with an ascending order.

In the following the efficiency of this algorithm is demonstrated using numerical examples and a comparison is made through the results of the present algorithm and the LU factorization method. The comparisons are confined to those of sparsity, condition number and computational time of the formation of the flexibility matrices. It should be noted that all the algebraic methods use LU decomposition approach for the formation of the null basis or controlling the independence of the columns of the equilibrium matrix.

6.2.6 *Numerical Examples*

In this section examples with different topological properties are studied. The models are assumed to be supported in a statically determinate fashion. The effect of the presence of additional supports can separately be included for each special case with no difficulty.

Example 1. A beam with one opening which is supported in a statically determinate fashion is depicted in Fig. 6.9. This structure is also discretized using rectangular and triangular finite elements. The properties of the model are:

Number of rectangular elements = 16, E = 2e + 8 kN/m^2, $\nu = 0.3$
Number of Triangular elements = 16, t = 0.02 m, $n_c = 1$
Number of type I self stress systems = 44 (76 %), Number of nodes = 36
Number of type II self stress systems = 12, $DSI_T = 59 = (44 + 12 + 3)$

Pattern of the equilibrium matrix, the null basis matrices and the corresponding flexibility matrices for the present algorithm are illustrated in Figs. 6.10, 6.11, and 6.12, respectively.

Comparison of the results of the displacement method and the present force method can be found in Ref. [4].

Example 2. A finite element model which is supported in a statically determinate fashion is depicted in Fig. 6.13. This structure is also discretized using quadrilateral and triangular finite elements. The properties of the model are:

Number of quadrilateral elements = 66, E = 2e + 8kN/m^2, $\nu = 0.3$
Number of Triangular elements = 44, t = 0.1 m
Number of type I self stress systems = 172, Number of nodes = 115
Number of type II self stress systems = 63, $DSI_T = 235 = (172 + 63)$

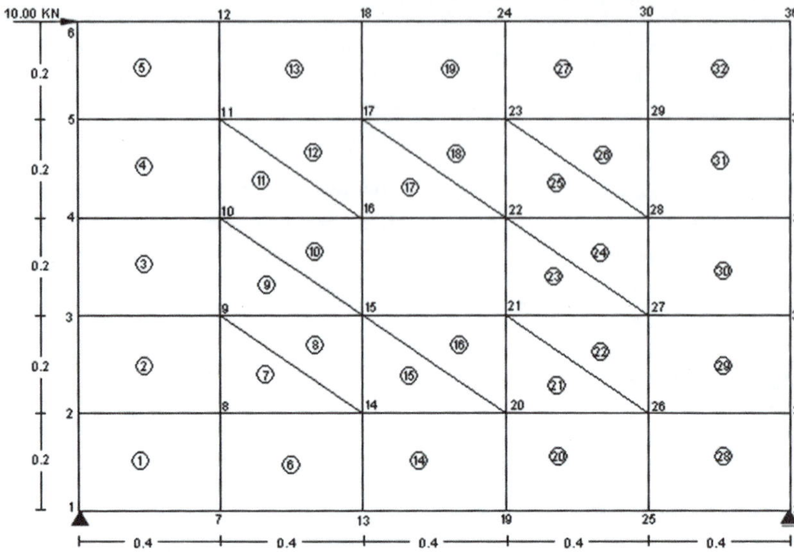

Fig. 6.9 A beam and the discretization of the selected part

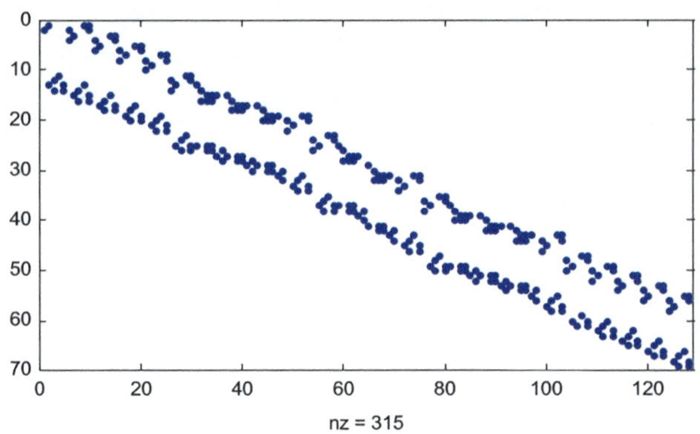

Fig. 6.10 Pattern of the equilibrium matrix

Pattern of the null basis matrix with 1,100 entries for the present method is shown in Fig. 6.14. Pattern of the flexibility matrix using the present algorithms are illustrated in Fig. 6.15.

Example 3. A circular disk, shown in Fig. 6.16, is analyzed using plane stress triangular elements with the following properties:

Diameter = 4.4 m, thickness = 0.05 m, E = 2e + 8 kN/m^2, $\nu = 0.3$,
Number of triangular elements = 312, Number of nodes = 169,
Number of members of the natural associate graph = 456 (type-1 S.E.Ss),
First Betti number of the natural associate graph = 145 (type-2 S.E.Ss),
DSI = 601 = (456 + 145).

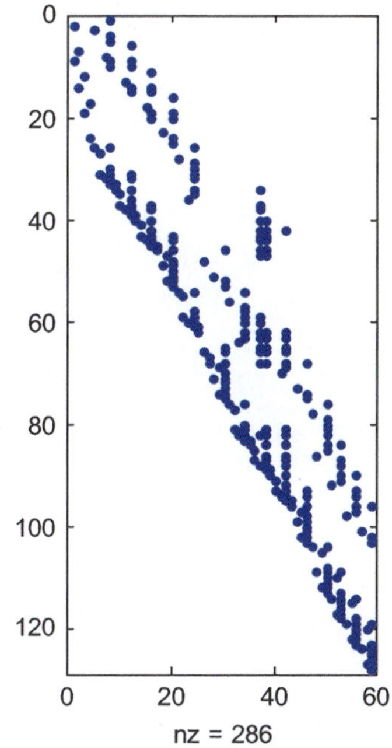

Fig. 6.11 Pattern of the null bases matrix

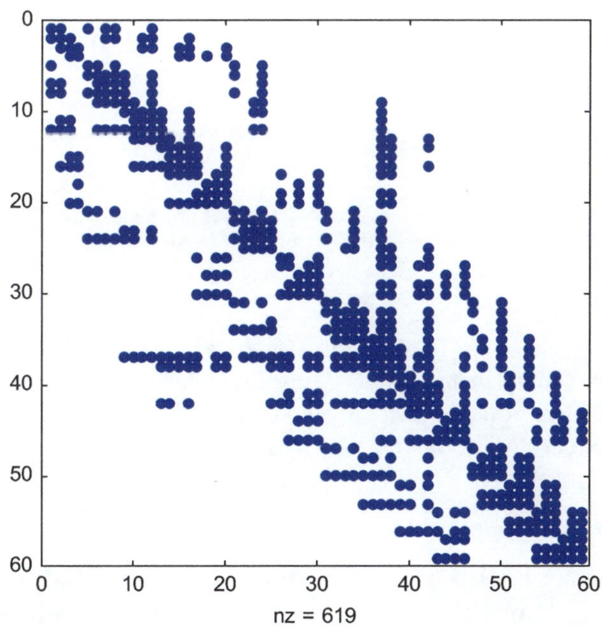

Fig. 6.12 Pattern of the flexibility matrix

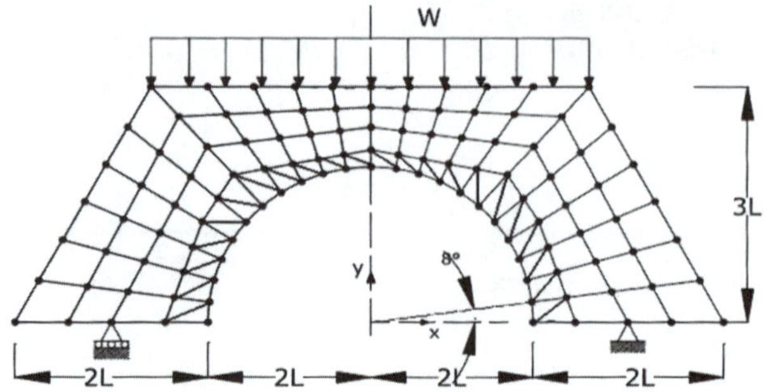

Fig. 6.13 A finite element model with qudrilateral and triangular elements

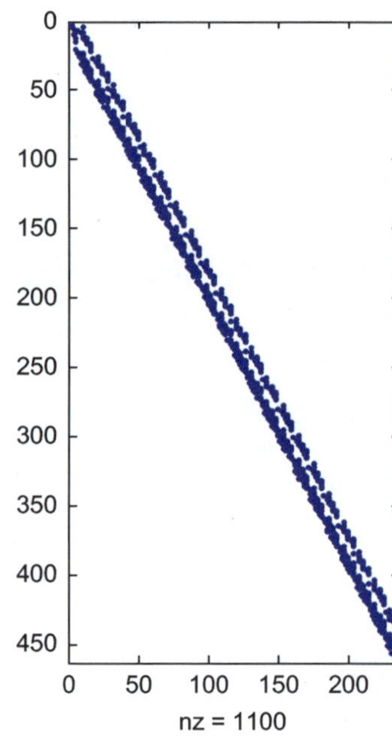

Fig. 6.14 Pattern of the null bases matrix

Patterns of the null basis matrices are shown in Fig. 6.17, and pattern of the flexibility matrix using the present algorithm is illustrated in Fig. 6.18. For LU factorization the null basis contains 11,014 entries, while the present method leads to only 2,623 entries.

6.3 Finite Element Analysis Force Method: Triangular and Rectangular Plate Bending Elements

In this section, an efficient algorithm is presented for the formation of null bases for finite element models consisting of triangular and rectangular plate bending elements [5]. The null bases obtained by this algorithm are highly sparse and narrowly

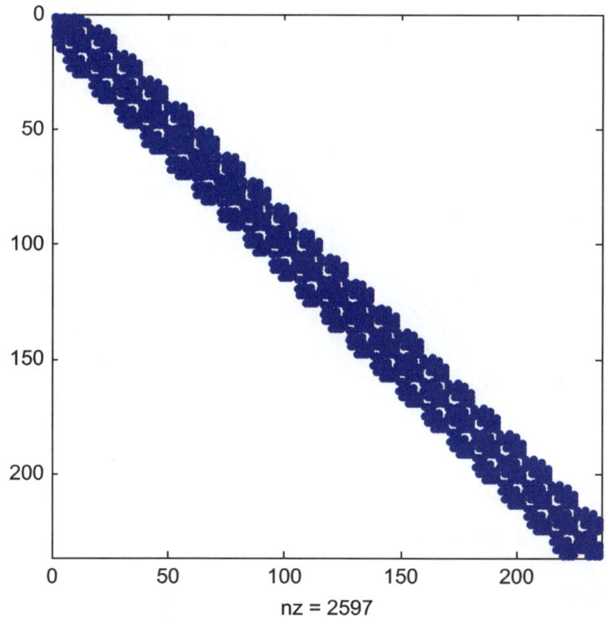

Fig. 6.15 Pattern of the flexibility matrix

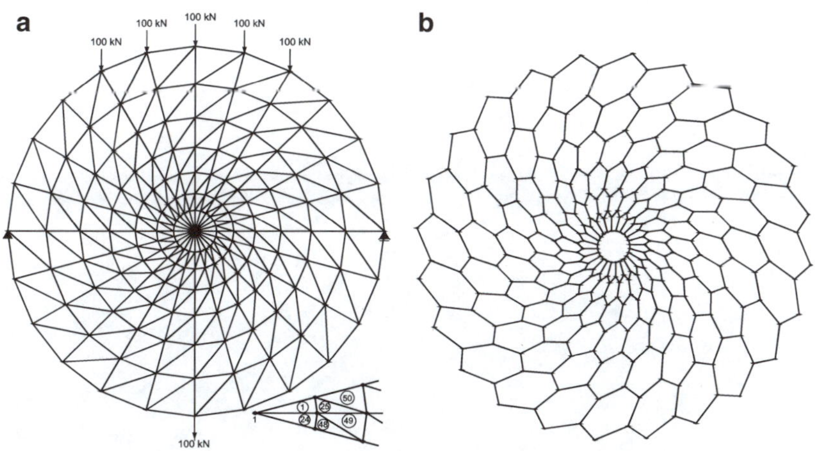

Fig. 6.16 (**a**) A circular disk with loading (**b**) its natural associate graph

banded and can be used for optimal finite element analysis by force method. In the present method, using topological transformations the non-zero patterns of null bases are identified and their numerical values are calculated by an algebraic process. For this purpose, associate digraph and interface graph are utilized.

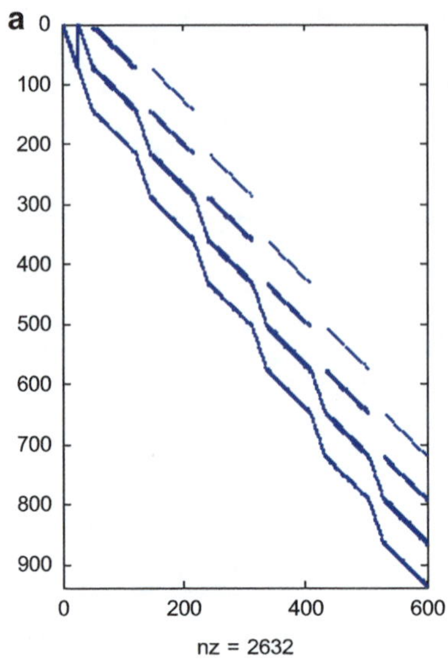

Fig. 6.17 Patterns of the null basis matrices; (**a**) The present approach (**b**) LU factorization approach

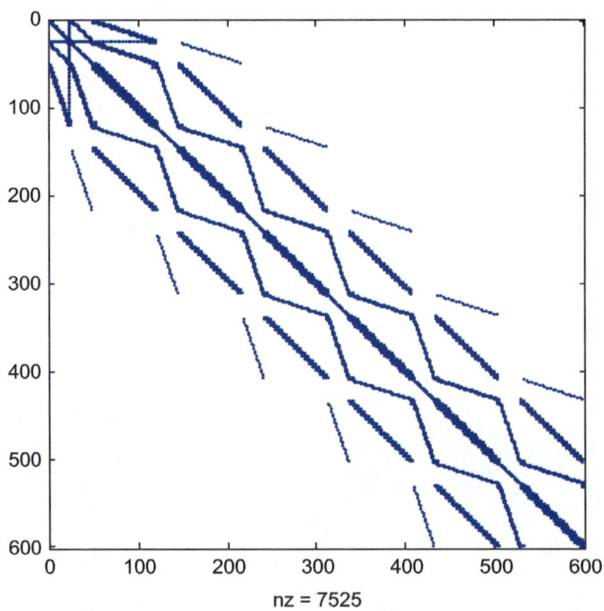

Fig. 6.18 Pattern of the flexibility matrix using the presented method

6.3.1 Graphs Associated with Finite Element Models

An Associate Digraph: In this graph one node is associated with each element of the FEM and two nodes are connected with a member if the elements have a common edge. A typical member of the graph is directed from the node with smaller number to the node with higher number. Except the node numbered as 1, all the other nodes have one or two negatively incident (one or two entering members) members defined as the negative incidence number of the node (if the nodes are badly numbered this number can be increased). Owing to the importance of these numbers in recognizing the types of SESs, the negative incidence numbers of the nodes of the graph should carefully be calculated. In Fig. 6.19, a rectangular and a triangular FEM with element numbering and their corresponding associate digraphs and negative incidence number of nodes are shown.

An Interface Graph. This graph can easily be constructed for triangular FEM using the following two rules:

1. This graph contains all the nodes of the FEM.
2. With each edge of an element of FEM, two graph members are associated. Therefore, in the interface of two elements, four members are present incident with the two end nodes of the common edge.

   For rectangular FEM the following additional rule should be used:

3. For each element a diagonal member is added in the interface graph. This member can be added between the first and third nodes of the element. These graphs for a rectangular and triangular FEM are shown in Fig. 6.20.

The member numbering of the interface graph should be performed according to the numbering of the FEM, taking into account the primary nodal numbering of considered element in the model. Thus for each triangular element six, and for rectangular element nine members of the interface graph will be numbered sequentially. In Fig. 6.20, such a numbering is shown for a typical element (a).

6.3.2 Subgraphs Corresponding to Self-Equilibrating Systems

6.3.2.1 Definitions of Independent Elements Forces

For the generation of equilibrium matrix **A** of a FEM, a system of independent force systems should be defined and also their relations with the element nodal forces should be established. The system of independent element forces for a rectangular finite element contains four symmetric moments (F_1,F_3,F_5,F_7) and four

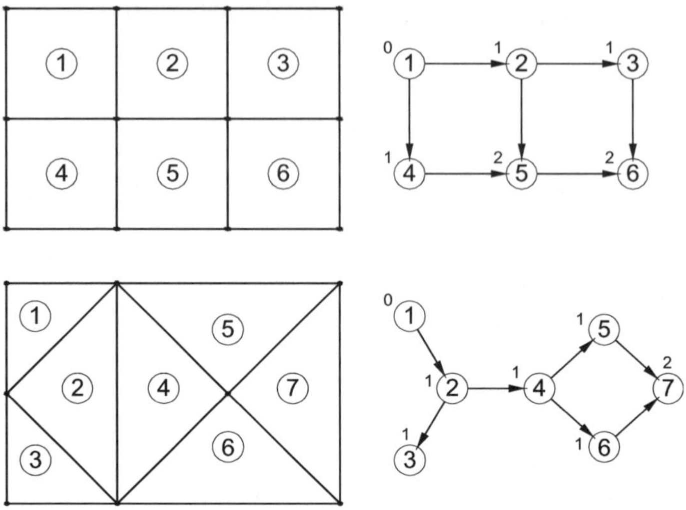

Fig. 6.19 A rectangular and a triangular FEM with their associate digraphs and nodes incidence numbers

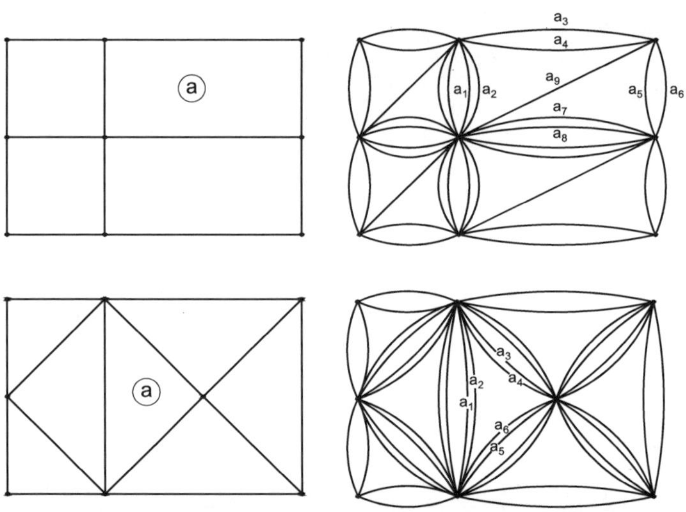

Fig. 6.20 A rectangular and a triangular FEM with their interface graphs and their numbering for a typical element (**a**)

anti-symmetric moments (F_2,F_4,F_6,F_8) and a set of four forces (F_9), which are applied at four corners of the element. These forces are related to the nodal forces (S_1 ~ S_{12}) by a 12 × 9 transformation matrix. A comprehensive study of these forces and their corresponding transformation matrix can be found in [12].

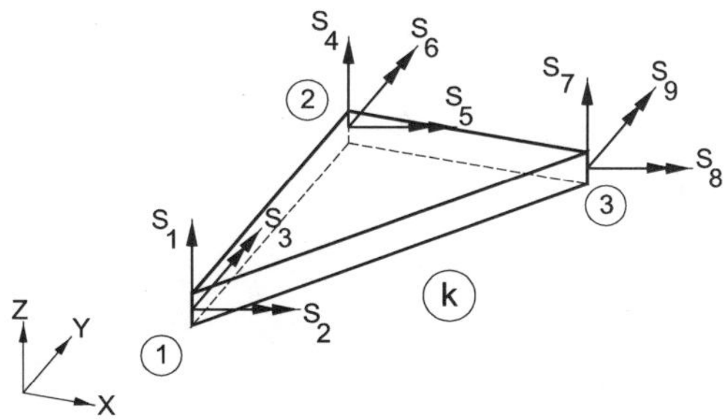

Fig. 6.21 Nodal forces for a triangular finite element

The system of independent element forces for a triangular finite element can be defined as three symmetric moments (F_1,F_3,F_5) and three anti-symmetric moments (F_2,F_4,F_6).

These forces can be related to element nodal forces ($S_1 \sim S_9$) using Eq. 6.16. The nodal forces are shown in Fig. 6.21, and the defined element forces for a triangular finite element are illustrated in Fig. 6.22. The interface graph defined in the preceding section is formed based on the way these element forces are considered and members of this graph have one-to-one correspondence with the element forces.

$$\mathbf{S} = \mathbf{TF} \tag{6.16}$$

$$\begin{bmatrix} S_1 \\ S_2 \\ S_3 \\ S_4 \\ S_5 \\ S_6 \\ S_7 \\ S_8 \\ S_9 \end{bmatrix} = \begin{bmatrix} 0 & \frac{2}{L_{12}} & 0 & 0 & 0 & -\frac{2}{L_{13}} \\ m_{12} & m_{12} & 0 & 0 & -m_{31} & m_{31} \\ -l_{12} & -l_{12} & 0 & 0 & l_{31} & -l_{31} \\ 0 & -\frac{2}{L_{12}} & 0 & \frac{2}{L_{23}} & 0 & 0 \\ -m_{12} & m_{12} & m_{23} & m_{23} & 0 & 0 \\ l_{12} & -l_{12} & -l_{23} & -l_{23} & 0 & 0 \\ 0 & 0 & 0 & -\frac{2}{L_{23}} & 0 & \frac{2}{L_{13}} \\ 0 & 0 & -m_{23} & m_{23} & m_{31} & m_{31} \\ 0 & 0 & l_{23} & -l_{23} & -l_{31} & -l_{31} \end{bmatrix} \begin{bmatrix} F_1 \\ F_2 \\ F_3 \\ F_4 \\ F_5 \\ F_6 \end{bmatrix}$$

In the transformation matrix **T**, L_{ij} is the length and l_{ij}, m_{ij} are direction cosines of the edge ij, which have the following definitions according to nodal coordinates:

$$l_{ij} = \frac{x_j - x_i}{L_{ij}} \quad m_{ij} = \frac{y_j - y_i}{L_{ij}}$$

Considering the above definitions, the degree of statical indeterminacy (DSI) for a rectangular and triangular plate bending FEM with determinate support conditions is as follows:

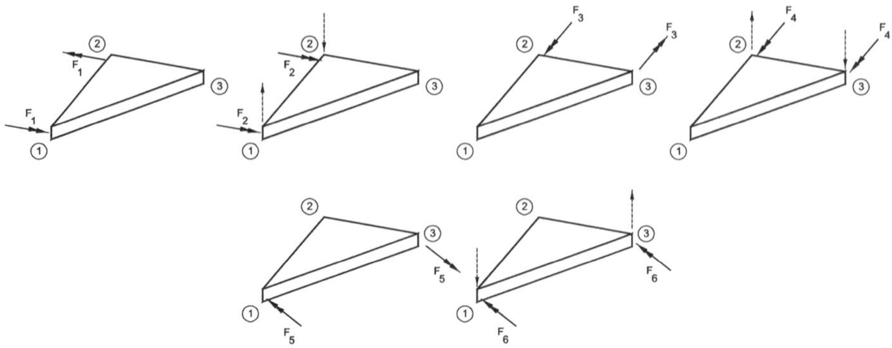

Fig. 6.22 Independent element forces for a triangular finite element

$$\text{DSI} = 9\text{m} - 3\text{n} + 3 \quad \text{(For rectangular FEM)} \tag{6.17}$$

$$\text{DSI} = 6\text{m} - 3\text{n} + 3 \quad \text{(For triangular FEM)} \tag{6.18}$$

where m is the total number of finite elements and n is the total number of nodes of FEM.

6.3.2.2 Self-Equilibrating Systems of Type I

Every set consisting of four members of interface graph, corresponding to two elements of the FEM with common edges, is called a self-equilibrating system of Type I.

The corresponding subgraph contains two SESs. Therefore, the set of four members corresponding to the common edges of the two elements i and j(i < j), has two members m_i and n_i(m < n), and r_j and s_j (r < s). The two SESs obtained from this set are (m,r) with (−1, 1) and (n,s) with (1, 1). On the other hand, a null vector with non-zero entries (−1, 1) in rows (m,r) and another null vector with non-zero entries (1, 1) at rows (n,s) are formed. Obviously, the number of such minimal SESs is twice the number of the members of associate digraph, since each member of this graph passes from interface of two elements. Nearly, two-third of null vectors for a rectangular or triangular FEM are of this type, corresponding to high sparsity for the null basis matrix.

6.3.2.3 Self-Equilibrating Systems of Type II

For each two adjacent finite element (two adjacent node in the associate digraph) such as r and j (r < j) in which j have negative incidence 1, another type of SES can be constructed which is called as the self-equilibrating system of Type II. In Fig. 6.23a, two adjacent rectangular and triangular finite element as well as their

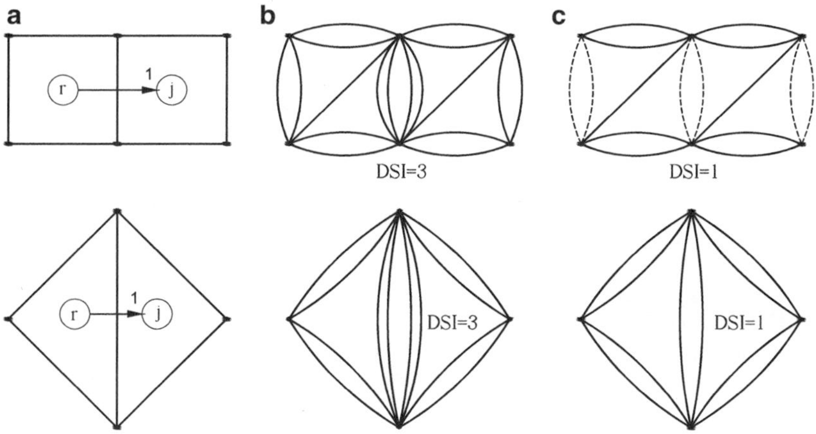

Fig. 6.23 (**a**) Two adjacent elements, (**b**) Related interface graphs, (**c**) Subgraphs of Type II SESs

associate digraphs are shown. Their corresponding interface graphs are also shown in Fig. 6.23b. The DSI of interface graph is 3 and thus correspond to three null vectors. Two null vectors are previously formed using four members in the interface of two elements. Therefore, in order to form the third SES, the generator of two type I SESs, should be removed from interface subgraph. Thus the DSI of remaining subgraph equals one and an independent null vector can simply be extracted.

It should be noted that the remaining subgraph corresponding to rectangular elements have still six ineffective members (hidden members in Fig. 6.23c) which can analytically be shown that always lead to zero entries in related null vector. Thus the subgraphs corresponding to Type II SESs, of rectangular and triangular finite elements have always ten members. In Fig. 6.23, the interface graph and the subgraphs corresponding to Type II SESs are shown for two rectangular and triangular adjacent elements.

For each node with a negative incidence two, a self equilibrating system of Type II can also be extracted. For each element k with negative incidence two which is adjacent to two elements i and j with $k > i, j$, pairs (i, k) or (j, k) can be used for the formation of a SES. Though both choices are valid, for maximum reduction in bandwidth of null basis matrix, the pair (max(i,j), k) should be selected.

6.3.2.4 Self-Equilibrating Systems of Type III

There are two elements i and j with $k > i, j$ in the adjacency of an element such as k with negative incidence two. Using these three elements and from their related interface graph, a subgraph corresponding to another minimal SES can be decomposed which is defined as the self equilibrating system of Type III. The interface graph related to these three elements has $\text{DSI} = 6$ and corresponds to six null vectors. Therefore, in order to maintain the independency of null vectors, one

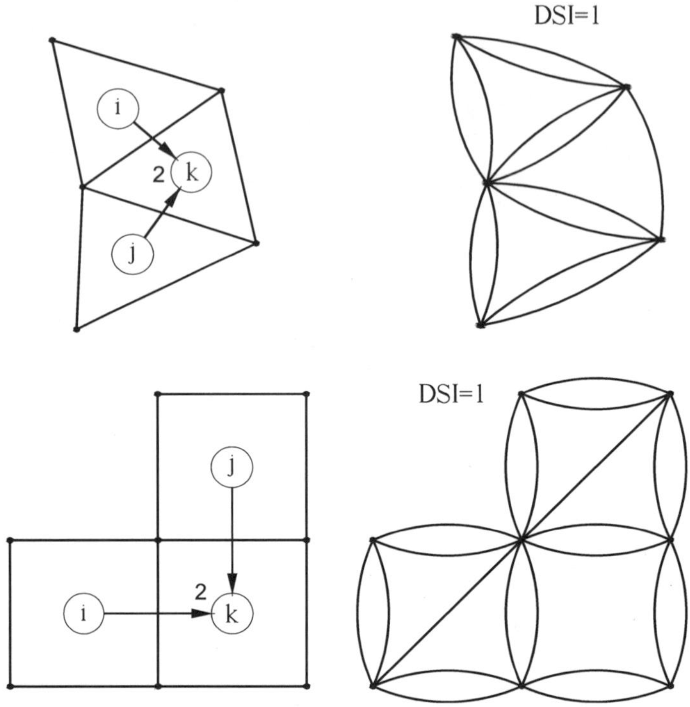

Fig. 6.24 Triangular and rectangular FEM and the corresponding Type III SESs

independent SES should be extracted from this graph. In this subgraph, there exist four and one member corresponding to te generators of Type I and II SESs, respectively, which using them five null vectors were previously formed. Thus the remaining subgraph after removing these members will be one degree statically indeterminate (DSI $=$ 1) and corresponds to an independent null vector.

This process can be used without any changes for rectangular and triangular FEM. However, in rectangular finite elements the remaining subgraphs have always some ineffective members. In Fig. 6.24, these subgraphs are shown for triangular and rectangular FEMs.

6.3.2.5 Self-Equilibrating Systems of Type IV

In the previous sections, three types of SESs were defined. These systems are sufficient for formation of null bases of finite element models without openings. However, if a FEM contains one or more openings, then another type of SESs can be identified which is called the self equilibrating system of Type IV, Fig. 6.25. In fact, from each opening in the FEM three independent SESs can be extracted. The subgraphs corresponding to these SESs have usually more members than the previous systems and also their related null vectors have more non-zero entries.

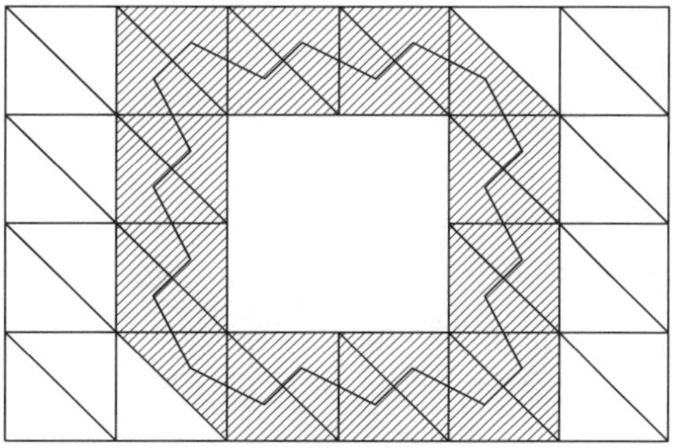

Fig. 6.25 A FEM with an opening and related cycle from the associate digraph

Every cycle of the associate digraph, if their related elements have no common in one node, corresponds to an opening. Since, every cycle has the same number of members as its nodes; therefore a cycle with m members passes through **m** finite elements of the FEM. As **m** triangular and rectangular finite elements surrounding an opening have **m** and 2**m** nodes respectively, therefore using Eqs. 6.17 and 6.18 the DSI of their related interface subgraphs will be 3**m** + 3. However, in these subgraphs 3**m** self equilibrating systems consisting 2**m** SESs of Type I and **m** SESs of Types II and III are previously selected. Then simply by removing the generators of these SESs from interface subgraph corresponding to an opening, a subgraph with DSI = 3 will be remained which corresponds to three null vectors. These three null vectors can simply be calculated using the remaining members of the interface graph as the columns of the related equilibrium submatrix and by utilizing an algebraic procedure. The null vectors related to openings which are calculated by the above process are subminimal. Finally, using the present procedures all minimal and subminimal SESs are simply calculated and the null basis matrix is generated. Due to the nature of present method, the calculated null bases are highly sparse and narrowly banded. However, for further reduction in bandwidth of null basis matrix (without any changes in sparsity) for each SES, an optimal list should be selected.

Algorithm

This algorithm consists of the following steps:

Step 1. In this step the associate digraph of the considered FEM is formed. In order to have a banded null basis, the nodes of this graph should be numbered by any efficient nodal ordering algorithms. Obviously, the effect of final numbering should be considered in FEM and rectangular equilibrium matrix. However, the ordering of the elements of FEM (nodes of associate digraph) is sufficient for formation of a banded null basis and there is no need for nodal ordering of FEM.

Step 2. The rectangular equilibrium matrix of the FEM is formed in this step.

Step 3. Formation of the interface graph of the FEM and the numbering their members according to nodal and element numbering of FEM is performed n this step.

Step 4. In this step the SESs of Type I are formed and the corresponding null vectors are obtained.

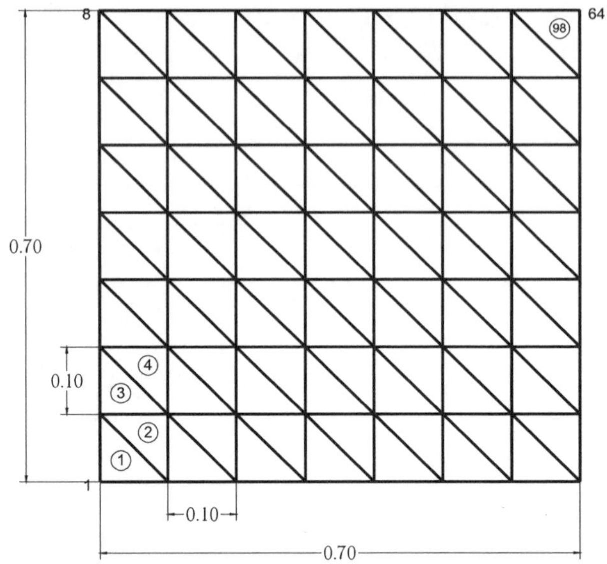

Fig. 6.26 A triangular FEM with numbering

Step 5. Formation of the SESs of Type II and calculation of the corresponding null vectors by using an algebraic process (such as LU factorization) on the related submatrices is performed in this step.

Step 6. Formation of the SESs of Type III, Numerical values of null vectors are found by a similar process in Step 5.

Step 7. The SESs of Type IV are formed and calculation of the numerical values of related null vectors is carried out if the model contains one or more openings, similar to Steps 5 and 6.

Step 8. The calculated null vectors are combined and ordered in a matrix in such a way that their generators make an ascending ordered list.

6.3.3 Numerical Examples

In this section three examples from triangular and rectangular FEM are studied. All of the models are assumed to be supported in a statically determinate fashion. The effect of indeterminate support conditions can separately be included with no difficulty [13]. However, the null basis matrices for each model are calculated using the present algorithm and **LU** factorization methods and the results are compared through computational time, sparsity, pattern of matrices and accuracy.

Example 1. In this example (Fig. 6.26), the null basis matrix ($\mathbf{B}_1$) for a triangular FEM with statically determinate support conditions is calculated and the sparsity, computational time and two norms of $\mathbf{AB}_1$ matrix, namely Frobenious and infinite

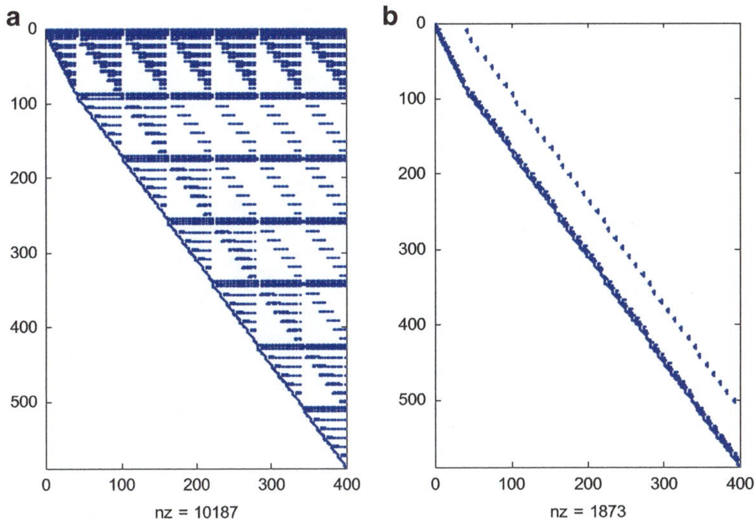

Fig. 6.27 Pattern of the null basis matrix: (**a**) LU factorization, (**b**) present method

Table 6.1 Comparison of the sparsity, computational time and accuracy of the present algorithm versus the **LU** factorization

	Number of non-zero entries (nz)	$\frac{\text{Time}}{\textbf{LU}\text{ Time}}$	$\|\mathbf{AB}_1\|_{fro}$	$\|\mathbf{AB}_1\|_{\infty}$
LU factorization	10,187	1.0000	7.39e−13	5.57e−12
Present algorithm	1,873	0.7885	2.95e−14	1.86e−14

norms are compared with **LU** factorization method, Fig. 6.27 (Table 6.1). The FEM properties are as follows:

Number of triangular elements = 98, Number of nodes = 64, DSI = 399, Thickness = 0.1 m, E = 2e + 8 kN/m^2, $\nu = 0.3$.

Example 2. In Fig. 6.28, a rectangular 1.6 m × 0.8 m plate which is discretized as 120 rectangular finite elements is shown. Patterns of the calculated null basis matrix using two methods are shown in Fig. 6.29. Also the results of the comparison are presented in Table 6.2. The properties of the model are as follows:

Number of rectangular finite elements = 128, Number of nodes = 153, DSI = 696, Thickness = 0.05 m, E = 2e + 8 kN/m^2, $\nu = 0.3$.

Example 3. In this example, a circular plate (with diameter 4) which is clamped at its center (determinate support condition) is studied, Fig. 6.30. Pattern of the null basis matrix for two methods and the comparison of results are shown in Fig. 6.31 and Table 6.3, respectively. The properties of the model are as follows:

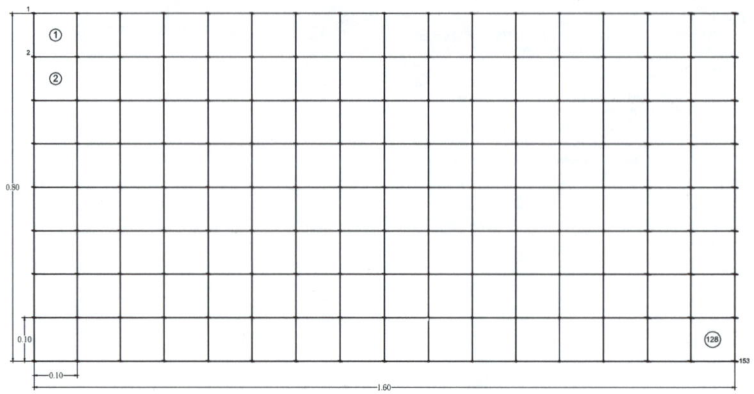

Fig. 6.28 The FEM of a rectangular plate and its numbering

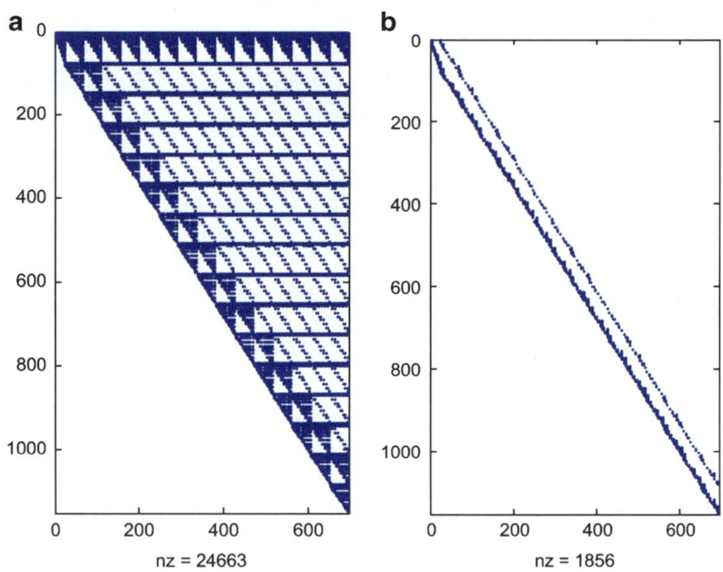

Fig. 6.29 Pattern of the null basis matrix: (**a**) LU factorization, (**b**) present method

Table 6.2 Comparison of the sparsity, computational time and accuracy of the present algorithm versus the **LU** factorization

	Number of non-zero entries (nz)	$\frac{\text{Time}}{\textbf{LU}\text{ Time}}$	$\lVert\mathbf{AB}_1\rVert_{fro}$	$\lVert\mathbf{AB}_1\rVert_{\infty}$
LU factorization	24,663	1.0000	3.76e−12	2.62e−11
Present algorithm	1,856	0.6539	1.74e−14	7.10e−15

Fig. 6.30 The FEM of a circular plate

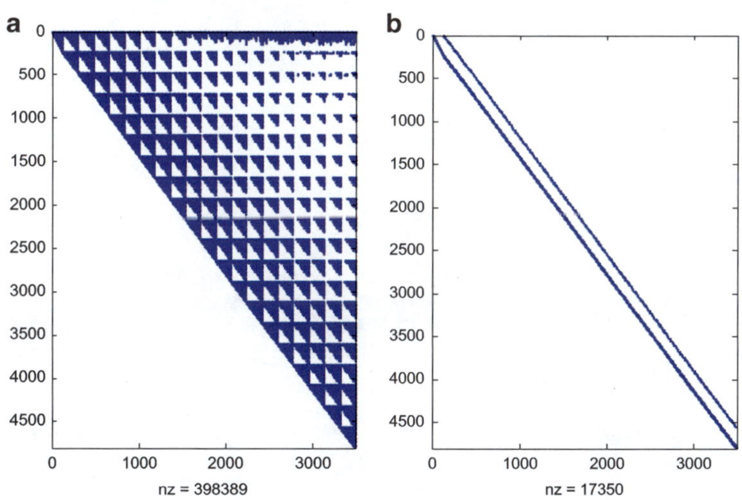

Fig. 6.31 Pattern of the null basis matrix: (**a**) LU method, (**b**) the present method

Table 6.3 Comparison of the sparsity, computational time and accuracy of the present algorithm versus the **LU** factorization

	Number of non-zero entries (nz)	$\frac{\text{Time}}{\textbf{LU}\ \text{Time}}$	$\|\|\mathbf{AB}_1\|\|_{fro}$	$\|\|\mathbf{AB}_1\|\|_{\infty}$
LU factorization	398,389	1.0000	4.15e−11	6.65e−10
Present algorithm	17,350	0.1852	7.51e−14	1.95e−14

Number of triangular finite elements = 800, Number of nodes = 441, DSI = 3,480, Thickness = 0.2 m, E = 2e + 7 kN/m^2, $\nu = 0.2$.

The results of examples clearly reveal the efficiency of the present method in reduction of non-zero entries and bandwidth. In Example 3, the difference of the computational time for two methods has been considerable which means that the complexity of present method is lower than **LU** method, and this difference becomes even more when the DSI is increased. The values of norms also indicate the higher accuracy of the present algorithm.

Finally, the results show that, the present method can be used as an efficient tool for null basis calculation of plate bending FEM and optimal finite element force method because in all aspects of comparisons, (sparsity, computational time and accuracy) the present algorithm has considerable priority versus the **LU** method and thus versus other algebraic algorithms which **LU** factorization is one of the primary steps of those methods.

6.4 Force Method for Three Dimensional Finite Element Analysis

In this section an efficient method is presented for the formation of null bases of finite element models comprised of tetrahedron elements, corresponding to highly sparse and banded flexibility matrices [7]. This is achieved by associating special graphs to the finite element model and selecting appropriate subgraphs and forming the self stress systems on these subgraphs.

6.4.1 Graphs Associated with Finite Element Model

Here, the natural associate graph and the interface graph are utilized as defined in the following:

The interface graph S_I. This graph can be constructed using the following two rules:

a. There is 1–1 correspondence between the nodes of the interface graph and the nodes of the FEM.
b. For each edge of the tetrahedron, one independent member is associated. Therefore, if k tetrahedrons have a common edge, then the corresponding member of the interface graph will consists of *k* members (multiple members). A FEM and the corresponding interface graph are shown in Fig. 6.32a, b, respectively.

The members of the interface graph should be numbered according to the FEM. For each tetrahedron element like *a*, six members of the interface graph should be numbered consequently. The numbering is performed according to the direction of the independent element forces. A typical numbering is shown in Fig. 6.32c.

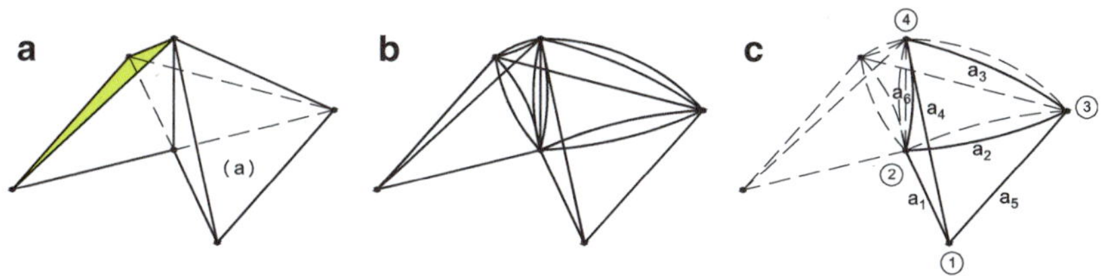

Fig. 6.32 (**a**) A 3D finite element model; (**b**) The interface graph; (**c**) Numbering for the skeleton of a typical tetrahedron *a*

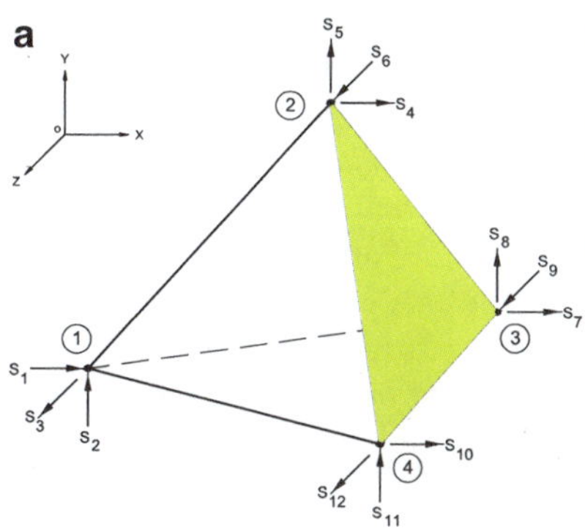

Force system or direction i	Location (nodes)
1	1,2
2	2,3
3	3,4
4	4,1
5	1,3
6	2,4

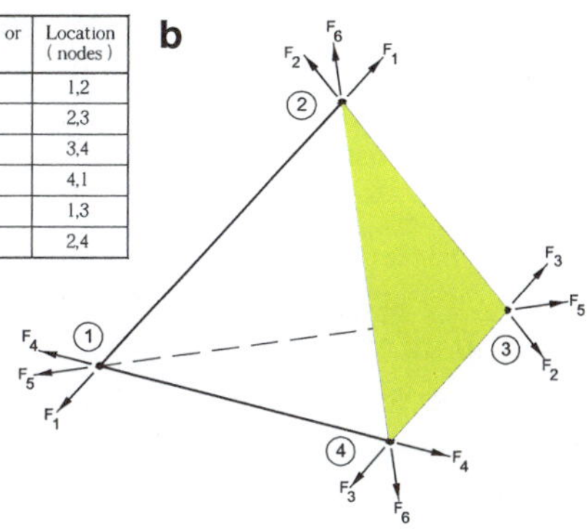

Fig. 6.33 The nodal and element forces of a tetrahedron element

6.4.2 *The Pattern Corresponding to the Self Stress Systems*

The nodal forces and independent element forces of a tetrahedron is defined as shown in Fig. 6.33. This is the same convention as that of the Przemieniecki [10].

Considering Fig. 6.32, in order to find the patterns corresponding to the self stress systems, the skeleton of tetrahedra are simulated as a space truss. This is possible since the independent element forces F_1 to F_6 are applied in the nodes and are along the edges of the tetrahedron, Fig. 6.33. The statical indeterminacy of a space truss with m members and n nodes is given as $\gamma(S) = m - 3n + 6$, therefore the Degree of Statical Indeteminacy (DSI) of the entire FEM, supported in a statically determinate fashion, can be calculated with same relationship as:

$$\mathrm{DSI_T} = 6\mathrm{M} - 3\mathrm{N} + 6 \tag{6.19}$$

where *M* is the number of tetrahedron elements and *N* is the total number of nodes of the FEM.

With the above simulation, the patterns of the self stress systems can be identified as follows:

6.4.2.1 Type I Self Stress Systems

Each k-multiple member of the interface graph is a subgraph on which $k - 1$ self stress systems can be generated. In other words, on a k-multiple member numbered as $(i_1, j_2, l_3, \ldots, m_{k-1}, n_k)$ with the condition $(i < j < l < \ldots < m < n)$, $k - 1$ self stress systems each formed on two single members can be constructed.

Each $k(k - 1)/2$ combination of double members from the above list is valid for a self stress system but obviously, for maximum reduction in bandwidth of the final null basis, $k - 1$ pairs of duplicate members should be selected as (i,j), (j,l), …, (m,n). Each pair (i,j) with $(i < j)$, corresponds to a null vector with their nonzero entries are located in rows i and j, and their numeric values are −1, 1, respectively. The member with bigger member number (j) is called the generator. Each pairs forms the underlying subgraph of a Type I self stress system.

For finite elements models with tetrahedron elements, more than 85 % of total self stress systems are of Type I. Thus a large percent of the minimal null vectors can be formed only by the determination of member numbers of these pairs. It should be noted that in the process of the formation of the interface graph, these pairs and their numbers can simply be identified.

6.4.2.2 Type II Self Stress Systems

There are other types of self stress systems which are topologically identical to the minimal self stress systems of the corresponding space truss. In the other words, if a k-multiple member from the interface graph is substituted by a member, or if the generators of the Type I self stress systems are removed from S_I, then the remaining subgraph is a graph, denoted by S. In general the self stress systems built on S are called Type II self stress systems.

In general, the self stress systems which can be selected from subgraph S are called Type II systems. In fact these systems are γ-cycles, which correspond to the cycles of minimal lengths of the associate graph of the finite element model. A connected rigid subgraph C_k of S with $\gamma(C_k) = 1$, which has no removable subgraph, is termed a *γ-cycle* of S, Ref. [11]. A *removable subgraph* S_j of a graph S_i, is the elementary subgraph for which $\gamma(S_i - S_j) = \gamma(S_i)$.

The associate graph of tetrahedron finite element models, denoted by A(S), is a graph in which to each tetrahedron element one node is associated and two such nodes are connected together by a member if their corresponding elements having a common face (3 common nodes). A finite element model with 24 tetrahedron elements is shown in Fig. 6.34a, its associate graph, which is the 1-skeleton of a polyhedron, is depicted in Fig. 6.34b.

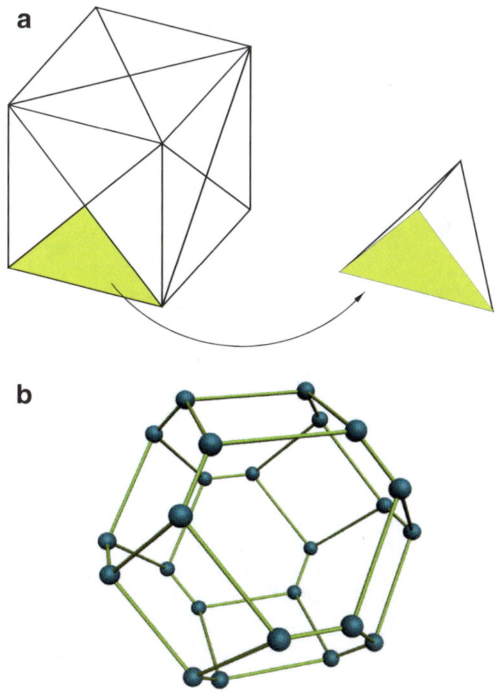

Fig. 6.34 A finite element model with 24 tetrahedron elements and the corresponding associate graph

Corresponding to the regional cycle basis of a planer graph (the set of cycles which are the boundaries of the internal regions [11]), in general, two types of minimal cycles can be extracted from the associate graph of a finite element model. These cycles are as follows.

6.4.2.3 Type I Minimal Cycles

In these cycles all the corresponding finite elements have two common nodes. Each cycle in this type passes through M finite elements for which its corresponding interface graph has N = M + 2 nodes, and (3M − 1) Type I self stress systems can be extracted. Therefore, by using Eq. 6.8, the degree of statical indeterminacy of equivalent γ-cycle is 1. Thus each Type I cycle corresponds to one null vector.

6.4.2.4 Type II Minimal Cycles

A minimal cycle which surrounds an opening (in the form of a hole through a structure), is called Type II minimal cycle. Such a cycle passes through M finite elements and its corresponding interface graph has N = M nodes, and 3M Type I self stress systems can be extracted. Again by using Eq. 6.8, the DSI of equivalent γ-cycle is 6. Thus each Type II cycle corresponds to six null vectors.

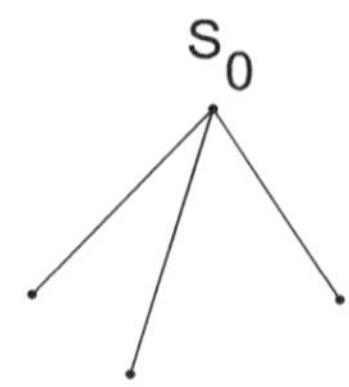

Fig. 6.35 A typical subgraph S_0

6.4.3 Relationship Between $\gamma(S)$ *and* $b_1(A(S))$

The goal of this section, which has theoretical importance, is to derive a relationship between the degree of statical indeterminacy of the 1-skeleton of a FE model *S* and the first Betti number of the associate graph of the model without openings (analogous to 2-dimensional fully triangulated trusses [11]). For this purpose an expansion is employed.

6.4.3.1 Number of Nodes of the Associate Graph

Consider a tetrahedron element. The 1-skeleton S corresponding to this element has six members and its associate graph is only a single node. Second tetrahedron element results in the addition of a typical subgraph S_0 as shown in Fig. 6.35. Each time by adding this subgraph to the previous graph leads to addition of three members to the main graph *S* and one node to its associate graph.

The associate graph which is formed using this process is a tree and therefore its number of nodes can simply be calculated as:

$$N' = \frac{M(S) - 3}{3} \tag{6.20}$$

where M(S) and N′ are the number of members and nodes of the 1-skeleton of the model *S* and its associate graph A(S), respectively. Obviously N′ is also equal to the number of tetrahedron elements.

Addition of one node and three members in each stage of expansion is the basic condition for validity of Eq. 6.20. Obviously, in this case the 1-skeleton can be viewed as a space truss having DSI equal 0. Some stages of the expansion process is shown in Figs. 6.36a–c. If the subgraph S_0 is joined to the previous 1-skeleton in a manner that only one new member is added to *S* without addition of a new node (equivalent to the addition of a new tetrahedron element, one new node to A(S) and formation of a cycle in A(S)), then the DSI of corresponding space truss will be increased by unity, Fig. 6.36d. In such a case, Eq. 6.9 is not valid and must be modified using a new parameter. Clearly, the DSI of the space truss, $\gamma(S)$, should be considered as this new parameter.

Considering the above mentioned point, Eq. 6.20 is modified as

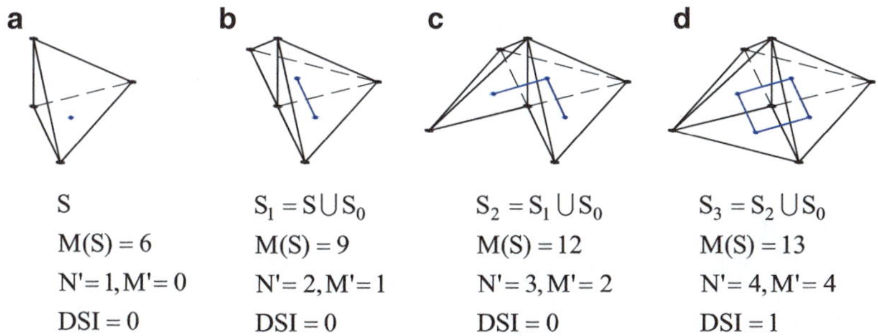

Fig. 6.36 The process of expansion for the formation of a γ-cycle with DSI equal to unity

$$N' = \frac{M(S) + 2\gamma(S) - 3}{3} \tag{6.21}$$

Substitution $\gamma(S) = M(S) - 3N(S) + 6$ in Eq. 6.21 leads to

$$N' = M(S) - 2N(S) + 3 \tag{6.22}$$

In fact, the number of nodes of the associate graph is equal to the DSI of the 1-skeleton *S*, when *S* is viewed as a two dimensional truss!

6.4.3.2 The Number of Members of the Associate Graph

Similar to the previous section, the number of members of the associate graph can also be determined. However, if the expansion process is in a manner that leads to an associate graph which is a tree, its number of members can be simply calculated using the property of a tree, i.e.

$$M' = N' - 1 \tag{6.23}$$

or

$$M' = \frac{M(S) - 6}{3} \tag{6.24}$$

Here again, if joining a subgraph S_0 leads to the addition of only one member to the graph S (this case corresponds to the addition of two members and one node to A(S), and one unit increase in the DSI of space truss), then Eq. 6.24 can simply be modified using $\gamma(S)$ as

$$M' = \frac{M(S) + 5\gamma(S) - 6}{3} \tag{6.25}$$

or

$$M' = 2M(S) - 5N(S) + 8 \tag{6.26}$$

Now it is possible to relate the number of independent cycles of the associate graph to the number of nodes and members of the 1-skeleton S. The dimension of the cycle space, or the first Betti number of the associate graph, can be calculated using

$$b_1(A(S)) = M' - N' + 1 \tag{6.27}$$

By substitution of M' and N' from Eqs. 6.11 and 6.26, the relation for $b_1(A(S))$ is obtained in terms of M(S) and N(S) as

$$b_1(A(S)) = M(S) - 3N(S) + 6 \tag{6.28}$$

and this is the relationship for the DSI of a three dimensional truss.

The right hand of the above formula is identical to the DSI of a space truss. Examining further models with tetrahedron finite elements and their corresponding associate graphs, it becomes obvious that the relationship presented in earlier sections are valid for all the cases where A(S) is not the 1-skeleton of a polyhedron. If A(S) is the 1-skeleton of a polyhedron (Fig. 6.34b), then internal nodes will be created in the finite element model or graph S (a node is called *internal* if it is not positioned on the surface of the FEM). This case corresponds to situations where in the process of expansion, adding one tetrahedron element leads to the addition of three members and one node for the graph A(S). For such cases, the present relationship must be modified considering the contribution of the number of internal nodes as:

$$N' = M(S) - 2N(S) + 3 + N_i(S) \tag{6.29}$$

$$M' = 2M(S) - 5N(S) + 8 + 3N_i(S) \tag{6.30}$$

$$b_1(A(S)) = M(S) - 3N(S) + 6 + 2N_i(S) \tag{6.31}$$

or

$$b_1(A(S)) = \gamma(S) + 2N_i(S) \tag{6.32}$$

In which $N_i(S)$ is the total number of internal nodes in the 1-skeleton S of the finite element model.

The above equations are general relationships for finding the number of nodes, members and the dimension of the cycle space of an associate graph. Equation 6.32

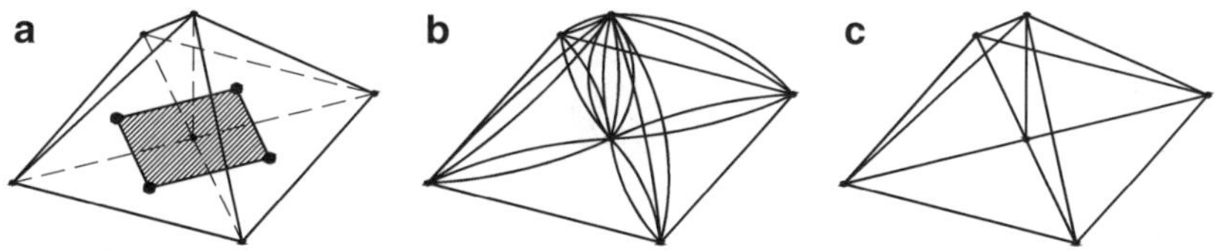

Fig. 6.37 (**a**) Finite element model with its associate graph; (**b**) corresponding s_I^c; (**c**) corresponding γ-cycle

shows that, if no internal node is created, then the dimension of the cycle space of the associate graph is equal to the DSI of the corresponding space truss, and therefore for each cycle, an independent null vector can be formed. If there is one or more internal nodes in the model, then the dimension of the cycle space of A (S) is greater than of the DSI of *S* and thus all the null vectors corresponding to cycles cannot be used in formation of the final null basis. In this case, $2N_i(S)$ of vectors must be selected and ignored. Some of the null vectors will have the same generators. In Sect. 6.4.2 a method is presented for the selection of these vectors. It should be noted that, in 2-dimensional trusses and plane stress and strain finite elements, the dimension of the cycle space of A(S) is always equal to the DSI of the 1-skeleton S.

6.4.4 Selection of Optimal γ-Cycles Corresponding to Type II Self Stress Systems

Thus far, it is found out that each γ-cycle corresponds to a minimal cycle of the associate graph. Also each minimal cycle with *n* nodes from A(S) such as *c* passes through n tetrahedron elements. The subgraph $s_I^c(s_I^c \subset S_I)$ which is relevant to these n elements and cycle c, is a base for the selection of an optimal γ-cycle. Such a subgraph may contain simple and multiple members, where each multiple member with k members corresponds to the overlap of *k* tetrahedron elements, and each simple member corresponds to the edge of a element in the boundary of the model. By applying a special condition to such subgraphs, lists corresponding to optimal γ-cycles can be obtained.

A finite element model with four tetrahedron elements and its associate graph are shown in Fig. 6.37a. The corresponding s_I^c which contains multiple and simple members is illustrated in Fig. 6.37b, and the corresponding γ-cycle is depicted in Fig. 6.37c.

In general, from each s_I^c many γ-cycles (self stress systems) can be extracted, since each simple member of s_I^c is included in a γ-cycle, and all the members of a multiple member can be used in the formation of final self stress system. Thus for obtaining an optimal self stress system, on each s_I^c, two basic selections should be performed:

Table 6.4 Lists corresponding to graph s_I^c

d_1^1	d_2^1	...	s_1	s_2	...	d_i^1	s_j	...	d_k^1
d_1^2	d_2^2	...			...	d_i^2		...	d_k^2
	d_2^3								
						d_i^n			

1. Selection of the generator or the last member of a self stress system, which is required for the null vectors to be independent.
2. Selection of a list of members from graph s_I^c with maximum possible number for the first member. Such a selection can reduce the bandwidth of the null basis matrix considerably.

The mathematical representation of this selection can be written as

$$\text{Minimize}\,(j - i) \tag{6.33}$$

where j is the generator's member number and *i* is the least member number of the current γ-cycle. In the following a simple and fast method is presented for these selections.

6.4.5 Selection of Optimal Lists

In Table 6.4, members of a graph s_I^c which are relevant to a Type I minimal cycle of A(S) are shown. In this table d_i, (i = 1, ..., k) are the member numbers of multiple members, where $d_i^1 < d_i^2 < \ldots < d_i^n$, and s_j; (j = 1, ..., t) are the member numbers of simple members. All d_i^m with $(m \neq 1)$ are already used as the generators of Type I self stress systems. Therefore, it is obvious that the max $\{d_i^1, s_j\} = d_L^1$, (i = 1, ..., k, j = 1, ..., t) must be selected as the generator of the current γ-cycle.

For maximizing the difference between the first member number and the generator of the current γ-cycle, the following condition can be used:

$$\text{find}\left(\max\left(d_i^j\right)\right)\Big|d_i^j < d_L^1 \quad \text{then} \quad d_i^1 = \max\left(d_i^j\right) \quad (i = 1, \ldots, k, j = 2, \ldots, n) \tag{6.34}$$

Equation 6.34 means that, in each multiple member the largest d_i^j, (j = 2, ..., n) which is also less than the generator's member number (d_L^1), should be substituted with d_i^1 for all the indices of *i*. After using the above process, the remained list, $\{d_i^1, s_j\}$, (i = 1, ..., k, j = 1, ..., t) is the desired optimal list which corresponds to subgraph of the current optimal self stress system.

After finding optimal lists corresponding to Type I minimal cycles, using relevant equilibrium submatrix, numerical values for each null vector are calculated. For Type II minimal cycles, finding an optimal list is a time consuming process and considering the fact that the number of openings is low in the real

Table 6.5 Schematic view of three null vectors with identical generator

v_1	v_2	0	0	0	v_6	v_7	v_8
0	0	v_3	v_4	0	0	v_7	v_8
v_1	0	v_3	0	v_5	0	0	v_8

structures, the use of this process is not economical for improvement of the final null basis. Thus for each cycle of this type, graph s_I^c is decomposed and all members corresponding to Type I self stress systems are removed. The remaining subgraph has the DSI equal to 6, and then 6 null vectors can be calculated from relevant equilibrium submatrix. Obviously, these null vectors will be suboptimal.

At this stage, considering the number of Type I self stress systems previously selected, decision for performing the rest of the process should be taken. Suppose t_1 Type I self stress systems are identified previously. Then $t_2 = DSI_T - t_1$ Type II self stress systems should be selected. Therefore, if we have $t_2 = b_1(A(S)) + 5n_c$ (with n_c being the total number of openings) meaning that there is no internal node in the model, then all $\gamma - cycles$ corresponding to cycles of $A(S)$ should be involved in the formation of final null basis. Otherwise, $t_2 < b_1(A(S)) + 5n_c$ means that there are one or more internal nodes. This case corresponds to the generation of null vectors with identical generator numbers. These vectors can usually be grouped in triplex sets and some of them should be deleted. It should be noted that all vectors which have unique generators are valid and independent. In the following, an algebraic procedure is presented for the formation of a desired list of vectors.

In Table 6.5, a schematic view for the patterns of three null vectors with identical generator is illustrated. These vectors correspond to three minimal cycles of A (S) which according to the process presented in Sect. 4.1, their corresponding optimal lists have identical generator as v_8.

According to Eq. 6.22, it is obvious that if $v_8 \neq 0$, then for the generator v_8, the second vector will be the desired vector from these three sets (rows). In such a case, after the selection of one optimal vector, one cannot simply delete the remaining vectors. For this purpose the following two controls should be performed.

a. Numerical cancellation control

   Each vector for which the numerical value of its generator is equal to zero has in fact another generator (closer nonzero entry to the generator). If there is no such a vector with identical generator among all the previously selected vectors, then this vector should be selected as a new and independent null vector.

b. New generator control

   All combinations of m vectors (m is usually equal 3) for possibility of the formation of vectors with new generator should be calculated. These combinations should be found in a manner that the common generator member of vectors is removed. Here again, from newly created vectors, those with new generators should be selected as valid and independent null vectors. As an example, in Table 6.5, the combination of the first and third rows leads to a new vector with new generator v_7. If there is no such a vector with this new generator among all the previously selected null vectors, then this new vector should be selected as a valid and independent null vector with the generator as v_7.

Finally, using the above mentioned two conditions, t_2 valid and independent vectors are identified and totally DSI_T null vectors will be left. Since this process is performed on all vectors with identical generators, therefore all the desired vectors are obtained automatically and there is no need to additional information about the number of internal nodes. In the following an efficient algorithm is presented for finding the null basis of tetrahedron finite element models.

Algorithm

Step 1: Generate the associate graph of finite element model and use an efficient method for its node numbering, Kaveh [11]. It is obvious that a suitable numbering of this graph corresponds to good numbering of elements of finite element model. This numbering leads to a banded adjacency matrix of the graph and correspondingly to a banded flexibility matrix.

Step 2: Setup the equilibrium matrix of finite elements model.

Step.3 Generate the interface graph and perform its numbering. The numbering of this graph should be performed according to the element numbering of the considered finite elements model.

Step 4: Find the Type I self stress systems. All multiple members of interface graph are identified and the values -1 and 1 are assigned to appropriate rows (corresponding to the member numbers). At the end of this step t_1 minimal null vectors are created.

Step 5: Find the Type II self stress systems. Using the Type I and Type II minimal cycles of the associate graph, relevant subgraphs are identified and their corresponding optimal lists are constructed.

Step 6: Calculate numerical values of the optimal lists. Using optimal lists selected in Step 5, null vectors corresponding to the Type I and Type II minimal cycles are calculated from the relevant equilibrium submatrix.

Step 7: Order the null vectors. At this step the constructed null vectors should be ordered such that their generators form a list with an ascending order.

In the following the efficiency of this algorithm is demonstrated using two numerical examples and a comparison is made through the results of the present algorithm and the LU factorization method. The comparisons are confined to those of sparsity, condition number and computational time of the formation of the flexibility matrices.

6.4.6 Numerical Examples

In this section two examples with different topological properties are studied. The models are assumed to be supported in a statically determinate fashion. The effect of the presence of additional supports can separately be included for each special case with no difficulty, Kaveh and Fazli [13]. The patterns of the null basis matrix

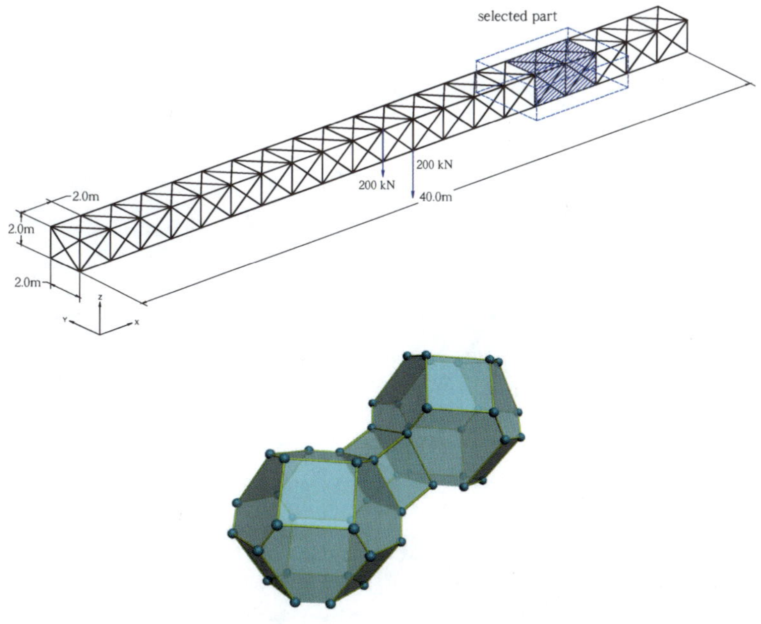

Fig. 6.38 A thick beam-type structure and the associate graph of the selected part

$\mathbf{B}_1$ and the flexibility matrix **G** are formed for two examples, and the number nonzero entries of these matrices are denoted by nz.

Example 1. A thick beam-type structure supported in a statically determinate fashion is depicted in Fig. 6.38. This structure is discretized using tetrahedron finite elements. The properties of the model are as follows:

Number of tetrahedron elements = 480, Number of nodes = 205
Elastic modulus E = 2e + 7 kN/m^2, Poisson's ratio $\nu = 0.2$
Number of Type I self stress systems = 2,032 (89.5 %)
First Betti number of the associate graph = 317 (independent cycles)
Number of Type II self stress systems = 239
Number of internal nodes (N_i) = 39, DSI_T = 2,271 = (2,032 + 239).

The sparsity of the final null basis obtained by the present algorithm is approximately 12 % of **LU** method, as shown in Fig. 6.39. The conditioning numbers, the ∞ norms and the Frobenius norms of $\mathbf{AB}_1$ are given for the present method and **LU** factorization approach, where **A** is the equilibrium matrix. The computational time is lower than 50 % for the present algorithm. The flexibility matrix shown in Fig. 6.40 is quite banded (Table 6.6).

In the above table, $\lambda_{max}/\lambda_{min}$ is the condition number, and $\|\|\|_\infty$, $\|\|\|_{fro}$ are the ∞ norm and Frobenius norm of $\mathbf{AB}_1$, respectively.

Example 2. A thick flat plate with 3D tetrahedra in a single layer is considered which is supported in a statically determinate fashion as depicted in Fig. 6.41. The 1-skeleton

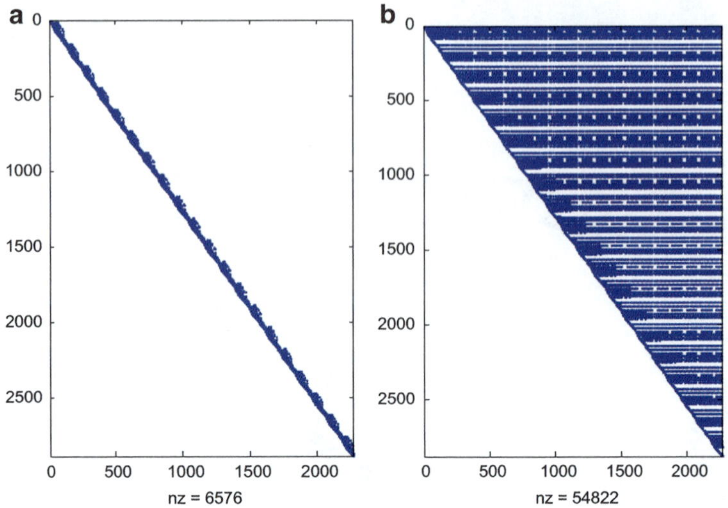

Fig. 6.39 Patterns of $\mathbf{B}_1(2{,}880 \times 2{,}271)$ and the number of nonzero entries, nz, of null basis; (**a**) Present algorithm; (**b**) **LU** factorization

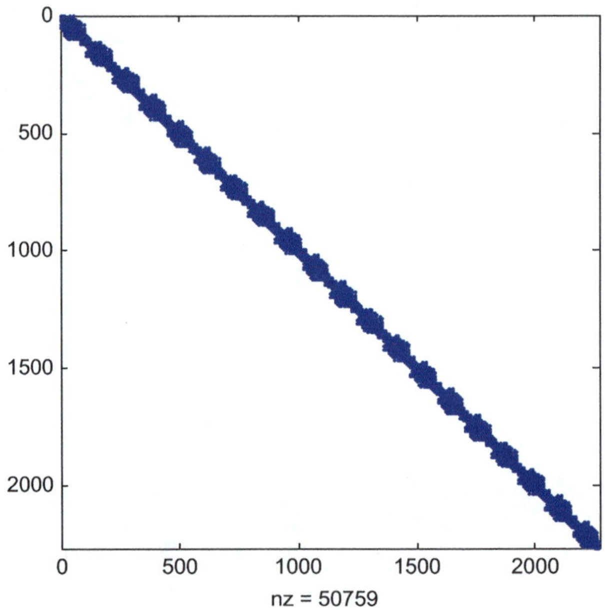

Fig. 6.40 Pattern of the flexibility matrix $\mathbf{G}(2{,}271 \times 2{,}271)$ and the number of its nonzero entries obtained by the present algorithm

Table 6.6 Comparison of the condition number of **G**, the norms and the computational time

	Time/**LU** time	$\lambda_{max}/\lambda_{min}$	$\|\mathbf{AB}_1\|_\infty$	$\|\mathbf{AB}_1\|_{fro}$
LU	1.00	1.67e+5	5.73e−12	1.29e−12
Present Algorithm	0.48	3.74e+5	0.00	0.00

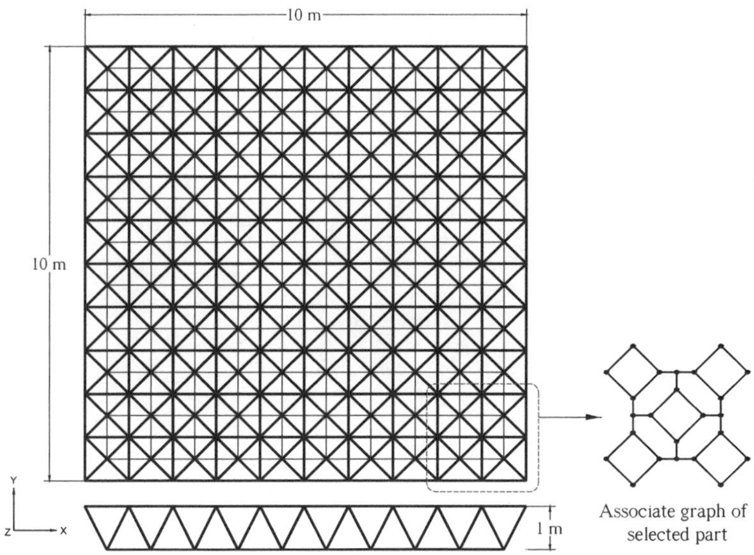

Fig. 6.41 A double layer grid, and the associate graph of the selected part of the grid

of this model is similar to a double layer space structure. The associate graph of this model is also shown in Fig. 6.41. The properties of the model are as follows:

Number of tetrahedron elements = 904, Number of nodes = 402
Elastic modulus E = 2e + 7 kN/m^2, Poisson's ratio $\nu = 0.2$
Number of Type I self stress systems = 3,719 (88.0 %)
First Betti number of the associate graph = 505 (independent cycles)
Number of Type II self stress systems = 505
Number of internal nodes (N_i) = 0, DSI_T = 4,224 = (3,719 + 505)

Here again, the sparsity of final null basis obtained by the present algorithm is approximately 10.5 % of **LU** method, as depicted in Fig. 6.42, while its computational time is nearly 11 % and also the condition number of **G** is improved, Table 6.7. The flexibility matrix **G** is also well structured as shown in Fig. 6.43.

In this chapter low order elements were presented. Higher order element will be discussed in subsequent chapter.

6.5 Efficient Finite Element Analysis Using Graph-Theoretical Force Method: Brick Element

In this section, an efficient graph theoretical method is presented for FEA of models composed of 3D brick elements. For this purpose first independent force systems and flexibility matrix of the element are presented, followed by the formation of the

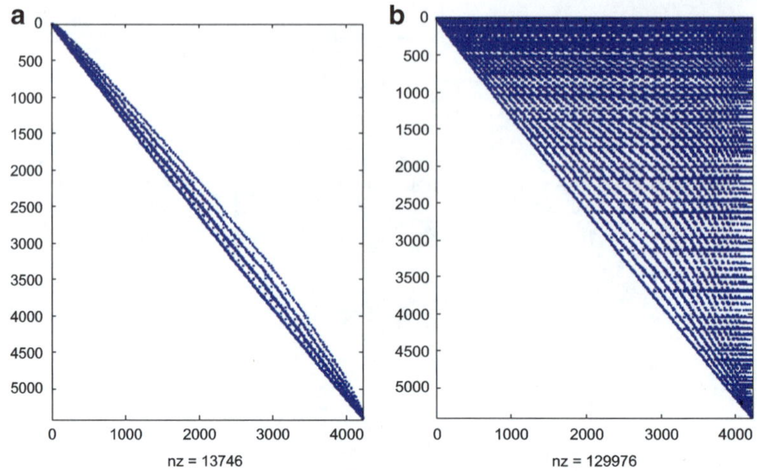

Fig. 6.42 Patterns of $\mathbf{B}_1(5{,}424 \times 4{,}224)$ and the number of nonzero entries, nz, of null basis; (**a**) present algorithm; (**b**) **LU** factorization

Table 6.7 Comparison of the condition number of **G**, the norms and the computational time

	Time/**LU** time	$\lambda_{max}/\lambda_{min}$	$\lVert \mathbf{AB}_1 \rVert_\infty$	$\lVert \mathbf{AB}_1 \rVert_{fro}$
LU	1.00	1.34e+6	1.64e−10	1.83e−11
Present Algorithm	0.11	1.68e+5	1.94e−15	1.18e−14

minimal subgraphs of the graph models of the considered FEMs. Then the self-equilibrating systems are constructed on these subgraphs forming a statical basis of the FEM corresponding to highly sparse and banded flexibility matrix.

6.5.1 Definition of the Independent Element Forces

In displacement method we use three forces at each node of the element, while in the force method, as shown in Fig. 6.44, it is preferable to select twelve edge force systems plus six diagonal force systems on six faces of the brick element between the second and third nodes of the current face. These element forces can be related to nodal forces using Eq. 6.35 as

$$\mathbf{S} = \mathbf{TF} \tag{6.35}$$

where l_{ij} is the length and m_{ij}, n_{ij}, p_{ij} are the direction cosines of the line between nodes i and j.

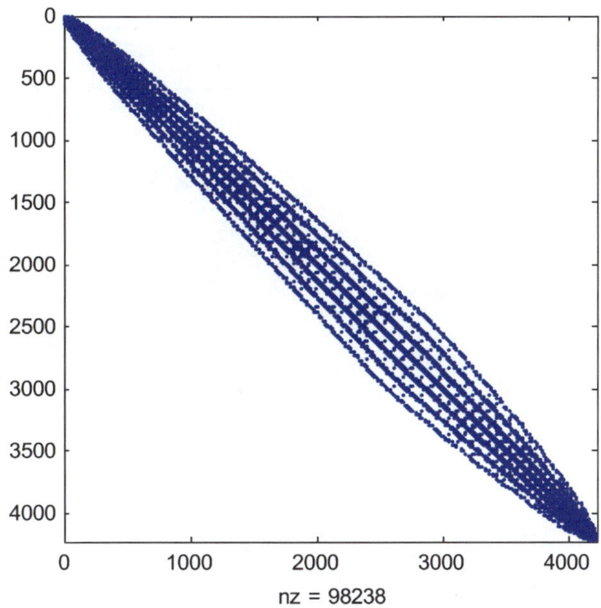

Fig. 6.43 Pattern of the flexibility matrix **G** (4,224 × 4,224) and the number of its nonzero entries obtained by the present algorithm

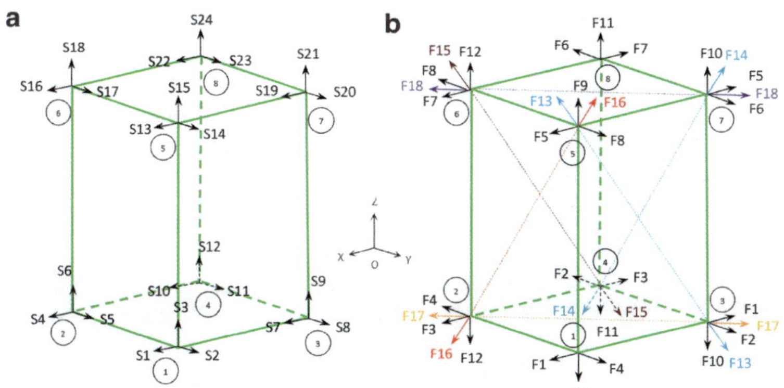

Fig. 6.44 Nodal and element force systems of a brick element

6.5.2 Flexibility Matrix of an Element

Formulation of a discrete element equivalent to the actual continuous structure is the first step in matrix structural analysis. For a linear system it can be assumed that the stresses $\boldsymbol{\sigma}$.are related to the forces **F** by linear equation as

$$\boldsymbol{\sigma} = \bar{\mathbf{c}}\mathbf{F} \tag{6.36}$$

The matrix $\bar{\mathbf{c}}$ represents statically equivalent stresses system due to the unit force **F**. The flexibility matrix of an element can be written as

$$
\begin{bmatrix} S_1\\ S_2\\ S_3\\ S_4\\ S_5\\ S_6\\ S_7\\ S_8\\ S_9\\ S_{10}\\ S_{11}\\ S_{12}\\ S_{13}\\ S_{14}\\ S_{15}\\ S_{16}\\ S_{17}\\ S_{18}\\ S_{19}\\ S_{20}\\ S_{21}\\ S_{22}\\ S_{23}\\ S_{24} \end{bmatrix}
=
\begin{bmatrix}
m_{13} & 0 & 0 & m_{12} & 0 & 0 & 0 & 0 & m_{15} & 0 & 0 & 0 & 0 & 0 & 0 & 0 & 0 & 0\\
n_{13} & 0 & 0 & n_{12} & 0 & 0 & 0 & 0 & n_{15} & 0 & 0 & 0 & 0 & 0 & 0 & 0 & 0 & 0\\
p_{13} & 0 & 0 & p_{12} & 0 & 0 & 0 & 0 & p_{15} & 0 & 0 & 0 & 0 & 0 & 0 & 0 & 0 & 0\\
0 & 0 & m_{24} & m_{21} & 0 & 0 & 0 & 0 & 0 & 0 & 0 & m_{26} & 0 & 0 & 0 & m_{25} & m_{23} & 0\\
0 & 0 & n_{24} & n_{21} & 0 & 0 & 0 & 0 & 0 & 0 & 0 & n_{26} & 0 & 0 & 0 & n_{25} & n_{23} & 0\\
0 & 0 & p_{24} & p_{21} & 0 & 0 & 0 & 0 & 0 & 0 & 0 & p_{26} & 0 & 0 & 0 & p_{25} & p_{23} & 0\\
m_{31} & m_{34} & 0 & 0 & 0 & 0 & 0 & 0 & 0 & m_{37} & 0 & 0 & m_{35} & 0 & 0 & 0 & m_{32} & 0\\
n_{31} & n_{34} & 0 & 0 & 0 & 0 & 0 & 0 & 0 & n_{37} & 0 & 0 & n_{35} & 0 & 0 & 0 & n_{32} & 0\\
p_{31} & p_{34} & 0 & 0 & 0 & 0 & 0 & 0 & 0 & p_{37} & 0 & 0 & p_{35} & 0 & 0 & 0 & p_{32} & 0\\
0 & m_{43} & m_{42} & 0 & 0 & 0 & 0 & 0 & 0 & 0 & m_{48} & 0 & 0 & m_{47} & m_{46} & 0 & 0 & 0\\
0 & n_{43} & n_{42} & 0 & 0 & 0 & 0 & 0 & 0 & 0 & n_{48} & 0 & 0 & n_{47} & n_{46} & 0 & 0 & 0\\
0 & p_{43} & p_{42} & 0 & 0 & 0 & 0 & 0 & 0 & 0 & p_{48} & 0 & 0 & p_{47} & p_{46} & 0 & 0 & 0\\
0 & 0 & 0 & 0 & m_{57} & 0 & 0 & m_{56} & m_{51} & 0 & 0 & 0 & m_{53} & 0 & 0 & m_{52} & 0 & 0\\
0 & 0 & 0 & 0 & n_{57} & 0 & 0 & n_{56} & n_{51} & 0 & 0 & 0 & n_{53} & 0 & 0 & n_{52} & 0 & 0\\
0 & 0 & 0 & 0 & p_{57} & 0 & 0 & p_{56} & p_{51} & 0 & 0 & 0 & p_{53} & 0 & 0 & p_{52} & 0 & 0\\
0 & 0 & 0 & 0 & 0 & 0 & m_{68} & m_{65} & 0 & 0 & 0 & m_{62} & 0 & 0 & m_{64} & 0 & 0 & m_{67}\\
0 & 0 & 0 & 0 & 0 & 0 & n_{68} & n_{65} & 0 & 0 & 0 & n_{62} & 0 & 0 & n_{64} & 0 & 0 & n_{67}\\
0 & 0 & 0 & 0 & 0 & 0 & p_{68} & p_{65} & 0 & 0 & 0 & p_{62} & 0 & 0 & p_{64} & 0 & 0 & p_{67}\\
0 & 0 & 0 & 0 & m_{75} & m_{78} & 0 & 0 & 0 & m_{73} & 0 & 0 & 0 & m_{74} & 0 & 0 & 0 & m_{76}\\
0 & 0 & 0 & 0 & n_{75} & n_{78} & 0 & 0 & 0 & n_{73} & 0 & 0 & 0 & n_{74} & 0 & 0 & 0 & n_{76}\\
0 & 0 & 0 & 0 & p_{75} & p_{78} & 0 & 0 & 0 & p_{73} & 0 & 0 & 0 & p_{74} & 0 & 0 & 0 & p_{76}\\
0 & 0 & 0 & 0 & 0 & m_{87} & m_{86} & 0 & 0 & 0 & m_{84} & 0 & 0 & 0 & 0 & 0 & 0 & 0\\
0 & 0 & 0 & 0 & 0 & n_{87} & n_{86} & 0 & 0 & 0 & n_{84} & 0 & 0 & 0 & 0 & 0 & 0 & 0\\
0 & 0 & 0 & 0 & 0 & p_{87} & p_{86} & 0 & 0 & 0 & p_{84} & 0 & 0 & 0 & 0 & 0 & 0 & 0
\end{bmatrix}
\begin{bmatrix} F_1\\ F_2\\ F_3\\ F_4\\ F_5\\ F_6\\ F_7\\ F_8\\ F_9\\ F_{10}\\ F_{11}\\ F_{12}\\ F_{13}\\ F_{14}\\ F_{15}\\ F_{16}\\ F_{17}\\ F_{18} \end{bmatrix}
$$

$m_{ij} = (x_i - x_j)/l_{ij}$, $n_{ij} = (y_i - y_j)/l_{ij}$, and $p_{ij} = (z_i - z_j)/l_{ij}$.

$$\mathbf{f}_{\mathrm{m}} = \int_{V} \bar{\mathbf{c}}^{t} \boldsymbol{\varphi} \bar{\mathbf{c}} \mathrm{dV} \tag{6.37}$$

The integration is taken over the volume of the element, where $\boldsymbol{\varphi}$ is the matrix relating the stresses to strains $\boldsymbol{\varepsilon} = \boldsymbol{\varphi\sigma}$ in three dimensional problems. The primary step in achieving the flexibility matrix of an element is determining the matrix $\bar{\mathbf{c}}$. It is obvious that the *i*th column of $\bar{\mathbf{c}}$ represents the resultant stresses due to unit element force $\mathbf{F}_i$ in force method and also stresses due to nodal forces $\mathbf{S}$ is equal to the *i*th column of $\mathbf{T}$ utilizing displacement method. Hence we can form matrix $\bar{\mathbf{c}}$ using stiffness properties of the brick element using the displacement method. Now the flexibility matrix of the element in the force method is formed from Eq. 6.37 using Gauss numerical integration method with eight Gauss points.

6.5.3 Graphs Associated with Finite Element Model

Here, topological properties of the FEM are transferred into the connectivity of its interface graph and natural associate graph.

6.5.3.1 Interface Graph

Interface graph of a FEM, denoted by IG(FEM), is constructed by the following rules:

1. Nodes of the IG(FEM) correspond to the nodes of FEM.
2. For each edge of a break element, one new member is added to the IG(FEM).
3. For each face of a break element, one new diagonal member is added to the IG (FEM). This member is located between second and third nodes of the current face of the element.

In fact there is one to one to one correspondence between element forces and member of the IG(FEM). The members of the interface graph are numbered according to the element numbers of the FEM. In this way for each element, corresponding members in interface graph are numbered consequently and then members of the next element are numbered. A FEM and the corresponding interface graph and the schematic numbering of the members corresponding to n*th* element in the interface graph are illustrated in Fig. 6.45.

6.5.3.2 Natural Associate Graph

The natural associate graph represented by NAG(FEM) is constructed by the following rules:

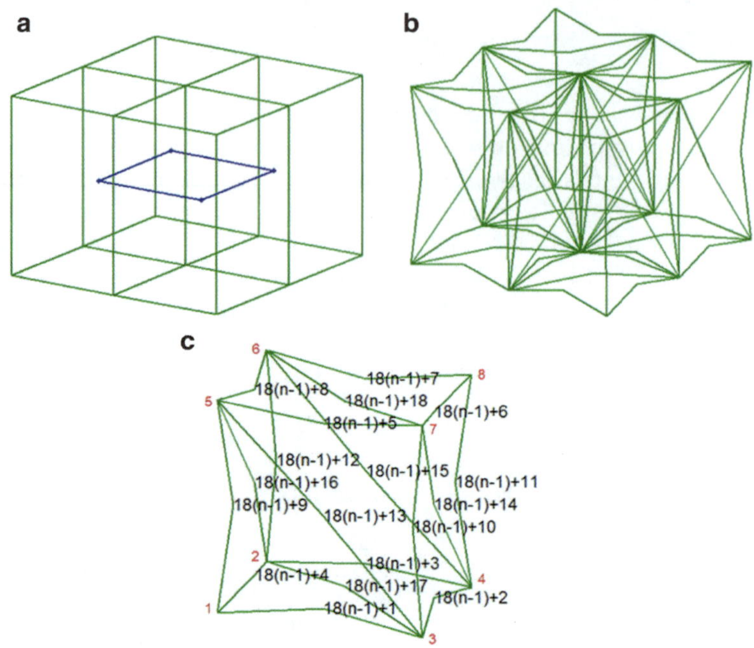

Fig. 6.45 (**a**) Finite element model (**b**) Interface graph of the FEM (**c**) Schematic numbering of the n*th* element

1. Nodes of the NAG(FEM) correspond to the elements of FEM.
2. For each pair of elements in FEM having four common nodes, one member is added between the corresponding two nodes in NAG(FEM).

NAG(FEM) can be constructed using the following procedure: One of the preliminary steps in FEA is defining the elements with their connected nodes. In this way the element connectivity matrix is constructed which contains the element-node incidence relationships. In the process of constructing the element connectivity matrix, another matrix which contains node-element incidence properties can be formed. This matrix is named the node connectivity matrix. Now using the element connectivity and the node connectivity matrices leads to an algorithm with complexity O(n) for an efficient generation of NAG.

In order to recognize the adjacent elements to the n*th* element which have common four nodes or one common face, first the connected nodes to the n*th* element are identified from the element connectivity matrix. In the subsequent step using the node connectivity matrix, elements which have at least one common node with the n*th* element are identified. Now it is convenient to seek for the adjacent elements in this reduced search space. A FEM and its corresponding NAG are illustrated in Fig. 6.46.

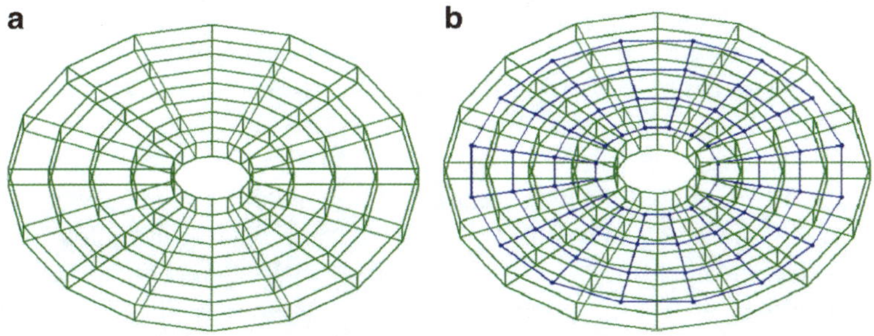

Fig. 6.46 Finite element model and Finite element model with natural associate graph

Fig. 6.47 Space truss model equivalent to a brick element

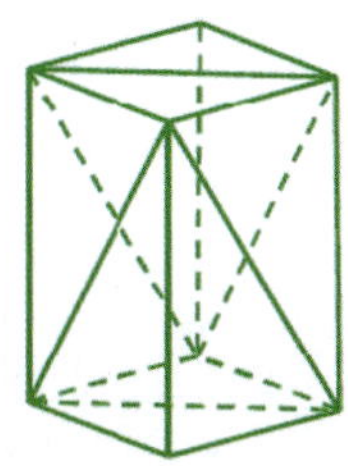

6.5.4 *Topological Interpretation of Static Indeterminacy*

6.5.4.1 Degree of Static Indeterminacy of the FEM

Each bi-action element force in a brick element can be considered as bi-action element forces in a bar element. In this way, the force system of the brick element will be equivalent to the force system of the corresponding space truss as indicated in Fig. 6.47.

Thus calculating the degree of static indeterminacy (DSI) and forming the self equilibrating systems of the FEM are replaced by the DSI and self equilibrating systems of the equivalent truss model. In this way using the DSI of a space truss with n nodes and m members as DSI = m − 3n + 6, the degree of indeterminacy of a FEM is obtained as.

$$DSI = 18E - 3N + 6 \tag{6.38}$$

where E is the number of brick elements and N is the total number of the nodes of the FEM.

6.5.4.2 Pattern of Type I Self-Equilibrating Systems

For each k multiple member in equivalent truss model of FEM, there are k unknown forces and one equilibrium equation in the member's direction. Thus DSI of the substructure is equal to k − 1 and k − 1 self equilibrating systems can be generated on each k multiple member of interface graph of the FEM. In this way, first each k multiple members are arranged in ascending order as ($m_1, m_2, m_3, \ldots, m_{k-1}, m_k$). where ($m_1 < m_2 < m_3 < \ldots < m_{k-1} < m_k$). Each selection of two members from this list is valid to construct a type I self-equilibrating system, but in order to achieve a better bandwidth reduction; selection of adjacent members from the defined list is preferable. Therefore k − 1 duplicate members are selected as (m_1, m_2), (m_2, m_3),. . ., (m_{k-1}, m_k). Each pair (m_i, m_j) with $i < j$ represents the numbers of corresponding self-equilibrating system. The member with bigger number is selected as the generator of the current SES and also as a redundant force. The null vectors corresponding to the type I SESs have two non-zero entries in rows i and j equal to −1 and 1, respectively.

Therefore by generating type I SESs, about three fourths of null basis is formed with maximum sparsity. These SESs are generated easily in the process of constructing natural associate graph of the FEM.

6.5.4.3 Relationship Between γ(S) and NAG(FEM)

By reducing the generators of the type I SESs from IG(FEM), the remaining subgraph is called graph *S*, with its associate graph A(S) being equivalent to NAG(FEM). In order to generate other types of the SESs, a relationship between the DSI of the equivalent truss of graph S and the natural associate graph of the FEM should be established. For achieving this aim an expansion process is employed.

Consider a brick element, as illustrated in Fig. 6.48a. The corresponding graph S is denoted by S_1 and NAG(FEM) is a single node. The equivalent structure is determinate. The graph S corresponding to two brick elements, denoted by S_2, is constructed by adding the subgraph $S_0^{\,1}$ (Fig. 6.49a) to the graph S_1 as indicated in Fig. 6.48b, and also one node and one member is added to NAG(FEM) with the DSI becoming one. Consequently by adding subgraph $S_0^{\,1}$ to the previous graph, it adds one node and member to the NAG(FEM) and it is growing as a tree and the DSI increases by unity.

In some stages of the expansion process adding subgraph $S_0^{\,2}$ (Fig. 6.49b) to the previous graph S from two faces, as shown in Fig. 6.48d, is equivalent to adding one node and two members to the NAG(FEM) and a cycle is formed in the NAG(FEM). In this case, the DSI of corresponding truss is increased by three.

Considering the above points, the number of the nodes and members of the NAG (FEM) can be calculated as

$$N' = \frac{M(S) + 2\gamma(S) - 3}{15} \tag{6.39}$$

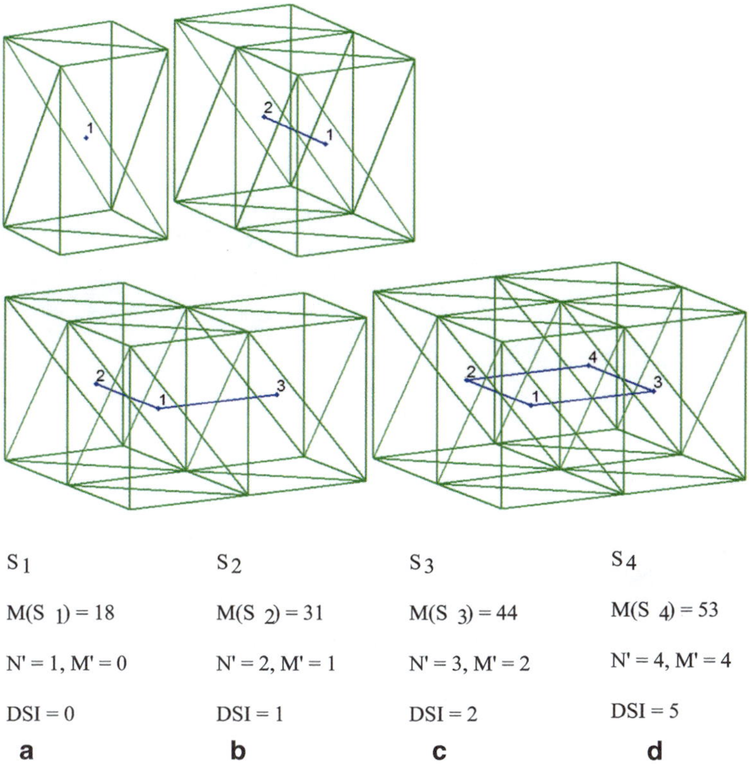

Fig. 6.48 The expansion process for the formation of a γ-cycle without internal node

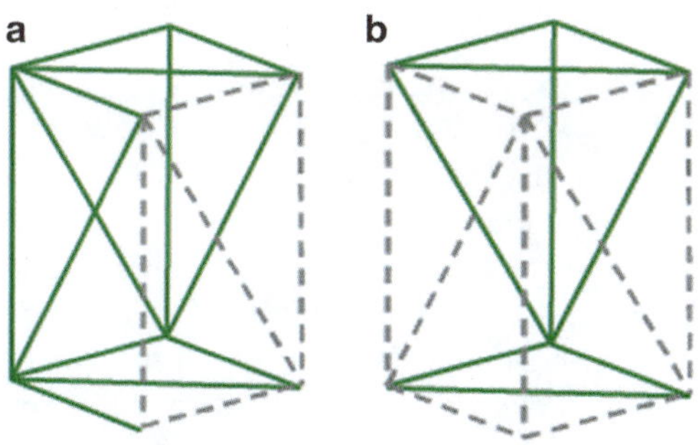

Fig. 6.49 A typical S_0 subgraphs. (**a**) S_0^1, (**b**) S_0^2

$$M' = \frac{M(S) + 17\gamma(S) - 18}{30} \tag{6.40}$$

Now, the relation between the DSI of the equivalent truss of the graph *S* and the independent cycles of the natural associate graph of FEM can be established. The

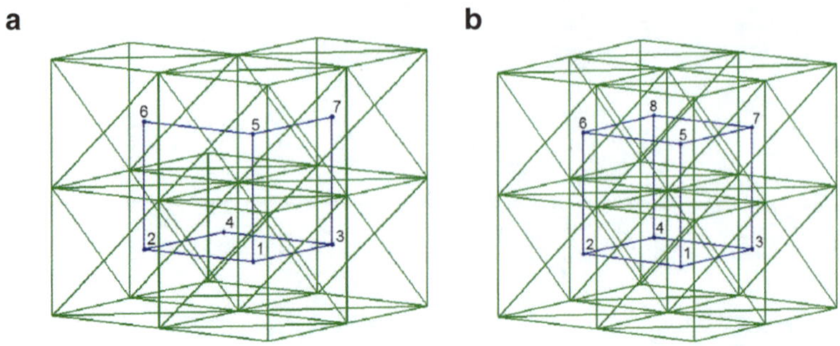

S_7
$M(S_7) = 84$
$N' = 7, M' = 9$
$DSI = 12$

S_8
$M(S_8) = 90$
$N' = 8, M' = 12$
$DSI = 15$

Fig. 6.50 The expansion process for the formation of a γ-cycle with internal node

first Betti number of the natural associate graph of the FEM, states the number of independent cycles of this graph which is expressed as

$$b_1(\mathrm{NAG(FEM)}) = M' - N' + 1 \tag{6.41}$$

Adding M' to both sides of the Eq. 6.41 leads to

$$b_1(\mathrm{NAG(FEM)}) + M' = 2M' - N' + 1 \tag{6.42}$$

Substituting Eqs. 6.39, 6.40, 6.41, and 6.42 results in

$$\gamma(S) = b_1(\mathrm{NAG(FEM)}) + M' \tag{6.43}$$

According to this equation, the DSI of the equivalent truss of the graph *S* can be expressed as the sum of the number of members and the first Betti number of NAG (FEM) that corresponds to type II and type III self-equilibrating systems.

The graph *S* corresponding to the eight brick elements denoted by S_8 is constructed by adding one node and six members to the graph S_7 as illustrated in Fig. 6.50. This process adds one node, three members and three minimal cycles to the NAG(FEM), and also the DSI of the equivalent graph increases by three.

When the FEM or corresponding graph *S* has an internal node and the NAG (FEM) becomes a polyhedral, then Eqs. 6.39 and 6.40 will be modified as

$$N' = \frac{M(S) + 2\gamma(S) + 3N_i(S) - 3}{15} \tag{6.44}$$

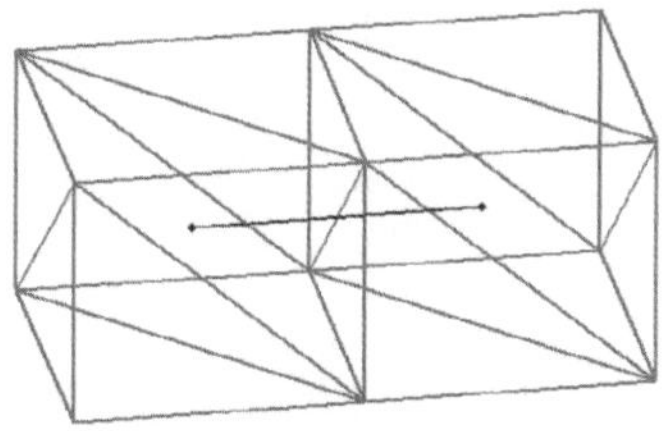

Fig. 6.51 Pattern of a Type II self-equilibrating system

$$M' = \frac{M(S) + 17\gamma(S) + 33N_i(S) - 18}{30} \tag{6.45}$$

Therefore the relation between the DSI of equivalent truss of graph *S* and the first Betti number of the NAG(FEM) is modified as

$$\gamma(S) = b_1(NAG(FEM)) + M' - 2N_i(FEM) \tag{6.46}$$

Comparing Eqs. 6.43 and 6.46 demonstrates that the FEM has no internal node the DSI of the equivalent truss of graph *S* is equal to the sum of the number of members and first Betti number of NAG(FEM), however, when the FEM has one or more internal nodes, $2N_i$(FEM) self-equilibrating systems are not independent and must be ignored.

6.5.4.4 Pattern of Type II Self-Equilibrating Systems

As mentioned, type II self-equilibrating systems as indicated in Fig. 6.51 are topologically identical to the subgraph of graph *S* which corresponds to the two connected nodes of the natural associate graph of the FEM.

The most important point in type II self-equilibrating systems is to select an appropriate generator. Because by eliminating these generators from graph *S*, the sub-structure of type III SESs and primary structure of the structure *S* must be stable. To achieve this, the following rule for appropriate selection of generators of type II SESs is suggested.

In this way avoiding instability of the subsequent type of the SESs, the following procedure is applied, as indicated in Fig. 6.52. For a type II SESs (in any coordinate system such as Cartesian, cylindrical or spherical) generators of the type II SESs in directions 1, 2 and 3 are the chosen members which are numbered as 8, 11, and 23.

6.5.4.5 Pattern of Type III Self-Equilibrating Systems

According to the expansion process in models without opening, sub-structures which are topologically identical to the minimal cycles of the natural associate

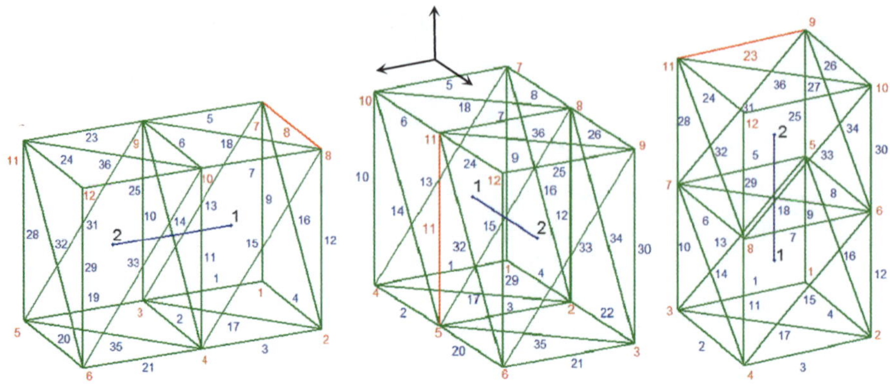

Fig. 6.52 Selected generators of the type II SES

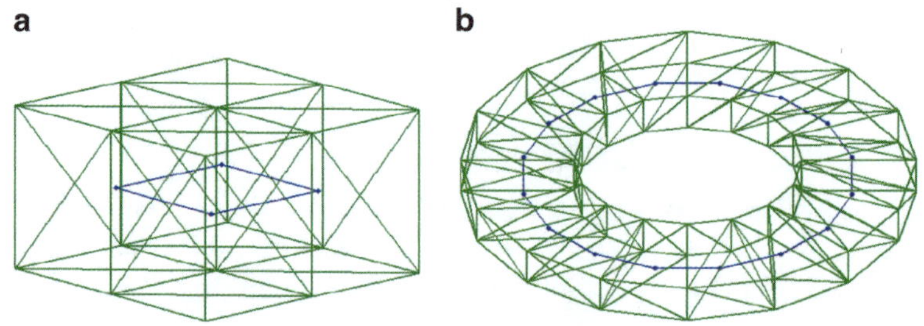

Fig. 6.53 Minimal cycles of the natural associate graph of the FEM

graph of FEM contains some type II and one type III self-equilibrating systems as indicated in Figs. 6.53a and 6.54.

6.5.4.6 Type I Minimal Cycles

These minimal cycles of the natural associate graph of the FEM pass through E elements which have two common nodes and one edge. Corresponding interface graph of these elements have $N = 4E + 2$ nodes. Therefore using Eq. 6.37 the DSI of the related sub-structure is equal to $6E$. Obviously $5E - 1$ and E type I and type II self-equilibrating systems can be extracted from the mentioned sub-structure. The DSI of the remaining sub-structure is 1. Thus each type I minimal cycle of the natural associate graph of the FEM contains a type III self-equilibrating system and one null vector.

Avoiding instability of the primary structure S, the procedure indicated in Fig. 6.55 is applied to selection of the generators of the type III SESs. For a type III SESs (in any coordinate system such as Cartesian, cylindrical or spherical) generators of SESs perpendicular to the directions 1, 2 and 3 are chosen members which are numbered as 70, 51 and 17, respectively.

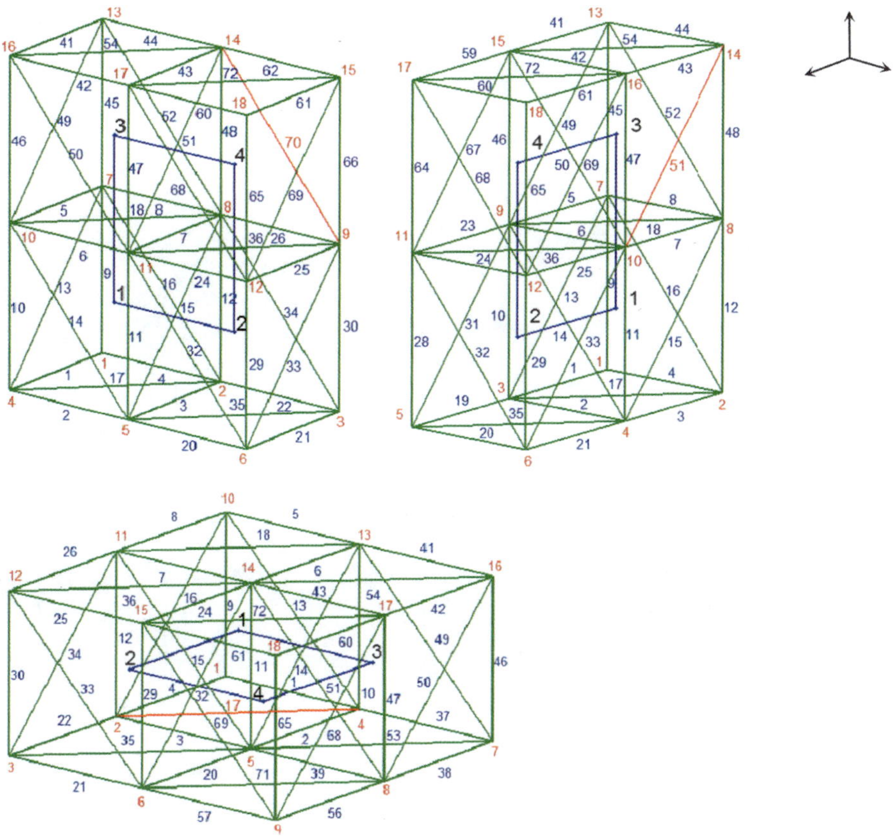

Fig. 6.54 Selected generators of the type III SESs

6.5.4.7 Type II Minimal Cycles

For models with openings, each independent cycle of the natural associate graph which surrounds an opening of the FEM is called type II minimal cycle of the natural associate graph. Considering that this cycle passes through E elements and its corresponding interface graph has $N = 4E$ nodes. Using Eq. 6.37, the DSI of the related sub-structure is equal to $6E + 6$. Obviously $5E$ and E type I and type II self-equilibrating systems can be extracted from the mentioned sub-structure. Therefore the DSI of the remaining sub-structure is 6. Thus each type II minimal cycle of the natural associate graph of the FEM contains six self-equilibrating systems of type III and six corresponding null vectors. These null vectors can easily be generated on the corresponding sub-structure utilizing an algebraic method.

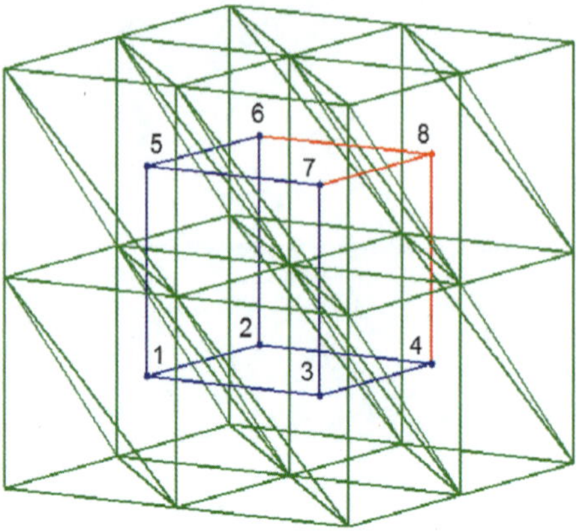

Fig. 6.55 Eight elements and the corresponding hexahedron natural associate graph

6.5.5 Models Including Internal Node

For N_i internal node in the FEM, $2N_i$ self-equilibrating systems are not independent from others and then should be selected and ignored. Since each type III SESs include some type I and type II SESs, therefore ignoring each type I and type II SESs causes the corresponding type III SESs not to be valid. Therefore for any internal node, two type III SESs should be selected and ignored. The following procedure should be applied to select dependent SESs.

Considering graph S_7 its equivalent truss is twelve times statically indeterminate. As it can be seen from Fig. 6.50a, b adding one node and six members consisting of three edge members and three diagonal ones, to graph S_7 forms graph S_8. Then the DSI of the equivalent truss is increased by three. Considering the equivalent truss of graph S_7 and by eliminating the restraints corresponding to the generators of SESs, the primary structure which is determinate and stable is obtained. Also corresponding primary structure of graph S_8 is obtained by eliminating the generators of SESs of graph S_7 plus the above mentioned three diagonal members, form graph S_8.

Eight elements correspond to each internal node and the natural associate graph corresponding to these elements is a hexahedron. At the beginning, from each hexahedron as illustrated in Fig. 6.55, node 8 is considered as the last node which makes up the hexahedron. These nodes must be distinct from each others.

From each hexahedron three type III SESs corresponding to three minimal cycles of NAG(FEM) which pass through the selected node, are ignored and also the mentioned three diagonal members used as generator of three type II SESs corresponding to three members of NAG(FEM) which pass through the selected node. In Figs. 6.55 and 6.56, the red members represent the modified type II SESs.

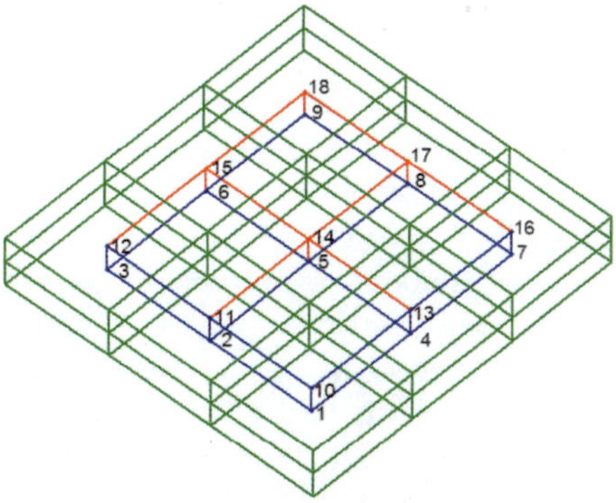

Fig. 6.56 A finite element model and the corresponding natural associate graph

6.5.6 Selection of an Optimal List Corresponding to Minimal Self-Equilibrating Stress Systems

The main goal of this section is the selection of a member list for each self-equilibrating systems which has the nearest member numbers to the generator of the system.

$$\text{Minimize}\,(\text{Abs}(i - j)) \tag{6.47}$$

where j is the number of the generator member, and i are the number of members of the self-equilibrating system.

Consider d_i^m; $(i = 1, 2, \ldots, k; \ m = 1, 2, \ldots, n)$, representing the member numbers of the multiple members where $d_i^1 < d_i^2 < \ldots < d_i^n$, and s_j; $(j = 1, 2, \ldots, t)$ representing the member numbers of the simple members. Since d_i^m with $m \neq 1$ is used as generator of type I of SESs, as illustrated in Figs. 6.52 and 6.53, the generator of type II and type III of SESs must be selected from $\{d_i^1, s_j\}$. For maximum bandwidth reduction, from each multiple member one member is selected which has the nearest number to the generator's number of the self-equilibrating system. In order to achieve this goal, Eq. 6.48 should be applied.

$$\text{find}\left(d_i^j\right)|\text{Abs}\left(d_i^j - d_G^1\right) = \text{Min}\left(\text{abs}\left(d_i^j - d_G^1\right)\right) \tag{6.48}$$

then

$$d_i^1 = d_i^j \ \ (i = 1, 2, \ldots, k; j = 1, 2, \ldots, n) \tag{6.49}$$

where, d_G^1 is the generator of the self-equilibrating system.

Algorithm. Step 1: Number the nodes of the FEM. Nodal numbering does not affect the pattern of the flexibility matrix of the FEM.

Step 2: Define the brick elements through its eight nodes. Use an efficient method for element numbering, for having small bandwidth for the null basis matrix $\boldsymbol{B}_1$ and the flexibility matrix.

Step 3: Generate the natural associate graph of the FEM.

Step 4: Generate the interface graph of the FEM in manner that its member numbering is according to the element numbering of the FEM. In this way for each element, corresponding members in interface graph are numbered consequently.

Step 5: Construct the equilibrium matrix of the FEM.

Step 6: Set up the type I self-equilibrating systems and calculate the corresponding null vectors which have two nonzero entries in the rows corresponding to the member numbers.

Step 7: Set up the type II self-equilibrating systems and calculate the corresponding null vectors form the relevant equilibrium sub-matrix.

Step 8: Set up the type III self-equilibrating systems and calculate the corresponding null vectors form the relevant equilibrium sub-matrix.

Step 9: Construct the statical basis (null basis) of the FEM by arranging the null vectors in the null basis in the ascending manner utilizing the highest member number of the corresponding self-equilibrating systems.

The efficiency of this algorithm is shown through two examples by comparing the required computational times for the construction of the null basis matrices, also non-zero patterns and condition numbers of the flexibility matrices. In this comparison (a) Present method (b) Turn-back method (c) Gauss-Jordan elimination method are considered.

6.5.7 Numerical Examples

In this section two FEMs are considered, one of these models is assumed to be supported in statically indeterminate fashion and the other supported in a determinate fashion. Null basis and flexibility matrices are formed and the required computational times, and the condition numbers are calculated. In the following examples, nz represents the number of non-zero entries and $\lambda_{max}/\lambda_{min}$ is the ratio of the extreme eigenvalues taken as the condition number of a matrix.

Example 1. A thick arch type structure, having internal radius of 8 m, discretized by brick elements. The corresponding FEM is supported in a statically indeterminate fashion as illustrated in Fig. 6.57. The mechanical and topological properties of the model are as follow:

Poisson's ratio = 0.2; Elastic modulus E = 2E + 10 N/m^2; Density ρ = 2,400 kg/m^3;
Number of nodes = 165; Number of internal nodes (N_i) = 27;
Number of elements = 80;
Number of members of the natural associate graph = 172;

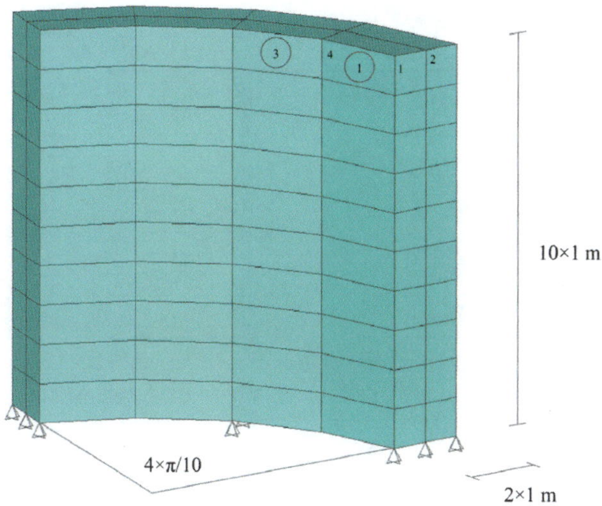

Fig. 6.57 A thick arch type structure discretized by brick elements

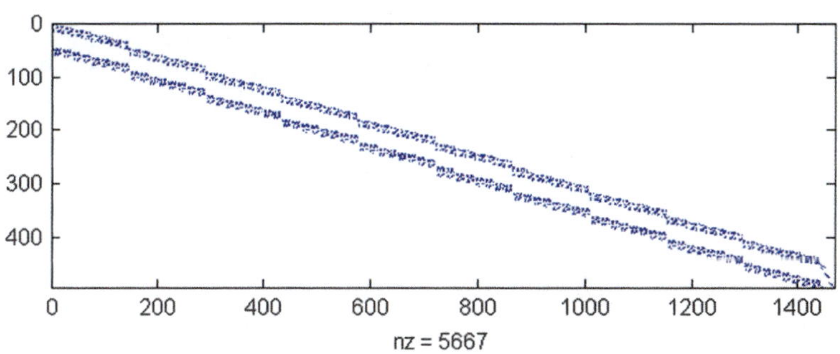

Fig. 6.58 Pattern of the equilibrium matrix of the FEM for Example 1

First Betti number of the natural associate graph (independent cycles) = 93;
Number of Type I self-stress systems = 740 (77.8 %);
Number of Type II self-stress systems = 172 (18.1 %);
Number of Type III self-stress systems = 93 − 2 × 27 = 39 (4.1 %);
$DSI_{Internal} = 951$; $DSI_{External} = 21$; $DSI_{Total} = 972$;

The pattern of the equilibrium matrix of the FEM is displayed in Fig. 6.58. The nodes and elements of the FEM are numbered in a way to produce a banded equilibrium matrix. This characteristic facilitates the Turn-back method to form a null basis with less required computational time and more banded form.

The null basis of the FEM can be constructed using a mixed algebraic-graph theoretical and pure algebraic methods. In mixed methods, graph theoretical

Table 6.8 Definition of the element forces

Force system	Location (nodes)	Force system	Location (nodes)
1	1,3	10	3,7
2	3,4	11	4,8
3	2,4	12	2,6
4	1,2	13	3,5
5	5,7	14	4,7
6	7,8	15	4,6
7	6,8	16	2,5
8	5,6	17	2,3
9	1,5	18	6,7

Table 6.9 Member list corresponding to the type II or type III self-equilibrating systems

d_1^1	d_2^1	…	S_1	S_2	…	d_G^1	…	d_i^1	…	S_j	…	d_k^1
d_1^1	d_2^2					d_G^2		d_i^2				d_k^2
	d_2^3					$\vdots$		$\vdots$				
						d_G^n		d_i^n				

approach is utilized to form columns of the null basis which are related to the internal indeterminacies where algebraic procedures form the columns corresponding to the external indeterminacies.

In this case, graph theoretical approach is employed together with the Turn-back method (Tables 6.8 and 6.9). As displayed in Fig. 6.59 and Table 6.10, applying the mixed graph theoretical with Turn-back method lead to a highly sparse and banded null basis; however it requires some additional computational time than using the mixed graph theoretical with QR decomposition method.

Pure algebraic methods are also used to form the null basis of the equilibrium matrix of the FEM. From Fig. 6.59 and Table 7.3 it can be observed that each pure algebraic method has better performance when they are used together with the present graph theoretical method (Fig. 6.60).

Example 2. An arch type structure with an internal radius of 8 m, discretized by brick elements. As shown in Fig. 6.61, this structure has two openings and is supported in a statically determinate fashion. Properties of the model are as follow:

Mechanical Properties

Poisson's ratio = 0.2; Elastic modulus E = 2E + 10 N/m^2; Density ρ = 2,400 kg/m^3;

Topological Properties

Number of nodes = 108; Number of internal nodes (N_i) = 0; Number of elements = 38;

Number of members of the natural associate graph = 59

First Betti number of the natural associate graph (independent cycles) = 22;

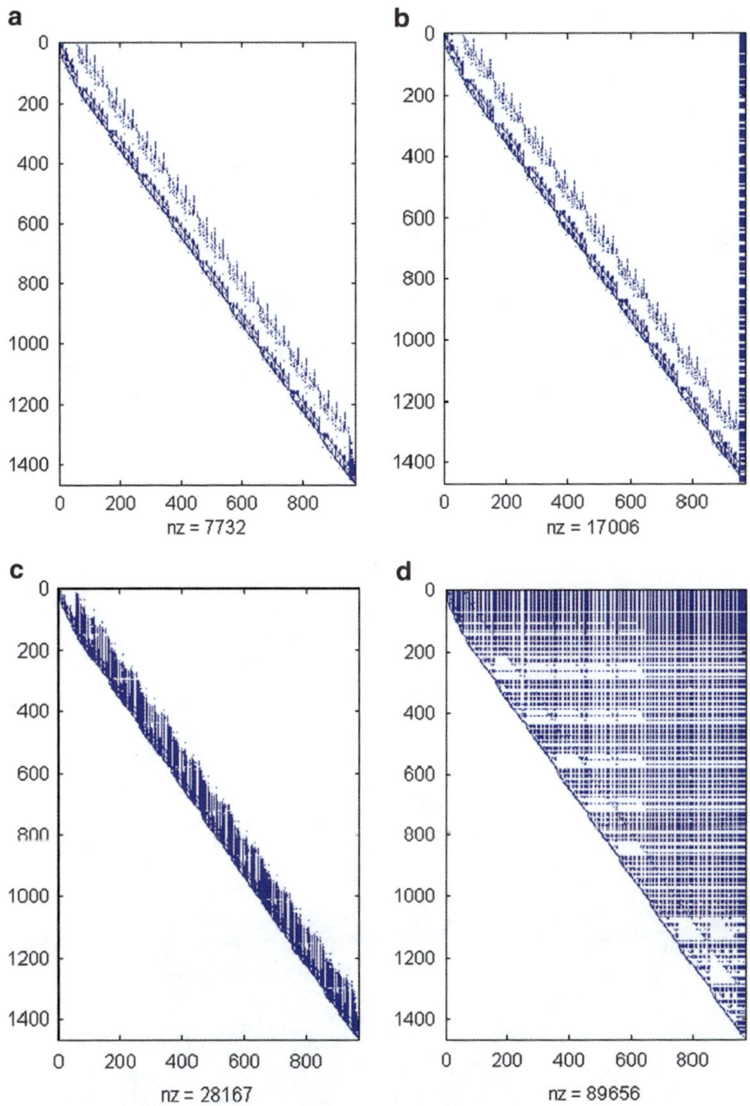

Fig. 6.59 Pattern of the null basis B_1 matrices corresponding to Example 1 utilizing: (**a**) Graph theoretical-Turn back method (**b**) Graph theoretical-QR method (**c**) Turn-back method (**d**) Gauss Jordan elimination method

Number of Type I self-stress systems = 275 (75.1 %);
Number of Type II self-stress systems = 59 (16.1 %);
Number of Type III self-stress systems = 20 + 2 × 6 = 32 (8.8 %);
$DSI_{Internal} = 366$; $DSI_{External} = 0$; $DSI_{Total} = 366$;

Table 6.10 Comparison of the optimality characteristics of the null basis matrices $\mathbf{B}_1$ and the flexibility matrices **G** for the FEM of Example 1

	Null basis $\mathbf{B}_1$		Flexibility matrix **G**	
	Time (sec)	$\frac{\text{Time}}{\text{Time}_{\text{Present method}}}$	$\frac{\lambda_{max}}{\lambda_{min}}$	$\frac{\text{nz entries}}{\text{nz entries}_{\text{Present method}}}$
Graph theoretical-Turn back method	2.464	1.000	2.475e+07	1.000
Graph theoretical-**QR** method	0.824	0.334	4.699e+08	1.692
Turn back method	70.358	28.554	2.776e+07	1.622
Gauss Jordan elimination method	25.921	10.520	5.655e+06	7.279

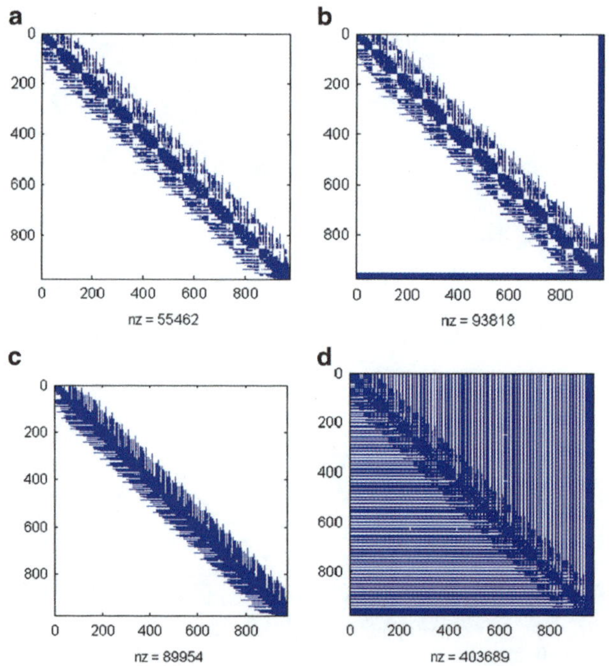

Fig. 6.60 Pattern of the flexibility matrices G corresponding to Example 1 utilizing: (**a**) Graph theoretical-Turn back method (**b**) Graph theoretical-QR method (**c**) Turn-back method (**d**) Gauss Jordan elimination method

Fig. 6.61 An arch type structure containing two openings, discretized by brick elements

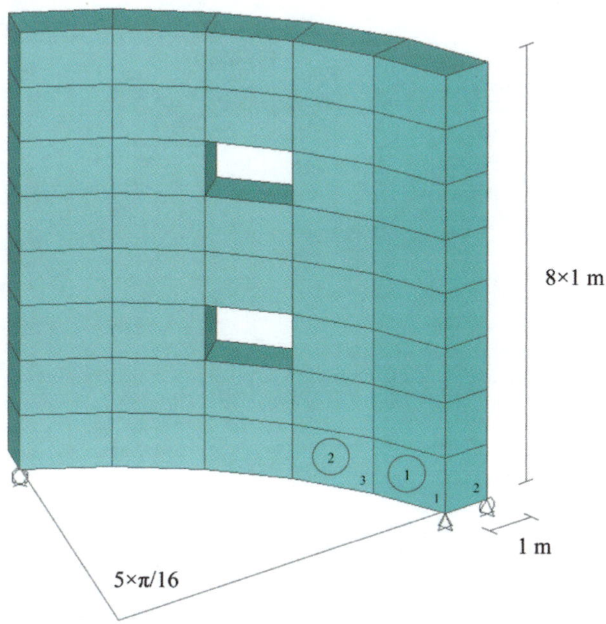

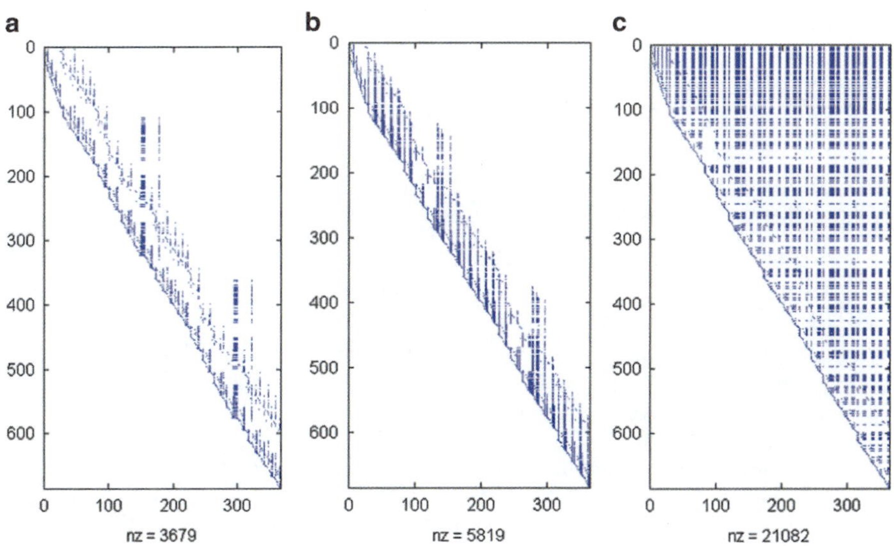

Fig. 6.62 Pattern of the null basis $\mathbf{B}_1$ matrices corresponding to Example 2 utilizing: (**a**) Graph theoretical method (**b**) Turn-back method (**c**) Gauss Jordan elimination method

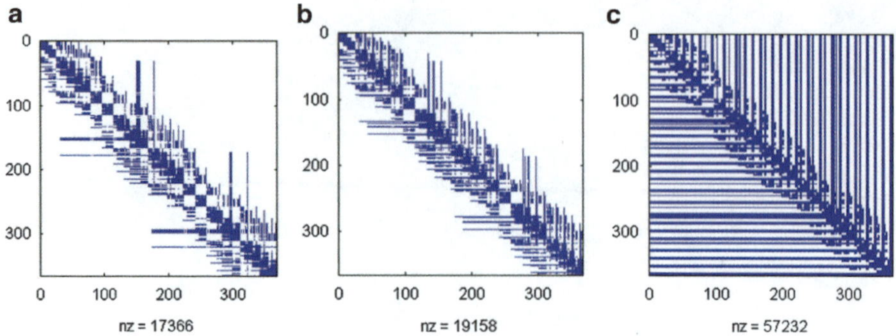

Fig. 6.63 Pattern of the flexibility matrices G corresponding to Example 2 utilizing (**a**) Graph theoretic method (**b**) Turn-back method (**c**) Gauss Jordan elimination method

Table 6.11 Comparison of the optimality characteristics of the null basis matrices $\boldsymbol{B}_1$ and the flexibility matrices $\boldsymbol{G}$ for the FEM of Example 2

	Null basis $\boldsymbol{B}_1$		Flexibility matrix $\boldsymbol{G}$	
	Time (sec)	Time / Time Present method	$\frac{\lambda_{max}}{\lambda_{min}}$	nz entries / nz entries Present method
Graph theoretical method	0.171	1.000	2.860e+4	1.000
Turn back method	16.816	98.339	1.033e+4	1.103
Gauss-Jordan elimination method	6.394	37.392	1.814e+5	3.296

The pattern of the null basis and flexibility matrices are illustrated in Figs. 6.62 and 6.63 and it can be easily seen that present graph theoretical method and Turn-back method lead to banded null basis and flexibility matrices. Table 6.11 containing the optimality characteristics of the applied methods reveals that the presented method requires acceptable computational time for constructing the null basis of the FEM.

As mentioned, the present method leads to a highly sparse and banded flexibility matrix requiring a low computational time with an acceptable condition number.

These examples are also analyzed by the standard displacement method and the integrated force method. Figure 6.64 illustrates the pattern of the reduced stiffness matrix, and its optimality characteristics are provided in Table 6.12. It can be seen that for these examples, the displacement method analyzes the model with less unknowns than the presented graph theoretical force method, and the reduced stiffness matrix $\boldsymbol{K}_r$ has less non-zero entries than the flexibility matrix $\boldsymbol{G}$.

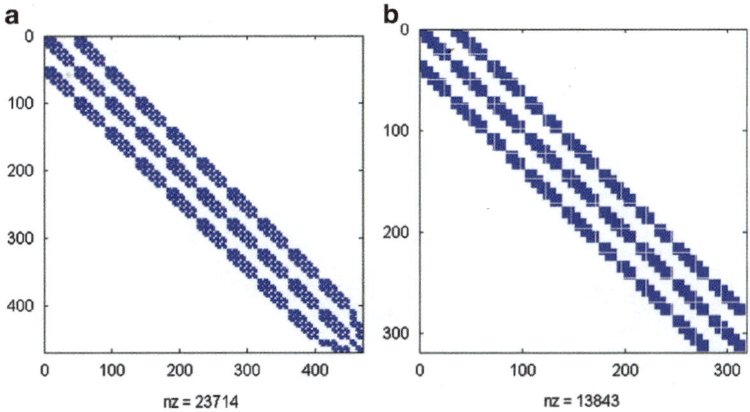

Fig. 6.64 Pattern of the reduced stiffness matrix K_r corresponding to (**a**) Example 1 (**b**) Example 2

Table 6.12 Optimality characteristics of the stiffness matrix $\boldsymbol{K_r}$

	$\frac{\lambda_{max}}{\lambda_{min}}$	$\frac{\text{nz entries}}{\text{nz entries}_{\text{Present method}}}$
Example 1	1.793e+4	0.428
Example 2	2.667e+5	0.797

References

1. Kaveh A, Koohestani K, Taghizadeh N (2007) Efficient finite element analysis by graph-theoretical force method. Finite Elem Anal Des 43(6–7):543–554
2. Kaveh A, Koohestani K (2009) Efficient graph-theoretical force method for two dimensional rectangular finite element analysis. Commun Numer Methods Eng 25(9):951–971
3. Kaveh A, Ebrahimi E (2012) Graph-theoretical force method of finite element models with triangular and rectangular elements. Asian J Civil Eng 13(5):597–616
4. Kaveh A (1974) Applications of topology and matroid theory to the analysis of structures. Ph. D. thesis, Imperial College, London University, London, UK
5. Kaveh A, Koohestani K (2008) Efficient finite element analysis by graph-theoretical force method; triangular and rectangular plate bending elements. Finite Elem Anal Des 44:646–654
6. Kaveh A, Koohestani K (2007) An efficient graph theoretical method for plate bending finite element analysis via force method. Eng Comput 24(7):679–698
7. Kaveh A, Koohestani K (2008) Efficient graph-theoretical force method for three dimensional finite element analysis. Commun Numer Methods Eng 24(11):1533–1551
8. Kaveh A, Tolou Kian SJ (2012) Efficient finite element analysis using graph-theoretical force method; brick element. Finite Elem Anal Des 54:1–15
9. Kaveh A, Roosta GR (1988) Comparative study of finite element nodal ordering methods. Eng Struct 20(1–2):86–96
10. Przemieniecki JS (1968) Theory of matrix structural analysis. McGraw-Hill, New York

11. Kaveh A (2004) Structural mechanics: graph and matrix methods, 3rd edn. Research Studies Press (Wiley), Somerset
12. Watkins DS (2002) Fundamentals of matrix computations, 2nd edn. Wiley, New York
13. Kaveh A, Fazli H (2008) Analysis of frames by substructuring technique based on using algebraic and graph methods. Commun Numer Methods Eng 24(10):867–874

Chapter 7
Optimal Force Method for FEMS: Higher Order Elements

7.1 Introduction

In this chapter force method for the analysis of finite element models comprising of higher order elements are studied.

In the first part, an efficient graph theoretical force method is presented for the analysis of FEMs comprising of higher order triangular elements, corresponding to highly sparse and banded flexibility matrices [1]. This is achieved by associating special graphs to a finite element model, and selecting subgraphs for the formation of localized self stress systems.

In second part, a method is described for the formation of null bases for FEMs comprised of higher order rectangular plane stress and plane strain elements (serendipity family elements) leading to highly sparse and banded flexibility matrices for optimal finite element analysis by force method [2].

In the third part, an competent method is described for the formation of null bases of finite element models (FEMs) consisting of hexahedron elements, corresponding to highly sparse and banded flexibility matrices. This is achieved by associating special graphs with the FEM and selecting appropriate subgraphs and forming the self-equilibrating systems on these subgraphs [3].

7.2 Finite Element Analysis of Models Comprised of Higher Order Triangular Elements

This part introduces an efficient method for the finite element analysis of models comprised of higher order triangular elements. The presented method is based on the force method and benefits graph theoretical transformations. For this purpose, minimal subgraphs of predefined special patterns are selected. Self-equilibrating systems (S.E.Ss) are then constructed on these subgraphs leading to sparse and

A. Kaveh, *Computational Structural Analysis and Finite Element Methods*, 281
DOI 10.1007/978-3-319-02964-1_7, © Springer International Publishing Switzerland 2014

banded null basis. Finally, well-structured flexibility matrices are formed for efficient finite element analysis.

7.2.1 Definition of the Element Force System

Defining appropriate structural elements is the first step of structural analysis. Based on the analysis approaches, structural elements are formulated in different manners. In case of higher order triangular elements (in-plane forces), in displacement method two forces are employed at each node of the element, while in force method the following force system is utilized.

Considering an O(n) element first, 3n sets of edge bi-action forces are described between adjacent side nodes. Then n(n − 1)/2 bi-action forces are added between adjacent nodes parallel to side 23. The same forces are added parallel to side 13. Finally n − 1 bi-action forces are added in the same manner, parallel and in the closest position to side 12. Force systems corresponding to the second, third and fourth order elements are shown in Fig. 7.1a–c. These independent element forces denoted by $\boldsymbol{F}$ are related to nodal forces $\boldsymbol{S}$ using Eq. 7.1.

$$\mathbf{S} = \mathbf{TF} \tag{7.1}$$

7.2.2 Flexibility Matrix of the Element

The flexibility matrices of higher order triangular elements can simply be formed using the stiffness matrices of such elements.

$$\mathbf{f}_{\mathrm{m}} = (\mathbf{T}_{\mathrm{r}})^{\mathrm{t}}(\mathbf{K}_{\mathrm{r}})^{-1}\mathbf{T}_{\mathrm{r}} \tag{7.2}$$

where the subscript r indicates that, corresponding orders of matrices to dependent forces are reduced.

7.2.3 Graphs Associated with Finite Element Model

In order to benefit topology in finite element analysis, first some topological transformations of FEM are needed. In this relation ten different graphs are presented in Ref. [4]. Here natural associate graph and interface graph are used that are defined in the following:

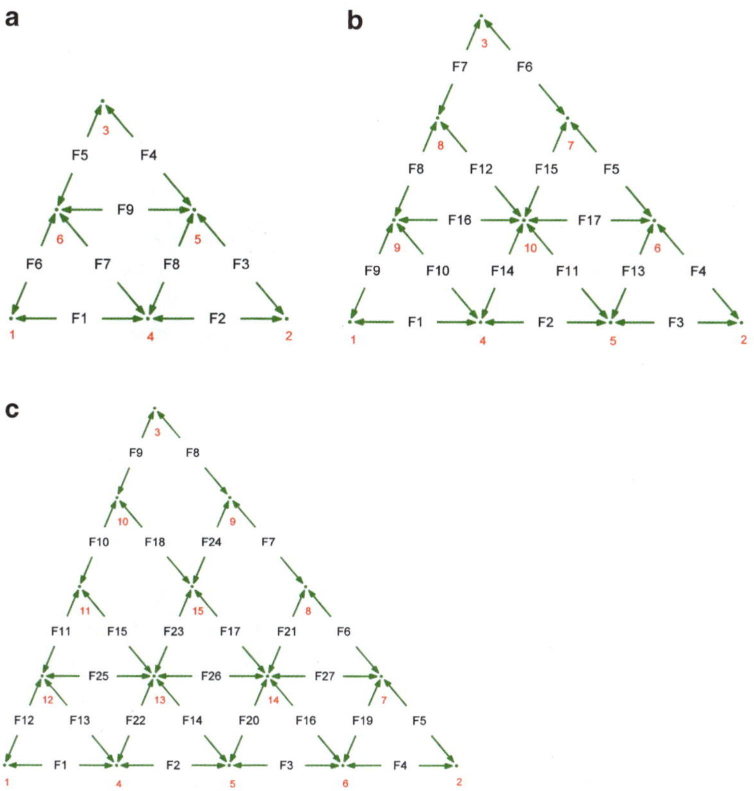

Fig. 7.1 Element force systems of higher order triangular elements

7.2.3.1 Natural Associate Graph

The natural associate graph of a FEM is represented by NAG(FEM). This graph reveals elements adjacency properties and as illustrated in Fig. 7.2a is constructed by following rules:

1. Each node of NAG(FEM) corresponds to each element of the FEM.
2. Two nodes of NAG(FEM) are connected with a member if two corresponding O(n) elements have n + 1 common nodes on a common edge.

Natural associate graph can easily be generated using the following procedure:

Connected nodes with a considered element are identified using element connectivity matrix.

1. Connected elements with these nodes are identified using node connectivity matrix.

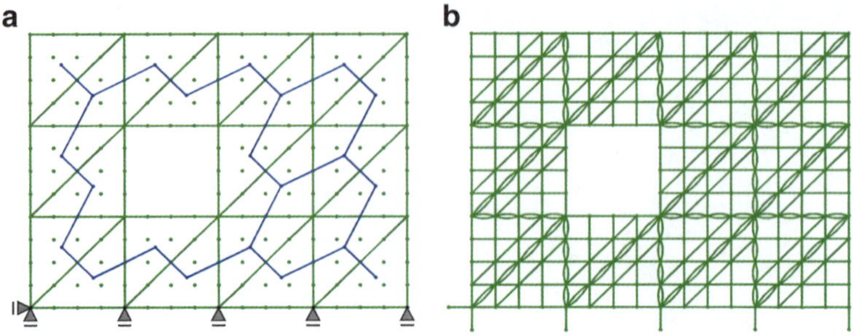

Fig. 7.2 Natural associate graph and interface graph of the corresponding FEM comprised of fourth order elements

2. All identified elements in step 2, at least have one common node with the considered element in step 1. Now among these identified elements the one which has n + 1 common nodes with the considered element, is desirable.

7.2.3.2 Interface Graph

The interface graph of a FEM is represented by IG(FEM). This graph corresponds to the force system of the FEM, and as indicated in Fig. 7.2b it is constructed by the following rules:

1. Each node of IG(FEM) corresponds to the each node of the FEM.
2. Members of the IG(FEM) correspond to the force system of FEM between their adjacent nodes.
3. Each support condition is considered as a member of IG(FEM).

Members of the interface graph corresponding to the element forces are numbered according to element numbering. Meantime corresponding members to support conditions are numbered before members of their connected elements.

7.2.4 Topological Interpretation of Static Indeterminacies

7.2.4.1 Degree of Static Indeterminacy of the FEM

As mentioned in Sect. 3.1, the introduced element force system is comprised of a number of bi-action forces. In accordance with Przemieniecki [5] each bi-action force can be considered as force system of a bar element, hence force system of the equivalent truss element can be employed instead of the force system of the original element. Therefore, the DSI of the FEM and self-equilibrating systems can be conveniently explored.

Each higher order triangular element has $n^2 + 3n - 1$ bi-action forces, and hence the DSI of a FEM which is comprised of *E* elements and contains *N* nodes is obtained as

$$\mathrm{DSI} = \left(n^2 + 3n - 1\right)E - 2N + 3 \tag{7.3}$$

Following self-equilibrating systems are found. Then null vectors which express the equilibrium conditions of the self-equilibrating systems are generated.

7.2.4.2 Pattern of the Type I Self-Equilibrating Systems

The interface graph of a FEM contains double members at the interface of two elements (Fig. 7.2b). Each double member of the interface graph correspond to a Type I self-equilibrating system. The self-equilibrating system consisting of two members numbered as i and j (j > i). The member with bigger number is selected as the generator of the self-equilibrating system and is considered as the redundant force of the FEM. Typical null vector corresponding to a Type I self-equilibrating system contains two nonzero entries in *ith* and *jth* rows equal to −1 and 1, respectively.

The above mentioned double members can conveniently be identified while natural associate graph is being generated.

7.2.4.3 Identification of Other Self-Equilibrating Systems Using an Expansion Process

This section adopts a method to identify other self-equilibrating systems and locate other redundant forces. Consider a graph *S* the same as the interface graph of the FEM with generators of Type I self-equilibrating systems being removed.

7.2.4.4 Models Excluding Openings

Consider a general triangular element as shown in Fig. 7.3a. The corresponding graph *S* contains $N_E = (n + 1)(n + 2)/2$ nodes and $M_E = (n + 1)(n + 2) - 3$ members, thus the equivalent truss is determinate. The NAG(FEM) is an isolated node. When another element is added (Fig. 7.3b, c), each time $N_E - n - 1$ nodes and $M_E - n$ members are added to the corresponding graph *S*, thus the indeterminacy is increased by $n - 1$. The NAG(FEM) grows with a node and member. This is true while The NAG(FEM) is growing like a tree (with no cycle).

In some steps of the expansion process adding an element grows NAG(FEM) by a node and two members, and a cycle is formed in the natural associate graph, as illustrated in Fig. 7.3d. In this situation $(N_E - 2n - 1)$ nodes and $(M_E - 2n)$ members are added to the corresponding graph *S*, hence the indeterminacy is increased by $2(n - 1) + 1$.

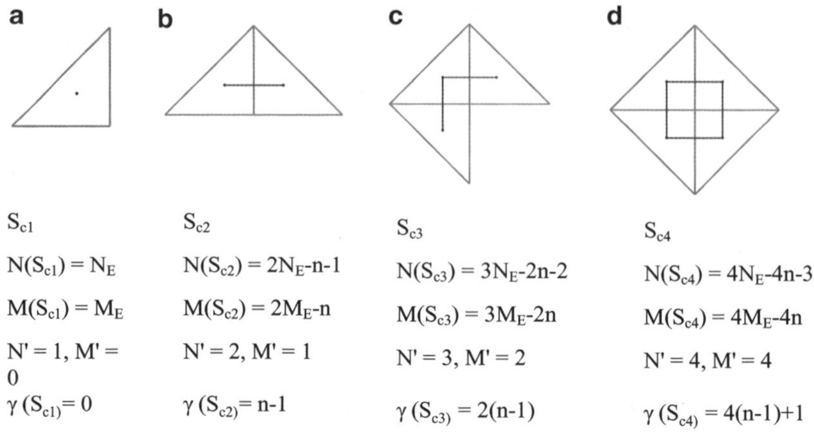

Fig. 7.3 Expansion process in FEMs comprising of O(n) elements and without opening

Based on the above mentioned remarks it can clearly be seen that each member and cycle of natural associate graph corresponds to n − 1 and one degrees of indeterminacy, respectively. This conclusion is theoretically validated as follows. Considering the above points, the number of nodes and members of the natural associate graph are derived as

$$M' = \frac{M(S) + M_E\gamma(S) - M_E}{n(M_E - 1)} \tag{7.4}$$

$$N' = \frac{M(S) + \gamma(S) - 1}{(M_E - 1)} \tag{7.5}$$

Now the first Betti number is epmloyed to calculate the number of independent cycles of the natural associate graph of the FEM

$$b_1(NAG(FEM)) = M' - N' + 1 \tag{7.6}$$

By substituting Eqs. 7.4 and 7.5 in Eq. 7.6, the degree of static indeterminacy of the equivalent truuss is obtained using the natural associate graph of the FEM:

$$\gamma(S) = b_1(NAG(FEM)) + (n - 1)M' \tag{7.7}$$

This equation shows that the subgraphs of S which correspond to memebers of NAG(FEM), represent n − 1 degree of indeterminacy and n − 1 self-equilibrating systems can be constructed which are called Type II self-equilibrating systems. Meantime subgraphs of S which correspond to independent cycles of NAG(FEM), represent one degree of indeterminacy and one self-equilibrating system, called Type III self-equilibrating systems, can be formed.

7.2.5 *Models Including Opening*

In this section, an expansion process is employed in the process of expanding a cycle which surounds an opening, to identiy new degrees of indeterminacy and corresponding the new self-equilibrating systems.

Consider the finite element model of Fig. 7.4a. Adding an element in manner shown in Fig. 7.4b, adds $N_E - n - 2$ nodes and $M_E - n$ members to the corresponding graph S, thus increasing the indeterminacy of equivalent truss by $(n - 1) + 2$.

When the final element is added as shown in Fig. 7.4c, $N_E - 2n - 1$ nodes and $M_E - 2n$ members are added to the corresponding graph S leading to an increase of DSI by $2(n - 1) + 1$.

In the step which is shown in Fig. 7.4b if the Type II self-equilibrating systems are ignored, two new self-equilibrating systems can be recognized and considering Fig. 7.4c there is one new self-equilibrating system. As pointed out, the truss corresponding to a mininal cycle of NAG(FEM) that surrounds an opening, contains three self-equilibrating systems. These self-equilibrating systems are classified in Type III self-equilibrating systems. Here, Eq. 7.7 is modified by adding the term $2n_c$. Each cycle of NAG(FEM) which surrounds an opening is considered as an independent cycle by the first Betti number as

$$\gamma(S) = b_1(\mathrm{NAG(FEM)}) + (n - 1)M' + 2n_c \tag{7.8}$$

where n_c is the number of openings in FEM.

7.2.5.1 Pattern of Type II Self-Equilibrating Systems

Subgraphs of the graph S which correspond to members of the NAG(FEM) are the underlying subgraphs of Type II self-equilibrating systems. If *n* is considered as the order of elements, $n - 1$ Type II self-equilibrating systems can be constructed on each subgraph.

Consider a triangular element; the second element can be attached from each three sides. Depending on the side to which the second element is attached, generators of Type II self-equilibrating systems are selected in different ways. Figure 7.5a shows a second order element indicated by bold nodes which is connected to three elements from three sides. In each case the corresponding generator is identified by dashed red line. The same is shown in Fig. 7.5b, c considering third and fourth order elements. Meantime as it can be noticed from Fig. 7.5, the pattern of the generators can conveniently be expanded for elements with higher orders.

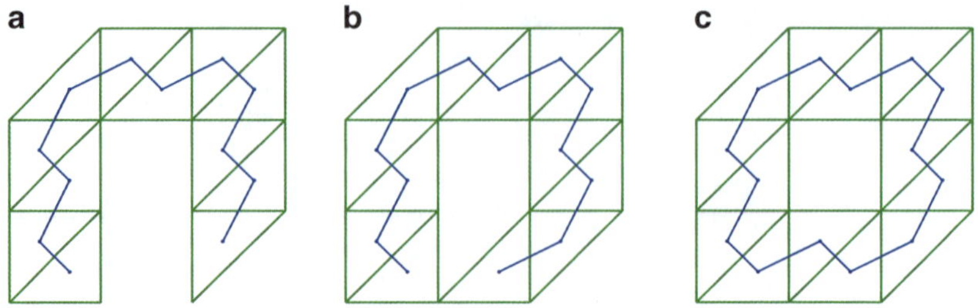

S_{o1}	S_{o2}	S_{o3}
$N(S_{o1}) = N_{o1}$	$N(S_{o2}) = N_{o1}+ N_E\text{-}n\text{-}2$	$N(S_{o3}) = N_{o1}+ 2N_E\text{-}3n\text{-}3$
$M(S_{o1}) = M_{o1}$	$M(S_{o2}) = M_{o1}+ M_E\text{-}n$	$M(S_{o3}) = M_{o1}+ 2M_E\text{-}3n$
$N' = N'_{o1}, M' = M'_{o1}$	$N' = N'_{o1}+1, M' = M'_{o1}+1$	$N' = N'_{o1}+2, M' = M'_{o1}+3$
$DSI = DSI_{o1}$	$DSI = DSI_{o1}+(n-1)+2$	$DSI = DSI_{o1}+3(n-1)+3$

Fig. 7.4 Expansion process in FEMs comprising of O(n) elements, containing an opening

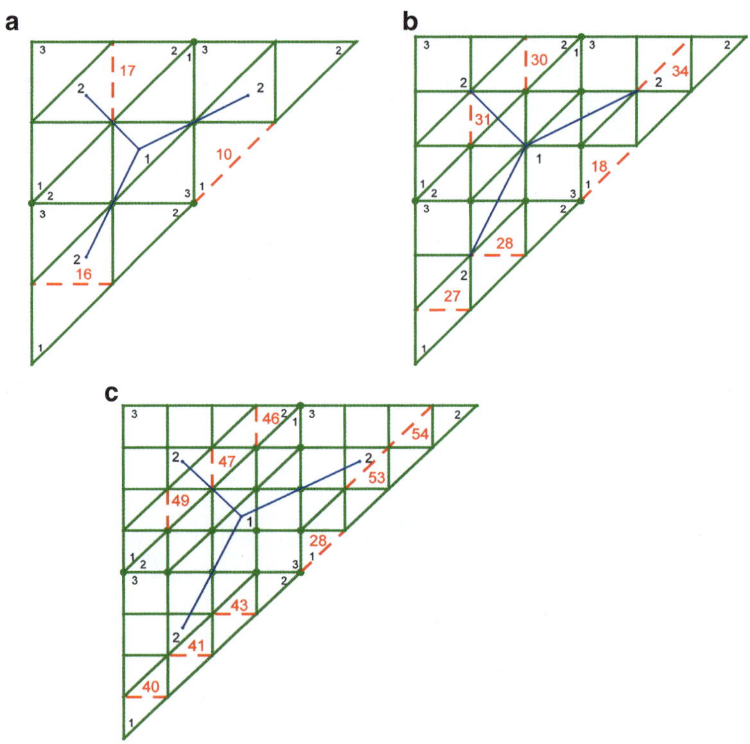

Fig. 7.5 Appropriate generators of Type II self-equilibrating systems from models comprised of second, third and fourth order elements

7.2.5.2 Pattern of Type III Self-Equilibrating Systems

Subgraphs of the graph S which correspond to minimal cycles of the NAG(FEM) are underlying subgraphs of Type III self-equilibrating systems. These minimal cycles can be categorized into two classes.

7.2.5.3 Type I Minimal Cycles

These cycles pass through elements which all have one certain node in common. As discussed in expansion process, subgraphs of graph S corresponding to Type I minimal cycles lead to one Type III self-equilibrating system. Figure 7.6a–c represent the underlying subgraphs and generators of Type III self-equilibrating systems corresponding to second, third and fourth order elements. The generators are indicated by dashed red lines.

Meantime Fig. 7.6 implies that the pattern of the generators can conveniently be expanded for the elements with higher orders.

7.2.5.4 Type II Minimal Cycles

These cycles pass through elements which surround an opening. According to the expansion process, subgraphs of the graph S corresponding to Type II minimal cycles contain three self-equilibrating systems of Type III. Consider Fig. 7.4b, in this situation, based on expansion process two Type III self-equilibrating systems are formed hence the two corresponding generators can simply be selected from members of the last added element.

The last Type III self-equilibrating systems is formed when the Type II minimal cycle is completed (Fig. 7.4c). Here again the corresponding generator is simply selected from the members of the added element.

7.2.5.5 Self-Equilibrating Systems Corresponding to the External Indeterminacies

These self-equilibrating systems are formed in relation with the external degrees of indeterminacy. For this purpose, each indeterminate restraint forces is considered as a redundant force. Here unlike the internal redundant forces, the external ones are not bi-action forces. Thus the corresponding self-equilibrating systems will require simple support conditions. A typical self-equilibrating system regarding to an external indeterminacy shown in Fig. 7.7b is formed based on a tree of NAG (FEM) which connects the external redundant force to a close simple support configurations. However, it is essential that these self-equilibrating systems remain independent from each other. After selecting two ends of the tree, it is desirable the above mentioned tree to pass through elements with close numbers.

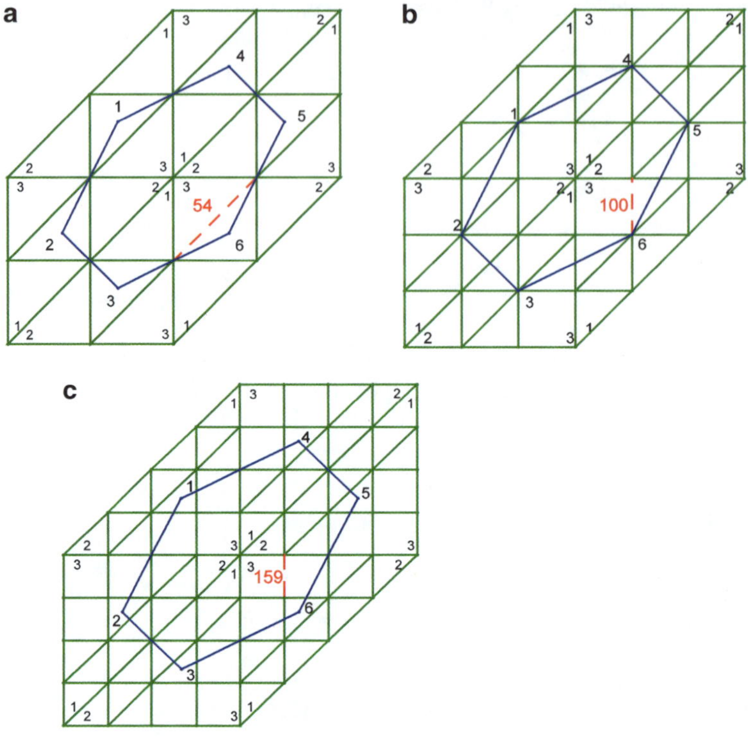

Fig. 7.6 Appropriate generators of Type III self-equilibrating systems from models comprised of second, third and fourth order elements

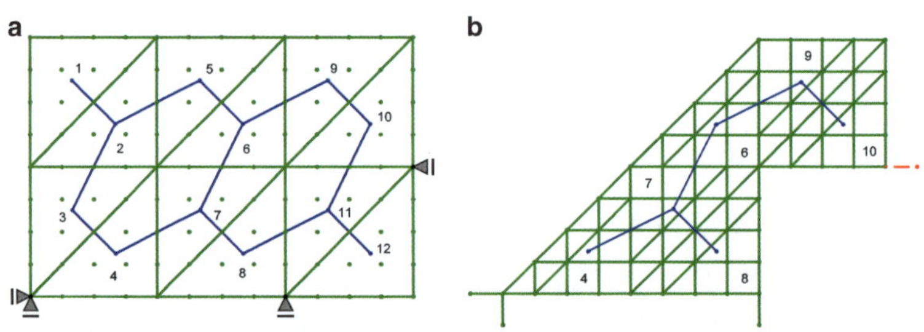

Fig. 7.7 A self-equilibrating system corresponding to an indeterminate support condition

7.2.6 Selection of an Optimal List Corresponding to Minimal Self-Equilibrating Stress Systems

Consider a general self-equilibrating system of Type II or Type III. According to the above procedure SESs consist of single members like s_j and double members d_i^l

Table 7.1 Member list of a typical Type II or Type III self-equilibrating system

d_1^1	d_2^1	...	s_1	s_2	...	g	...	d_i^1	...	s_j	...	d_k^1
d_1^2	d_2^2							d_i^2				d_k^2

which have a twin member as d_i^2. It is clear that replacing each d_i^1 member with its twin member does not affect the topology of the corresponding SES, Table 7.1. Thus except the generators which are unique and necessary for the independency of the SESs, other double members can be replaced by their twin members.

Here, considering the bandwidth reduction of the null basis, the following procedure is utilized to select members with closer numbers to the generator number.

$$\textit{If}\ \mathrm{abs}\left(g - d_i^2\right) < \mathrm{abs}\left(g - d_i^1\right)\ \ \text{then}\ \ d_i^2 \rightarrow d_i^1 \tag{7.9}$$

where, g is the generator number.

Algorithm

Step 1: Define and number the nodes of the FEM.

Step 2: Define the triangular elements and use an efficient numbering method to reduce the bandwidth of the null basis and flexibility matrice.

Step 3: Generate the natural associate graph of the FEM based on the adjacency of the elements.

Step 4: Generate the interface graph of the FEM in a manner that its members are numbered according to the element numbering of the FEM.

Step 5: Select the Type I self-equilibrating systems and form the corresponding null vectors.

Step 6: Set up the Type II and Type III self-equilibrating systems, and form the corresponding null vectors consisting of the members' forces when the generator's force is equal to unity.

Step 7: Finally assemble the null basis (static basis) of the FEM by arranging the null vectors in the ascending order of the highest member number of the self-equilibrating systems.

The above algorithm is implemented in MATLAB and is used to analyze three structures, and the efficiency of the present method is illustrated through these examples.

7.2.7 Numerical Examples

In this section three examples are studied. In each case, first the structure is idealized using second order triangular elements. The null basis matrices are constructed utilizing the present method and two algebraic procedures, namely the Gauss-Jordan elimination method and QR factorization. Then the results are contrasted through normalized computational time for the formation of the null basis matrices and nonzero pattern and condition numbers of the flexibility matrices. In the following examples, nz represents the number of non-zero entries and, the ratio of the extreme eigenvalues, $\lambda_{max}/\lambda_{min}$, is taken as the condition number of the matrices.

In the second step, each example is idealized using second, third and fourth order elements. Then the properties of the flexibility matrices obtained from the present method are compared to those of the corresponding stiffness matrices. For this purpose condition s and the number of unknowns, namely the DSI for the force method and the DKI degree of kinematic indeterminacy for the displacement method are utilized.

Example 1. Consider a beam structure with determinate support conditions. The beam is bent under a uniformly distributed load of intensity q = 10 kN/m. The structure is idealized using plain stress triangular elements. As indicated in Fig. 7.8, three types of elements are generated using second, to fourth order elements. All models have the same nodes with the following mechanical properties:

Thickness = 0.01 m, E = 2e + 8 kN/m^2, $\nu = 0.3$. Topological properties of the models are collected in Table 7.2.

Figure 7.9 displays pattern of the null basis matrices employing the present method and two other algebraic procedures for the model comprising of second order elements. In this relation Table 7.3 contains other optimality characteristics of the force method procedures. It is clear that the graph theoretical method forms the most well-structured null basis in smallest computational time.

Figure 7.10 shows the pattern of the flexibility matrices for the models comprising of the second to fourth order elements. It is noticeable that as the order of elements increases the DSI of the model decreases. Meanwhile, with identical number of nodes for the models, the DKI stays the same, as illustrated in Fig. 7.11. Table 7.4 contains the ratio of the DKI/DSI.

Finally, for a model with second order elements, the average σ_{xx} stresses at nodes of Path 1 are compared through the results of the present force method and displacement method in [1].

Example 2. A beam structure which depicted in Fig. 7.12 is bent under a uniformly distributed load of intensity q = 10 kN/m. The structure is analyzed three times using second, third and fourth order elements. Plane stress elements are considered with the following mechanical properties: thickness = 0.01 m, E = 2 e + 8 kN/m^2, $\nu = 0.3$. Topological properties of the models are collected in Table 7.5.

Figure 7.13 and Table 7.6 reveal the optimality characteristics of the present graph theoretical and the two algebraic force methods. The present method leads to a sparse and banded null basis using the smallest computational time. **QR** factorization method forms a null basis in reasonable computational time and leads to a well-conditioned flexibility matrix but the flexibility matrix is not well-structured at all.

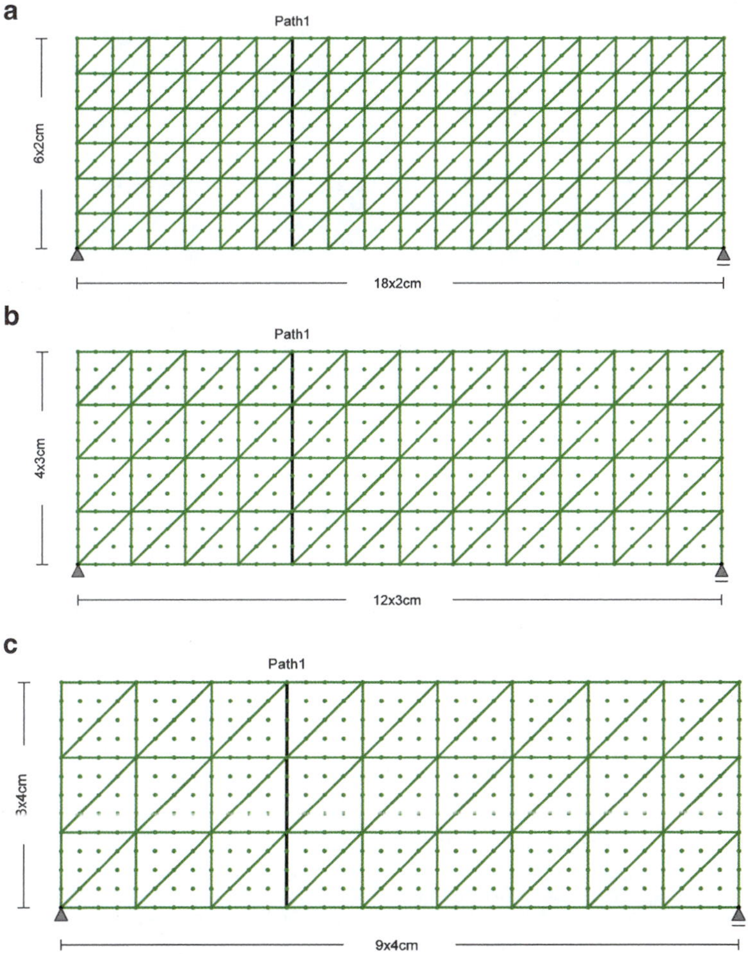

Fig. 7.8 FEM of the beam of Example 1

Table 7.2 Topological properties for the FEM of Example 1

	FEM		Self-equilibrating systems			Force system	
Element type	Nodes	Elements	Type I	Type II	Type III	Int.	DSI
Second order	481	216	600	300	85	1,944	985
Third order	481	96	384	256	33	1,632	673
Fourth order	481	54	276	207	16	1,458	499

Comparing Figs. 7.14 and 7.15 shows that, by using higher order triangular elements there will be fewer compatibility conditions than equilibrium equations. Thus utilizing the graph theoretical force method can be attractive and economical.

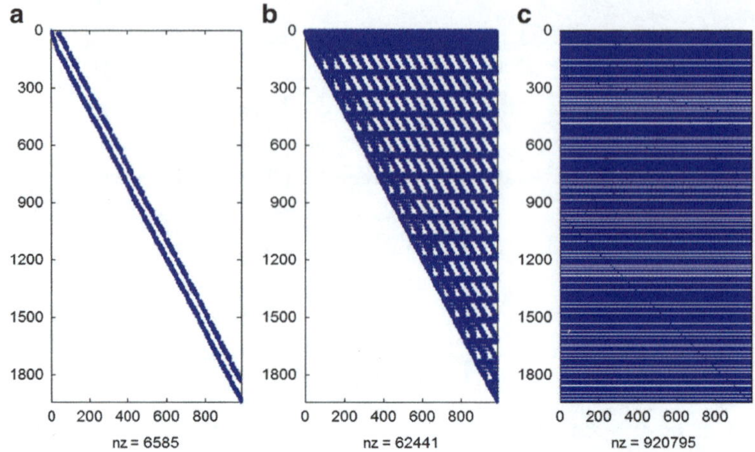

Fig. 7.9 Pattern of the null basis $\boldsymbol{B}_1$ corresponding to Example 1 with second order elements and utilizing (**a**) Graph theoretical method (**b**) Gauss Jordan elimination method (**c**) **QR** decomposition method

Table 7.3 Comparison of the optimality characteristics of the null basis matrices $\boldsymbol{B}_1$ and flexibility matrices $\boldsymbol{G}$ for the FEM of Example 1

	Null basis $\mathbf{B}_1$	Flexibility matrix $\mathbf{G}$	
	Time	nz entries	λ_{max}
	Time Present method	nz entries Present method	λ_{min}
Graph theoretical method	1.00	1.00	1.21e+4
Gauss Jordan elimination method	106.48	17.01	3.08e+4
QR factorization method	3.06	43.27	8.22e+3

Table 7.7 contains the ratios of the DKI/DSI and also condition numbers of the stiffness matrices.

Example 3. A cross section of a retaining wall is idealized using triangular elements, as illustrated in Fig. 7.16. The model is analyzed three times using second, third and fourth order elements. Here plane strain elements are utilized with the following mechanical properties:

Thickness = 1 m, E = 2e + 7 kN/m^2, $\nu = 0.2$. Topological properties of the models are collected in Table 7.8.

Optimality characteristics of the employed force methods can be seen in Fig. 7.17 and Table 7.16. The patterns of the flexibility matrices utilizing present method are illustrated in Fig. 7.18. The difference between nonzero numbers of the

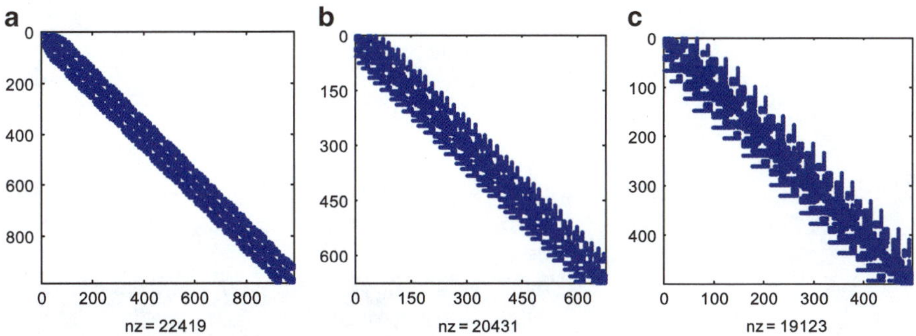

Fig. 7.10 Pattern of the flexibility matrices G corresponding to Example 1, considering (**a**) second order (**b**) third order (**c**) fourth order elements

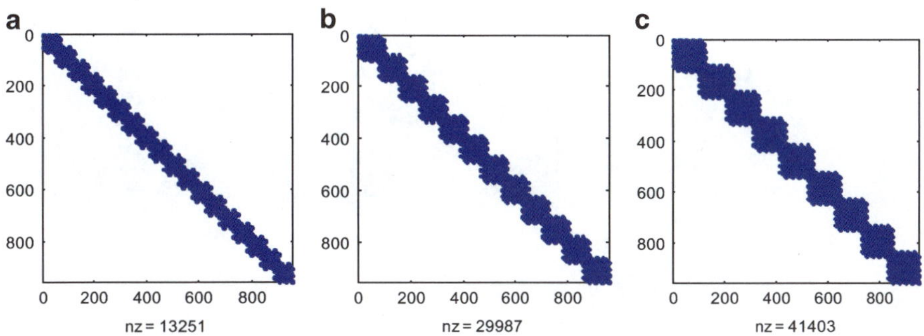

Fig. 7.11 Pattern of the reduced stiffness matrices K_r corresponding to Example 1 considering (**a**) second order (**b**) third order (**c**) fourth order elements

Table 7.4 Optimality characteristics of the reduced stiffness matrices $\mathbf{K_r}$ for Example 1.

Element type	$\frac{\lambda_{max}}{\lambda_{min}}$	$\frac{DKI}{DSI}$
Second order	6.22e+4	0.97
Third order	9.45e+4	1.42
Fourth order	1.59e+5	1.92

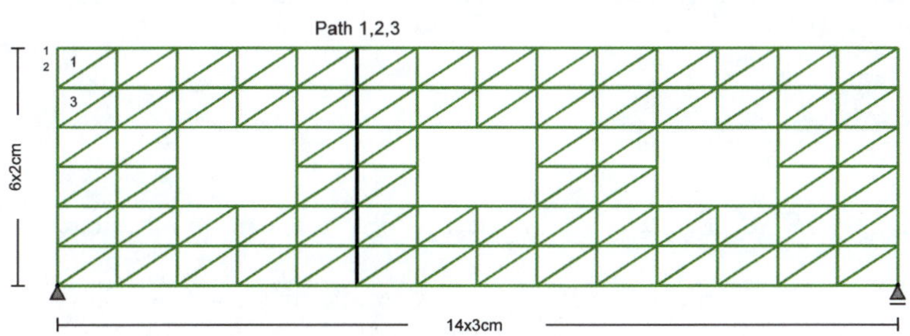

Fig. 7.12 Beam structure of Example 2

Table 7.5 Topological properties for the FEM of Example 2

	FEM		Self-equilibrating systems			Force system	
Element type	Nodes	Elements	Type I	Type II	Type III	Int.	DSI
Second order	350	144	368	184	47	1,296	599
Third order	742	144	552	368	47	2,448	967
Fourth order	1,278	144	736	552	47	3,888	1,335

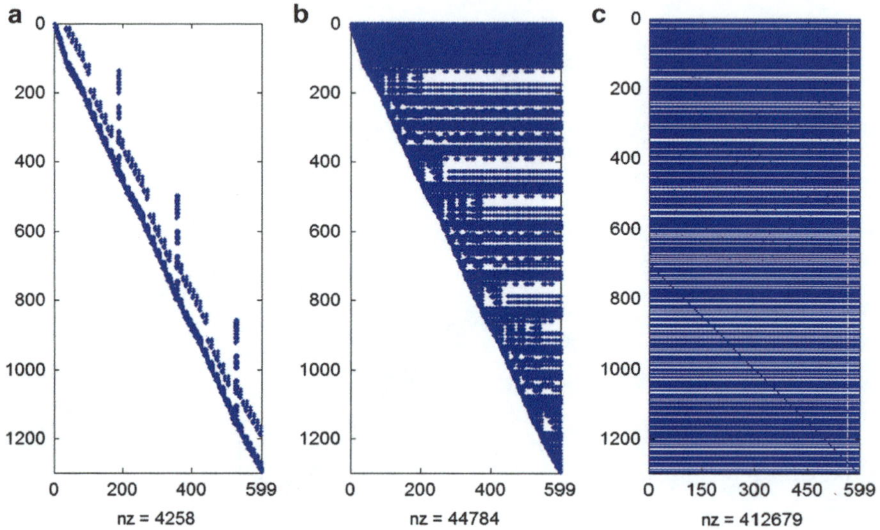

Fig. 7.13 Pattern of the null basis $\mathbf{B_1}$ corresponding to Example 2 with second order elements and utilizing (**a**) Graph theoretical method (**b**) Gauss Jordan elimination method (**c**) **QR** decomposition method

Table 7.6 Comparison of the optimality characteristics of the null basis matrices $\mathbf{B_1}$ and flexibility matrices **G** for the FEM of Example 2

	Null basis $\mathbf{B}_1$	Flexibility matrix **G**	
	$\frac{\text{Time}}{\text{Time}_{\text{Present method}}}$	$\frac{\text{nz entries}}{\text{nz entries}_{\text{Present method}}}$	$\frac{\lambda_{max}}{\lambda_{min}}$
Graph theoretical method	1.00	1.00	4.24e+4
Gauss Jordan elimination method	60.33	11.50	1.27e+5
QR factorization method	1.67	26.21	1.66e+3

matrices is due to the use of different SESs for external redundant forces (Table 7.9).

The patterns of reduced stiffness matrices and the corresponding condition numbers are provided in Fig. 7.19 and Table 7.10, respectively.

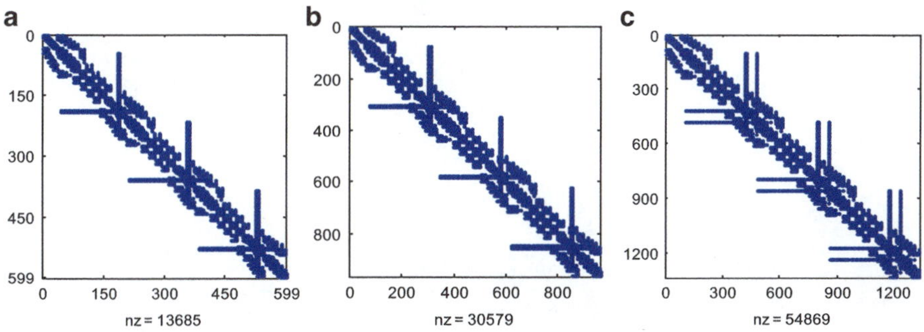

Fig. 7.14 Pattern of the flexibility matrices G corresponding to Example 2 considering (**a**) second order (**b**) third order (**c**) fourth order elements

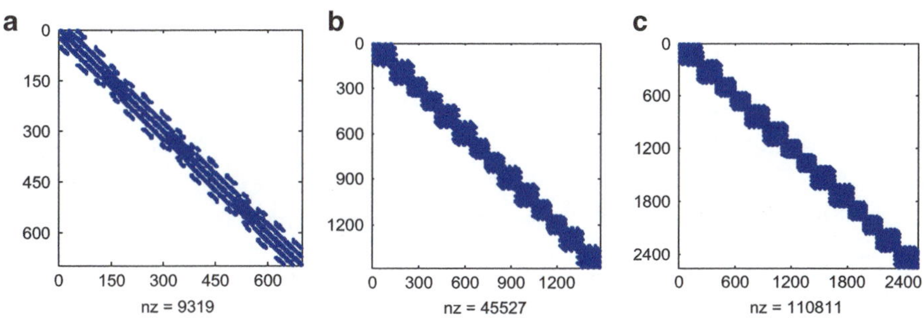

Fig. 7.15 Pattern of the reduced stiffness matrices $\mathbf{K}_r$ corresponding to Example 2 considering (**a**) second order (**b**) third order (**c**) fourth order elements

Table 7.7 Optimality characteristics of the reduced stiffness matrices $\mathbf{K}_r$ for Example 2

	$\frac{\lambda_{max}}{\lambda_{min}}$	$\frac{DKI}{DSI}$
Second order	5.94e+4	1.16
Third order	2.08e+5	1.53
Fourth order	6.25e+5	1.91

7.3 Finite Element Analysis of Models Comprised of Higher Order Rectangular Elements

In this section, an efficient method is developed for the formation of null bases of finite element models (FEMs) consisting of rectangular plane stress and plane strain serendipity family elements, corresponding to highly sparse and banded flexibility matrices. This is achieved by associating special graphs with the FEM and selecting appropriate subgraphs and forming the self-equilibrating systems (SESs) on these subgraphs. The efficiency of the present method is illustrated through three examples.

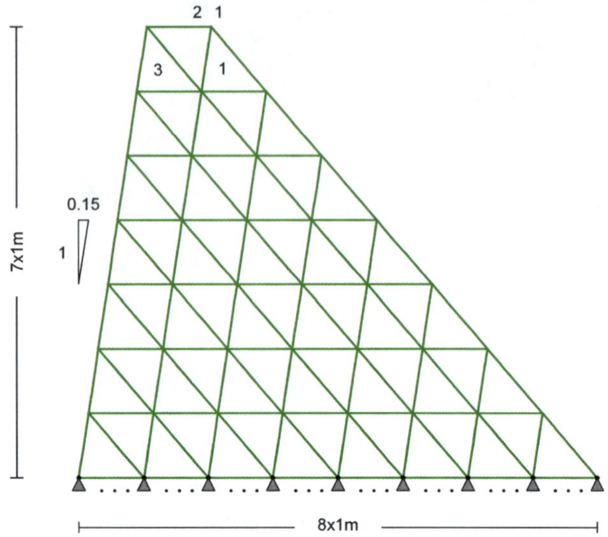

Fig. 7.16 A section of the retaining wall of Example 3

Table 7.8 Topological properties for the FEM of Example 3

	FEM		Self-equilibrating systems			Force system		
Element type	Nodes	Elements	Type I	Type II	Type III	Int.	Ext.	DSI
Second order	150	63	166	83	21	567	34	301
Third order	319	63	249	166	21	1,071	50	483
Fourth order	551	63	332	249	21	1,701	66	665

7.3.1 Definition of Element Force System

For the generation of the equilibrium matrix **A** of a FEM, a set of independent forces system should be defined and also their relations with the element nodal forces should be established.

In displacement method we have two forces at each node of the element. For an element with N nodes, 2 × N nodal forces can be defined. Using three equilibrium equations, 2N − 3 independent forces will remain. In other words, there are 2N − 3 independent element forces in an element with N nodes. The nodal forces and element forces systems are shown in Fig. 7.20 for rectangular plane stress and plane strain serendipity family elements with various numbers of boundary nodes. For the higher order elements, the element forces system can be obtained with the same procedure.

These element forces can be related to the nodal forces for a rectangular element by a (2N) × (2N − 3) transformation matrix using Eq. 7.10 as

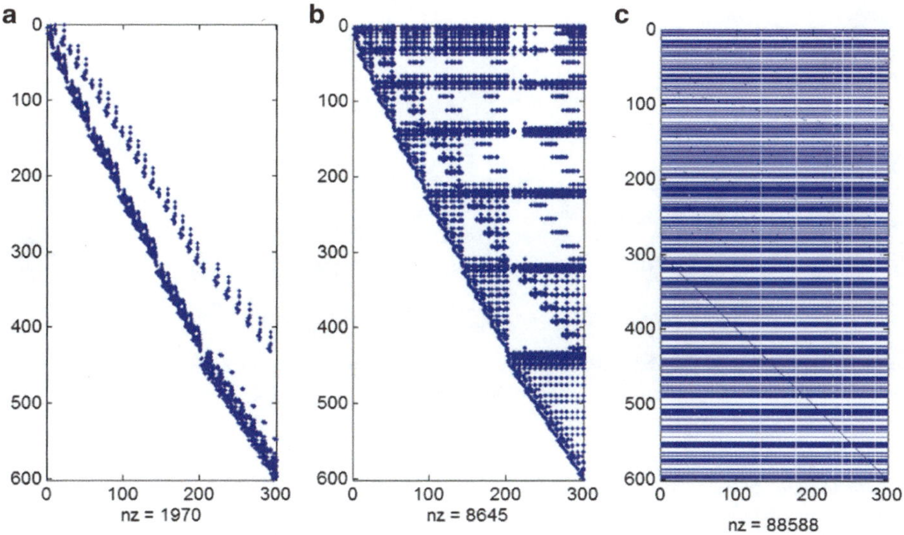

Fig. 7.17 Pattern of the null basis $\mathbf{B}_1$ corresponding to Example 3 with second order elements and utilizing (**a**) Graph theoretical method (**b**) Gauss Jordan elimination method (**c**) **QR** decomposition method

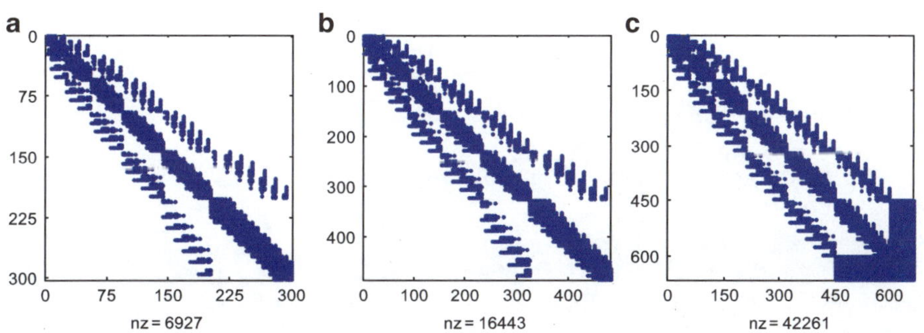

Fig. 7.18 Pattern of the flexibility matrices G corresponding to Example 3 considering (**a**) second order (**b**) third order (**c**) fourth order elements

Table 7.9 Comparison of the optimality characteristics of the null basis matrices $\mathbf{B}_1$ and flexibility matrices **G** for the FEM of Example 3

	Null basis $\mathbf{B}_1$	Flexibility matrix **G**	
	$\frac{\text{Time}}{\text{Time}_{\text{Present method}}}$	$\frac{\text{nz entries}}{\text{nz entries}_{\text{Present method}}}$	$\frac{\lambda_{max}}{\lambda_{min}}$
Graph theoretical method	1.00	1.00	5.02e+5
Gauss Jordan elimination method	48.17	5.54	1.19e+5
QR factorization method	1.62	13.07	4.80e+3

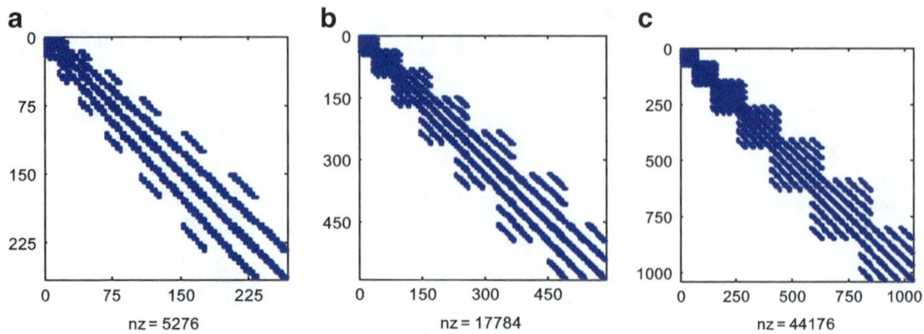

Fig. 7.19 Pattern of the reduced stiffness matrices $\mathbf{K_r}$ corresponding to Example 3 considering (**a**) second order (**b**) third order (**c**) fourth order elements

Table 7.10 Optimality characteristics of the reduced stiffness matrices $\mathbf{K}_r$ for Example 3

	$\frac{\lambda_{max}}{\lambda_{min}}$	$\frac{DKI}{DSI}$
Second order	4.35e+3	0.88
Third order	1.31e+4	1.21
Fourth order	3.61e+4	1.55

$$\mathbf{S} = \mathbf{TF} \tag{7.10}$$

Transformation matrix can be formed simply as

$$
\begin{array}{l}
(n_1, n_2) = \text{ end nodes of element force } F_j \\
\text{For} \quad i = 1 : N \\
\quad \text{For} \quad j = 1 : 2N - 3 \\
\quad\quad \text{If} \quad i == n_1 \quad T(2i-1,\ j) = m_{n_1 n_2} \text{ and } \quad T(2i, j) = n_{n_1 n_2} \\
\quad\quad \text{If} \quad i == n_2 \quad T(2i-1, j) = m_{n_2 n_1} \text{ and } \quad T(2i, j) = n_{n_2 n_1} \\
\quad \text{End} \\
\text{End}
\end{array}
$$

Where x_i and y_i are the Cartesian coordinates of node i, $m_{ij} = \frac{x_i - x_j}{l_{ij}}$, $n_{ij} = \frac{y_i - y_j}{l_{ij}}$ are the direction cosines and l_{ij} is the length of the line between nodes i and j.

7.3.2 *Flexibility Matrix of the Element*

In this case flexibility matrices of higher order triangular elements can simply be formed using the stiffness matrices of such elements.

Element	Nodes number	Size of T	Nodal forces	Element forces	Element forces
Linear	$N=4$	8×5	$S_{8\times1}=[S_1 \ldots S_8]^t$	$F_{5\times1}=[F_1 \ldots F_5]^t$	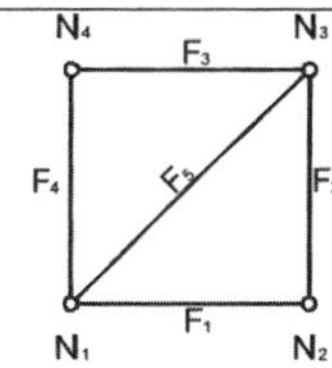
Quadratic	$N=8$	16×13	$S_{16\times1}=[S_1 \ldots S_{16}]^t$	$F_{13\times1}=[F_1 \ldots F_{13}]^t$	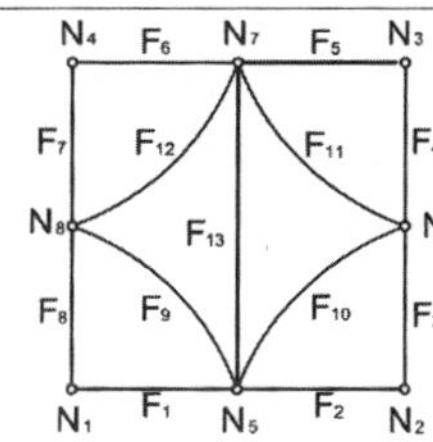
Cubic	$N=12$	24×21	$S_{24\times1}=[S_1 \ldots S_{24}]^t$	$F_{21\times1}=[F_1 \ldots F_{21}]^t$	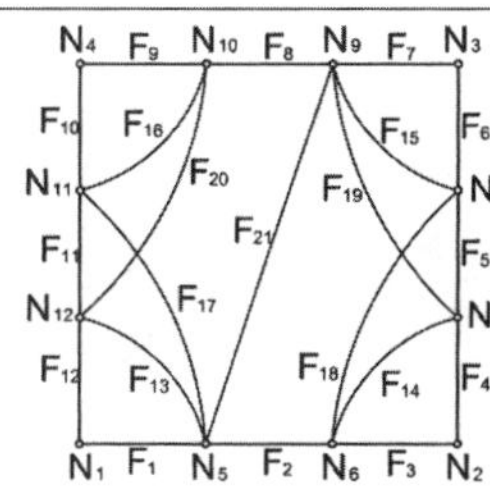
Quartic	$N=16$	32×29	$S_{32\times1}=[S_1 \ldots S_{32}]^t$	$F_{29\times1}=[F_1 \ldots F_{29}]^t$	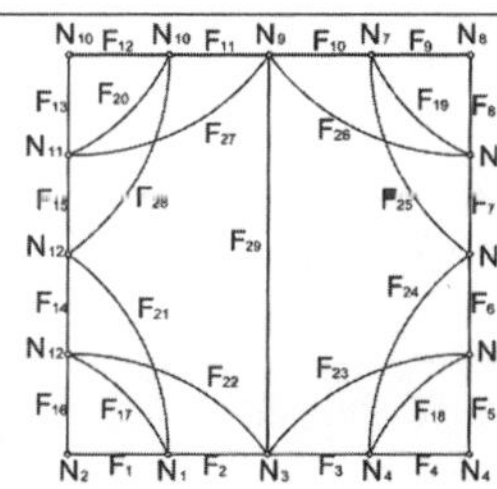

Fig. 7.20 A set of rectangular serendipity family elements

$$\mathbf{f}_{\mathrm{m}} = (\mathbf{T}_{\mathrm{r}})^{\mathrm{t}}(\mathbf{K}_{\mathrm{r}})^{-1}\mathbf{T}_{\mathrm{r}} \tag{7.11}$$

where the subscript r indicates that, corresponding orders of matrices to dependent forces are reduced.

7.3.3 Graphs Associated with Finite Element Model

In order to transfer the topological property of a finite element model to the connectivity of a graph ten different graphs are previously introduced in Chap. 4.

Here natural associate graph and interface graph are used that are defined in the following:

7.3.3.1 Natural Associate Graph

The natural associate graph represented by NAG(FEM) is constructed by the following rules:

1. Nodes of the NAG(FEM) correspond to the elements of FEM.
2. For each pair of elements in FEM having (N + 4)/4 common nodes, one member is added between the corresponding two nodes in NAG(FEM).

NAG can be constructed using the following procedure: One of the preliminary steps in FEM is defining the elements with their connected nodes. In this way the element connectivity matrix is constructed which contains the element-node incidence relationships. In the process of constructing the element connectivity matrix, another matrix which contains node-element incidence properties can be formed. This matrix is named the node connectivity matrix. Now using the element connectivity and the node connectivity matrices leads to an algorithm with complexity O(n) for an efficient generation of NAG.

In order to recognize the adjacent elements to the nth element which have (N + 4)/4 common nodes or one common face, first the connected nodes to the nth element are identified from the element connectivity matrix. In the subsequent step using the node connectivity matrix, elements which have at least one common node with the nth element are identified. Now it is convenient to seek for the adjacent elements in this reduced search space. A FEM and its corresponding NAG are illustrated in Fig. 7.21.

7.3.3.2 An Interface Graph

The interface graph of a finite element model denoted by IG (FEM) can easily be constructed for rectangular FEM using the following rules:

1. This graph contains all the nodes of the FEM.
2. With the all edges of an element of FEM, N graph members are associated. Therefore, in the interface of two elements, 2-multiple members are presented.
3. For each element with N nodes, 2N − 3 members should be considered in the interface graph. Thus, N − 3 = (2N − 3) − N) diagonal members should be added. This graph for a quadratic and cubic FEM is shown in Fig. 7.22.

The member numbering of the interface graph should be performed according to the numbering of the FEM, taking into account the primary nodal numbering of a considered element in the model. Thus, for each rectangular element 2N − 3 members of the interface graph will be numbered sequentially according to the patterns which were illustrated in Fig. 7.20.

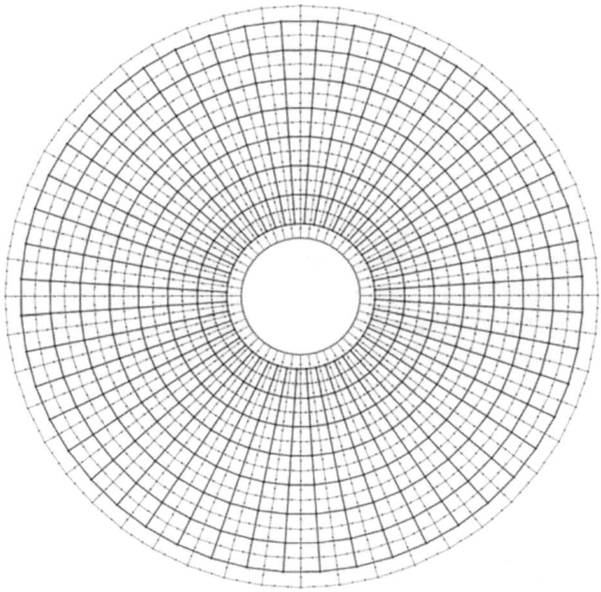

Fig. 7.21 A quadratic rectangular FEM with its natural associate graph (bold lines) for a circular plate

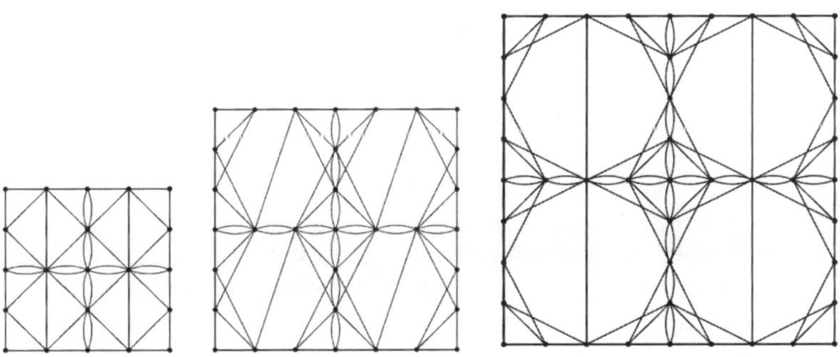

Fig. 7.22 A quadratic, cubic and quartic rectangular FEM with their interface graphs

7.3.4 Topological Interpretation of Static Indeterminacies

7.3.4.1 Degree of Static Indeterminacy of the FEM

Considering Fig. 7.20, in order to find the patterns corresponding to the self-equilibrating systems, a rectangular element is simulated as a planar truss formed as the 1-skeleton of the rectangular element together with some diagonal members. This is possible since the independent element forces applied at in the nodes and are along the edges of the rectangular element. The statical indeterminacy of a planar truss with m members and n nodes is given as $\gamma(S) = m - 2n + 3$; therefore, the degree of statical indeterminacy (DSI) of the entire model supported in a statically determinate manner can be calculated with the same relationship as

$$\mathrm{DSI} = (2N - 3) \times M - 2n + 3 \tag{7.12}$$

Where M is the total number of elements, N is the number of nodes within an element and n is the total number of nodes of the FEM.

Following self-equilibrating systems are found. Then null vectors which express the equilibrium conditions of the self-equilibrating systems are generated.

7.3.4.2 Type I Self-Equilibrating Systems

Each multiple member of the interface graph is a subgraph on which one self-equilibrating system can be generated. In other words, on a 2-multiple member numbered as (i, j) with the condition $(i < j)$, one self-equilibrating system can be constructed (extracted).

Each pair such as (i, j) for which $(i < j)$ corresponds to a null vector with their non-zero entries being located in rows i and j, and their numeric values are $(-1, 1)$, respectively. The member with bigger member number (j) is called the generator. These pairs are called Type I self-equilibrating systems.

For a FEM we have $\frac{N}{4} \times M'$ self-equilibrating systems of Type I, where M' is the number of members of the associate graph of the model.

7.3.4.3 Type II Self-Equilibrating Systems

There are other types of self-equilibrating systems which are extracted from two adjacent elements of FEM. In other words, for two adjacent elements with N nodes, the DSI can be calculated as:

$$\begin{aligned}
&\mathrm{DSI} = (2N - 3) \times M - 2n + 3 \\
&\Rightarrow \mathrm{DSI} = (2N - 3) \times 2 - 2 \times \left(2N - \frac{N+4}{4}\right) + 3 = \quad N - 7
\end{aligned} \tag{7.13}$$

$\frac{N}{4}$ self-equilibrating systems were generated as Type I systems. Thus the number of remaining self-equilibrating systems is

$$\text{Type II} = \frac{N}{2} - 1 - \frac{N}{4} = \frac{N}{4} - 1 \tag{7.14}$$

In other words, $\frac{N}{4} - 1$ SESs should be extracted from two adjacent elements. This number is equal to the number of internal nodes of the remaining subgraph after

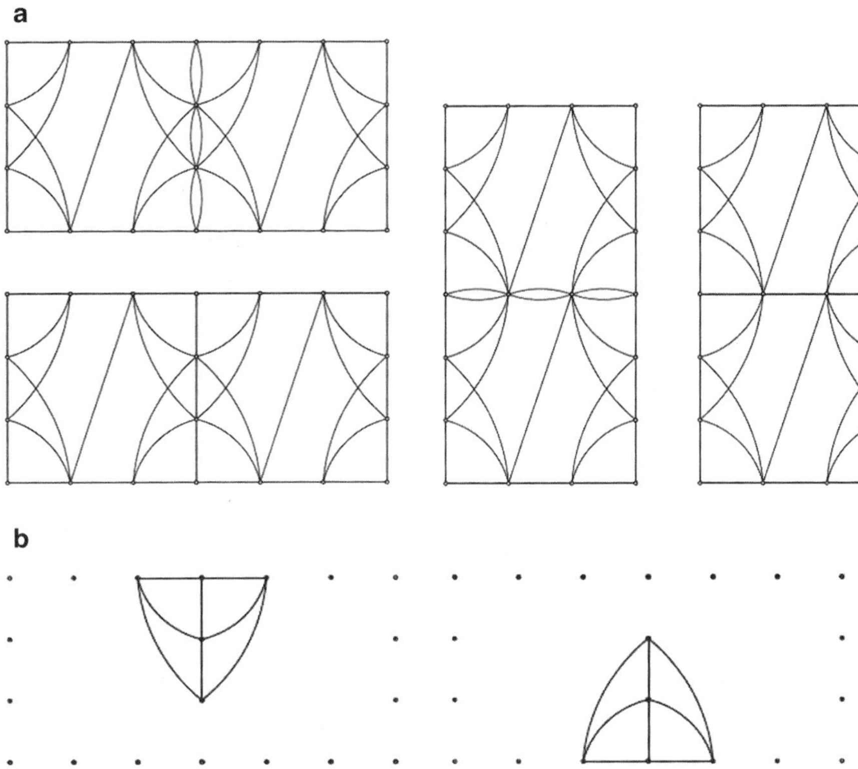

Fig. 7.23 (a) Subgraph corresponding to SESs of Type II, (b) Pattern of Type II self-equilibrating systems in horizontal direction

deleting the generators of SESs of Type I. For example, the remaining subgraphs for two adjacent cubic elements are shown in Fig. 7.23a in two directions. In this figure, the diagonal members are curved for better illustration. After deleting the generators corresponding to Type I SESs, the null vectors should be calculated from the remaining subgraph. These null vectors can easily be generated on the corresponding sub-structure utilizing an algebraic method. For instance, results SESs in horizontal direction are shown in Fig. 7.23b.

In a FEM, the total number of Type II SESs can be calculated as:

$$\text{Type II} = \text{M}' \times \left(\frac{\text{N}}{4} - 1\right) \tag{7.15}$$

where M′ is the number of members of the associate graph of the model and N is the number of nodes of an element.

The most important point in Type II self-equilibrating systems is to select an appropriate generator. In fact by eliminating these generators from graph S, the sub-structure of Type III SESs and the primary structure of the structure S must remain stable.

7.3.4.4 Type III Self-Equilibrating Systems

Sub-structures which are topologically identical to the minimal cycles of the natural associate graph of FEM contains some Type I, II and one Type III self-equilibrating systems.

Type I Minimal Cycles of NAG(S)

These minimal cycles of the natural associate graph of the FEM pass through four elements which have one common node. Corresponding interface graph of these elements have n nodes and m edges for a FEM with N-node elements.

$$m = 4 \times (2N - 3) \tag{7.16}$$

$$n = 4N - 4 \times \left(\frac{N+4}{4}\right) + 1 = 3 \times (N - 1) \tag{7.17}$$

Subsequently, the DSI of the interface graph is

$$\begin{aligned} &DSI = m - 2n + 3 \\ &\Rightarrow DSI = 4 \times (2N - 3) - 2 \times (3 \times (N - 1)) + 3 = 2N - 3 \end{aligned} \tag{7.18}$$

The N, $\left(N = \frac{N}{4} \times M' = \frac{N}{4} \times 4\right)$, SESs are Type I and there are $N - 4$, $\left(N - 4 = M' \times \left(\frac{N}{4} - 1\right) = 4 \times \left(\frac{N}{4} - 1\right)\right)$, SESs of Type II.

$$DSI - (TypeI\&II) = (2N - 3) - (N + (N - 4)) = 1 \tag{7.19}$$

Therefore, one independent SES should be extracted. This SES with eight members can be formed for any types of rectangular elements around the common node as is indicated bold in Fig. 7.24.

Type II Minimal Cycles of NAG(S)

Each minimal cycle that surrounds an opening is called the Type II minimal cycle. Such a cycle passes through M′, ($M' \geq 8$), finite elements and its corresponding interface graph has $\left(\frac{3N}{4} - 1\right) \times M'$ nodes and $M' \times (2N - 3)$ members. The DSI of subgraph is

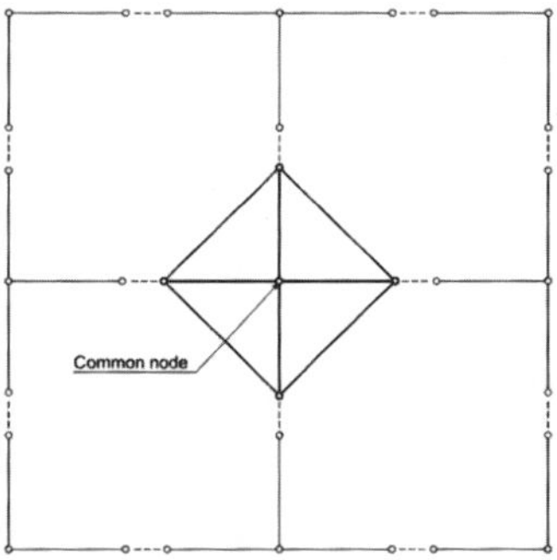

Fig. 7.24 The SES of Type III corresponding to the common node of four rectangular elements

$$\mathrm{DSI} = \mathrm{M}' \times (2\mathrm{N} - 3) - 2 \times \left(\frac{3\mathrm{N}}{4} - 1\right) \times \mathrm{M}' + 3$$

$$\Rightarrow \mathrm{DSI} = \mathrm{M}' \times \left(\frac{\mathrm{N}}{2} - 1\right) + 3 \qquad (7.20)$$

that $\frac{\mathrm{N}}{4} \times \mathrm{M}'$ SESs of Type I and Type II $= \mathrm{M}' \times \left(\frac{\mathrm{N}}{4} - 1\right)$ SESs of Type II can be extracted.

$$\mathrm{DSI} - \mathrm{Type\,I}\ \&\ \mathrm{Type\,II} = \mathrm{M}' \times \left(\frac{\mathrm{N}}{2} - 1\right) - \left[\mathrm{M}' \times \frac{\mathrm{N}}{4} + \mathrm{M}' \times \left(\frac{\mathrm{N}}{4} - 1\right)\right] + 3 = 3 \quad (7.21)$$

Therefore, each Type II minimal cycle corresponds to three null vectors which are calculated utilizing an algebraic method.

7.3.5 Selection of Generators for SESs of Type II and Type III

The most important point in Type II and Type III self-equilibrating systems is to select appropriate generators. This is by eliminating these generators from graph S, the sub-structure of primary structure of the structure S must remain stable. To achieve this, the following rule for appropriate selection of generators of Type II SESs is suggested.

For quadratic and rectangular element the generators of SESs Type II and Type III are illustrated in Figs. 7.25 and 7.26, respectively. It should be noted that the generators corresponding to Type I were chosen previously. In addition, the generators corresponding to an opening are the last non-zero entries of its columns which are not common with the previously selected generators.

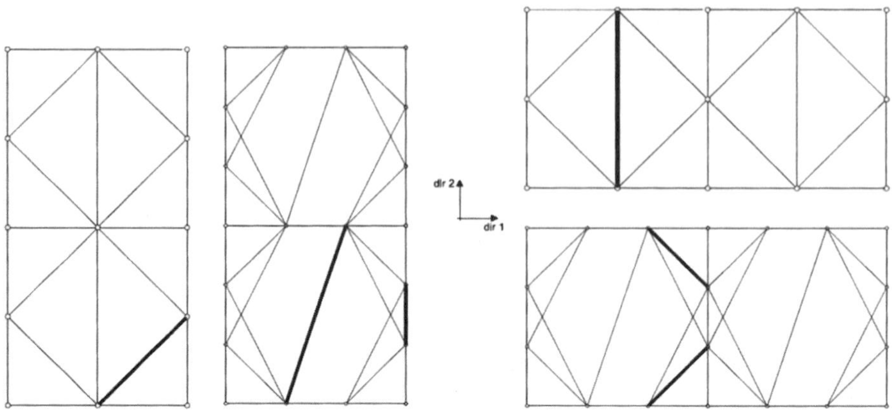

Fig. 7.25 Selected generators of the Type II SES

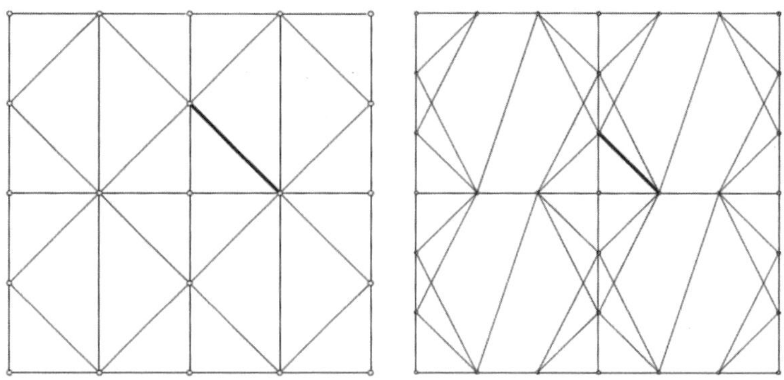

Fig. 7.26 Selected generators of the Type III SES

7.3.6 Algorithm

Step 1: Generate the associate graph of the FEM and use an efficient method for its nodal numbering see Chaps. 4 and 5. It is obvious that good numbering of this graph corresponds to good numbering of elements of the FEM. This numbering leads to a banded adjacency matrix of the graph and correspondingly to a banded flexibility matrix. Since the numbering of the members of the interface graphs corresponds to the element numbering of the finite elements, such a numbering is the only parameter for controlling the bandwidth of the flexibility matrix.

Step 2: Set up the equilibrium matrix of the FEM.

Step 3: Generate the interface graph and perform its numbering. The numbering of this graph should be performed according to the element numbering of the considered FEM. After this numbering the interface graph can easily be formed and its members can be numbered.

Step 4: Find the Type I self-equilibrating systems. All multiple members of the interface graph are identified and the values -1 and 1 are assigned to appropriate rows (corresponding to the member numbers) and the corresponding minimal null vectors are created.

Step 5: Find the Type II self-equilibrating systems. The $\frac{N}{4} - 1$ SESs of Type II should be extracted from two adjacent elements.

Step 6: Find the Type III self-equilibrating systems. For each minimal cycle of natural associate graph of FEM with four members (one common node), one SES and with eight or more members (Opening), three SESs should be extracted.

Step 7: Order the null vectors. At this step the constructed null vectors should be ordered such that their last non-zero entries form a list with an ascending order.

7.3.7 Numerical Examples

In this section three FEMs are considered, one of these models is assumed to be supported in statically indeterminate fashion and the other two supported in a determinate fashion. The effect of the presence of additional supports can separately be included for each special case with no difficulty. The equilibrium matrices are formed. Null bases and the flexibility matrices are constructed and the required computational times, and the condition numbers are calculated. In all the following examples, nnz represents the number of non-zero entries and $\lambda_{max}/\lambda_{min}$ is the ratio of the extreme eigenvalues taken as the condition number of a matrix. The comparison between present algorithm and algebraic force method is shown in Table 7.11 for all three examples. Finally the present method is validated through comparison of resulting stresses using the present graph-theoretical force method and the displacement method.

Example 1. The lining of a tunnel is considered supported in a statically determinate manner, and its applied load is depicted in Fig. 7.27. This structure is discretized using rectangular 8-node finite elements. The properties of the model are as follows:

Poisson's ratio = 0.3; Elastic modulus E = 2e + 7 kN/m^2; Thickness t = 1.00 m
Number of rectangular 8-node elements = 100
Number of nodes = 405
DSI = 100 × 13 − 2 × 405 + 3 = 493
Number of Type I self-equilibrating systems = 296 (60 %)
Number of Type II self-equilibrating systems = 148 (30 %)
Number of Type III self-equilibrating systems = 49 (10 %)

Table 7.11 The comparison between present algorithm and algebraic force method for all three examples

Example	$\frac{\text{Computational time}}{\text{LU time}}$	Condition number (flexibility matrices) Present method	$\frac{\lambda_{max}}{\lambda_{min}}$ LU factorization	Norms max$\lvert \mathbf{A} \times \mathbf{B}_1 \rvert$ Present method	LU factorization
Tunnel lining	1.21	47.65	1.63e+5	1.08e−15	4.04e−14
Circulate beam	0.45	9.38e+5	8.73e+7	5.51e−14	1.76e−13
Retaining wall (8-node)	0.84	2.68e+4	4.28e+7	7.43e−12	2.67e−13
Retaining wall (12-node)	0.78	3.59e+5	8.01e+7	1.22e−14	1.90e−13

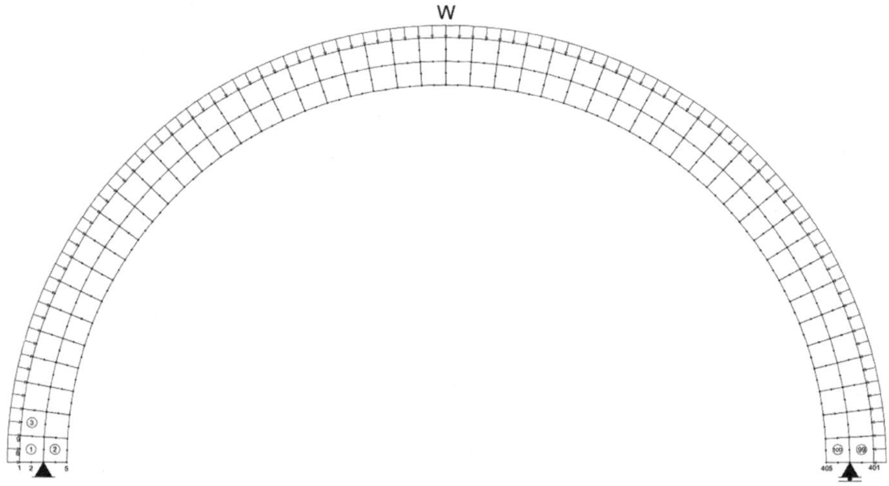

Fig. 7.27 A lining of a tunnel, the discretization and loading of the structure

The interface and natural associate graphs of the FEM model are illustrated in Fig. 7.28. The pattern of the equilibrium matrix is shown in Fig. 7.29. The sparsity of the final null basis obtained by the present method is approximately 6.7 % of that of QR method and 6.07 % of the LU method as depicted in Fig. 7.30. The flexibility matrix, G, is also well-structured as shown in Fig. 7.31. The results are verified by standard displacement method in Table 7.12.

Example 2. A circular plate and its applied load are shown in Fig. 7.32. The internal and external diameters are 1.00 and 5.00 m, respectively. This structure is discretized using 12-node rectangular finite elements. The properties of the model are as follows:

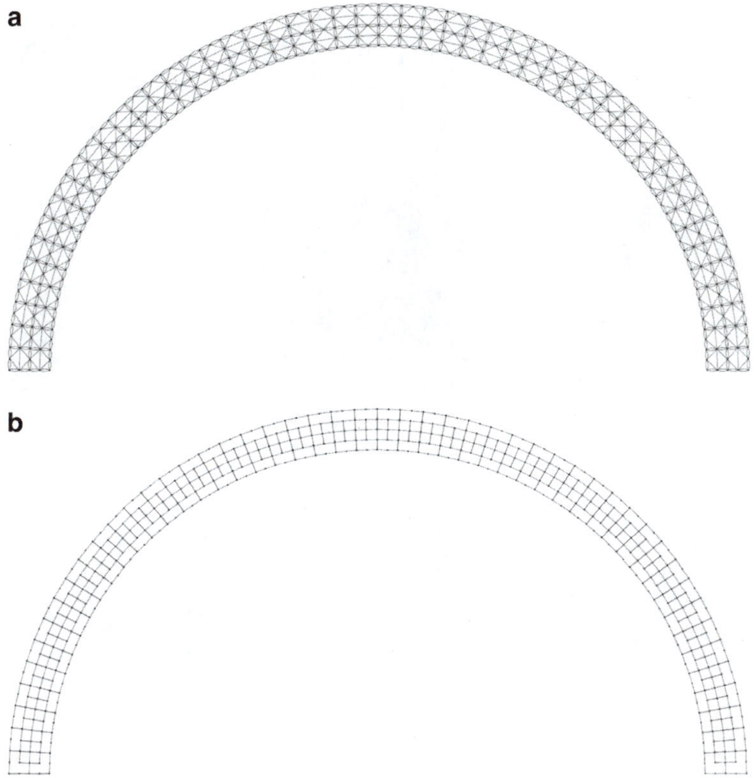

Fig. 7.28 Interface and natural associate graphs of Example 1. (**a**) Interface graph, (**b**) natural associate graph

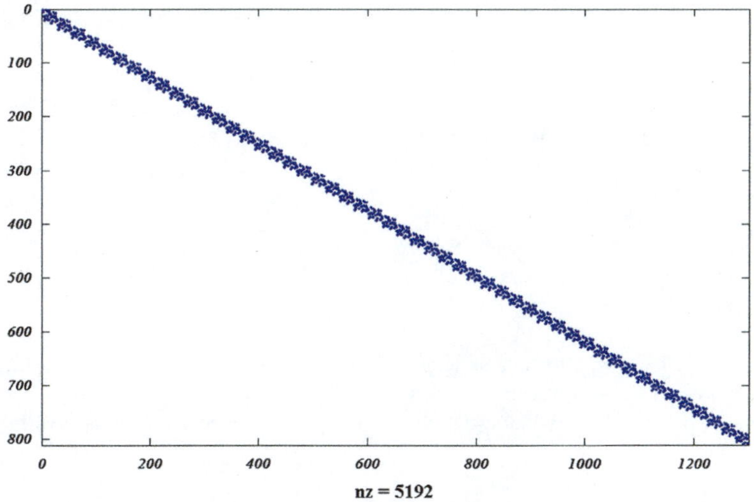

Fig. 7.29 Pattern of the equilibrium matrix for Example 1

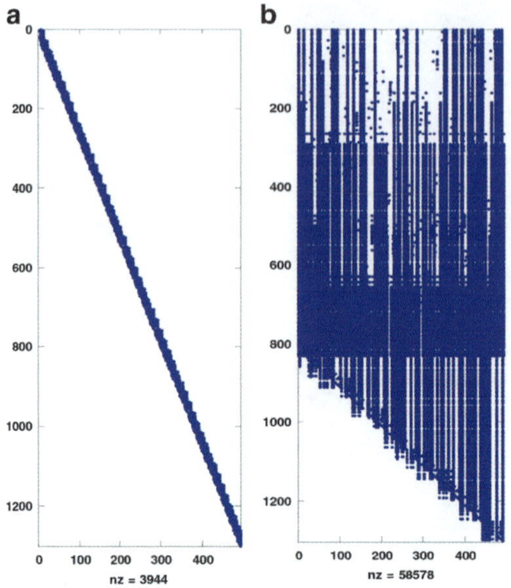

Fig. 7.30 Patterns and number of non-zero entries of the null bases of Example 1: (**a**) present algorithm, (**b**) QR factorization and (**c**) LU factorization

Fig. 7.31 Patterns of the flexibility matrix $\mathbf{G} = \mathbf{B}_1^t \mathbf{F}_m \mathbf{B}_1$ for Example 1 using the proposed method

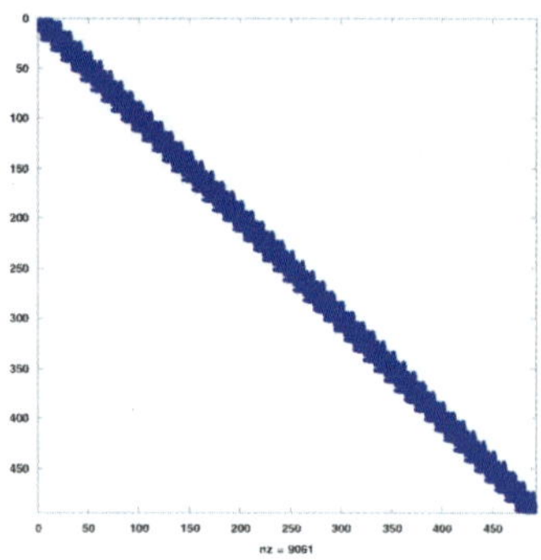

Poisson's ratio = 0.3; Elastic modulus E = 2e + 7 kN/m^2; Thickness t = 1.00 m
Number of rectangular 12-node elements = 384
Number of nodes = 2,064
DSI = 384 × 21 − 2 × 2064 + 3 = 3939
Number of Type I self-equilibrating systems = 2,160 (≈55 %)
Number of Type II self-equilibrating systems = 1,440 (≈36 %)
Number of Type III self-equilibrating systems = 336 (internal nodes) + 3 (an opening) = 339 (≈8.5 %)

Table 7.12 Comparison of the displacement method and the present force method for Example 1

	Element stresses					
Method	Displacement method			The present force method		
Element	σ_{xx}	σ_{yy}	σ_{xy}	σ_{xx}	σ_{yy}	σ_{xy}
	kN/cm^2			kN/cm^2		
1	−0.1806	−0.7815	0.2763	−0.1806	−0.7815	0.2763
10	−0.0918	−0.7379	−0.2186	−0.0918	−0.7379	−0.2186
20	−0.3097	−0.6082	−0.4040	−0.3097	−0.6082	−0.4040
30	−0.6168	−0.3721	−0.4470	−0.6168	−0.3721	−0.4470
40	−0.8943	−0.1416	−0.3060	−0.8943	−0.1416	−0.3060
50	−1.0196	−0.0346	−0.0306	−1.0196	−0.0346	−0.0306
60	−0.9346	−0.1073	0.2586	−0.9346	−0.1073	0.2586
70	−0.6790	−0.3214	0.4333	−0.6790	−0.3214	0.4333
80	−0.3672	−0.5666	0.4265	−0.3672	−0.5666	0.4265
90	−0.1244	−0.7240	0.2627	−0.1244	−0.7240	0.2627
100	−0.1361	−0.7739	0.3446	−0.1361	−0.7739	0.3446

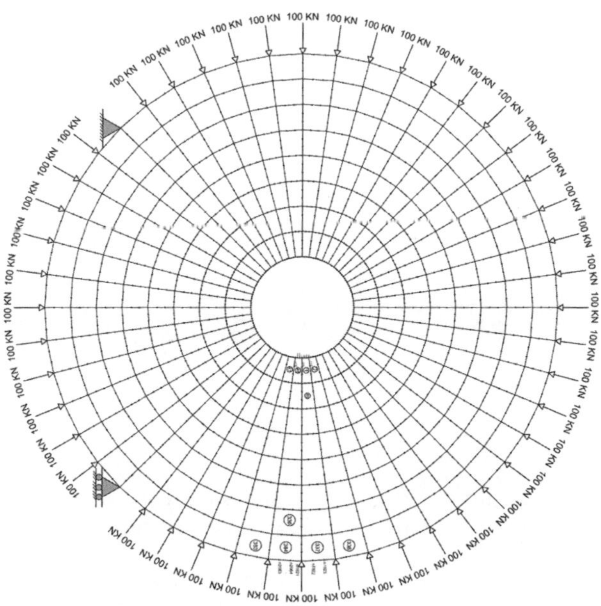

Fig. 7.32 A circulate plate with an opening

The interface and natural associate graph of the FEM model is illustrated in Figs. 7.33 and 7.21. The pattern of the equilibrium matrix is shown in Fig. 7.34. The sparsity of the final null basis obtained by the present method is approximately 0.46 % of the QR method and 1.9 % of the LU approach as depicted in Fig. 7.35.

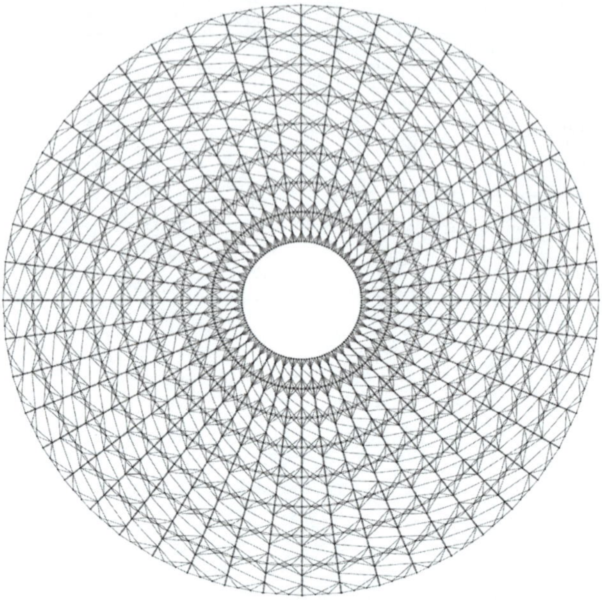

Fig. 7.33 The interface graph of Example 2

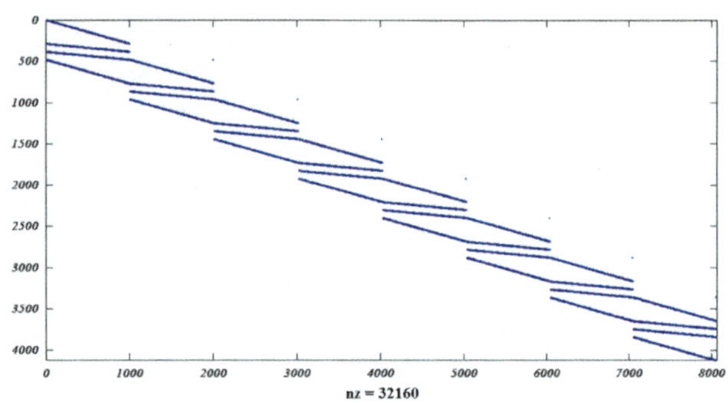

Fig. 7.34 Pattern of the equilibrium matrix for Example 2

The flexibility matrix is also well-structured as shown in Fig. 7.36. The results are verified by the standard displacement method in Table 7.13.

Example 3. The FEM of a dam which is supported in a statically indeterminate fashion is depicted in Fig. 7.37. This structure is discretized using 8-node and 12-node rectangular finite elements separately. It should be noted that the number of support elements depends on the choice of 8 or 12 nodes per finite element. The properties of the models are:

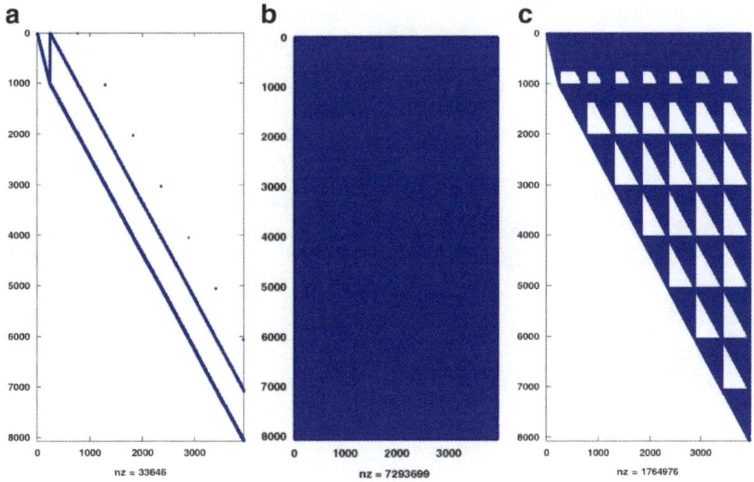

Fig. 7.35 Patterns and the number of non-zero entries of the null bases of Example 2: (**a**) present algorithm, (**b**) QR factorization and (**c**) LU factorization

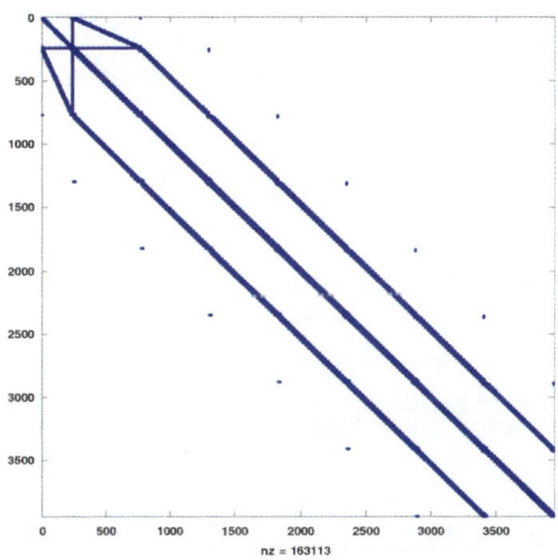

Fig. 7.36 Patterns of flexibility matrix $\mathbf{G} = \mathbf{B}_1^t \mathbf{F}_m \mathbf{B}_1$ for Example 2 using the proposed method

Poisson's ratio = 0.3; Elastic modulus E = 2e + 7 kN/m^2; Thickness t = 1.00 m
Case 1: Number of rectangular 8-node elements = 192, Number of nodes = 681
Case 2: Number of rectangular 12-node elements = 192, Number of nodes = 1,117
$DSI_{8-\text{node}} = 192 \times 13 - 2 \times 681 + 82 = 1,216$
$DSI_{12-\text{node}} = 192 \times 21 - 2 \times 1117 + 122 = 1,920$
Number of Type I self-equilibrating systems, Case 1 = 664 (58.5 %)
Number of Type II self-equilibrating systems, Case 1 = 332 (29 %)
Number of Type III self-equilibrating systems, Case 1 = 141 (12.5 %)
Number of Type I self-equilibrating systems, Case 2 = 996 (55 %)
Number of Type II self-equilibrating systems, Case 2 = 664 (36.8 %)
Number of Type III self-equilibrating systems, Case 2 = 141 (8.2 %)

Table 7.13 Comparison of the displacement method and the present force method for Example 2

	Element stresses					
Method	Displacement method			The present force method		
Element	σ_{xx}	σ_{yy}	σ_{xy}	σ_{xx}	σ_{yy}	σ_{xy}
	kN/cm^2			kN/cm^2		
337	−2.6962	−2.8935	0.0188	−2.6962	−2.8935	0.0188
340	−2.7314	−2.8329	0.0701	−2.7314	−2.8329	0.0701
343	−2.7879	−2.7646	0.0793	−2.7879	−2.7646	0.0793
346	−2.8348	−2.7121	0.0464	−2.8348	−2.7121	0.0464
349	−2.8483	−2.6972	−0.0098	−2.8483	−2.6972	−0.0098
352	−2.8220	−2.7262	−0.0612	−2.8220	−2.7262	−0.0612
255	−2.7686	−2.7870	−0.0813	−2.7686	−2.7870	−0.0813
358	−2.7158	−2.8547	−0.0572	−2.7158	−2.8547	−0.0572
361	−2.6939	−2.9112	0.0059	−2.6939	−2.9112	0.0059
364	−2.7288	−2.9929	0.1074	−2.7288	−2.9929	0.1074
367	−0.8483	−2.4765	−0.0155	−0.8483	−2.4765	−0.0155
370	−2.4803	−2.7025	−0.0552	−2.4803	−2.7025	−0.0552
373	−2.5319	−2.7105	0.0084	−2.5319	−2.7105	0.0084
376	−2.3806	−2.6839	0.0927	−2.3806	−2.6839	0.0927
379	−2.6693	−2.7841	−0.4591	−2.6693	−2.7841	−0.4591
382	−2.7092	−2.9526	−0.0679	−2.7092	−2.9526	−0.0679

The interface and natural associate graphs of the FEM model are illustrated in Fig. 7.38 for FEM with 8-node elements. The interface graph for other cases can simply be obtained. The final null basis obtained for both cases by the present method are depicted in Figs. 7.39 and 7.40. The flexibility matrix is also well-structured as shown in Fig. 7.41. The results are verified by the standard displacement method in Table 7.14.

7.4 Efficient Finite Element Analysis Using Graph-Theoretical Force Method: Hexa-Hedron Elements

Formation of a suitable null basis for equilibrium matrix is the main problem of finite elements analysis via force method. For an optimal analysis, the selected null basis matrices should be sparse and banded corresponding to sparse, banded and well-conditioned flexibility matrices. In this section, an efficient method is developed for the formation of null bases of finite element models (FEMs) consisting of hexahedron elements, corresponding to highly sparse and banded flexibility matrices. This is achieved by associating special graphs with the FEM and selecting appropriate subgraphs and forming the self-equilibrating systems (SESs) on these subgraphs.

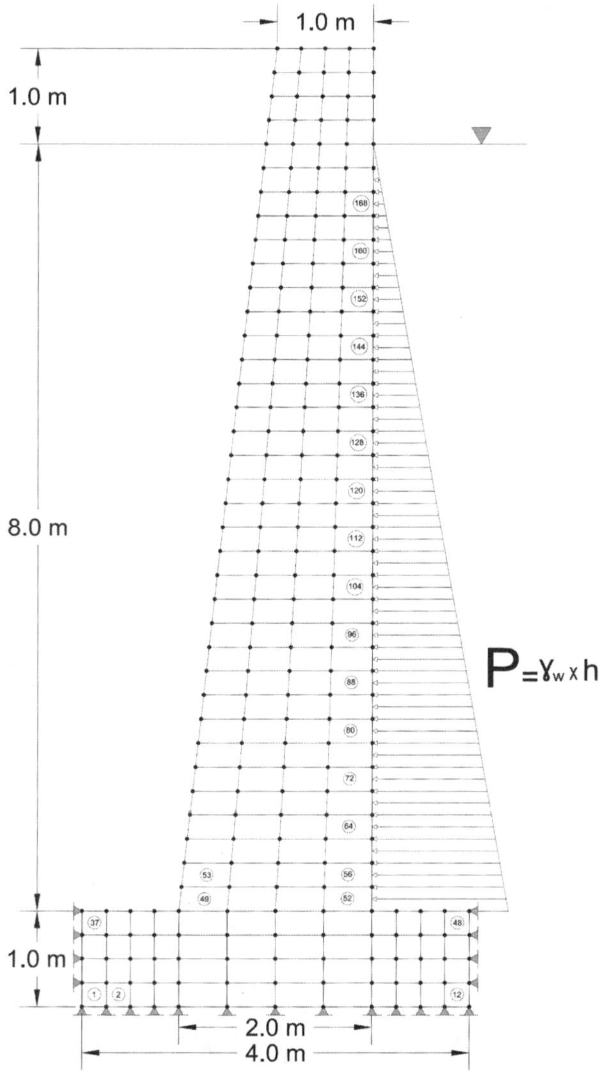

Fig. 7.37 A retaining wall and the corresponding rectangular meshes

7.4.1 *Independent Element Forces and Flexibility Matrix of Hexahedron Elements*

In the force method the efficiency of this analysis depends on the required time for the formation of the matrix. $\mathbf{G} = \mathbf{B}_1^t \mathbf{F}_m \mathbf{B}_1$ and its characteristics, i.e. sparsity and bandedness together with its conditioning. For the formation of a well-structured matrix G, one should select a well-structured $\boldsymbol{B}_1$ matrix.

For the generation of the equilibrium matrix **A** of a FEM, a system of independent force systems should be defined and also their relations with the element nodal forces should be established.

In displacement method we have three forces at each node of the element. For an element with N nodes, $3 \times N$ nodal forces can be defined. Using six equilibrium equations, $3N - 6$ independent forces will be remained. In other words, there are $3N - 6$ independent element forces in an element with *N* nodes. The nodal forces and element forces systems are shown in Fig. 7.42 for hexahedron elements with various numbers of boundary nodes. For the higher order elements, the element forces system can be obtained with the same procedure.

These element forces **F** can be related to the nodal forces **S** for a N-node element by a $(3N) \times (3N - 6)$ transformation matrix using Eq. 7.22 as

$$\mathbf{S} = \mathbf{TF} \tag{7.22}$$

Transformation matrix can be formed simply as
where x_i, y_i and z_i are the Cartesian coordinates of node i, $m_{ij} = (x_i - x_j)/l_{ij}$, $n_{ij} = (y_i - y_j)/l_{ij}$, and $p_{ij} = (z_i - z_j)/l_{ij}$, are the direction cosines and l_{ij} is the length of the line between nodes i and j.

Formulation of a discrete element equivalent to the actual continuous structure is the first step in matrix structural analysis. For a linear system it can be assumed that the stresses $\boldsymbol{\sigma}$ .are related to the forces **F** by linear equation as

$$\boldsymbol{\sigma} = \bar{\mathbf{c}}\mathbf{F} \tag{7.24}$$

The matrix $\bar{\mathbf{c}}$ represents a statically equivalent stress system due to the unit force F. The flexibility matrix of an element can be written as

$$\mathbf{f}_{\mathrm{m}} = \int_V \bar{\mathbf{c}}^t \boldsymbol{\varphi} \bar{\mathbf{c}} \mathrm{d}V \tag{7.25}$$

The integration is taken over the volume of the element, where $\boldsymbol{\varphi}$ is the matrix relating the stresses to strains $\boldsymbol{\varepsilon} = \boldsymbol{\varphi}\boldsymbol{\sigma}$ in three dimensional problems. The primary step in the formation of the flexibility matrix of an element is determining the matrix $\bar{\mathbf{c}}$. It is obvious that the *i*th column of $\bar{\mathbf{c}}$ represents the resultant stresses due to unit element force F_i in the force method and also stresses due to nodal forces **S** is equal to the *i*th column of **T** utilizing the displacement method. Hence, we can form matrix $\bar{\mathbf{c}}$ using the stiffness properties of the hexahedron element using the displacement method. Now the flexibility matrix of the element in the force method is formed from Eq. 7.25 using Gauss numerical integration method with sixty four Gauss points ($4 \times 4 \times 4$ Gauss Points Integration).

$$
\begin{bmatrix} S_1 \\ S_2 \\ S_3 \\ S_4 \\ S_5 \\ S_6 \\ S_7 \\ S_8 \\ S_9 \\ S_{10} \\ S_{11} \\ S_{12} \\ S_{13} \\ S_{14} \\ S_{15} \\ S_{16} \\ S_{17} \\ S_{18} \\ S_{19} \\ S_{20} \\ S_{21} \\ S_{22} \\ S_{23} \\ S_{24} \end{bmatrix}
=
\begin{bmatrix}
m_{13} & 0 & 0 & m_{12} & 0 & 0 & 0 & 0 & m_{15} & 0 & 0 & 0 & 0 & 0 & 0 & 0 & 0 & 0 \\
n_{13} & 0 & 0 & n_{12} & 0 & 0 & 0 & 0 & n_{15} & 0 & 0 & 0 & 0 & 0 & 0 & 0 & 0 & 0 \\
p_{13} & 0 & 0 & p_{12} & 0 & 0 & 0 & 0 & p_{15} & 0 & 0 & 0 & 0 & 0 & 0 & 0 & 0 & 0 \\
0 & 0 & m_{24} & m_{21} & 0 & 0 & 0 & 0 & 0 & 0 & 0 & m_{26} & 0 & 0 & 0 & m_{25} & m_{23} & 0 \\
0 & 0 & n_{24} & n_{21} & 0 & 0 & 0 & 0 & 0 & 0 & 0 & n_{26} & 0 & 0 & 0 & n_{25} & n_{23} & 0 \\
0 & 0 & p_{24} & p_{21} & 0 & 0 & 0 & 0 & 0 & 0 & 0 & p_{26} & 0 & 0 & 0 & p_{25} & p_{23} & 0 \\
m_{31} & m_{34} & 0 & 0 & 0 & 0 & 0 & 0 & 0 & m_{37} & 0 & 0 & m_{35} & 0 & 0 & 0 & m_{32} & 0 \\
n_{31} & n_{34} & 0 & 0 & 0 & 0 & 0 & 0 & 0 & n_{37} & 0 & 0 & n_{35} & 0 & 0 & 0 & n_{32} & 0 \\
p_{31} & p_{34} & 0 & 0 & 0 & 0 & 0 & 0 & 0 & p_{37} & 0 & 0 & p_{35} & 0 & 0 & 0 & p_{32} & 0 \\
0 & m_{43} & m_{42} & 0 & 0 & 0 & 0 & 0 & 0 & 0 & m_{48} & 0 & 0 & m_{47} & m_{46} & 0 & 0 & 0 \\
0 & n_{43} & n_{42} & 0 & 0 & 0 & 0 & 0 & 0 & 0 & n_{48} & 0 & 0 & n_{47} & n_{46} & 0 & 0 & 0 \\
0 & p_{43} & p_{42} & 0 & 0 & 0 & 0 & 0 & 0 & 0 & p_{48} & 0 & 0 & p_{47} & p_{46} & 0 & 0 & 0 \\
0 & 0 & 0 & 0 & m_{57} & 0 & 0 & m_{56} & m_{51} & 0 & 0 & 0 & m_{53} & 0 & 0 & m_{52} & 0 & 0 \\
0 & 0 & 0 & 0 & n_{57} & 0 & 0 & n_{56} & n_{51} & 0 & 0 & 0 & n_{53} & 0 & 0 & n_{52} & 0 & 0 \\
0 & 0 & 0 & 0 & p_{57} & 0 & 0 & p_{56} & p_{51} & 0 & 0 & 0 & p_{53} & 0 & 0 & p_{52} & 0 & 0 \\
0 & 0 & 0 & 0 & 0 & 0 & m_{68} & m_{65} & 0 & 0 & 0 & m_{62} & 0 & 0 & m_{64} & 0 & 0 & m_{67} \\
0 & 0 & 0 & 0 & 0 & 0 & n_{68} & n_{65} & 0 & 0 & 0 & n_{62} & 0 & 0 & n_{64} & 0 & 0 & n_{67} \\
0 & 0 & 0 & 0 & 0 & 0 & p_{68} & p_{65} & 0 & 0 & 0 & p_{62} & 0 & 0 & p_{64} & 0 & 0 & p_{67} \\
0 & 0 & 0 & 0 & m_{75} & m_{78} & 0 & 0 & 0 & m_{73} & 0 & 0 & 0 & m_{74} & 0 & 0 & 0 & m_{76} \\
0 & 0 & 0 & 0 & n_{75} & n_{78} & 0 & 0 & 0 & n_{73} & 0 & 0 & 0 & n_{74} & 0 & 0 & 0 & n_{76} \\
0 & 0 & 0 & 0 & p_{75} & p_{78} & 0 & 0 & 0 & p_{73} & 0 & 0 & 0 & p_{74} & 0 & 0 & 0 & p_{76} \\
0 & 0 & 0 & 0 & 0 & m_{87} & m_{86} & 0 & 0 & 0 & m_{84} & 0 & 0 & 0 & 0 & 0 & 0 & 0 \\
0 & 0 & 0 & 0 & 0 & n_{87} & n_{86} & 0 & 0 & 0 & n_{84} & 0 & 0 & 0 & 0 & 0 & 0 & 0 \\
0 & 0 & 0 & 0 & 0 & p_{87} & p_{86} & 0 & 0 & 0 & p_{84} & 0 & 0 & 0 & 0 & 0 & 0 & 0
\end{bmatrix}
\begin{bmatrix} F_1 \\ F_2 \\ F_3 \\ F_4 \\ F_5 \\ F_6 \\ F_7 \\ F_8 \\ F_9 \\ F_{10} \\ F_{11} \\ F_{12} \\ F_{13} \\ F_{14} \\ F_{15} \\ F_{16} \\ F_{17} \\ F_{18} \end{bmatrix}
\tag{7.23}
$$

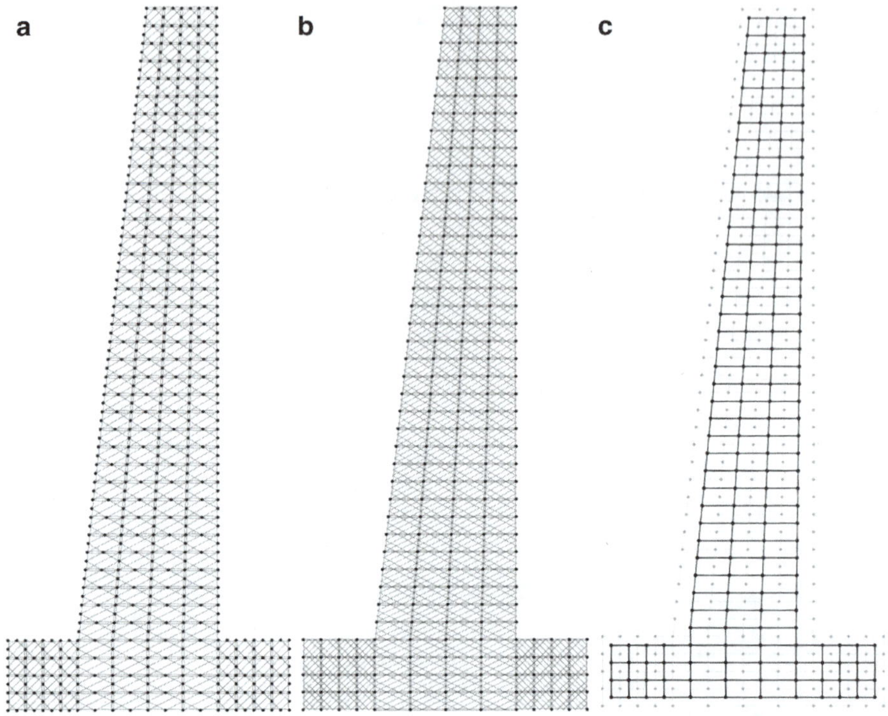

Fig. 7.38 Interface graph and natural associate graph for both cases of Example 3, (**a**) Interface graph for 8-node element, (**b**) Interface graph for 12-node element and (**c**) Associate graph for both cases

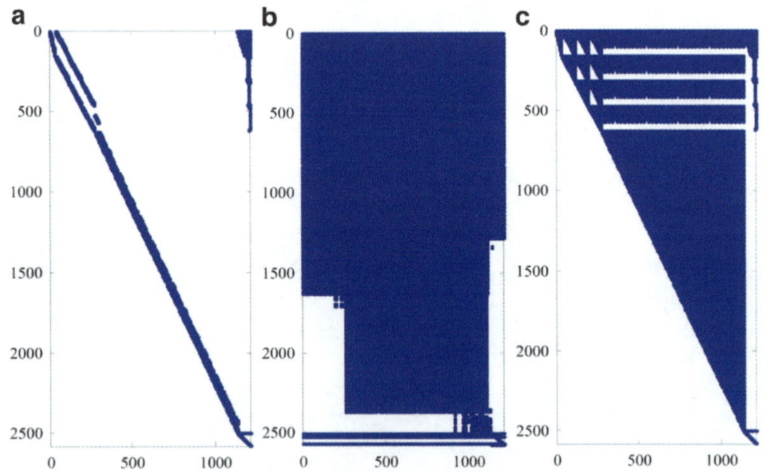

Fig. 7.39 Patterns and number of non-zero entries of null bases of Example 3 (8-node element): (**a**) present algorithm, (**b**) QR factorization and (**c**) LU factorization

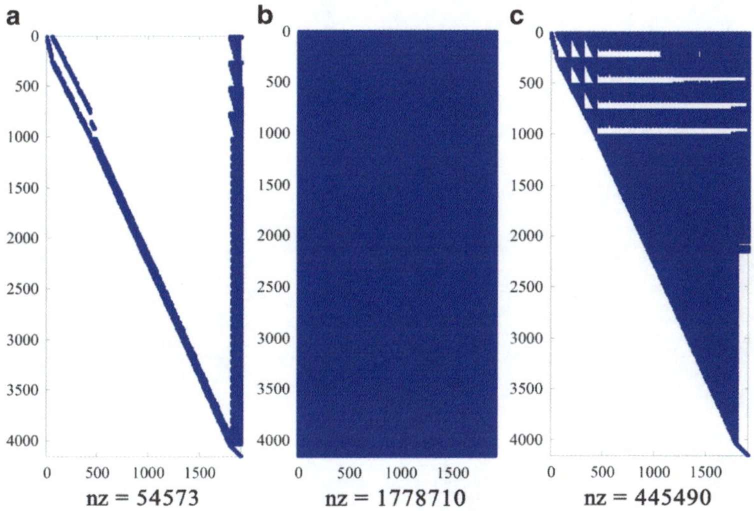

Fig. 7.40 Patterns and number of non-zero entries of null bases of Example 3 (12-node element): (**a**) present algorithm, (**b**) QR factorization and (**c**) LU factorization

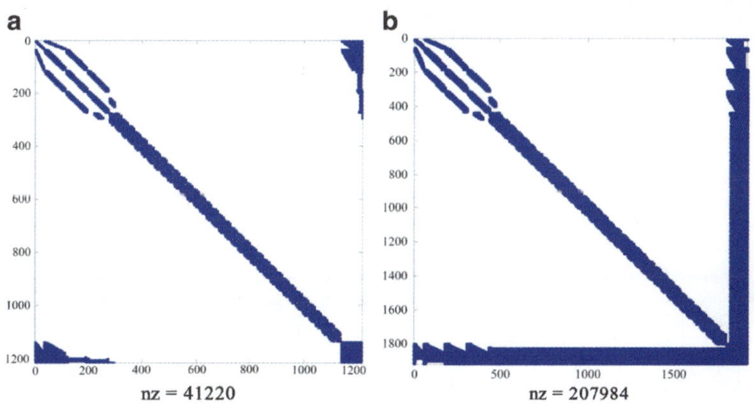

Fig. 7.41 Patterns of flexibility matrix $\mathbf{G} = \mathbf{B}_1^t \mathbf{F}_m \mathbf{B}_1$ of Example 3, (**a**) 8-node element, (**b**) 12-node element

7.4.2 Graphs Associated with Finite Element Models

7.4.2.1 An Interface Graph

The interface graph of a finite element model denoted by IG (FEM) can easily be constructed for hexahedron FEM using the following rules:

1. This graph contains all the nodes of the FEM.

Table 7.14 The comparison of the displacement method and the present force method for Example 3

	Element stresses											
Method	8-node						12-node					
	Displacement method			The present force method			Displacement method			The present force method		
Element	σ_{xx}	σ_{yy}	σ_{xy}	σ_{xx}	σ_{yy}	σ_{xy}	σ_{xx}	σ_{yy}	σ_{xy}	σ_{xx}	σ_{yy}	σ_{xy}
	kN/cm^2						kN/cm^2					
8	0.0560	0.9029	0.0436	0.0560	0.9029	0.0436	0.0322	0.8158	0.0426	0.0322	0.8158	0.0426
16	−0.0430	−1.0513	−0.2987	−0.0430	−1.0513	−0.2987	0.0346	−0.6760	−0.3087	0.0346	−0.6760	−0.3087
24	0.2060	0.0529	−0.0512	0.2060	0.0529	−0.0512	0.2399	0.1925	−0.0526	0.2399	0.1925	−0.0526
32	0.0655	1.4381	0.1071	0.0655	1.4381	0.1071	0.1210	1.6845	0.0629	0.1210	1.6845	0.0629
40	−1.0339	−1.4361	−0.9329	−1.0339	−1.4361	−0.9329	−0.9501	−1.1784	−0.9567	−0.9501	−1.1784	−0.9567
48	0.0694	0.0060	0.0006	0.0694	0.0060	0.0006	0.0736	0.0340	0.0035	0.0736	0.0340	0.0035
56	−0.1939	1.2847	0.0540	−0.1939	1.2847	0.0540	−0.1916	1.2772	0.0490	−0.1916	1.2772	0.0490
64	−0.1546	1.1539	−0.0439	−0.1546	1.1539	−0.0439	−0.1572	1.1521	−0.0411	−0.1572	1.1521	−0.0411
72	−0.1114	1.0027	−0.0788	−0.1114	1.0027	−0.0788	−0.1124	1.0021	−0.0764	−0.1124	1.0021	−0.0764
80	−0.0979	0.8474	−0.0779	−0.0979	0.8474	−0.0779	−0.0975	0.8475	−0.0767	−0.0975	0.8475	−0.0767
88	−0.0891	0.6999	−0.0703	−0.0891	0.6999	−0.0703	−0.0884	0.7002	−0.0695	−0.0884	0.7002	−0.0695
96	−0.0809	0.5645	−0.0616	−0.0809	0.5645	−0.0616	−0.0802	0.5648	−0.0608	−0.0802	0.5648	−0.0608
104	−0.0729	0.4431	−0.0527	−0.0729	0.4431	−0.0527	−0.0722	0.4434	−0.0521	−0.0722	0.4434	−0.0521
112	−0.0650	0.3363	−0.0441	−0.0650	0.3363	−0.0441	−0.0643	0.3366	−0.0436	−0.0643	0.3366	−0.0436
120	−0.0570	0.2446	−0.0359	−0.0570	0.2446	−0.0359	−0.0564	0.2449	−0.0354	−0.0564	0.2449	−0.0354
128	−0.0491	0.1683	−0.0281	−0.0491	0.1683	−0.0281	−0.0485	0.1686	−0.0278	−0.0485	0.1686	−0.0278
136	−0.0412	0.1073	−0.0210	−0.0412	0.1073	−0.0210	−0.0407	0.1075	−0.0207	−0.0407	0.1075	−0.0207
144	−0.0333	0.0612	−0.0145	−0.0333	0.0612	−0.0145	−0.0329	0.0614	−0.0143	−0.0329	0.0614	−0.0143
152	−0.0254	0.0294	−0.0089	−0.0254	0.0294	−0.0089	−0.0251	0.0296	−0.0088	−0.0251	0.0296	−0.0088
160	−0.0176	0.0104	−0.0044	−0.0176	0.0104	−0.0044	−0.0173	0.0105	−0.0044	−0.0173	0.0105	−0.0044
168	−0.0097	0.0019	−0.0012	−0.0097	0.0019	−0.0012	−0.0096	0.0020	−0.0012	−0.0096	0.0020	−0.0012

Element	Nodes number	Size of T	Nodal forces	Element forces	Element shape
Linear	$N = 8$	24×18	$S_{24\times1} = [S_1 \ldots S_{24}]^t$	$F_{18\times1} = [F_1 \ldots F_{18}]^t$	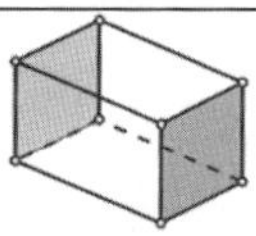
Quadratic	$N = 20$	60×54	$S_{60\times1} = [S_1 \ldots S_{60}]^t$	$F_{54\times1} = [F_1 \ldots F_{54}]^t$	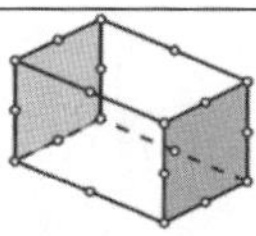
Cubic	$N = 32$	96×90	$S_{96\times1} = [S_1 \ldots S_{96}]^t$	$F_{90\times1} = [F_1 \ldots F_{90}]^t$	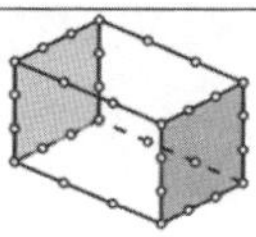
Quartic	$N = 44$	132×126	$S_{132\times1} = [S_1 \ldots S_{132}]^t$	$F_{126\times1} = [F_1 \ldots F_{126}]^t$	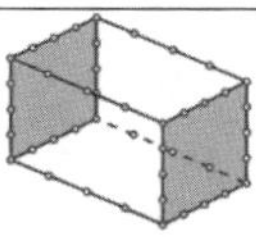

Fig. 7.42 A set of hexahedron elements

2. With each edge of an element of FEM, (N + 4)/12 graph elements are associated.
3. For each element with N nodes, $3N - 6$ members should be considered in the interface graph. Thus, $2N - 10 = (3\ N - 6) - (N + 4)$ diagonal members should be added.

Therefore, in the interface of two elements (common side), $4 \times \frac{(N+4)}{12} + \frac{(2N-10)}{6}$ multiple members are present. The member numbering of the interface graph should be performed according to the numbering of the FEM, taking into account the primary nodal numbering of a consider element in the model. Thus, for each hexahedron element $3N - 6$ edges of the interface graph will be numbered sequentially according to the patterns which were illustrated in Fig. 7.43. In this figure, quartic element numbering was neglected and just the element forces are displayed. Numbering of this type can be easily obtained according to the pattern of other elements.

7.4.2.2 Natural Associate Graph

The natural associate graph represented by NAG(FEM) is constructed by the following rules:

1. Nodes of the NAG(FEM) correspond to the elements of FEM.
2. For each pair of elements in FEM having $\frac{(2N+8)}{6}$ common nodes ($N =$ the number of nodes of an element), one member is added between the corresponding two nodes in NAG(FEM).

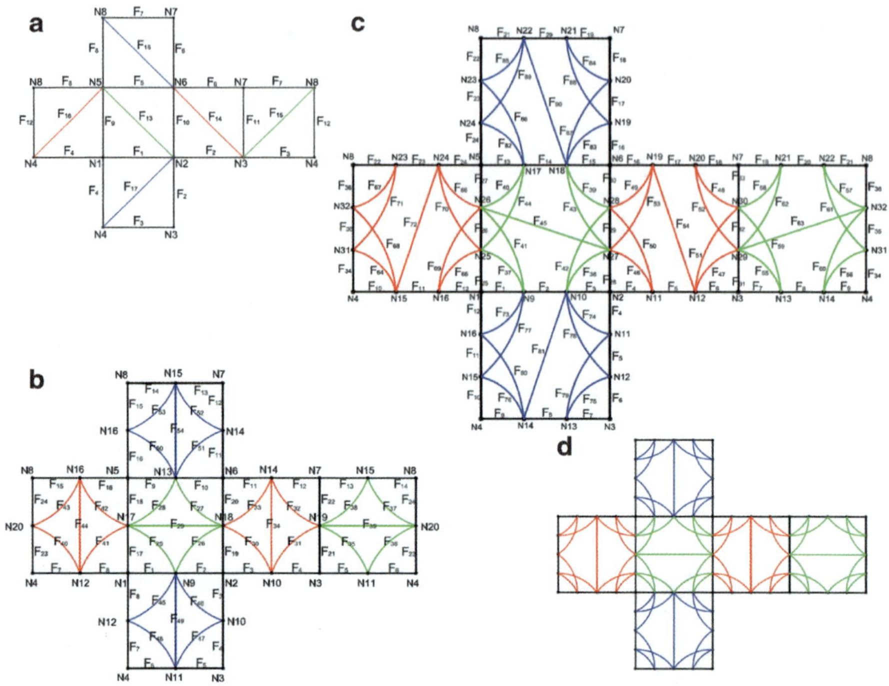

Fig. 7.43 Nodal numbering and element forces for hexahedron elements; (**a**) linear, (**b**) quadratic, (**c**) cubic and (**d**) quartic

NAG(FEM) can be constructed using the following procedure: One of the preliminary steps in FEA is defining the elements with their connected nodes. In this way the element connectivity matrix is constructed which contains the element-node incidence relationships. In the process of constructing the element connectivity matrix, another matrix which contains node-element incidence properties can be formed. This matrix is named the node connectivity matrix. Now using the element connectivity and the node connectivity matrices leads to an algorithm with complexity O(n) for an efficient generation of NAG.

In order to recognize the adjacent elements to the *n*th element which have common $\frac{(2N+8)}{6}$ nodes or one common face, first the connected nodes to the *n*th element are identified from the element connectivity matrix. In the subsequent step using the node connectivity matrix, elements which have at least one common node with the *n*th element are identified. Now it is convenient to seek for the adjacent elements in this reduced search space. A FEM and its corresponding NAG are illustrated in Fig. 7.44.

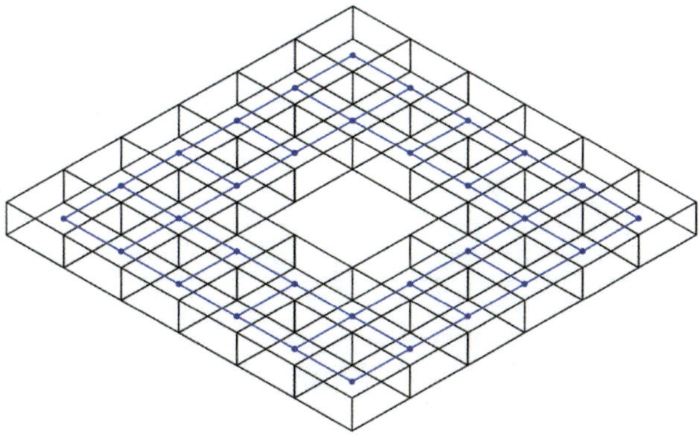

Fig. 7.44 Finite element model (black part) with natural associate graph (blue part)

7.4.3 Negative Incidence Number

Negative incidence number (NIN) is necessary for each node of NAG(FEM). This number can be found as following:

After generation of natural associate graph of the FEM, use an efficient method for its nodal numbering. A typical edge of the graph connects smaller number to the node with higher number. Negative incidence number of each node is the number of its adjacent nodes with smaller nodal number. Except the node numbered as 1, all the other nodes have one, two or three negatively incident edges defined as the negative incidence number of the node. Owing to the importance of these numbers in recognizing the types of SESs, the negative incidence numbers of the nodes of the graph should carefully be calculated. In Fig. 7.45, a hexahedron FEM with element numbering, its corresponding associate graph and negative incidence number of nodes are shown. The nodes should be numbered such that the incidence numbers do not become large. Any simple nodal ordering will lead to a logical ordering.

7.4.4 Pattern Corresponding to Self-Equilibrating Systems

Considering Fig. 7.43, in order to find the patterns corresponding to the self-equilibrating systems, a hexahedron element is simulated as a spatial truss formed as the 1-skeleton of the hexahedron element together with some diagonal members. This is possible since the independent element forces are applied in the nodes and are along the edges of the element. In Fig. 7.46, an IG(FEM) with four quadratic elements is shown which is simulated as a spatial truss containing some multiple members.

The statical indeterminacy of a spatial truss with *m* members and *n* nodes is given as $\gamma(S) = m - 3n + 6$; therefore, the degree of statical indeterminacy (DSI)

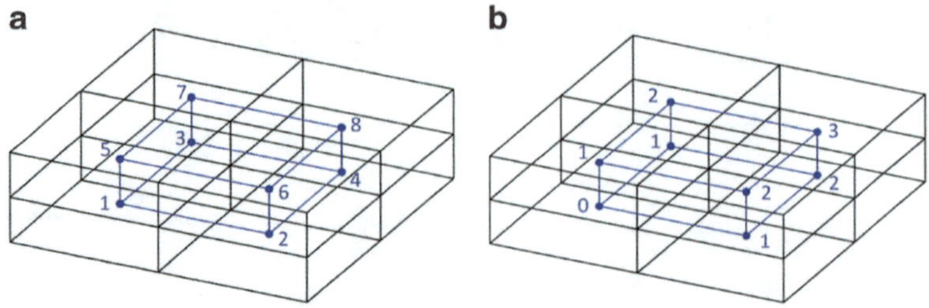

Fig. 7.45 Finite element model (black part) with natural associate graph (blue part); (**a**) nodal numbering of NAG; (**b**) negative incidence numbers of NAG

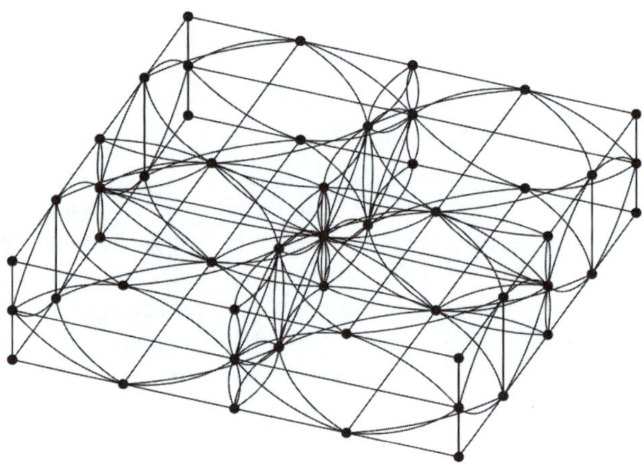

Fig. 7.46 An IG(FEM) with four quadratic elements is shown which is simulated as a spatial truss

of the entire model supported in a statically determinate manner can be calculated with the same relationship as

$$\mathrm{DSI} = (3\mathrm{N} - 6) \times \mathrm{M} - 3\mathrm{n} + 6 \tag{7.26}$$

where M is the total number of finite elements, N is the number of nodes of one element and n is the total number of nodes of the FEM.

With the above simulation, the patterns of the self-equilibrating systems can be identified as follows:

7.4.4.1 Type I Self-Equilibrating Systems

For each k multiple member in equivalent truss model of FEM, there are k unknown forces and one equilibrium equation in the member's direction. Thus DSI of the substructure is equal to k − 1 and k − 1 self-equilibrating systems can be generated on each k multiple member of interface graph of the FEM. In this way, first each k multiple members are arranged in ascending order as (m_1, m_2, m_3,. . ., m_{k-1}, m_k). where ($m_1 < m_2 < m_3 < \ldots < m_{k-1} < m_k$). Each selection of two members from this list is valid to construct a Type I self-equilibrating system, but in order to achieve a better bandwidth reduction; selection of adjacent members from the defined list is preferable. Therefore k − 1 duplicate members are selected as (m_1, m_2), (m_2, m_3),. . ., (m_{k-1}, m_k). Each pair (m_i, m_j) with $i < j$ represents the numbers of corresponding self-equilibrating system. The member with bigger number is selected as the generator of the current SES and also as a redundant force. The null vectors corresponding to the Type I SESs have two non-zero entries in rows i and j equal to −1 and 1, respectively.

For FEMs with hexahedron elements, more than 75 % of the total self-stress systems are of Type I. Thus, a large percent of the minimal null vectors can be formed only by the determination of member numbers of these pairs. It should be noted that in the process of the formation of the interface graph, these pairs and their numbers can simply be identified.

7.4.4.2 Type II Self-Equilibrating Systems

There are other types of self-equilibrating systems which are extracted from two adjacent elements of FEM. In other words, for two adjacent elements with N nodes, the DSI can be calculated as:

$$\mathrm{DSI} = (3\mathrm{N} - 6) \times \mathrm{M} - 3\mathrm{n} + 6$$

$$\Rightarrow \mathrm{DSI} = (3\mathrm{N} - 6) \times 2 - 3 \times \left(2\mathrm{N} - \frac{2\mathrm{N} + 8}{6}\right) + 6 = \mathrm{N} - 2 \tag{7.27}$$

$$\underbrace{\frac{2\mathrm{N} - 10}{6}}_{\substack{\text{diagonal members} \\ \text{of one side}}} + \underbrace{\frac{2\mathrm{N} + 8}{6}}_{\substack{\text{other members} \\ \text{of one side}}} = \underbrace{\frac{4\mathrm{N} - 2}{6}}_{\substack{\text{number of members} \\ \text{of common side of} \\ \text{two adjacent elements}}} \quad \text{self-equilibrating}$$

systems were generated as Type I systems. Thus the number of remaining self-equilibrating systems is

$$\text{Type II} = (\mathrm{N} - 2) - \left(\frac{4\mathrm{N} - 2}{6}\right) = \frac{2\mathrm{N} - 10}{6} \tag{7.28}$$

In other words, $\frac{2N-10}{6}$ SESs should be extracted from two adjacent elements. For example, the remaining subgraphs for two adjacent quadratic elements are shown in

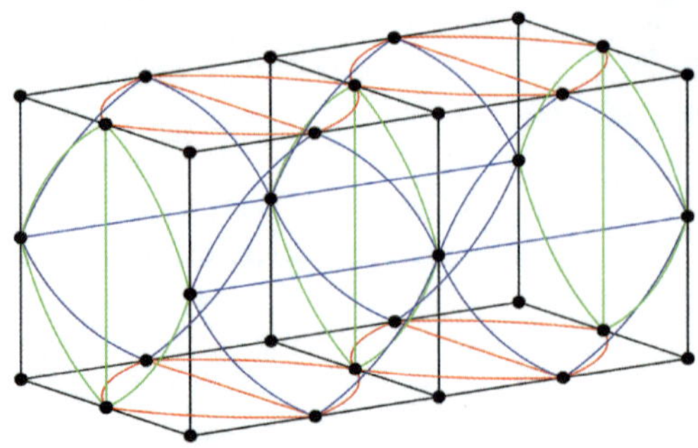

Fig. 7.47 Subgraph corresponding to SESs of Type II

Fig. 7.47. After deleting the generators corresponding to Type I SESs, the null vectors should be calculated from the remaining subgraph. These null vectors can easily be generated on the corresponding sub-structure utilizing an algebraic method.

Apart from the aforementioned about generating the SESs of Type II, if there is at least a negative incidence number higher than one in a FEM, another important point should be considered which is explained below:

Some of the calculated SESs of Type II are not independent of the others. For example, for a FEM with four quadratic elements M' is equal to four, where M' is the number of members of the associate graph of the model. The number of SESs of Type II is 18 instead of $20 = 4 \times 5$. In other words, two SESs are dependent and should not be selected. For determining the independent SESs, an appropriate approach is proposed. In this approach, independent SESs will be recognized utilizing negative incidence number of elements.

The SESs of Type II are extracted from two adjacent elements in a FEM which are the same as members of NAG(FEM). If a member of NAG(FEM) connects two elements M_i and M_j where $i < j$, the number of independent SESs of Type II which can be extracted from the subgraph corresponding to these two adjacent elements is equal to:

$$\text{Type II} = \left(\alpha - \text{NIN}_j \times \frac{\text{mod}(N, 8)}{4}\right) \tag{7.29}$$

Where NIN_j is the negative incidence number of jth element and α is 1,6,8 and 13 for linear, quadratic, cubic and quartic elements, respectively. For linear element, a SES Type II can be generated on each two adjacent elements on a FEM [6]. For other types of element, after deleting the generators corresponding to Type I SESs, the main diagonal member (longer diagonal member) of the jth element which is located in the common side with other adjacent elements with smaller number than j, the null vectors should be calculated from the remaining subgraph.

The most important point in Type II self-equilibrating systems is to select an appropriate generator. In fact by eliminating these generators and the generators corresponding to Type I SESs from IG(FEM), the sub-structure of Type III SESs and the primary structure of the IG(FEM) must remain stable.

7.4.4.3 Type III Self-Equilibrating Systems

Sub-structures which are topologically identical to the minimal cycles of the natural associate graph of FEM contain some Type I, Type II and one or six Type III self-equilibrating systems.

(a) *Type I minimal cycles of NAG(FEM)*

These minimal cycles of the natural associate graph of the FEM pass through four elements which have $\frac{\frac{2N+8}{6}}{4}+1=\frac{N+16}{12}$ common nodes or $\frac{N+16}{12}-1=\frac{N+4}{12}$ common edges. Corresponding interface graph of these elements have n nodes and m edges for a FEM with *N* node elements.

$$\mathrm{m} = 4 \times (3\mathrm{N} - 6) \tag{7.30}$$

$$\mathrm{n} = 4\mathrm{N} - 4 \times \left(\frac{2\mathrm{N}+8}{6}\right) + \frac{\mathrm{N}+16}{12} = \frac{11\mathrm{N}-16}{4} \tag{7.31}$$

Subsequently, the DSI of the interface graph is

$$\begin{aligned} \mathrm{DSI} &= \mathrm{m} - 3\mathrm{n} + 6 \Rightarrow \mathrm{DSI} = 4 \times (3\mathrm{N} - 6) - 3 \times \frac{11\mathrm{N}-16}{4} + 6 \\ &= \frac{15\mathrm{N}}{4} - 6 \end{aligned} \tag{7.32}$$

The $4 \times \frac{4N-2}{6} - \frac{2N+8}{24} = \frac{31N-20}{12}$ SESs are Type I and there are 4, 18, 32 and 46 SESs of Type II for linear, quadratic, cubic and quartic elements, respectively.

$$\mathrm{DSI} - (\mathrm{TypeI\&II}) = \begin{cases} \mathrm{N} = 8 \Rightarrow \mathrm{DSI} - (\mathrm{Type\ I\ \&\ II}) = 1 \\ \mathrm{N} = 20 \Rightarrow \mathrm{DSI} - (\mathrm{Type\ I\ \&\ II}) = 1 \\ \mathrm{N} = 32 \Rightarrow \mathrm{DSI} - (\mathrm{Type\ I\ \&\ II}) = 1 \\ \mathrm{N} = 44 \Rightarrow \mathrm{DSI} - (\mathrm{Type\ I\ \&\ II}) = 1 \end{cases} \tag{7.33}$$

Therefore, one independent SES should be extracted. This SES with thirteen members can be formed for any types of hexahedron elements around the common edge as is indicated bold in Fig. 7.48.

It should be noted that in a FEM, all of the SESs of Type III which are extracted from any four elements around one common edge, are not independent with all previous selected SESs. Independent ones should be selected utilizing NINs of elements. For this purpose, NIN of four elements with common edge should not be more than 2. In Fig. 7.45, a FEM with eight elements is shown. The independent SESs of Type III should be selected utilizing these three sets of elements:

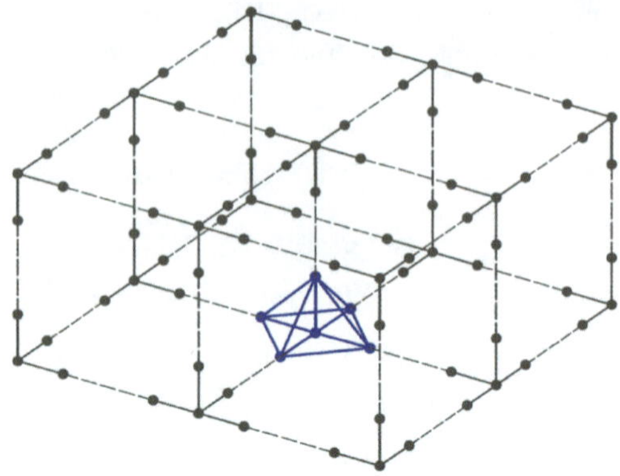

Fig. 7.48 The SES of Type III corresponding to the common edge of four elements

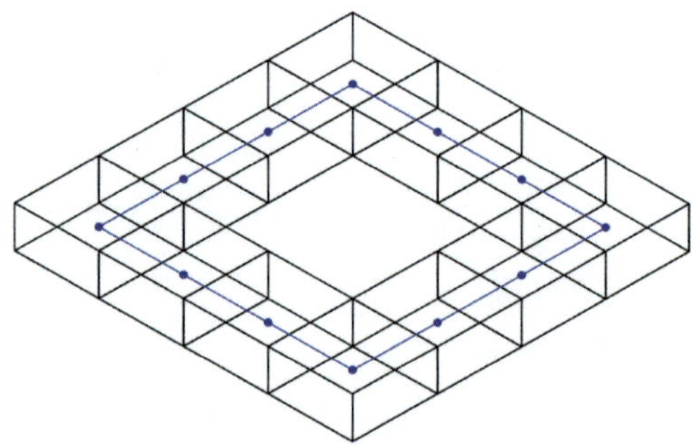

Fig. 7.49 A FEM with an opening and its NAG

{E1 E2 E3 E4}, {E1 E2 E5 E6} and {E1 E3 E5 E7}. In other words, E8 should not be in selections, because the NIN of E8 is 3.

(b) *Type II minimal cycles of NAG(FEM)*

Each minimal cycle that surrounds an opening is called the Type II minimal cycle (Fig. 7.49). Such a cycle passes through M′ (M′ ≥ 8) finite elements and its corresponding interface graph has $\left(N - \frac{2N+8}{6}\right) \times M'$ nodes and $M' \times (3N - 6)$ members. The DSI of subgraph is

$$DSI = M' \times (3N - 6) - 3 \times \left(N - \frac{2N+8}{6}\right) \times M' + 6 \Rightarrow DSI$$

$$= M' \times (N - 2) + 6 \tag{7.34}$$

and $M' \times \underbrace{\frac{4N-2}{6}}_{\substack{\text{number of members}\\ \text{of common side of}\\ \text{two adjacent elements}}}$ SESs of Type I and $M' \times \left(\frac{2N-10}{6}\right)$ SESs of Type II (Eq. 7.28) can be extracted.

$$\begin{aligned} \mathrm{DSI}-(\mathrm{Type\,I\&II}) &= \mathrm{M}' \times (\mathrm{N}-2)+6 \\ &\quad - \left[\mathrm{M}' \times \frac{4\mathrm{N}-2}{6} + \mathrm{M}' \times \left(\frac{2\mathrm{N}-10}{6}\right)\right] \\ &= 6 \end{aligned} \tag{7.35}$$

Therefore, each Type II minimal cycle corresponds to six null vectors which are calculated utilizing an algebraic method.

7.4.5 Selection of Generators for SESs of Type II and Type III

The most important point in Type II and Type III self-equilibrating systems is to select appropriate generators, because by eliminating these generators from IG (FEM), the sub-structure of primary structure of the IG(FEM) must remain stable. To achieve this, the following rule for appropriate selection of generators of Type II SESs is suggested.

For quadratic hexahedron element the generators of SESs Type II and III are illustrated in Tables 7.15 and 7.16, respectively. Directions 1, 2 and 3 are shown in Fig. 7.50. In these Tables, $N_{\alpha,\beta}$ indicates the βth node of element α and $NIN(d)_j$ is the negative incidence number of element j in direction d. In other words, $NIN(d)_j$ is one if j has an adjacent element i where $i < j$ in direction d. It should be noted that the generators corresponding to Type I were chosen previously. In addition, the generators corresponding to an opening are the last six non-zero entries of its columns which are not common with the previously selected generators. For other element types, generators corresponding to Type II and Type III can be obtained following aforementioned patterns.

Algorithm. Step 1: Generate the associate graph of the FEM and use an efficient method for its nodal numbering [4]. It is obvious that good numbering of this graph corresponds to good numbering of elements of the FEM. This numbering leads to a banded adjacency matrix of the graph and correspondingly to a banded flexibility matrix. Since numbering the members of the interface graphs corresponds to the element numbering of the finite elements, such a numbering is the only parameter for controlling the bandwidth of the flexibility matrix. Negative incidence number of the NAG(FEM) should be calculated in this step.

Step 2: Set up the equilibrium matrix of the FEM.

Step 3: Generate the interface graph and perform its numbering. The numbering of this graph should be performed according to the element numbering of the considered FEM. After this numbering the interface graph can easily be formed and its members can be numbered.

Step 4: Find the Type I self-equilibrating systems. All multiple members of the interface graph are identified and the values −1 and 1 are assigned to appropriate

Table 7.15 Generators of Type II SESs in directions 1, 2 and 3 ($k < i < j$)

NIN_j	Pattern and direction of elements i and j			Selected generators
$NIN_j = 1$	$dir1\ or\ N_{i,2} = N_{j,1}$			$54 \times (i-1) + [13,25,36,46,51]$
	$dir2\ or\ N_{i,4} = N_{j,1}$			$54 \times (i-1) + [12,31,40,45,50]$
	$dir3\ or\ N_{i,5} = N_{j,1}$			$54 \times (i-1) + [22,27,30,38,41]$
$NIN_j = 2$	$dir1$ or $N_{i,2} = N_{j,1}$	$NIN(2)_i = 1\ \&\ NIN(3)_i = 0$	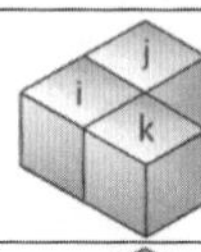	$54 \times (i-1) + [13,36,46,51]$
		$NIN(2)_i = 0\ \&\ NIN(3)_i = 1$		$54 \times (i-1) + [13,25,36,51]$
	$dir2$ or $N_{i,4} = N_{j,1}$	$NIN(1)_i = 1\ \&\ NIN(3)_i = 0$	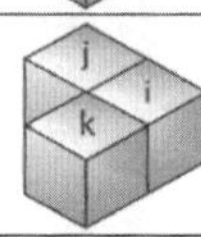	$54 \times (i-1) + [12,31,45,50]$
		$NIN(1)_i = 0\ \&\ NIN(3)_i = 1$	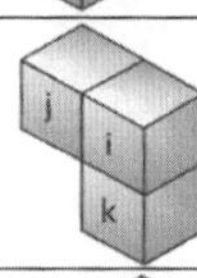	$54 \times (i-1) + [12,31,40,50]$
	$dir3$ or $N_{i,5} = N_{j,1}$	$NIN(1)_i = 1\ \&\ NIN(2)_i = 0$		$54 \times (i-1) + [22,27,30,38]$
		$NIN(1)_i = 0\ \&\ NIN(2)_i = 1$	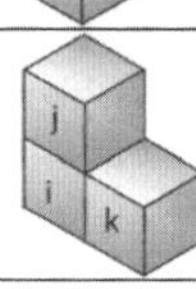	$54 \times (i-1) + [22,30,38,41]$
$NIN_j = 3$	$dir1\ or\ N_{i,2} = N_{j,1}$			$54 \times (i-1) + [13,36,51]$
	$dir2\ or\ N_{i,4} = N_{j,1}$			$54 \times (i-1) + [12,31,50]$
	$dir3\ or\ N_{i,5} = N_{j,1}$			$54 \times (i-1) + [22,30,38]$

rows (corresponding to the member numbers) and the corresponding minimal null vectors are created.

Step 5: Find the Type II self-equilibrating systems. The $\frac{(2N-10)}{6}$ SESs of Type II should be extracted from two adjacent elements and independent ones should be selected among these SESs utilizing the approach which is explained in Sect. 5.2. Calculate

Table 7.16 Generators of Type III SESs in planes 1-2, 2-3 and 1-3 ($i < j < k < l$)

	Four elements with common edges		Selected generators
Plane 1-2	$N_{i,7} = N_{j,6}$ $N_{j,6} = N_{k,8}$ $N_{k,8} = N_{l,5}$	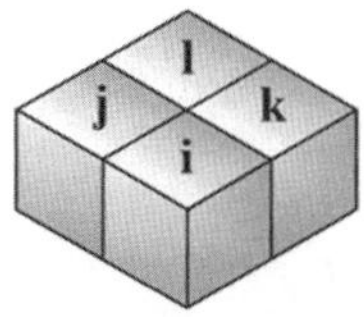	$54 \times (i-1) + [\,4\,]$
	For $\alpha = [i,j,k,l]$, $NIN_\alpha \neq 3$		
Plane 2-3	$N_{i,7} = N_{j,6}$ $N_{j,6} = N_{k,3}$ $N_{k,3} = N_{l,2}$		$54 \times (i-1) + [\,24\,]$
	For $\alpha = [i,j,k,l]$, $NIN_\alpha \neq 3$		
Plane 1-3	$N_{i,7} = N_{j,8}$ $N_{j,8} = N_{k,3}$ $N_{k,3} = N_{l,4}$		$54 \times (i-1) + [\,20\,]$
	For $\alpha = [i,j,k,l]$, $NIN_\alpha \neq 3$		

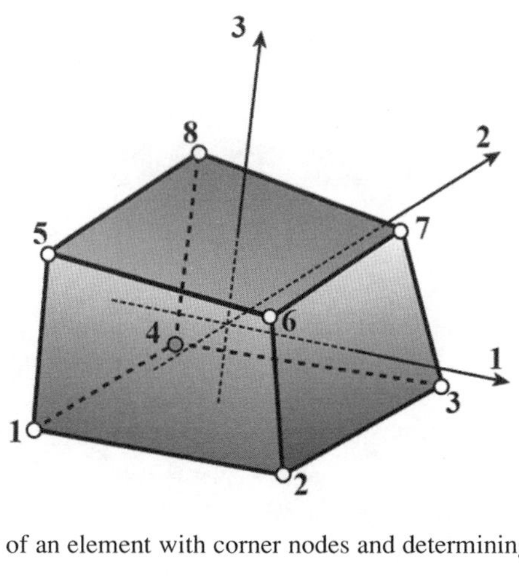

Fig. 7.50 Typical view of an element with corner nodes and determining directions 1, 2 and 3

the corresponding null vectors from the relevant equilibrium sub-matrix in this step.

Step 6: Find the Type III self-equilibrating systems. For each minimal cycle of natural associate graph of FEM with four members ($\frac{N+16}{12}$ common nodes or $\frac{N+4}{12}$

common edges and $NIN_j < 3$ for j = 1, 2, 3, 4), one SES and with eight or more members (opening) six SESs should be extracted. Calculate the corresponding null vectors from the relevant equilibrium sub-matrix.

Step 7: Order the null vectors. At this step the constructed null vectors should be ordered such that their last non-zero entries form a list with an ascending order.

7.4.6 Numerical Examples

In this section two FEMs are considered, which are assumed to be supported in statically indeterminate fashion. The translations of each support node are fixed in all three directions. The equilibrium matrices are formed. Null bases and the flexibility matrices are constructed and the required computational times, and the condition numbers are calculated. In all the following examples, *nz* represents the number of non-zero entries and $\lambda_{max}/\lambda_{min}$ is the ratio of the extreme eigenvalues taken as the condition number of a matrix. The comparison between present algorithm and algebraic force method will be shown in the conclusion section.

Example 1. An arch wall structure which is supported in a statically indeterminate fashion is illustrated in Fig. 7.51. This structure is discretized using 20-node hexahedron elements. The properties of the model are as follows:

Poisson's ratio = 0.2; Elastic modulus E = 2E + 10 N/m^2; Density ρ = 2,400 kg/m^3;
Internal radius = 8.0 m;
Number of 20-node hexahedron elements = 80
Number of nodes = 557
$DSI_{Internal} = 80 \times 54 - 3 \times 557 + 6 = 2655$; $DSI_{External} = 15 \times 3 - 6 = 39$
Number of Type I self-equilibrating systems = 1,996 (75.0 %)
Number of Type II self-equilibrating systems = 620 (23.3 %)
Number of Type III self-equilibrating systems = 39 (1.7 %)

The pattern of the equilibrium matrix is shown in Fig. 7.52. The sparsity of the final null basis obtained by the present method is approximately 22.41 % of the **LU** method as depicted in Fig. 7.53. The flexibility matrix **G** is also well-structured as shown in Figs. 7.54 and 7.55.

It should be added that the total DSI for the force method of this structure is 2,655 + 39 = 2,694, while the DOFs for the displacement method is 557 × 3 = 1,671, indicating less number of equations for the latter approach. However, since in the force method nearly 75 % of the null vectors are found by simple graph theoretical approach, one should compare 1,671 with approximately 25 % of 2,694 = 673, showing the superiority of the force method.

Example 2. A dome with an opening which is supported in a statically indeterminate fashion is illustrated in Fig. 7.56. The internal and external diameters are 5.00

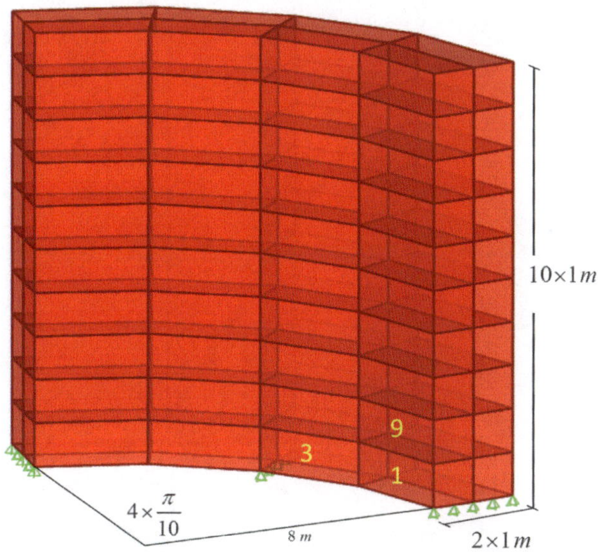

Fig. 7.51 An arch wall structure which is supported in a statically indeterminate fashion

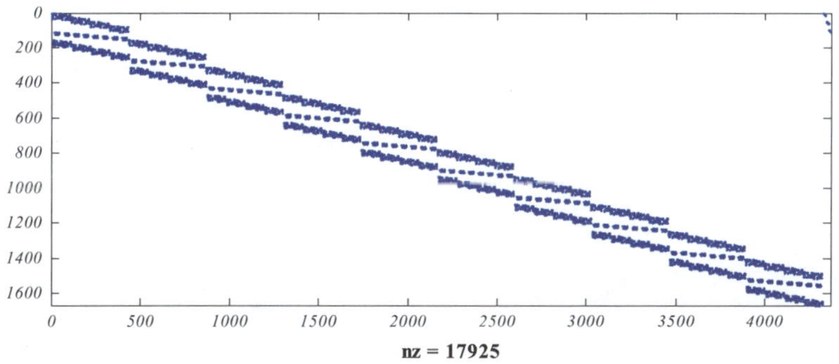

Fig. 7.52 Pattern of the equilibrium matrix for Example 1

and 5.50 m, respectively. This structure is discretized using 20-node hexahedron elements. The properties of the model are as follows:

Poisson's ratio = 0.2; Elastic modulus E = 2E + 10 N/m^2; Density ρ = 2,400 kg/m^3;
Number of 20-node hexahedron elements = 84,
Number of nodes = 648
$DSI_{Internal} = 84 \times 54 - 3 \times 648 + 6 = 2598$; $DSI_{External} = 60 \times 3 - 6 = 174$
Number of Type II self-equilibrating systems = 636 (24.4 %)
Number of Type III self-equilibrating systems = 72 (four elements with common edges) + 6 (an opening) = 78 (3.0 %)

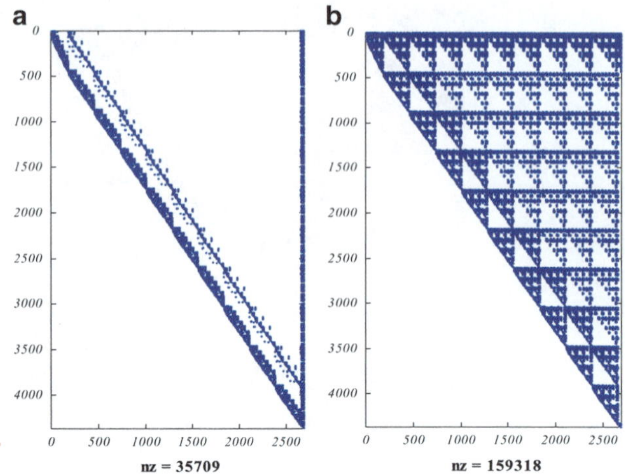

Fig. 7.53 Patterns and number of non-zero entries of the null bases of Example 1: (**a**) present algorithm, (**b**) LU factorization

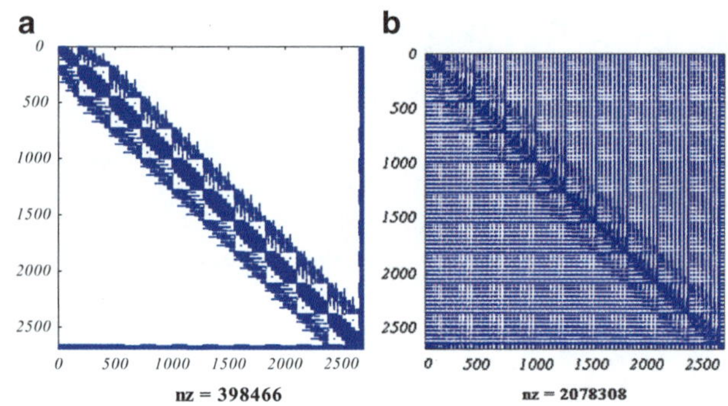

Fig. 7.54 Patterns of the flexibility matrix $\mathbf{G} = \mathbf{B}_1^t \mathbf{F}_m \mathbf{B}_1$ for Example 1: (**a**) present algorithm, (**b**) LU factorization

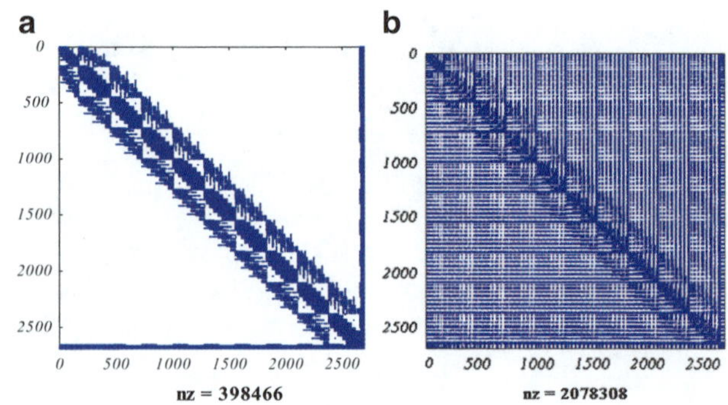

Fig. 7.55 Patterns of the flexibility matrix $\mathbf{G} = \mathbf{B}_1^t \mathbf{F}_m \mathbf{B}_1$ for Example 1: (**a**) present algorithm, (**b**) LU factorization

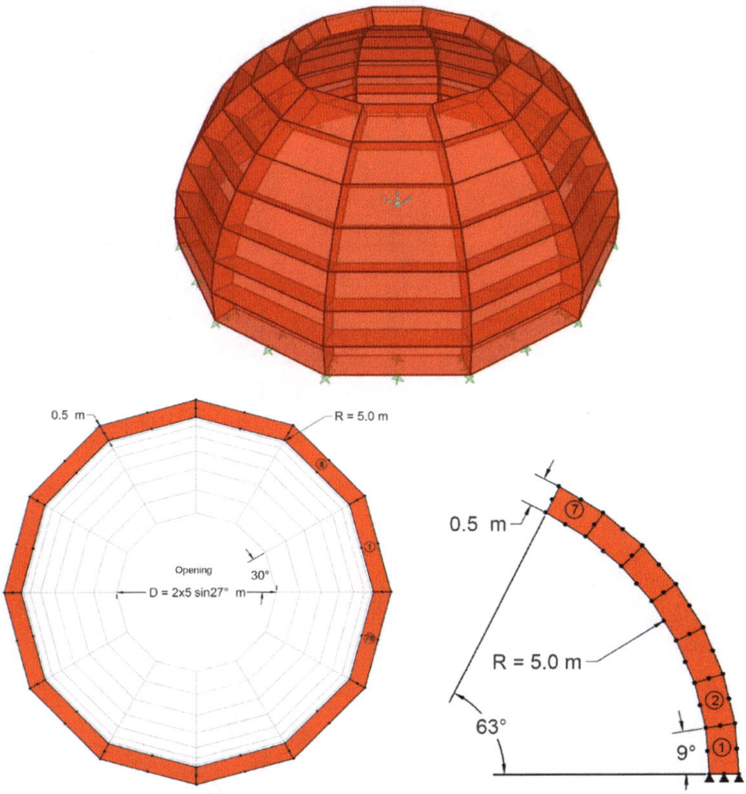

Fig. 7.56 A dome with an opening: 3D view, bottom view and a section

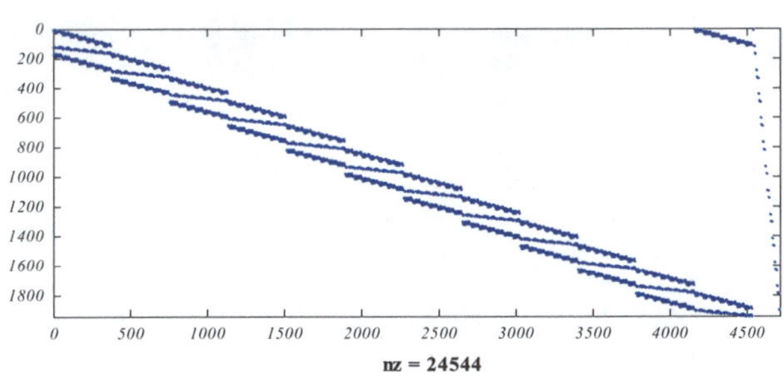

Fig. 7.57 Pattern of the equilibrium matrix for Example 2

The pattern of the equilibrium matrix is shown in Fig. 7.57. The sparsity of the final null basis obtained by the present method is approximately 28.1 % of the LU approach as depicted in Fig. 7.16. The flexibility matrix is also well-structured as shown in Figs. 7.58 and 7.59.

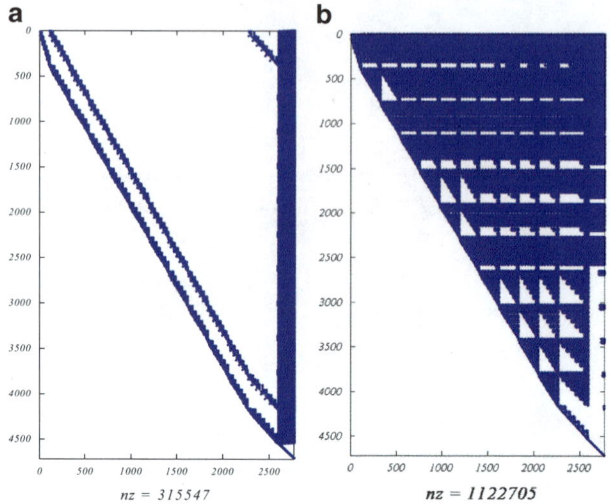

Fig. 7.58 Patterns and the number of non-zero entries of the null bases of Example 2: (**a**) present algorithm, (**b**) **LU** factorization

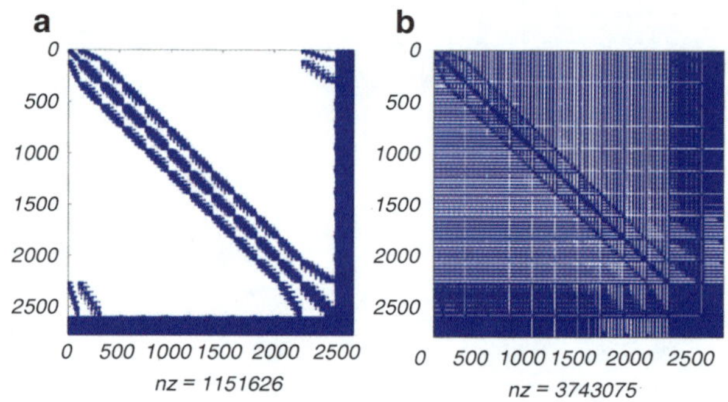

Fig. 7.59 Patterns of flexibility matrix $\mathbf{G} = \mathbf{B}_1^t \mathbf{F}_m \mathbf{B}_1$ for Example 2: (**a**) present algorithm, (**b**) **LU** factorization

Finally, it is hoped that the extension of elements for the force method continues similar to those of the displacement method, to enable these dual approaches to be efficiently utilized in the analysis of large-scale finite element models.

References

1. Kaveh A, Tolou Kian MJ (2013) Efficient finite element analysis of models comprised of higher order triangular elements. Acta Mech 224(9):1957–1975

2. Kaveh A, Massoudi MS, Massoudi MJ (2014) Efficient finite element analysis using graph-theoretical force method; rectangular plane stress and plane strain serendipity family elements. Periodica Polytechnica in print
3. Kaveh A, Massoudi MS, Massoudi MJ (2013) Efficient finite element analysis using graph-theoretical force method; hexahedron elements. Comput Struct 128:175–188
4. Kaveh A, Roosta GR (1998) Comparative study of finite element nodal ordering methods. Eng Struct 20(1–2):86–96
5. Przemieniecki JS (1968) Theory of matrix structural analysis. McGraw-Hill, New York
6. Kaveh A, Koohestani K (2007) An efficient graph theoretical method for plate bending finite element analysis via force method. Eng Comput 24(7):679–698

Chapter 8
Decomposition for Parallel Computing: Graph Theory Methods

8.1 Introduction

In the last decade, parallel processing has come to be widely used in the analysis of large-scale structures. This chapter is devoted to the optimal decomposition of structural models using graph theory approaches. First, efficient graph theory methods are presented for the optimal decomposition of space structures. The subdomaining approaches are then presented for partitioning of finite element models. A substructuring technique for the force method of structural analysis is discussed.

Several partitioning algorithms are developed for solution of multi-member systems, which can be categorised as graph theory methods and algebraic graph theory approaches.

For the graph theory method, Farhat [1] proposed an automatic finite element domain decomposer, which is based on a Greedy type algorithm and seeks to decompose an FEM into balanced domains, sharing a minimum number of common nodal points. In order to avoid domain splitting, Al-Nasra and Nguyen [2] incorporated geometrical information of the FEM into an automatic decomposition algorithm similar to the one proposed by Farhat [1]. The Sparpak uses nested dissection due to George and Liu [3], which uses a level tree for dissecting a model. Kaveh and Roosta [4] employed different expansion processes for decomposing space structures and finite element meshes.

Applications of the methods of this chapter are by no means confined to structural systems; these methods can equally be applied to other large-scale problems like the analysis of hydraulic systems and electrical networks.

A. Kaveh, *Computational Structural Analysis and Finite Element Methods*,
DOI 10.1007/978-3-319-02964-1_8,

8.2 Earlier Works on Partitioning

8.2.1 Nested Dissection

The term "nested dissection" was introduced by George [5], following a suggestion of Birkhoff. Its roots lie in finite element substructuring, and it is closely related to the tearing and interconnecting method of Kron [6].

The central concept for nested dissection is the removal of a set of nodes from the graph (separator) of a symmetric matrix (or the model of a structure) that leaves the remaining graph in two or more disconnected parts. In nested dissection, these parts are themselves further divided by the removal of sets of nodes, with the dissection nested to any depth.

If the variables of each subgraph are grouped together, by ordering the nodes of their nodes contiguously followed by numbering the nodes, in the separator, then the following block form will be obtained:

$$\begin{bmatrix} \mathbf{A}_{11} & \mathbf{0} & \mathbf{A}_{13} \\ \mathbf{0} & \mathbf{A}_{22} & \mathbf{A}_{23} \\ \mathbf{A}_{31} & \mathbf{A}_{32} & \mathbf{A}_{33} \end{bmatrix}. \tag{8.1}$$

The blocks $\mathbf{A}_{11}$ and $\mathbf{A}_{22}$ may themselves be ordered to such a form by using dissection sets. This way every level defines a nested dissection order.

The significance of the above partitioning of the matrix is twofold: first, the zero blocks are preserved in the factorisation, thereby limiting fill; second, factorisation of the matrices $\mathbf{A}_{11}$ and $\mathbf{A}_{22}$ can proceed independently, thereby enabling parallel execution on separate processors.

When a complicated design is assembled from simpler substructures, it makes sense to exploit these natural substructures. The resulting ordering is likely to be good, simply because, when each variable is eliminated, only the other variables of its substructures are involved.

8.2.2 A Modified Level-Tree Separator Algorithm

The separator routine in Sparspak, FNDSEP, finds a pseudo-peripheral node in the graph and generates a level structure from it. It then chooses the median level in the level structure as the node separator. However, this choice may separate the graph into widely disparate parts. In a modification made by Pothen et al. [7], the node separator is selected to the smallest level k, such that the first k levels together contain more than half of the nodes. A node separator is obtained by removing from the nodes in level k those nodes that are not adjacent to any node in level k − 1, and therefore these are added to the part containing the nodes in the first k − 1 levels. The other part has nodes in levels k + 1 and higher. Although such a method is simple; however, the spectral bisection method computes a smaller node separator than the Sparspak algorithm.

8.3 Substructuring for Parallel Analysis of Skeletal Structures

8.3.1 Introduction

In many engineering applications, particularly in the analysis and design of large systems, it is convenient to allocate the design of certain components (substructures) to individual design groups. The study of each substructure is carried out more or less independently, and the dependencies between the substructures are resolved after the study of individual substructures is completed. The dependencies among the components may of course require redesign of some of the substructures, so the above procedure may be iterated several times.

As an example, suppose for a structural model, we choose a set of nodes I and their incident members which, if removed, disconnect it into two substructures. If the variables associated with each substructure are numbered consecutively, followed by the variables associated with I, then the partitioning of the stiffness matrix **A** will be as that of Eq. 8.1.

The Cholesky factor **L** of **A**, correspondingly, will be partitioned as,

$$\mathbf{L} = \begin{bmatrix} \mathbf{L}_{11} & \mathbf{0} & \mathbf{0} \\ \mathbf{0} & \mathbf{L}_{22} & \mathbf{0} \\ \mathbf{W}_{13}^{t} & \mathbf{W}_{23}^{t} & \mathbf{L}_{33} \end{bmatrix}, \tag{8.2}$$

where

$$\mathbf{A}_{11} = \mathbf{L}_{11}\mathbf{L}_{11}^{t}, \mathbf{A}_{22} = \mathbf{L}_{22}\mathbf{L}_{22}^{t}, \mathbf{W}_{13} = \mathbf{L}_{11}^{t}\mathbf{A}_{13}, \quad \mathbf{W}_{23} = \mathbf{L}_{22}\mathbf{L}_{23}^{t},$$

and

$$\mathbf{L}_{33}^{t}\mathbf{L}_{33} = \mathbf{A}_{33} - \mathbf{A}_{13}^{t}\mathbf{A}_{11}^{-1}\mathbf{A}_{23} - \mathbf{A}_{23}^{t}\mathbf{A}_{22}^{-1}\mathbf{A}_{23}. \tag{8.3}$$

Therefore, $\mathbf{A}_{11}$ and $\mathbf{A}_{22}$ correspond to each substructure, and the matrices $\mathbf{A}_{13}$ and $\mathbf{A}_{23}$ represent the "glue" which relates the substructures through the nodes of I.

Since the factors of $\mathbf{A}_{11}$ and $\mathbf{A}_{22}$ are independent, they can be computed in either order, or in parallel if two processors are available. Finally, in some design applications, several substructures may be identical, for example, have the same configuration and properties, and each substructure may be regarded as a super-element, which is constructed once and used repeatedly in the design of several structures. In the above example, $\mathbf{A}_{11}$ and $\mathbf{A}_{22}$ could be identical.

8.3.2 Substructuring Displacement Method

For the analysis of skeletal structures and for the finite element method, using the displacement approach, an appropriate formulation such as the Galerkian method reduces to solving the following matrix equation,

$$\mathbf{K}\mathbf{v} = \mathbf{p}, \tag{8.4}$$

where **K** is the global stiffness matrix, and **v** and **p** are the nodal displacement and nodal force vectors, respectively. To distribute the computation after decomposing the model into q subdomains, each subdomain can be treated as a super element and mapped onto the processors. Various methods for decomposition will be presented in this chapter. The global stiffness matrix and nodal force vector are equivalent to the assembly of its components for q subdomains:

$$\mathbf{K} = \sum_{j=1}^{q} \mathbf{k}_j \text{ and } \mathbf{p} = \sum_{j=1}^{q} \mathbf{p}_j. \tag{8.5}$$

Equation 8.4 can be written in the following partitioned form:

$$\begin{bmatrix} \mathbf{K}_{ii} & \mathbf{K}_{ib} \\ \mathbf{K}_{bi} & \mathbf{K}_{bb} \end{bmatrix} \begin{bmatrix} \mathbf{v}_i \\ \mathbf{v}_b \end{bmatrix} = \begin{bmatrix} \mathbf{p}_i \\ \mathbf{p}_b \end{bmatrix}. \tag{8.6}$$

In the above equation, a boundary node is defined as a node which is part of more than one subdomain and degrees of freedom at the boundary nodes are treated as boundary degrees of freedom. The vectors $\mathbf{v}_i$ and $\mathbf{v}_b$ are displacements, and $\mathbf{p}_i$ and $\mathbf{p}_b$ are forces, corresponding to internal and boundary nodes, respectively.

Each subdomain requires solution of an equation, similar to Eq. 8.4:

$$[\mathbf{k}]_j[\mathbf{d}]_j = [\mathbf{p}]_j. \tag{8.7}$$

For the full domain, Eq. 8.7 can be written in partitioned form as:

$$\begin{bmatrix} \mathbf{k}_{ii} & \mathbf{k}_{ib} \\ \mathbf{k}_{bi} & \mathbf{k}_{bb} \end{bmatrix} \begin{bmatrix} \mathbf{v}_i \\ \mathbf{v}_b \end{bmatrix} = \begin{bmatrix} \mathbf{p}_i \\ \mathbf{p}_b \end{bmatrix}. \tag{8.8}$$

Using static condensation for eliminating the interior degrees of freedom of each subdomain, the effective stiffnesses and load vectors on the interface boundaries are obtained.

For internal nodes we have,

$$[\mathbf{k}_{ii}][\mathbf{v}_i] + [\mathbf{k}_{ib}][\mathbf{v}_b] = [\mathbf{p}_i], \tag{8.9}$$

or

$$[\mathbf{v}_i] = [\mathbf{k}_{ii}]^{-1}\{[\mathbf{p}_i] - [\mathbf{k}_{ib}][\mathbf{v}_b]\}. \tag{8.10}$$

Substituting in Eq. 8.8 leads to,

$$[\mathbf{k}_{bi}][\mathbf{k}_{ii}]^{-1}\{[\mathbf{p}_i] - [\mathbf{k}_{ib}][\mathbf{v}_b]\} + [\mathbf{k}_{bb}][\mathbf{v}_b] = [\mathbf{p}_b], \tag{8.11}$$

or

$$[\mathbf{k}^*][\mathbf{v}_b] = [\mathbf{p}_b] - [\mathbf{k}_{ib}]^{-1}[\mathbf{k}_{ii}][\mathbf{p}_i], \tag{8.12}$$

where

$$[\mathbf{k}^*] = [\mathbf{k}_{bb}] - \left\{[\mathbf{k}_{bi}][\mathbf{k}_{ii}]^{-1}[\mathbf{k}_{ib}]\right\}, \tag{8.13}$$

is the condensed super element stiffness matrix and

$$[\mathbf{p}^*] = [\mathbf{p}_b] - [\mathbf{k}_{bi}][\mathbf{k}_{ii}]^{-1}[\mathbf{p}_i], \tag{8.14}$$

is the modified load vector. A summation of the interface conditions for the subdomains leads to the formation of the global interface stiffness matrix $\mathbf{K}^*$ and the global interface load vector $\mathbf{p}^*$ as follows:

$$\mathbf{K}^* = \sum_{j=1}^{q} \mathbf{k}_j^* \text{ and } \mathbf{p}^* = \sum_{j=1}^{q} \mathbf{p}_j^*. \tag{8.15}$$

$\mathbf{K}$ is symmetric and positive definite, and $\mathbf{K}^*$ has the same properties. The following interface system can now be solved:

$$[\mathbf{K}^*][\mathbf{v}_b] = [\mathbf{p}^*]. \tag{8.16}$$

Once $\mathbf{v}_b$ is found, the internal degrees of freedom for a subdomain can be evaluated employing Eq. 8.10.

A natural route to parallelism now is to provide it through domain decomposition by distributing the substructures onto the processors available. Several approaches can be used to solve Eq. 8.4. In the following, three broad classifications are briefly discussed:

8.3.3 Methods of Substructuring

8.3.3.1 Direct Methods

A substructuring method can be used to obtain the condensed stiffness matrix on each subdomain in parallel on the different processors. In order to create matrix $\mathbf{K}^*$, it is necessary to condense the stiffness matrix of each substructure (subdomain), i.e. from Eq. 8.13 the product $[\mathbf{k}_{bi}][\mathbf{k}_{ii}]^{-1}[\mathbf{k}_{ib}]$ should be calculated. The explicit formation of $[\mathbf{k}_{ii}]^{-1}[\mathbf{k}_{ib}]$ requires NB_{DOF} triangular system resolutions, where Mb_{DOF} is the number of subdomain boundary degrees of freedom (DOF). This step can be considered as follows:

Each internal DOF makes its contribution to the stiffness of each boundary DOF, such that the behaviour of the condensed boundary is equivalent to the behaviour of the entire domain. This step can be executed step by step, so that only the internal DOF connected to the boundary DOF updates the boundary stiffness matrix. This requires the internal DOF to appear at the bottom of the internal stiffness matrix $\mathbf{k}_{ii}$, so that they are modified by the elimination of all other internal DOF.

A frontal method can be used, which has the advantage of allowing very flexible strategies concerning the sequence of elimination of equations. When this method is applied to subdomain condensation, it is necessary to assemble the boundary DOF in the frontal matrix, and to retain them until all the internal DOF have been eliminated. At the end of the frontal elimination process, the frontal matrix is exactly the condensed matrix $[\mathbf{k}_{bi}][\mathbf{k}_{ii}]^{-1}[\mathbf{k}_{ib}]$.

The interface system of equations is then solved employing a direct approach (e.g. skyline method) on a single machine. Although the direct methods are simple and terminate in a fixed number of steps, the interface solution dominates the overall computational cost when the interface system is large, thus limiting the overall efficiency. In such a case, however, a distributed algorithm can be used for factorisation of the direct method to overcome this difficulty.

8.3.3.2 Iterative Methods

A different method to avoid the explicit inverse of $\mathbf{k}_{ii}$ in Eq. 8.13 is the use of an iterative approach. Among the iterative solutions, the conjugate gradient method is a promising candidate, because of its inherent parallelism and its rate of convergence. The theory of the conjugate gradient method is well known [8]. One iteration of this method for solving a system of equations $\mathbf{Kv} = \mathbf{p}$ is given as:

$$\{\mathbf{u}\} = [\mathbf{K}]\{\mathbf{f}\}, \tag{8.17a}$$

$$\alpha = \{\mathbf{r}\}^t\{\mathbf{r}\}/\{\mathbf{f}\}^t\{\mathbf{u}\}, \tag{8.17b}$$

$$\{\mathbf{v}_{\text{new}}\} = \{\mathbf{v}\} + \alpha\{\mathbf{f}\}, \tag{8.17c}$$

$$\{\mathbf{r}_{\text{new}}\} = \{\mathbf{r}\} + \alpha\{\mathbf{u}\}, \tag{8.17d}$$

$$\lambda = \{\mathbf{r}_{\text{new}}\}^{t}\{\mathbf{r}_{\text{new}}\}/\{\mathbf{r}\}^{t}\{\mathbf{r}\}, \tag{8.17e}$$

$$\{\mathbf{f}_{\text{new}}\} = \{\mathbf{r}_{\text{new}}\} + \lambda\{\mathbf{f}\}. \tag{8.17f}$$

Before each iteration, the vectors $\{\mathbf{v}\}$, $\{\mathbf{f}\}$ and $\{\mathbf{r}\}$ are set to $\{\mathbf{v}_{\text{new}}\}$, $\{\mathbf{f}_{\text{new}}\}$ and $\{\mathbf{r}_{\text{new}}\}$, respectively.

The vectors are initialised as,

$$\{\mathbf{r}\} = \{\mathbf{p}\} - [\mathbf{K}]\{\mathbf{v}_0\}, \tag{8.18a}$$

And

$$\{\mathbf{f}\} = \{\mathbf{r}\}, \tag{8.18b}$$

where $\{\mathbf{v}_0\}$ is usually taken as null, unless some approximation to the solution is known. Iteration is terminated when the residual is small. One criterion for handling the iteration is,

$$\|\mathbf{r}\|/\|\mathbf{p}\| < \varepsilon, \tag{8.19}$$

where ε is the tolerance specified for the problem.

In structural analysis, the vector $\mathbf{r}$ is the potential gradient and is identical to the residual force vector, $(\mathbf{p} - \mathbf{Kv})$ in the linear case. The vector $\mathbf{f}$ is the gradient direction to generate the displacement vector $\mathbf{v}$. For discussion and further details, the reader may refer to Law [9].

Preconditioned Conjugate Gradient (PCG) methods form a large class of the many iterative methods that have been suggested to reduce the cost of forming condensed stiffness matrices. A saving in total time may be achieved, since the predominant matrix-vector product at each iteration is computed in parallel. For further detailed discussion, the interested reader may refer to Keyes and Gropp [10].

8.3.3.3 Hybrid Methods

These methods use a combination of the direct and iterative methods. For instance, the components of the condensed matrix $\mathbf{k}^*$ may be obtained for the substructures using the direct method, and the resulting interface can be solved using an iterative approach.

A comparative study of direct, iterative and hybrid methods is made by Chadha and Baugh [11].

In the following sections, algorithms are presented for partitioning of the nodes of structural graph models, which can be incorporated in any program available for the analysis of skeletal structures. Domain decomposition algorithms are presented in Chap. 8.

8.3.4 Main Algorithm for Substructuring

Let S be the graph model of a structure. The following algorithm decomposes S into q subgraphs with equal or near equal number of nodes (support nodes are not counted) having the least number of interface nodes:

Step 1: Delete all the support nodes with their incident members, and denote the remaining subgraphs by S_r.
Step 2: Determine the distance between each pair of nodes of S_r, and evaluate the eccentricities of its nodes.
Step 3: Sort the remaining nodes (RN) in ascending order of their eccentricities.
Step 4: Select the first node of RN as the representative node of the subgraph S_1 to be determined and find a second node as the representative node of subgraph S_2 with a maximum distance from S_1.
Step 5: Find the third representative node with the maximum least distance from S_1 and S_2, and denote it with S_3.
Step 6: Subsequently, select a representative node of subgraph S_k for which the least distance from $S_1, S_2, \ldots, S_{k-1}$ is maximum. Repeat this process until q representative nodes of the subgraphs to be selected are found.
Step 7: For each subgraph S_j ($j = 1, \ldots, q$), add an unselected node n_i of RN, if it is adjacent only to S_j and its least distance from all nodes of other subgraphs is maximum.
Step 8: Continue the process of Step 7, without the restriction of transforming one node to each subgraph S_j, until no further node can be transferred. The remaining nodes in RN are interface nodes.
Step 9: Transfer the support nodes to the nearest subgraph.

Once the nodes for each subgraph S_j are found, the incidence members can easily be specified.

The algorithm is recursively applied to the selected substructures, decomposing each substructure into smaller ones, resulting in a further refinement.

8.3.5 Examples

Example 1. A double-layer grid supported at four corner nodes is considered and partitioned into $q = 2, 4$ substructures, Fig. 8.1. The corresponding node adjacency matrices (pattern of their stiffness matrices) are illustrated in Fig. 8.2a, b. For the case $q = 2$, the selected substructures are further refined with $q' = 2$ and 3, and the corresponding matrices are shown in Fig. 8.3a, b.

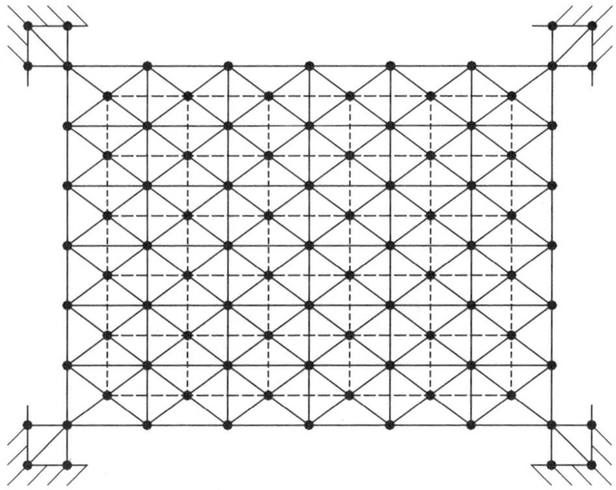

Fig. 8.1 A double-layer grid S

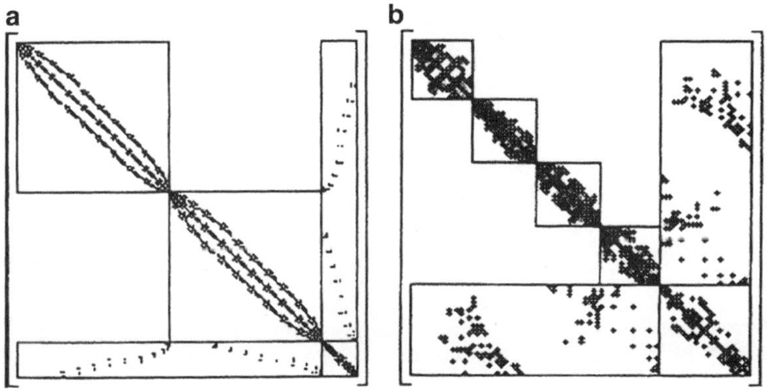

Fig. 8.2 Patterns of the adjacency matrices for different values of q. (**a**) q = 2. (**b**) q = 4

Example 2. A dome-type space structure supported at six nodes is considered and partitioned into q = 2, 3, 4 and 5 substructures, Fig. 8.4.

The corresponding node adjacency matrices are illustrated in Fig. 8.5a–d. For the case q = 2, the selected substructures are further refined with q′ = 2 and 3, and the corresponding matrices are shown in Fig. 8.6a, b.

Once the subgraphs and the interface nodes are specified, ordering the nodes of each subgraph reduces the bandwidth of each block, and appropriate numbering of the interface nodes, results in banded bordered for the entire matrix.

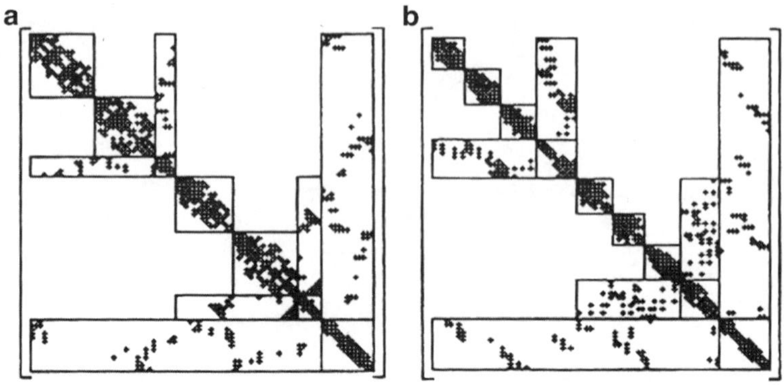

Fig. 8.3 Patterns of the adjacency matrices for q = 2 and q′ = 2 and 3. (**a**) q = 2 and q′ = 2. (**b**) q = 2 and q′ = 3

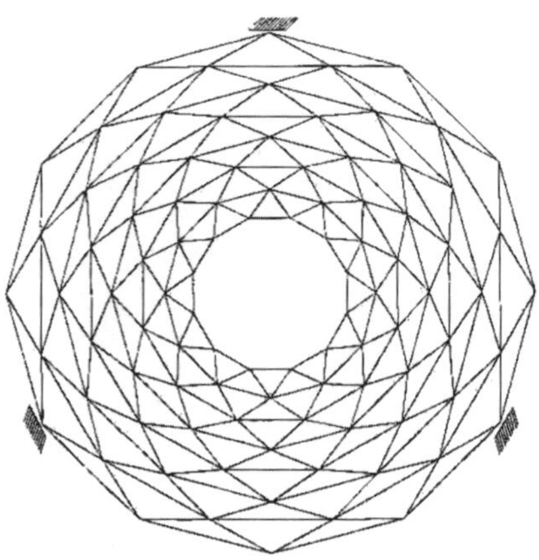

Fig. 8.4 A dome-type space structure

8.3.6 Simplified Algorithm for Substructuring

In the following, a simplified algorithm is presented which requires less storage and computer time than the main algorithm, at the expense of selecting subgraphs with a slightly higher number of interface nodes for some structural models. In this approach, the number of distances to be considered and compared for finding the nodes of substructures is far less than when the main algorithm is used, where the distances between each pair of nodes of S are required. This simplified algorithm consists of the following steps:

Step 1: Form an SRT rooted from an arbitrary node, in order to find a representative node of S_1 with maximum distance from the root. The selected node is also denoted by S_1.

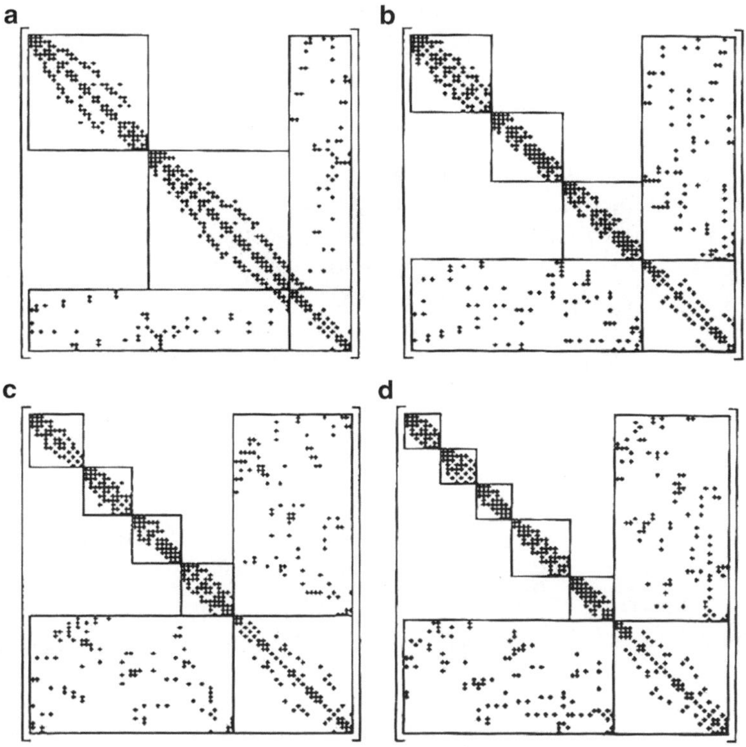

Fig. 8.5 Patterns of the adjacency matrices for different values of q. (a) q = 2. (b) q = 3. (c) q = 4. (d) q = 5

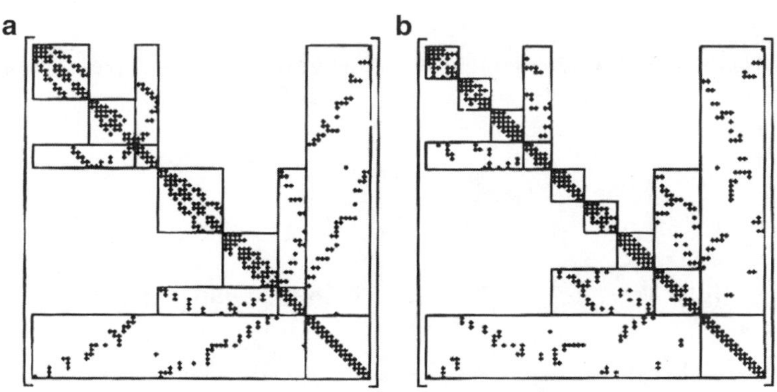

Fig. 8.6 Patterns of the adjacency matrices for q = 2 and different values of q′. (a) q = 2 and q′ = 2. (b) q = 2 and q′ = 3

Step 2: Form an SRT rooted from S_1, to calculate the distance between each node of S and S_1, and find the representative node S_2 in a maximum distance from S_1.
Step 3: Form an SRT rooted from S_2, to calculate the distance between each node of S and S_2 and find the representative node S_3 in a maximum least distance from the selected nodes. Repeat this process until q representative nodes S_1, S_2, . . . , S_q, forming a transversal , are selected.
Step 4: For each subgraph S_i, find a node adjacent to the previously formed S_i only, with maximum least distance from other representative nodes, in turn.
Step 5: Continue the process of Step 4, without the restriction of transforming one node to each subgraph S_i, until no further node can be transferred.

8.3.7 Greedy Type Algorithm

In this algorithm, the *weight* of a node is taken as the number of elements incident with that node. The *interior boundary* of a subdomain D_i is defined as the subset of its boundary that will interface with another subdomain D_j. The total number of elements in a given mesh is denoted by M(FEM).

Step 1: Start with a node and add incident elements one by one having the least current weight. The *current weight* is taken as the number of unselected elements at that stage incident with that node. Continue this process until M(FEM)/q elements are selected as D_1.
Step 2: Select an interior node of D_1, and repeat Step 1 to form D_2.
Step k: Repeat Step 2 for k = 3, 4, . . . , q with an interior node of D_{k-1} and form subdomain D_k.

This process is a Greedy type algorithm, which selects one element of minimal current weight at a time and completes a domain when N(FEM)/q (+1 if remainder $\neq$ 0) elements are selected for the formation of that subdomain. The current weight of an element is updated when an incident element is joined to the expanding subdomain.

8.4 Domain Decomposition for Finite Element Analysis

In this section, efficient algorithms are developed for automatic partitioning of unstructured meshes for the parallel solution of problems in the finite element method. These algorithms partitions a domain into subdomains with approximately equal loads and good aspect ratios, while the interface nodes are confined to the smallest possible. Examples are included to illustrate the performance and efficiency of the presented algorithms.

8.4.1 Introduction

Domain decomposition is attractive in finite element computations on parallel architectures, because it allows individual subdomain operations to be performed concurrently on separate processors and serial solutions on a sequential computer to overcome limitation of core storage capacity. Given a number of available processors q, an arbitrary finite element model (FEM) is decomposed into q subdomains, where formation of element matrices, assembly of global matrices, partial factorisation of the stiffness matrix and state determination or evaluation of generalised stresses can be carried out independently of similar computations for the other subdomains, and hence can be performed in parallel.

In parallel processing of subdomains, the time to complete a task will be the time to compute the longest subtask. An algorithm for domain decomposition will be efficient if it yields subdomains that require an equal amount of execution time. In other words, the algorithm has to achieve a load balance among the processors. In general, this will be particularly ensured if each subdomain contains an equal number of elements or an equal total number of degrees of freedom. However, for some numerical techniques based on domain decomposition, a balanced number of elements or total degrees of freedom among the subdomains does not imply balancing of the subdomain calculations themselves. The use of a frontal subdomain solver provides a relevant example. In this case, the computing load within a domain is not only a function of the number of elements within the subdomain, but also the element numbering. Thus, the optimal number of elements is a priori unknown and can vary significantly from one subdomain to another.

In order to reduce the cost of synchronisation and message passing between the processors in a parallel architecture, the amount of interface nodes should be minimised, because the parallel solution for the generalised displacements usually requires explicit synchronisation on a shared-memory multiprocessor and message passing on local-memory ones. In a domain decomposition method, another significant mesh partitioning factor which should be considered is the subdomain aspect ratio. This ratio has a vital impact on the convergence rate of the iterative approaches for the finite element tearing and interconnecting method.

The above features suggest that an automatic finite element domain decomposer should meet four basic requirements in order to be efficient:

1. It should be able to handle irregular geometry and arbitrary discretisation in order to be general purpose.
2. It must yield a set of balanced subdomains in order to ensure that the overall computational load be as evenly distributed as possible among the processor.
3. It should minimise the amount of interface nodes in order to reduce the cost of synchronisation and/or message passing between the processors.
4. It must result in subdomains with proper aspect ratios, in order to improve the convergence rate of the domain decomposition based iterative method.

Methods of subdomaining are well documented in the literature, see for example Farhat and Wilson [12], Farhat [1], Dorr [13] Malone [14], Farhat and Roux [15], Farhat and Lesoinne [16], Topping and Khan [17], Topping and Sziveri [18], Vanderstraeten and Keunings [19], and Kaveh and Roosta [4]. Several automatic domain decomposition methods that address the load balance and minimum interprocessor computation problems have already been reported in the literature. In general, these algorithms can be grouped into two categories: engineering based and graph theory based methods. For engineering based approaches, one can refer to those of Ref. [20], and for graph theory based methods the algorithms of Ref. [21] can be referred to.

In this section, two efficient algorithms are presented to decompose one- to three-dimensional finite element models of arbitrary shapes. The first method is a *graph based* method and uses a general expansion process. The second is an *engineering based* approach. In these algorithms the resulted subdomains generally have good aspect ratios, especially when the elements have this property originally.

8.4.2 A Graph Based Method for Subdomaining

In this algorithm, first the associate or incidence graph model G of the FEM is generated. Then a good starting node R_1 of G is selected. R_1 is taken as the first node of the first subgraph G_1. Next G_1 is expanded from R_1. The process of expansion is continued such that the equality of the total degrees of freedoms of subdomains is provided. G_2 is formed similar to G_1, but it is expanded from R_2, which is an unselected node in a maximum distance from R_1. R_2 should contain no node of G_1. The process of expansion is executed in a manner that provides the connectedness of the subgraph being formed (if it is possible). A similar approach is employed and $G_3, \ldots, G_q$ are generated, and the subdomains of the FEM corresponding to the selected subgraphs of G are identified. The steps of the algorithm are as follows:

Step 1: Use the associate or incidence graph G of the considered FEM and form an SRT rooted from an arbitrary node of G, in order to find a node R_1 with maximum distance from the root.

Step 2: Generate subgraph G_i ($i = 1$ to q) as follows:

(a) Form an SRT rooted from R_i in order to calculate the distance between each node of G and R_i (R_i is taken as the first selected node of G_i), and find an unselected node R_{i+1} with maximum distance from R_i.

(b) Find all the unselected boundary nodes of G_i, and denote them by UBN.

(c) Associate an integer with each node n_i of UBN which is the same as its distance from R_i plus the number of unselected nodes adjacent to n_i minus the number of selected nodes adjacent to n_i. Then detect the node with minimum integer and add it to G_i.

(d) If the total degrees of freedom of the corresponding subdomain is less than $[TDOF + W_0(q - 1)]/q$, then repeat the above steps from Step (b);

otherwise, execute Step 2 to generate subgraph G_{i+1}. TDOF is the total degrees of freedom of the FEM and W_0 is the total degrees of freedom for the nodes of the corresponding subdomain which are also contained in unselected elements.

In the above algorithm, only the connectivity of the nodes of G is considered, and no labels for edges of G, list or matrices of edges are needed. Therefore, the formation of SRTs of G and data keeping will be more simple and efficient. Since valencies of the nodes of an associate or incidence graph of an FEM are not generally very different, the adjacency list is an efficient means of keeping the connectivity data of G. The adjacency list of a graph G is a matrix containing N(G) rows and Δ columns, where Δ is the maximum degree of the nodes of G. The *i*th row contains the labels of the nodes adjacent to the node i.

Step 1 is carried out to select a good starting node in the generated associate or incidence graph G. Using the adjacency list of G, Step (a) can be performed as follows; however, any other type of list may also be used:

1. Select all the nodes of the R_ith row of the adjacency list of G. The distance between these nodes and the root is equal to unity.
2. Select all the unselected nodes of the rows j (j is an element of the set of the selected nodes in the previous step). The distance of these nodes from the root is one more.
3. Repeat Step 2 until all nodes are selected.

The last instruction of Step (a) is carried out to select the first node of the next subgraph. This node should not be included in the previously generated subgraphs (i.e. it should be an unselected node). In Step (b), UBN contains unselected nodes which are adjacent to selected nodes of G_i. In order to extend G_i, a node of UBN will be added to G_i in every execution of Step (c). In this step an integer will be associated with each node of UBN which defines the best possible node, having the following properties:

1. It is near to the root.
2. It does not make the next UBN very large.
3. It is connected to G_i with more nodes, which leads to a desirable configuration for G_i.

This integer is equal to the distance from R_i plus the valency of the node minus the number of selected adjacent nodes multiplied by 2. The value of $[TDOF + W_0(q-1)]/q$ is not needed to be calculated in every execution of Step (d). Since every subdomain should have at least TDOF/q degrees of freedom, W_0 can be calculated when the degrees of freedom of a subdomain becomes more than TDOF/q. Additional value, $W_0(q-1)/q$, is considered, since the degrees of freedom of the interface nodes of subdomains are calculated in two or more subdomains and the degrees of freedom of the subdomains should be equal or nearly equal.

In this algorithm, a disconnected subdomain may be generated. This happens when no node can be found in Step (b). In such a case, an unselected node with minimum distance should be added to the considered subgraph. In order to avoid such situations, one should avoid decomposing a small FEM into many subdomains. However, the following modifications can always be used:

1. Formation of a single SRT from an arbitrary node to find a good starting node, may not lead to the best node; however, the existing good starting node algorithms can be used to select a better node.
2. If a subgraph G_i contains two components G_i' and G_i'', one can exchange nodes of G_i' or G_i'' with the adjacent subgraphs to provide connectedness for G_i.
3. Use a non-deterministic heuristic of combinatorial optimisation such as Simulated Annealing to improve the initial partitioning to avoid the formation of multiconnected subdomains.

8.4.3 Renumbering of Decomposed Finite Element Models

Once the subdomains and interface nodes are specified, the nodes and/or elements of each subdomain and the interface nodes can be renumbered for bandwidth, profile or frontwidth reduction, depending on whether a band, profile or frontal solver is exploited, respectively. The process of renumbering includes the following steps:

(I) Renumber the internal nodes/elements of the subdomains $M_1, \ldots, M_q$ using an available algorithm.

(II) Select an interface node connected to M_1 which is contained in a minimum number of elements as the starting node, and number the interface nodes using a nodal ordering algorithm. In the process of renumbering, when possible, priority is given to the nodes connected to lower numbered subdomains.

It should be noted that, for a specified solver such as a frontal solver, the resulted subdomains and interface nodes should also satisfy additional conditions. For example in a frontal solver, a necessary condition for the applied domain decomposition approach to be feasible is that the number of degrees of freedom lying on the interface of any subdomain be smaller than the frontwidth associated with the direct (one domain) approach. However, such conditions cannot always be satisfied using the existing decomposition heuristics, because they generally depend on the shape and the connectivity of FEMs, see Lesoinne et al. [22].

8.4.4 Computational Results of the Graph Based Method

Example 1. A finite element model is considered with $\lambda = 606$, $\alpha = 1961$; each element has 4 corner nodes and 4 mid-side nodes, and each node has 2 degrees of

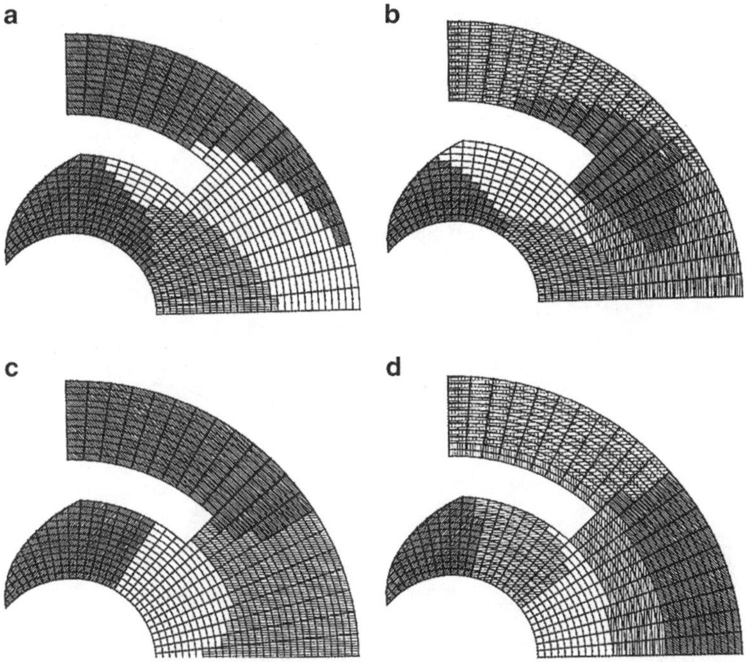

Fig. 8.7 A finite element model and its decompositions. (**a**) q = 4 using the associate graph. (**b**) q = 6 using the associate graph. (**c**) q = 4 using the incidence graph. (**d**) q = 6 using the incidence graph

freedom and is decomposed into 2, . . . ,6 subdomains, as shown in Fig. 8.7a–d for q = 4 and 6, where λ and α denote the numbers of elements and nodes, respectively. The degrees of freedom of the selected subdomains and interface nodes for q = 2, . . . ,6 are illustrated in Table 8.1, when associate and incidence graphs are used.

Example 2. An L-shaped finite element model is considered with $\lambda = 2{,}400$, $\alpha = 1{,}281$, and each node has degrees of freedom equal to 2. The model is decomposed into 6 and 12 subdomains, as shown in Fig. 8.8a–d. The degrees of freedom of the subdomains and interface nodes using associate and incidence graphs are illustrated in Table 8.2.

Example 3. A finite element model is considered with $\lambda = 528$, $\alpha = 307$, and each node has 2 degrees of freedom. The model is decomposed into 2, 3 and 4 subdomains, and the decomposed models for q = 4 are shown in Fig. 8.9a–b. The degrees of freedom of the subdomains and interface nodes using associate and incidence graphs are illustrated in Table 8.3. The patterns of the node adjacency matrices employing the associate graph for the model, after ordering, are shown in Fig. 8.10a–c.

Table 8.1 Results of Example 1

q	Type of graph	DOFs of subdomains; interface nodes
2	Associate	2002, 1994; 74
	Incidence	2016, 2008; 102
3	Associate	1370, 1360, 1352; 160
	Incidence	1352, 1370, 1352; 150
4	Associate	1048, 1052, 1060, 1044; 280
	Incidence	1022, 1030, 1030, 1022; 182
5	Associate	856, 860, 868, 852, 842; 352
	Incidence	828, 844, 848, 826, 816; 240
6	Associate	724, 730, 744, 748, 672, 706; 394
	Incidence	700, 728, 714, 692, 692, 694; 296

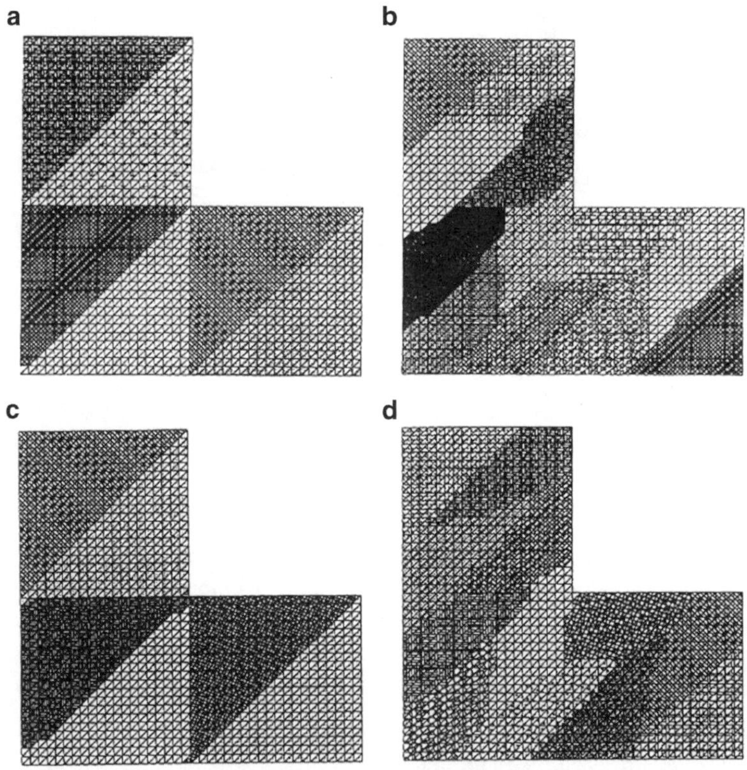

Fig. 8.8 An L-shaped finite element model and its decompositions. (**a**) q = 6 using the associate graph. (**b**) q = 12 using the associate graph. (**c**) q = 6 using the incidence graph. (**d**) q = 12 using the incidence graph

Table 8.2 Results of Example 2

q	Type of graph	DOFs of subdomains; interface nodes
8	Associate	462,462,462,462,462,462; 210
	Incidence	462,462,462,462,462,462; 216
12	Associate	244,244,248,248,246,246,242,252,246,232,236,232; 342
	Incidence	252,252,250,250,268,268,246,268,260,232,234,242; 440

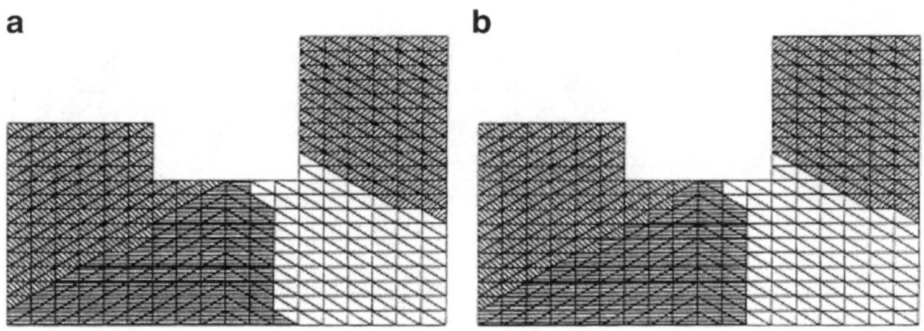

Fig. 8.9 A finite element model and its decompositions. (**a**) q = 4 using the associate graph. (**b**) q = 4 using the incidence graph

Table 8.3 Results of Example 3

q	Type of graph	DOFs of subdomains; interface nodes
2	Associate	324,324;34
	Incidence	322,322;30
3	Associate	224,216,218;42
	Incidence	222,222,220;50
4	Associate	170,164,170,168;58
	Incidence	176,164,170,170;66

8.4.5 Discussions on the Graph Based Method

This algorithm has low time complexity and is simple to program and leads to efficient partitioning of a finite element model into subdomains with the required properties; therefore it can also be considered as a good educational approach. The finite element model that should be partitioned can contain meshes with different dimensions, types and sizes. Although the problem of aspect ratios of the subdomains is not dealt with explicitly in this section, the algorithm has the feature of expansion in all directions, leading to good aspect ratios.

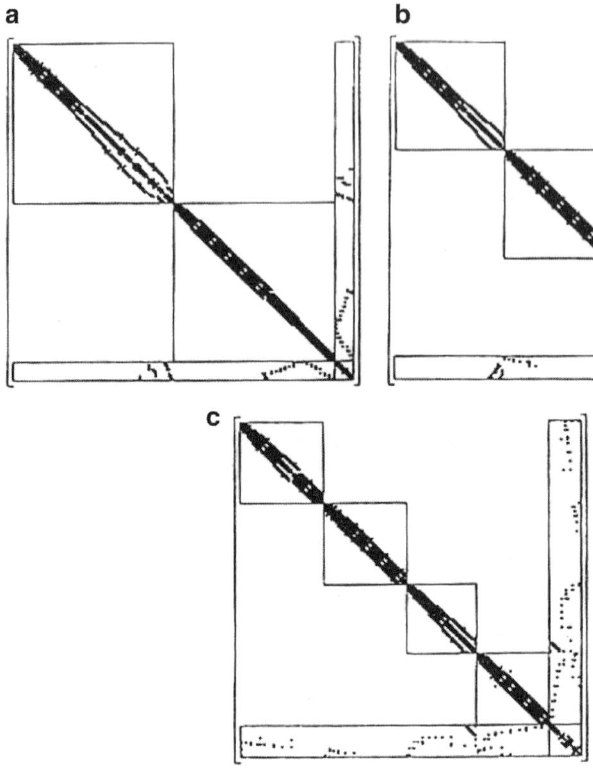

Fig. 8.10 The patterns of the ordered node adjacency matrices

8.4.6 Engineering Based Method for Subdomaining

Definitions. A *level structure* L(r) of a finite element model rooted from an element r (as the root), is defined as a partitioning of the set of elements into levels $l_1(r)$, $l_2(r)$, ..., $l_d(r)$ such that:

1. $l_1(r) = \{r\}$.
2. All elements adjacent to elements in level $l_i(r)$ $(1 < i < d)$ are in levels $l_{i-1}(r)$, $l_i(r)$ and $l_{i+1}(r)$.
3. All elements adjacent to elements in level $l_{di}(r)$ are in levels $l_{d-1}(r)$ and $l_d(r)$.

The overall level structure may be expressed as the set $L(r) = \{l_1(r), l_2(r), \ldots, l_d(r)\}$, where d is the *depth* of the level structure and is simply the total number of levels, and two elements are adjacent if they share a common node.

The *element adjacency list* of a finite element mesh contains the lists of elements adjacent to each element. The element-node list of an FEM contains the lists of nodes of each element and is generally employed as an input for data connectivity of finite element models. Following Webb and Froncioni [23], the *node-element list* contains the lists of elements containing each node of the finite element mesh.

A *genre structure* is a level structure in which each level is divided into one or more genres, and the index of each genre, as defined below, simply shows the pseudo-distance between the root and its elements. The overall genre structure rooted from an element r may be expressed as the set $G(r) = \{g_0(r), g_1(r), g_2(r), \ldots, g_s(r)\}$, in which the pseudo-distance between r and the elements of genre $g_i(r)$ is equal to *i*. The *index vector* $IV_r(i)$ of a genre structure rooted from an element r is an (n + 1)-dimensional vector whose *i*th array (i = 0, . . ., n) defines the total number of elements of g_i (j = 0, . . . , i), i.e.

$$IV_r(i) = \sum_{j=0}^{i} |\, g_i(r)|. \tag{8.20}$$

Thus, the cardinality of genre i $(0 < i \leq n)$ is simply equal to $IV_r(i) - IV_r(i - 1)$, and the cardinality of $g_0(r)$ is equal to 1. The following scheme (in pseudo code) should be used to form a genre structure from an arbitrary starting element r, to generate its index vector and to find the pseudo-distances $pd(r,e_i)$ between the root r and all elements $e_i(i = 1, \ldots, \lambda$, where λ denotes the number of elements) of the considered finite element model. In this scheme, $D \in \{1,2,3\}$ denotes the highest dimension of the elements in the model, and $CCN(g_i(r),e)$ denotes the set of common corner nodes between the elements of genre $g_i(r)$, and the element e.

```
1. Set g0(r) = {r}, IVr(0) = 1, pd(r,r) = 0 and mask r.
2. Set i = 1, a = 0 and b = 0.
3. for j = D to 1 step 1
   for k = a to b
     (I) put each unmasked element e with |CCN(gk(r),
e)| ≥ j into gi(r).
     (II) if |g1 (r) ≠ 0| then set IVr(i) = IVr(i − 1) + |g1
(r)|, pd(r,e) = i(e ∈ gi(r)), i = i + 1 and mask the ele-
ments of gi(r).
     end for
   end for
4. If Ivr(i − 1) < λ then set a = b + 1, b = i and repeat
Step 3.
```

8.4.7 Genre Structure Algorithm

Step 1: Form a genre structure rooted from an arbitrary element, and select an element e_s^1 from its last genre.

Step 2: Calculate the pseudo-distance between e_s^1 and each element, and select an element e_s^2 with maximum pseudo-distance from e_s^1.

Step 3: Calculate the pseudo-distance between e_s^2 and each element. If $q = 2$, then go to Step 5.

Step 4: Find an unselected element e_s^i ($i = 3,4, \ldots ,q$) contained in genres $g_{j1}(e_s^1)$, $g_{j2}(e_s^2)$, $g_{j3}(e_s^3)$, …, $g_{ji-1}(e_s^{i-1})$, such that the least value of $IV_r(j_k - 1)$ be maximum, where $j_k > 0$, $k = 1, \ldots, i - 1$, then calculate the pseudo-distances between e_s^i and the elements.

Step 5: For each selected element e_s^j ($j = 1, \ldots, q$) and each element e_k ($k = 1, \ldots, \lambda$), assign an integer in (e_s^j, e_k) as follows,

$$\mathrm{in}\left(e_s^j, e_k\right) = \lambda + \mathrm{mpd}\left(e_s^j, e_k\right) - \mathrm{pd}\left(e_s^j, e_k\right), \tag{8.21}$$

where

$$\mathrm{mpd}\left(e_s^j, e_k\right) = \min\left\{\mathrm{pd}\left(e_s^i, e_k\right) | 1 \le 1 \le q, 1 \ne j\right\}.$$

Step 6: Let e_s^j be the first element of the subdomain M_i, calculate the weight of M_i and mask e_s^i, where $i = 1, \ldots, q$.

Step 7: Find an expandable subdomain M_i with minimum weight, add an unmasked element e_k with maximum non-zero priority number $P_i = CN \times \mathrm{in}\,(e_s^i, e_k)$ to M_i, update the weight M_i and mask e_k, where if $|CCN(M_j,e_k)| \le 3$ then $CN = |CCN(M_i,e_k)|$, else $CN = 3$. If there is no element to be added to M_i, this subdomain is not expandable and should be masked. If there are several elements with the maximum priority number P_i, then select the one with the minimum sum of integers corresponding to e_s^i ($i = 1, \ldots, q$). Repeat this step until all the elements are masked.

In this algorithm, the weight of a subdomain M_i can be taken as an arbitrary single number such as the number of the elements of M_i, the total degrees of freedom of the nodes of M_j, a function of the number and labels of the elements of M_i, and so on. However, here the total degrees of freedom of the nodes of a subdomain, is considered as the *weight* of the subdomain.

Obviously, in this method only the corner nodes of a finite element mesh should be provided; i.e. mid-side nodes and interior nodes are not needed. This increases the efficiency of the algorithm and results in saving computer storage space for finite element models with high order elements.

An important problem which should be contemplated in a domain decomposition method is the connectedness of the elements of a single subdomain. In this algorithm, multicomponent subdomains are avoided. Since the integers which are calculated in Step 5 are more than zero, hence the priority number P_i of an element e_i corresponding to the subdomain M_i will be zero if $|CCN(M_i,e_i)| = 0$, i.e. the element is not connected to M_i with a corner node. As stated in Step 7, an element with priority number $P_i = 0$ cannot be added to M_i. This provides the connectedness of M_i ($i = 1, \ldots, q$); however, it leads to differences between the weights of the subdomains, because when a subdomain cannot be expanded and is masked, the other unmasked subdomains are still expanded. This problem has been nearly remedied in the present algorithm by Steps 2 and 4. In these steps, the first elements of the subdomains are selected in such a manner that there are enough elements to be added to them for further expansion of the subdomains. For complete balanced loads for subdomains, one can let elements with zero priority numbers be also added to a subdomain, in which case multicomponent subdomains will be generated. However, there are several non-deterministic heuristics used in combinatorial optimisation such as Simulated Annealing, Stochastic Evolution and Tabu Search which can be used for better load balancing of subdomains and reduction in the number of interface nodes, see for example Reference 13. These combinatorial optimisation methods are normally included in an FEM decomposition algorithm as follows:

Step I: Invoke a direct partitioning scheme to produce an initial decomposition of reasonable quality.

Step II: Use an optimisation procedure to improve the initial partitioning.

The second step generally needs high computer time, hence this algorithm is designed for careful partitioning of the finite element meshes in order to avoid (as far as possible) the use of optimisation procedures for general cases. However, this method can be applied as a direct method in Step I. This will be efficient, since the more the load balancing of subdomains and the less the number of interface nodes produced by a direct partitioning scheme, the less cost for the applied optimisation method.

The Step 1 of the algorithm presented in this section is carried out to find a good starting element e_s^1 for the first subdomains M_1. Step 2 is executed in order to calculate the pseudo-distance between e_s^1 and each element, and to find an element e_s^2 as the good starting element of the second subdomain M_2. It should be noted that, when $q > 2$, it is needed to know the index of genres containing a specified element because it is needed for the selection of the starting elements of subdomains M_i ($i = 3, \ldots, q$). Step 3 should be carried out to calculate the pseudo-distance between e_s^2 and each element of the considered finite element mesh. Also in this step, the index of genres containing a specified element should be defined for $q > 2$. Step 4 is executed in order to find good starting elements for subdomains M_i ($i = 3, \ldots, q$). The condition contained in this step is included in order to provide the starting elements of subdomains to be unobtrusive when the process of expansion is performed in Step 7. This condition increases the probability that a subdomain will remain expandable while the other subdomains are being expanded. Step 5 is carried out in order to calculate an integer for each selected (starting) element and each element of the finite element mesh. This integer is always more than zero since λ is always more than or equal to a pseudo-distance between two elements, and a pseudo-distance is always equal to or more than zero. The integers calculated in this step affect the priority number of elements in two ways when the process of expansion is performed: (1) the elements which are added to a subdomain have lower priority numbers for other subdomains, (2) the elements of a subdomain do not have flange positions in relation to the region of the subdomain (loosely speaking). These effects make the number of boundary interior nodes of a subdomain low and its aspect ratio a desired value. The less the differences between the geometrical dimensions of a subdomain with a given area/volume, the smaller the boundary and the better aspect ratio of the subdomain. However, this remark is true when the elements have originally good aspect ratios. For more details about the aspect ratio of a subdomain, see the recent paper of Farhat et al. [24] in which their final choice has been to compute the aspect ratio AR of a subdomain M_i as follows:

$$\mathrm{AR}(M_i) = c_2 \times \frac{\text{Surface } (M_i)}{\text{Surface of circumscribed circle}} \text{(two dimensional problems)}$$

$$\mathrm{AR}(M_i) = c_3 \times \frac{\text{Volume } (M_i)}{\text{Volume of circumscribed sphere}} \text{(three dimensional problems)} \tag{8.22}$$

where c_2 and c_3 are scaling constants designed such that $0 < \mathrm{AR} \leq 1$.

Step 6 is executed in order to initialise the subdomains M_i ($i = 1, \ldots, q$) and their weights and to mask their first (starting) elements. The elements of a subdomain are masked only in order to forbid their repeated selection. Step 7 contains the expansion process of the algorithm. In every execution of this step, an element with maximum priority number corresponding to a subdomain M_i is added to M_i, where M_i is the subdomain with current minimum weight. This way of expansion leads to equal loads for subdomains such that the subdomains remain expandable, and this condition is provided in the process of selecting e_s^i ($i = 1, \ldots, q$) and giving a priority number to an element corresponding to the subdomain being formed. The priority number defined in this step is simply designed to give more priority to an element connected to a subdomain M_k with more corner nodes in comparison with an element connected to M_k with less corner nodes having the same integers.

8.4.8 Example

Consider the simple finite element mesh, as shown in Fig. 8.11a, with each node having 2 degrees of freedom, and suppose it to be decomposed into three subdomains. The steps of the present algorithm are performed as follows:

Step 1: A genre structure is rooted from an arbitrary element such as the element 15. The elements of each genre are recognised with the index of the genre as illustrated in Fig. 8.11b. The last genre, $g_8(15)$, contains the element 6; hence $e_s^1 = 6$.

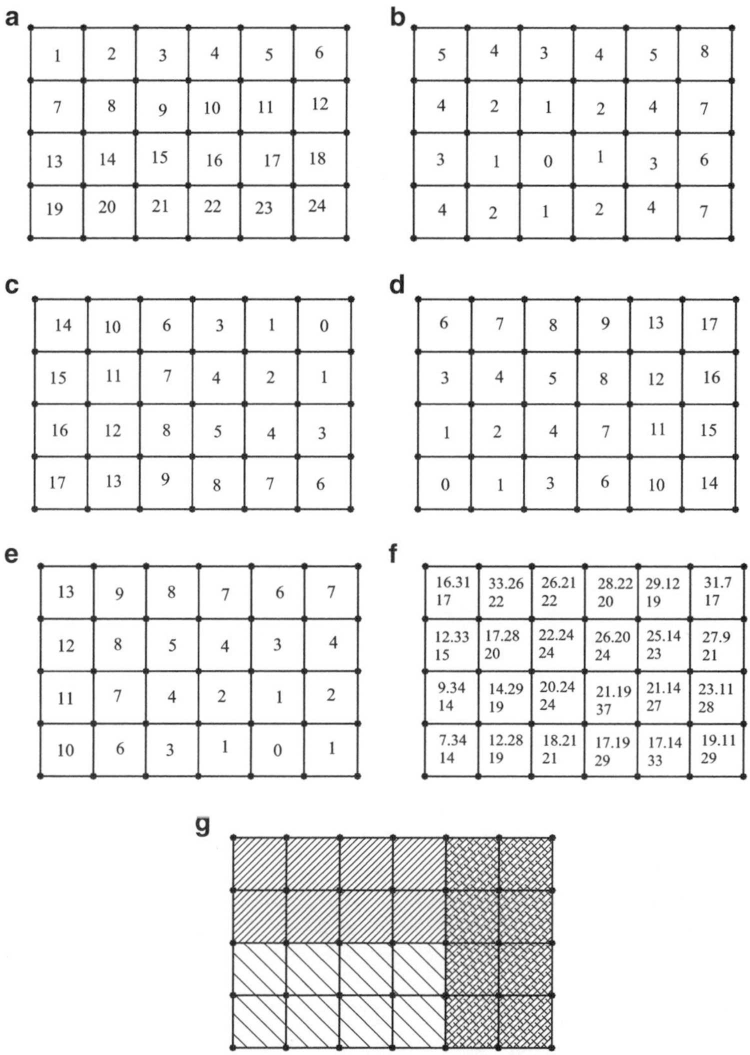

Fig. 8.11 Illustration of the steps for the example. (**a**) A simple two-dimensional FEM. (**b**) Genres of G(15). (**c**) Genres of G(6). (**d**) Genres of G(19). (**e**) Genres of G(23). (**f**) Integers of the elements. (**g**) Decomposition of the FEM for q = 3

Step 2: G(6) is formed to calculate the pseudo-distance between the element 6 and other elements. The elements of each genre are assigned with the index of the genre; this index is same as the pseudo-distances between the root and the elements of the genre. In Fig. 8.11c the pseudo-distance between the root (element 6) and other elements are depicted; the element 19 belongs to the last genre of G(6), having the highest pseudo-distance from the root, and thus $e_s^2 = 19$.

Step 3: G(19) is generated, and the pseudo-distances between the root and the elements are shown in Fig. 8.11d. Since $q > 2$, therefore Step 4 should be executed.

Step 4: Two elements 2 and 23 satisfy the condition of this step, since

$$2 \in g_{10}(6) \text{ and } g_7(19)$$
$$IV_6(9) = 16, IV_{19}(6) = 11$$
$$23 \in g_7(6) \text{ and } g_{10}(19)$$
$$IV_6(6) = 11, IV_{19}(9) = 16,$$

and

$$\min\{IV_6(i), IV_{19}(j)\} < 11,$$

where

$$0 \leq i, j \leq 16 \text{ and } (i, j) \neq (9, 6) \text{ and } (6, 9).$$

Element 2 or 23 can be selected for e_s^3 arbitrarily; suppose $e_s^3 = 23$. Figure 8.11e shows the pseudo-distances between e_s^3 and the other elements.

Step 5. For each element, three integers are assigned corresponding to e_s^3, e_s^2 and e_s^3. These integers are respectively illustrated in Fig. 8.11f for each element.

Step 6: Execution of this step leads to $M_1 = \{6\}$, $M_2 = \{19\}$ and $M_3 = \{23\}$. The weights of M_1, M_2 and M_3 are the same and equal to 8, and their elements are masked.

Step 7. This step is carried out $\lambda - q = 21$ times, and in each execution one element with maximum priority number is added to a subdomain with current minimum weight as follows:

All subdomains have the same weight; hence the subdomain M_1 is selected arbitrarily to be expanded. The elements with non-zero priority numbers which are connected to M_1 are 5, 11 and 12, and their priority numbers are 2×29, 1×25 and 2×27, respectively. Thus element 5 is added to M_1 and is masked. The weight of M_1 is now equal to 12. The subdomains M_2 and M_3 have minimum current weight. The subdomain M_2 is selected arbitrarily to be expanded. The elements 13, 14 and 20 are connected to M_2, and their priority numbers are 2×34, 1×29 and 2×29, respectively. Hence the element 13 is added to M_2 and is masked. The current weight of M_2 is now equal to 12. The subdomain M_3 has the least current weight. The elements 16, 17, 18, 22 and 24 are connected to M_3, and their priority numbers are 1×27, 2×27, 1×25, 2×29 and 2×29, respectively. The priority numbers of the elements 22 and 24 are maximum; however, element 24 is added to M_3 because the sum of its integers is less than that of the element 22. The element 24 is masked. The weight of the subdomain M_3 is now equal to 12. The repetitions of this step lead to the decomposition as illustrated in Fig. 8.11g.

8.4.9 Computational Results of the Engineering Based Method

Two examples are studied in this section, using the direct method for the formation of their element adjacency list.

Example 1. A multiconnected finite element mesh is shown in Fig. 8.12a, and decomposed into 2, 3, 4, 8 and 16 subdomains as illustrated in Fig. 8.12b–f. In this example, each node has 2 degrees of freedom. Computational time is provided in Table 8.4.

Example 2. A multiconnected H-shaped finite element mesh with each node having 2 degrees of freedom is shown in Fig. 8.13a, and decomposed into 2, 4, 5, 8, 16 and 32 subdomains as illustrated in Fig. 8.13b–g. Computational time is provided in Table 8.5.

8.4.10 Discussions

The algorithm developed in this section is designed as a pre-processor for concurrent finite element computations. It may also serve as an automatic decomposer for serial solutions on a sequential computer, to overcome limited core storage capacity. This algorithm has low time complexity and leads to efficient partitioning of a finite element mesh into subdomains with required properties. A finite element mesh to be partitioned, may contain various meshes with different dimensions, types and sizes. The algorithm uses a simultaneous expansion process which is an improved version of the algorithm presented in the previous section for substructuring. In this algorithm the method for selecting the first (representative) element for each subdomain is improved, and the better priority numbers for elements to be added to the expanding subdomains are defined in order to form subdomains with more appropriate properties.

This algorithm is designed to have properties required for an efficient decomposition and leads to subdomains with the following properties:

1. Low computer space and time requirements. In the present algorithm only the corner nodes are needed to be given, and this leads to a large space saving in FEMs with high order elements. The time complexity of the algorithm is independent of the number of nodes for the considered FEM, and the critical step of the algorithm takes $O(\lambda^2\theta)$ operations in worst-case.
2. General in use. The algorithm can be employed to decompose unstructured FEMs without any restriction, and an arbitrary parameter can be considered as the loads of the subdomains.
3. Balance loads for subdomains. Selection of the starting elements of subdomains and the expansion process are performed in a manner which leads to an efficient

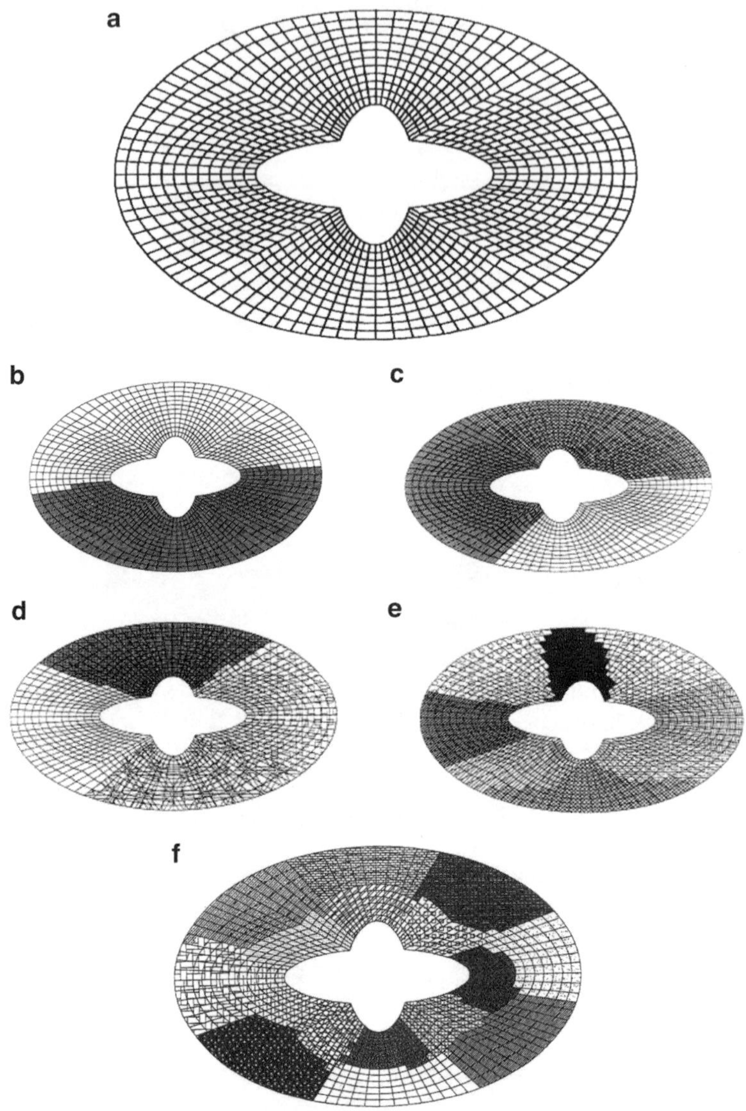

Fig. 8.12 Decompositions of the multiconnected finite element mesh. (**a**) A multiconnected FEM with 1,152 elements and 1,248 nodes. (**b**) q = 2. (**c**) q = 3. (**d**) q = 4. (**e**) q = 8. (**f**) q = 16

Table 8.4 Computational time

q	2	3	4	8	16
Time (sec.)	29.00	29.28	29.44	30.59	33.39

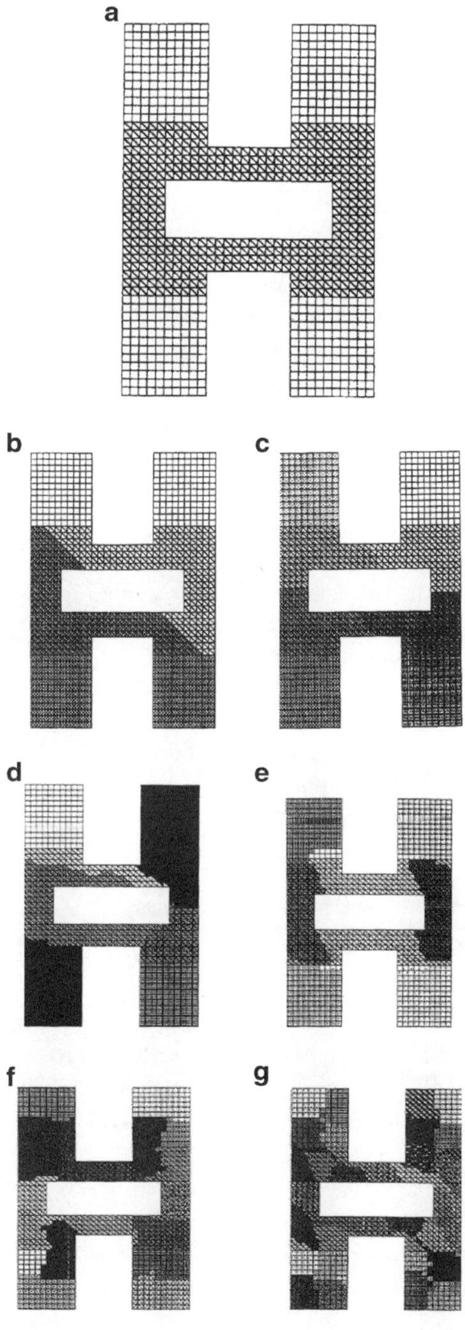

Fig. 8.13 Decompositions of a multiconnected H-shaped finite element mesh. (**a**) A multiconnected H-shaped FEM with 1,340 elements and 1,042 nodes. (**b**) q = 2. (**c**) q = 4. (**d**) q = 5. (**e**) q = 8. (**f**) q = 16. (**g**) q = 32

Table 8.5 Computational time

q	2	4	5	8	16	32
Time (sec.)	33.95	31.75	32.90	37.14	40.70	52.79

balancing of loads. However, in order to decrease the differences between the loads of subdomains, the following steps are included which should be executed in place of Steps 1–4 of the original algorithm:

(a) Find the pseudo-distance between each element and all the element of the finite element mesh.
(b) Find q elements $e_s^1, e_s^2, \ldots, e_s^q$, provided that e_s^i $(i = 1, \ldots, q)$, which is contained in genres $g_{j1}(e_s^1), g_{j1}(e_s^2), \ldots, g_{j1}(e_s^q)$, is selected in such a way that the least value of $IV(j_k - 1)$ is maximum where $j_k \neq 0$ $(k = 1, \ldots, q)$.

However, this takes more operations than those of Steps 1–4.

4. Close to minimum number of interface nodes. In this algorithm, the number of interface nodes is kept to the least possible by selecting the elements to be added to a subdomain which have not high priority numbers for the other subdomains, and have a proper position in relation to the previously selected elements of the subdomains.
5. Good aspect ratios for subdomains. When the elements of the considered finite element mesh have aspect ratios with proper values, the algorithm leads to a decomposition with subdomains having reasonable aspect ratios. This is because the subdomains are expanded in all directions, which makes the denominators of the equations introduced by Farhat et al. [24] to be increased.

8.5 Substructuring: Force Method

The force method can be employed in parallel analysis of structures. In this section, the formulation of substructuring is provided, and an algorithm is presented for such analysis. The computational process is illustrated using simple examples.

In this section, the notations and formulations presented in Chap. 3 will be used.

8.5.1 Algorithm for the Force Method Substructuring

Once a structural model has been decomposed using any of the methods presented in the previous sections, the following approach can be used for the analysis employing the force method:

In order to support a substructure in a statically determinate fashion, cuts are introduced at members incident with the interface nodes contained in the corresponding substructure, except at one arbitrary node where the substructure is connected to the previous one.

For a given substructure S_i, let the external forces be denoted by $\mathbf{p}_i$, and redundant forces by $\mathbf{q}_i$. Then the substructure S_i can be analysed for the internal forces in the substructure (not coupling redundants) $\mathbf{q}_i$ in the aforementioned manner, i.e.

$$\begin{bmatrix} \mathbf{v}_{0i} \\ \mathbf{v}_{1i} \end{bmatrix} = \begin{bmatrix} \mathbf{D}_{00} & \mathbf{D}_{01} \\ \mathbf{D}_{10} & \mathbf{D}_{11} \end{bmatrix}_i \begin{bmatrix} \mathbf{p}_i \\ \mathbf{q}_i \end{bmatrix}. \tag{8.23}$$

For continuity within the substructure:

$$\mathbf{q}_i = -\left(\mathbf{D}_{11}^{-1}\mathbf{D}_{10}\right)_i \mathbf{p}_{i.} \tag{8.24}$$

Deflections corresponding to the nodal force are,

$$\mathbf{v}_{0i} = \left(\mathbf{D}_{00} - \mathbf{D}_{10}^{t}\mathbf{D}_{11}^{-1}\mathbf{D}_{10}\right)_i \mathbf{p}_i, \tag{8.25}$$

that is

$$\mathbf{v}_{0i} = \mathbf{F}_i \mathbf{p}_i, \tag{8.26}$$

where $\mathbf{F}_i$ is the flexibility transformation matrix for the *i*th substructure. Internal forces are obtained as,

$$\mathbf{r}_i = \left(\mathbf{B}_0 - \mathbf{B}_1\mathbf{D}_{11}^{-1}\mathbf{D}_{10}\right)_i \mathbf{p}_i, \tag{8.27}$$

or

$$\mathbf{r}_i = \mathbf{B}_i \mathbf{p}_i \tag{8.28}$$

and

$$\mathbf{B}_i = \left(\mathbf{B}_0 - \mathbf{B}_1\mathbf{D}_{11}^{-1}\mathbf{D}_{10}\right)_i, \tag{8.29}$$

where $\mathbf{B}_i$ is the force transformation matrix in the redundant substructure. The matrices $\mathbf{F}_i$ and $\mathbf{B}_i$ are formed for each substructure, in turn.

For the complete structure S composed of q substructures (S_1,S_2, . . ., S_q), the force vector $\mathbf{p}_i$ acting on a substructure "s" is given by,

$$\mathbf{p}_{si} = [\mathbf{a}_e \quad \mathbf{b}_e] \begin{bmatrix} \mathbf{p}_e \\ \mathbf{q}_c \end{bmatrix}, \tag{8.30}$$

where $\mathbf{q}_c$ are the coupling redundants. On a particular substructure, there will be three different types of forces: $\mathbf{p}_{ee}$ is the external force vector, $\mathbf{p}_{ec}$ is the coupling redundant forces vector, and $\mathbf{p}_{eb}$ contains the statically determinate connection forces.

For the entire structure, the following matrices $\mathbf{A}_{ee}$ and $\mathbf{B}_{ec}$ are defined:

$$\mathbf{A}^t = \{a_{e(1)}, a_{e(2)}, \ldots, a_{e(q)}\}, \tag{8.31}$$

and

$$\mathbf{B}^t = \{b_{e(1)}, b_{e(2)}, \ldots, b_{e(q)}\}. \tag{8.32}$$

Then:

$$\mathbf{p}_s = \begin{bmatrix} \mathbf{p}_{s(1)} \\ \mathbf{p}_{s(2)} \\ \cdots \\ \cdots \\ \mathbf{p}_{s(q)} \end{bmatrix} = [\mathbf{A}_{ee} \quad \mathbf{B}_{ec}] \begin{bmatrix} \mathbf{p}_c \\ \mathbf{q}_c \end{bmatrix}. \tag{8.33}$$

The forces $\mathbf{p}_s$ can be partitioned according to three types of forces $\mathbf{p}_{ei}$, $\mathbf{p}_{eb}$, and $\mathbf{p}_{ec}$ as mentioned before. Then:

$$\begin{bmatrix} \mathbf{p}_{ei} \\ \mathbf{p}_{eb} \\ \mathbf{p}_{ec} \end{bmatrix} = \begin{bmatrix} \mathbf{a}_{ei} & \mathbf{0} \\ \mathbf{a}_{eb} & \mathbf{b}_{eb} \\ \mathbf{a}_{ec} & \mathbf{b}_{ec} \end{bmatrix} \begin{bmatrix} \mathbf{p}_e \\ \mathbf{q}_c \end{bmatrix}. \tag{8.34}$$

It is obvious that, whereas $\mathbf{q}_c$ may produce $\mathbf{p}_{eb}$ and $\mathbf{p}_{ec}$ forces, it does not produce $\mathbf{p}_{ei}$ forces.

The flexibility matrix of the entire structure corresponding to $\mathbf{p}_e$ and $\mathbf{q}_c$ can be formed using Eqs. 8.25 and 8.33 as:

$$\begin{bmatrix} \mathbf{f}_{ee} & \mathbf{f}_{ec} \\ \mathbf{f}_{ce} & \mathbf{f}_{cc} \end{bmatrix} = \begin{bmatrix} \mathbf{A}_{ee}^t \\ \mathbf{A}_{ec}^t \end{bmatrix} \begin{bmatrix} \mathbf{F}_{e(1)} & & \\ & \cdots & \\ & & \mathbf{F}_{e(q)} \end{bmatrix} [\mathbf{A}_{ee} \quad \mathbf{B}_{ec}], \tag{8.35}$$

and

Fig. 8.14 A singlebay four-storey frame with geometric and connectivity properties

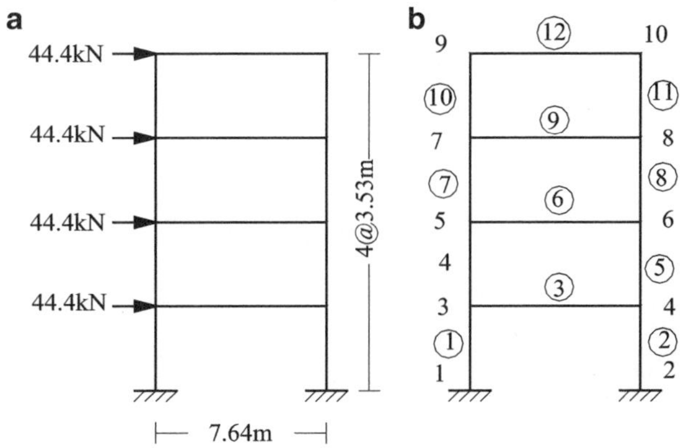

$$\begin{bmatrix} \mathbf{v}_e \\ \mathbf{v}_c \end{bmatrix} = \begin{bmatrix} \mathbf{f}_{ee} & \mathbf{f}_{ec} \\ \mathbf{f}_{ce} & \mathbf{f}_{cc} \end{bmatrix} \begin{bmatrix} \mathbf{p}_e \\ \mathbf{p}_c \end{bmatrix}. \tag{8.36}$$

For continuity across the cut sections of the structure,

$$\mathbf{v}_c = 0, \tag{8.37}$$

hence:

$$\mathbf{q}_c = -\mathbf{f}_{cc}^{-1}\mathbf{f}_{ce}\mathbf{p}_e \tag{8.38}$$

Deflections of the structure are then given as,

$$\mathbf{v}_e = \left(\mathbf{f}_{ee} - \mathbf{f}_{ec}\mathbf{f}_{cc}^{-1}\mathbf{f}_{ce}\right)\mathbf{p}_e, \tag{8.39}$$

making the complete analysis of the structure feasible.

8.5.2 Examples

Example 1. A single-bay four-storey frame is considered, as shown in Fig. 8.14. The forces are depicted in Fig. 8.14a, and the nodal and element orderings are given in Fig. 8.14.b. For this frame, I = 41,623.14 cm^4 (for all members) and E = 2.1 $\times$ 10^5 N/m^2.

The model is decomposed into two substructures as illustrated in Fig. 8.15. The analysis is performed and the bending moments are obtained as provided in Table 8.6.

Example 2. A three-bay pitched-roof frame together with material properties and dimensions are shown in Fig. 8.16.

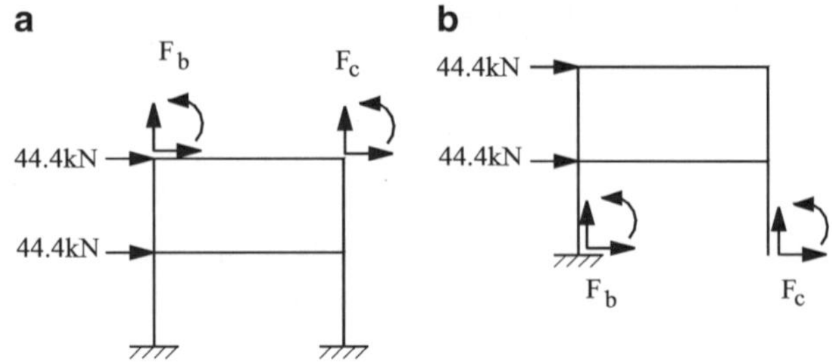

Fig. 8.15 Decomposition of the structural model

Table 8.6 Bending moments of Example 1

Nodes	End nodes of members	Bending moments(kN.m)
1	1–3	−219.17
	3–1	78.93
3	3–4	185.07
	3–5	−106.14
	5–3	0.34
5	5–6	2.94
	5–7	7.34
	7–5	96.2
7	7–8	112.7
	7–9	−16.48
9	9–7	58.04
	9–10	58.04
10	10–9	58.04
	10–8	−58.04
	8–10	−16.48
8	8–7	112.7
	8–6	−96.2
	6–8	−52.84
6	6–5	170.28
	6–4	−117.43
	4–6	−78.93
4	4–3	185.07
	4–2	−106.14
2	2–4	−219.17

This model is partitioned into two substructures, as illustrated in Fig. 8.17, where different groups of loads on each substructure are shown. For all the members, $I = 0.2\ m^4$ and $E = 2.1 \times 10^5\ N/m^2$. The bending moments for members of this frame are presented in Table 8.7.
The substructuring analysis, using the force method for frame structures, can be generalised to the analysis of other types of structures when the algebraic force method is employed, Plemmons and White [25]. In this method, appropriate partitioning of the incidence matrices of the structural graph models is performed,

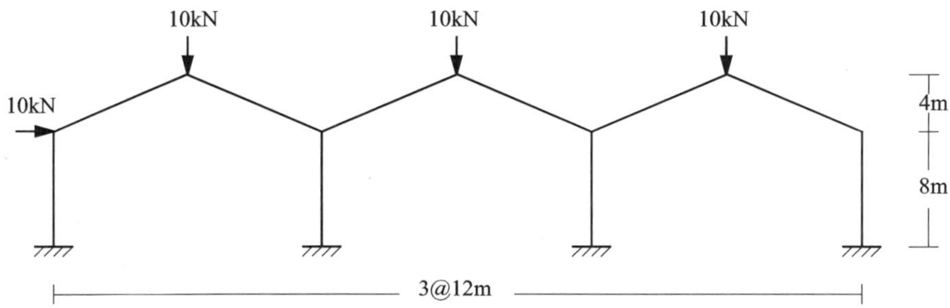

Fig. 8.16 A three bay pitched-roof frame

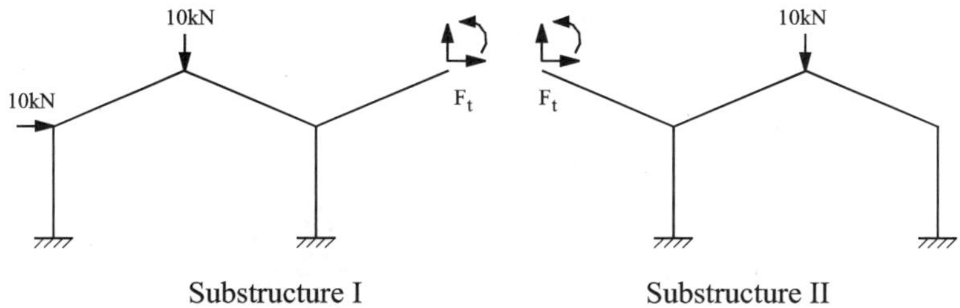

Fig. 8.17 Decomposition of the structural model

Table 8.7 Bending moments of Example 2

Nodes	End nodes of members	Bending moments(kN.m)
1	1–5	−72
5	5–1	−30
	5–6	30
6	6–5	3
	6–7	3
	7–6	26
7	7–8	65
	7–2	−91
2	2–7	−116
8	8–7	45
	8–8	45
	9–8	28
9	9–10	3.03
	9–3	60
3	3–9	40.02
10	10–9	52
	10–11	52
11	11–10	79

leading to well structured equilibrium equations. It is proved that sparse null bases can then be constructed in parallel, using the proposed decomposition. The performance of the method is illustrated by some examples from skeletal structures.

References

1. Farhat C (1988) A simple and efficient automatic FEM domain decomposer. Comput Struct 28:579–602
2. Al-Nasra M, Nguyen DT (1991) An algorithm for domain decomposition in finite element analysis. Comput Struct 39:277–289
3. George A, Liu JWH (1978) Algorithms for partitioning and numerical solution of finite element systems. SIAM J Numer Anal 15:297–327
4. Kaveh A, Roosta GR (1995) Graph-theoretical methods for substructuring, subdomaining and ordering. Int J Space Struct 10(2):121–131
5. George A (1971) Computer implementation of the finite elements. Report STAN-CS-71-208, Ph.D. thesis, Computer Science Department, Stanford University, CA
6. Kron G (1959) Diakoptics, piecewise solution of large-scale systems. A series of 20 chapters in the Electrical Journal, London
7. Pothen A, Simon H, Liou KP (1990) Partitioning sparse matrices with eigenvectors of graphs. SIAM J Matrix Anal Appl 11:430–452
8. Jennings A, McKeown JJ (1992) Matrix computation. Wiley, Chichester
9. Law KH (1986) A parallel finite element solution method. Comput Struct 23:845–858
10. Keyes DE, Gropp WD (1987) A comparison of domain decomposition techniques for elliptic partial differential equations and their parallel implementation. SIAM J Sci Statist Comput 8:166–202
11. Chadha HS, Baugh JW Jr (1996) Network-distributed finite element analysis. Adv Eng Softw 25:267–280
12. Farhat C, Wilson E (1987) A new finite element concurrent computer program architecture. Int J Numer Methods Eng 24:1771–1792
13. Dorr MR (1988) Domain decomposition via Lagrange multipliers. Report No. UCRL-98532, Lawrence Livermore National Laboratory
14. Malone JG (1988) Automated mesh decomposition and concurrent finite element analysis for hypercube multiprocessor computers. Comput Methods Appl Mech Eng 70:27–58
15. Farhat C, Roux FX (1991) A method of finite element tearing and interconnecting and its parallel solution algorithm. Int J Numer Methods Eng 32:1205–1227
16. Farhat C, Lesoinne M (1993) Automatic partitioning of unstructured meshes for the parallel solution of problems in computational mechanics. Int J Numer Methods Eng 36:745–764
17. Topping BHV, Khan AI (1995) Parallel finite element computations. Saxe-Coburg Publications, Edinburgh
18. Topping BHV, Sziveri J (1995) Parallel sub-domain generation method. In: Proceedings of Civil Comp 95, Edinburgh, pp 449–457
19. Vanderstraeten D, Zone O, Keunings R (1993) Non-deterministic heuristic for automatic domain decomposition in direct parallel finite element calculation. In: Proceedings of 6th SIAM conference on parallel processing, SIAM, pp 929–932
20. Felippa CA (1975) Solution of linear equations with skyline-stored symmetric matrix. Comput Struct 5:13–29
21. Kaveh A, Roosta GR (1994) A graph theoretical method for decomposition in finite element analysis. In: Proceedings of Civil-Comp 94, Edinburgh, pp 35–42
22. Lesoinne M, Farhat C, Geradin M (1991) Parallel/vector improvements of the frontal method. Int J Numer Methods Eng 32:1267–1281
23. Webb JP, Froncioni A (1986) A time-memory trade-off frontwidth reduction algorithm for finite element analysis. Int J Numer Methods Eng 23:1905–1914
24. Farhat C, Maman N, Brown GW (1995) Mesh partitioning for implicit computations via iterative domain decomposition: impact and optimization of the subdomain aspect ratio. Int J Numer Methods Eng 38:989–1000
25. Plemmons RJ, White RE (1990) Substructuring methods for computing the null space of equilibrium matrices. SIAM J Matrix Anal 11:1–22

Chapter 9
Analysis of Regular Structures Using Graph Products

9.1 Introduction

In this chapter, an efficient method is presented for the analysis of non-regular structures which are obtained by addition or removal of some members to regular structural models. Here a near-regular structure is divided into two sets, namely "the regular part of the structure" and "the excessive members". Regular part refers to the structure for which the inverse of the stiffness matrix can be obtained by the previously developed simplified methods, and excessive members refer to those which cause the non-regularity of the regular structure [1].

9.2 Definitions of Different Graph Products

Many structures have regular patterns and can be viewed as the Cartesian product, strong Cartesian product, or direct product of a number of simple graphs. These subgraphs, used in the formation of the entire model, are called the *generators* of that model. Graph products were developed in the past 50 years (see e.g. Imrich and Klavzar [2]) for mathematical aspects, and Kaveh [3] for extensive applications.

9.2.1 Boolean Operation on Graphs

In order to explain the products of graphs, let us consider a graph S as a subset of all unordered pairs of its nodes. The node set and member set of S are denoted by N(S) and M(S), respectively. The nodes of S are labelled as $v_1,v_2, \ldots, v_M$, and the resulting graph is a *labelled graph*. Two distinct adjacent nodes, v_i and v_j, form a member, denoted by $v_iv_j \in M(S)$.

A. Kaveh, *Computational Structural Analysis and Finite Element Methods*,
DOI 10.1007/978-3-319-02964-1_9, © Springer International Publishing Switzerland 2014

A Boolean operation on an ordered pair of disjoint labelled graphs K and H results in a labelled graph S, which has N(K) × N(H) as its nodes. The set M(S) of members of S is expressed in terms of the members in M(K) and M(H), differently for each Boolean operations. Three different operations are discussed in this chapter, corresponding to Cartesian product, strong Cartesian product and direct product of two graphs.

9.2.2 Cartesian Product of Two Graphs

Many structures have regular patterns and can be viewed as the Cartesian product of a number of simple graphs. These subgraphs, which are used in the formation of a model, are called the *generators* of that model.

The simplest Boolean operation on a graph is the Cartesian product K × H introduced by Sabidussi [4]. The Cartesian product is a Boolean operation S = K × H, in which, for any two nodes $u = (u_1,u_2)$ and $v = (v_1,v_2)$ in N(K) × N(H), the member uv is in M(S) whenever

$$u_1 = v_1 \text{ and } u_2v_2 \in M(H), \tag{9.1a}$$

or

$$u_2 = v_2 \text{ and } u_1v_1 \in M(K). \tag{9.1b}$$

As an example, the Cartesian product of $K = P_2$ and $H = P_3$ is shown in Fig. 9.1.

In this product, the two nodes (u_1,v_2) and (v_1,v_2) are joined by a member, since the condition (9.1b) is satisfied.

The Cartesian product of two graphs K and H can be constructed by taking one copy of H for each node of K and joining copies of H corresponding to adjacent nodes of K by matching of size N(H).

The graphs K and H will be referred to as the *generators* of S. The Cartesian product operation is symmetric, i.e. $K \times H \cong H \times K$. For other useful graph operations, the reader may refer to the work by Gross and Yellen [5].

Examples. In the first example, the Cartesian product $C_7 \times P_5$ of the path graph with five nodes denoted by P_5 and a cycle graph shown by C_7 is illustrated in Fig. 9.2.

Two representations of the Cartesian product $C_3 \times P_4$ are illustrated in Fig. 9.3.

The Cartesian product $P_{m1} \times P_{m2} \times P_{m3}$ of three paths forms a three-dimensional mesh. As the second example, the Cartesian product of $P_6 \times P_4 \times P_5$, resulting in a 5 × 3 × 4 mesh, is shown in Fig. 9.4.

A graph can be the product of more than two specific graphs, such as paths and cycles. As the third example, the product of three graphs, $P_2 \times K_3 \times P_4$, is shown

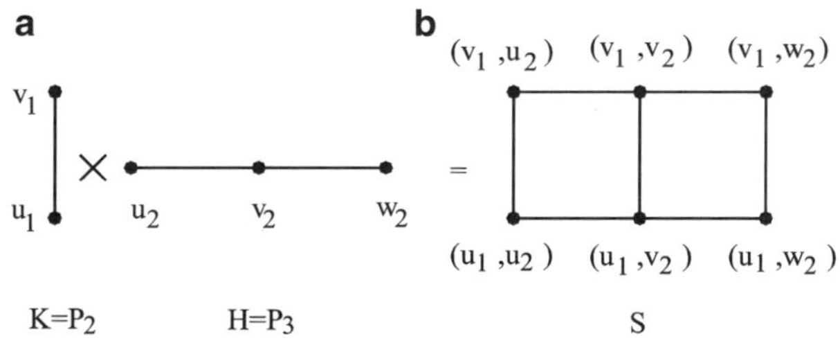

Fig. 9.1 The Cartesian product of two simple graphs

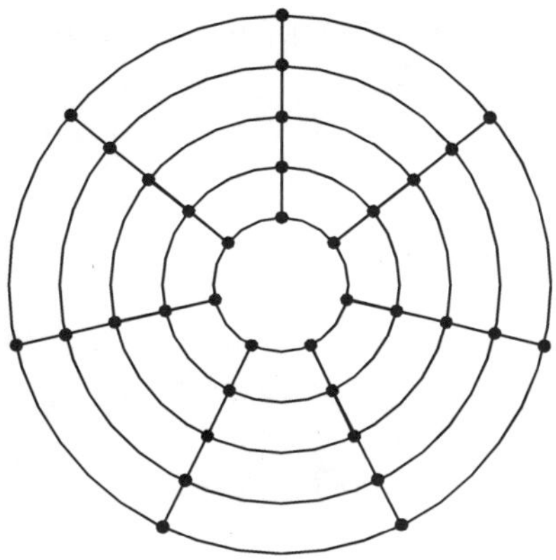

Fig. 9.2 Representation of $C_7 \times P_5$

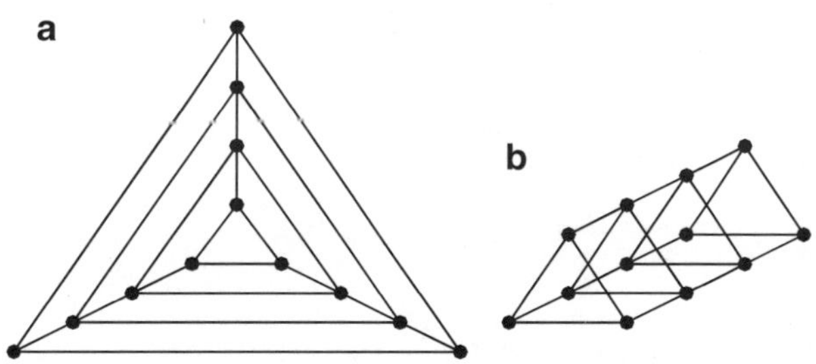

Fig. 9.3 Two different representations of $C_3 \times P_4$

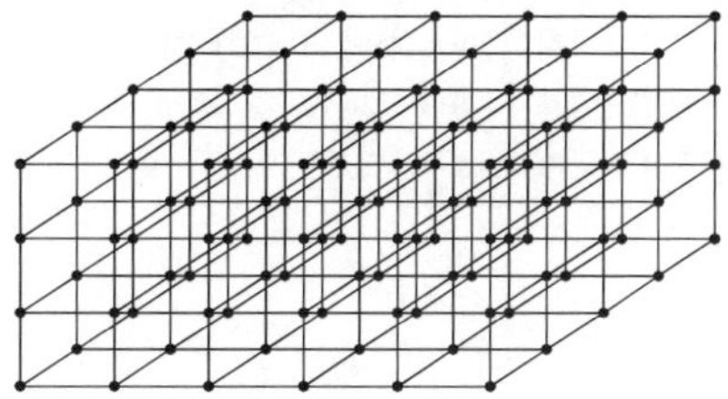

Fig. 9.4 Representation of a $5 \times 3 \times 4$ mesh

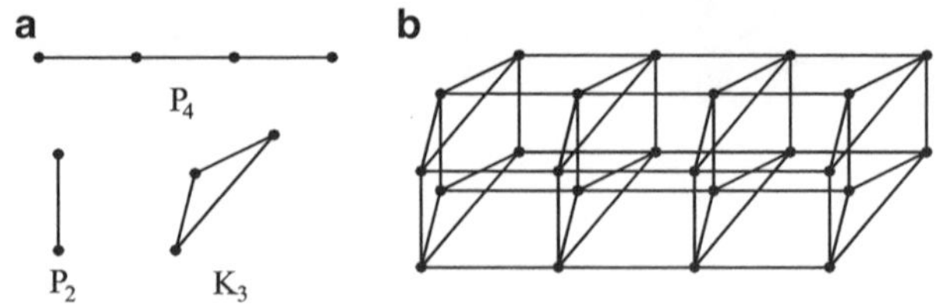

Fig. 9.5 The Cartesian product of three graphs $P_2 \times K_3 \times P_4$. (**a**) Generators. (**b**) Product

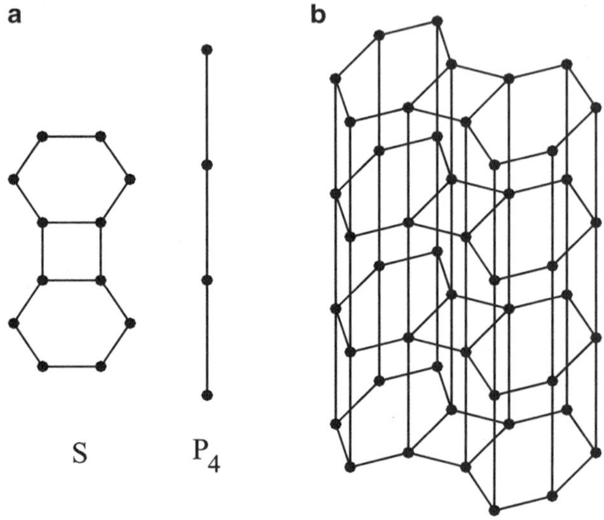

Fig. 9.6 The Cartesian product of S by P_4. (**a**) Generators. (**b**) Product

in Fig. 9.5. The product of a general graph and a path, $S \times P_4$, is illustrated in Fig. 9.6.

9.2.3 *Strong Cartesian Product of Two Graphs*

This is another Boolean operation, known as the *strong Cartesian product*. The strong Cartesian product is a Boolean operation $S = K \boxtimes H$ in which, for any two distinct nodes $u = (u_1, u_2)$ and $v = (v_1, v_2)$ in $N(K) \times N(H)$, the member uv is in M (S) if:

$$u_1 = v_1 \text{ and } u_2 v_2 \in M(H), \tag{9.2a}$$

or

$$u_2 = v_2 \text{ and } u_1 v_1 \in M(K), \tag{9.2b}$$

or

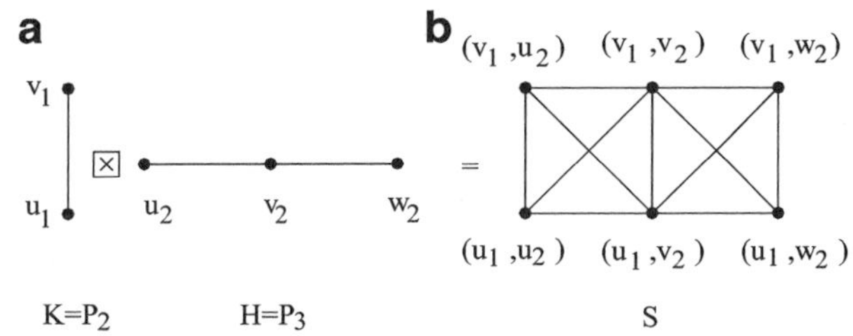

Fig. 9.7 The strong Cartesian product of two simple graphs. (**a**) Generators. (**b**) S = K⊠H

$$u_1v_1 \in M(K) \text{ and } u_2v_2 \in M(H). \tag{9.2c}$$

As an example, the strong Cartesian product of K = P_2 and H = P_3 is shown in Fig. 9.7.

In this example, the nodes (u_1,u_2) and (v_1,v_2) are joined, since the condition (9.2c) is satisfied.

Examples. In the first example, the strong Cartesian product $P_7 \boxtimes P_5$ of a path graph with seven nodes, denoted by P_7 and the path graph P_5 is illustrated in Fig. 9.8.

As the second example, the strong Cartesian product $C_7 \boxtimes P_4$ is shown in Fig. 9.9.

9.2.4 *Direct Product of Two Graphs*

This is another Boolean operation, known as the *direct product*, introduced by Weichsel [6], who called it the *Kronecker Product*. The direct product is a Boolean operation S = K*H, in which, for any two nodes u = (u_1,u_2) and v = (v_1,v_2) in N (K) × N(H), the member uv is in M(S) if:

$$u_1v_1 \in M(K) \text{ and } u_2v_2 \in M(H). \tag{9.3}$$

As an example, the direct product of K = P_2 and H = P_3 is shown in Fig. 9.10.

Here, the two nodes (u_1,u_2) and (v_1,v_2) are joined, since the condition (9.3) is satisfied.

Examples. The direct product $P_7{*}P_5$ of the path graph P_7 and path graph P_5 is illustrated in Fig. 9.11.

As the second example, the direct product $C_7{*}P_4$ is shown in Fig. 9.12.

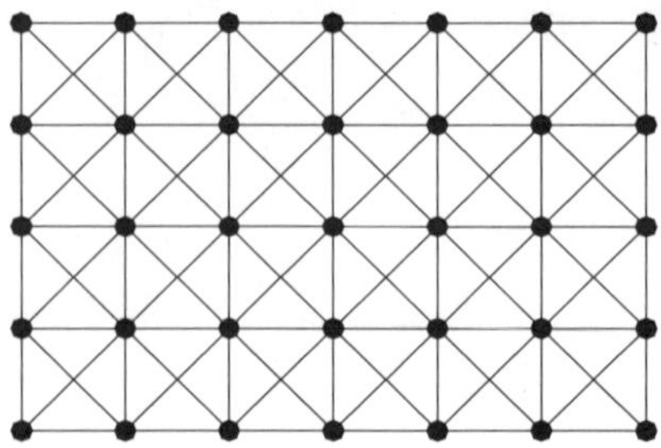

Fig. 9.8 Strong product representation of $P_7 \boxtimes P_5$

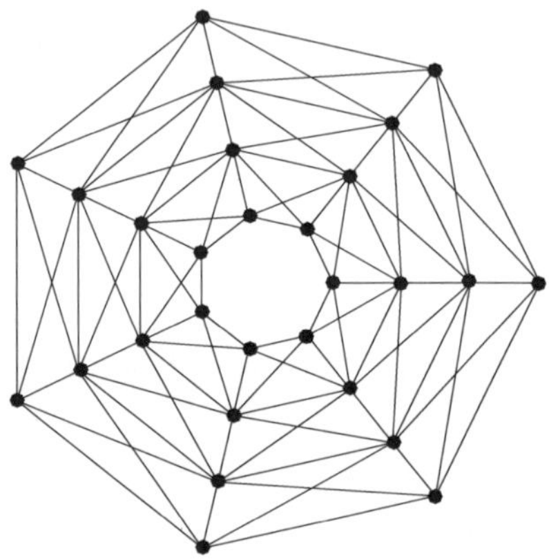

Fig. 9.9 Strong product representation of $C_7 \boxtimes P_4$

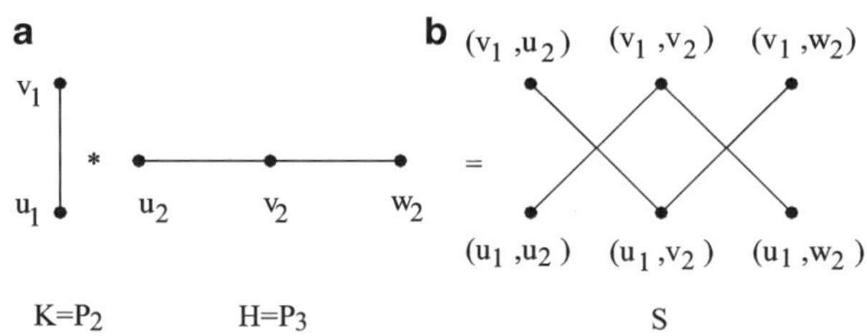

Fig. 9.10 The direct product of two simple graphs

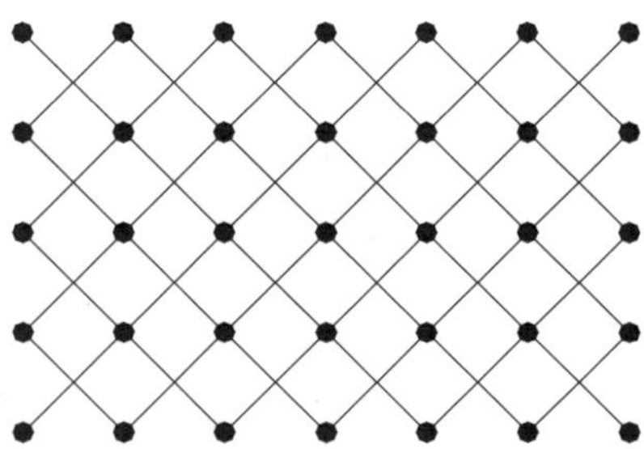

Fig. 9.11 Direct product representation of $P_7{*}P_5$

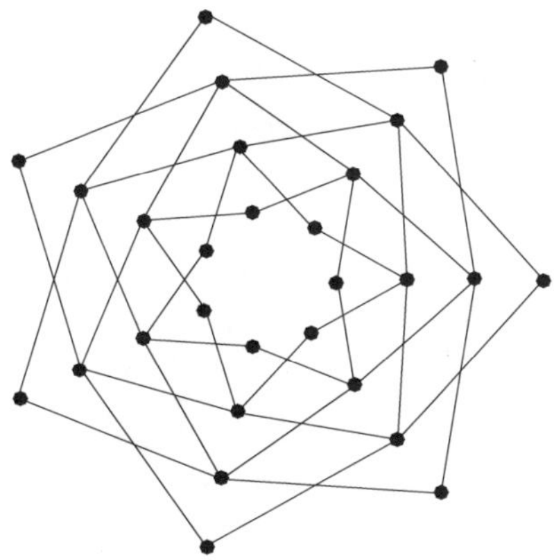

Fig. 9.12 Direct product representation of $C_7{*}P_4$

9.3 Analysis of Near-Regular Structures Using Force Method

Different simple and efficient methods for the analysis of structures are provided in Kaveh [3]. In the analysis of some near-regular structures one can solve the regular part independently and then superimpose the effect of the additional part. For such models, the matrices corresponding to regular part have canonical forms and their eigensolution or inversion can easily be performed [1]. The effect of member changing the regular to a near-regular structure can then be added. In this method, linear behaviour is assumed for the structures.

Here we use the force method, and instead of selecting a statically determinate basic structure (standard method) we employ the regular part of the structure as the basic structure [7].

A new algebraic method is introduced for the force method of analysis for efficient analysis of large near-regular structures.

In this part, we use the force method, however, instead of selecting a statically determinate basic structure we employ the regular part of the structure as the basic structure. Those additional elements are considered as redundant elements. This method is applied to truss and frame structures. In the present approach we can have missing elements instead of additional elements.

In order to demonstrate this problem, consider the truss shown in Fig. 9.13a. This structure consists of a regular part $P_4 \boxtimes P_{10}$ as shown in Fig. 9.13b and has become a near-regular because of having additional 10 bars. The main aim is to decompose these two parts in order to arrive at the analysis of the near-regular structure using the results of the analysis of the regular part. In Fig. 9.13c the positions of the excessive members are highlighted, where the regular part is shown in broken lines.

It should be mentioned that for some regular structures the stiffness matrices can be formed in special block forms, known as the canonical forms, Kaveh [3]. Here we assume that only the members cause irregularity and no additional nodes are

a

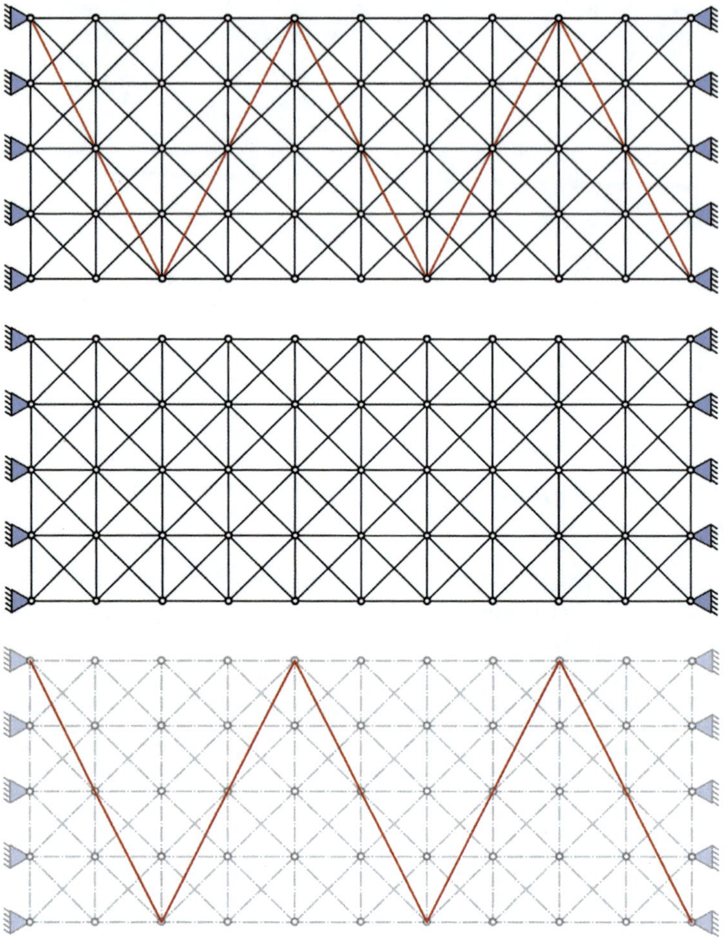

Fig. 9.13 (**a**) An irregular truss. (**b**) The regular part as the strong Cartesian product $P_4 \boxtimes P_{10}$. (**c**) The excessive members being highlighted [1]

present except those of the regular part, i.e. the nodes of the two ends of each excessive members are in the regular part of the structure.

At the beginning, methods suggested are presented for the formation of the matrices required in the force method. Obviously one can also obtain these matrices by other approaches.

The present method consists of two groups of structures as described in the following:

The first group is related to the analysis of those structures in which the excessive members have caused the irregularity. The second group is about those structures which require addition of some members to alter the near-regular structure to a regular one. In this case, by assuming pairs of members with two identical modulus of elasticity having positive and negative signs are added to those places where we need to have members to make the near-regular structure into regular one. In this case, the members with negative sign will be treated as the excessive members.

In the above force method, the internal forces of the excessive members will be considered as redundants, and the corresponding forces will be applied at the regular part of the structure as external loads to incorporate the effect of such members. Thus the regular part will be the main structure to be analyzed. This means we analyze the near-regular structure by considering the regular part and adding the effect of the internal forces of the complementary members as external loads.

Here we assume that the removal of the excessive members will leave the structure geometrically stable, and considering the topology of the regular structures this assumption is quite logical.

In each remaining section first the formulation will be presented, and then through a simple example the process of analysis will be described in a step by step manner. Then by some practical examples, the efficiency of the method will be demonstrated.

First the formation of the flexibility matrix is described. It should be mentioned that this matrix can be formed using any other available method.

9.3.1 *Formulation of the Flexibility Matrix*

In this section a method is presented for the formation of the matrix **B** in the following form:

$$\mathbf{B} = [\mathbf{B}_0 \quad \mathbf{B}_1] \tag{9.4}$$

where the ith column of $\mathbf{B}_0$ is a vector of internal force of the structure under a unit value of a load applied at the ith DOF of the structure ($P_i = 1$), and the ith column of $\mathbf{B}_1$ is a vector containing the internal forces of the structure under the unit load applied at the position of the ith redundant of the ($X_i = 1$) structure.

According to the above definitions for the formation a matrix **B** we are looking for a method by means of which having the externally applied loads of the structure we find the internal forces of the members. In the following a method is presented for this problem using the equilibrium matrix, though one can also find this employing the existing traditional method.

In order to calculate the internal forces of the regular structure under the external loading we proceed as the following:

In the global coordinate system we have

$$\mathbf{S\Delta} = \mathbf{P} \Rightarrow \mathbf{\Delta} = \mathbf{S}^{-1}\mathbf{P} \tag{9.5}$$

where $\mathbf{S}^{-1}$ is the inverse of the stiffness matrix of the DOFs of the regular part of the structure. Using the theorems previously developed for the block matrices, $\mathbf{S}^{-1}$ can be formed using the blocks constituting **S**. This matrix can be obtained using some concepts of graph products or employing concepts from group theory.

According to the definition of equilibrium matrices of the members of the structure and Eq. 9.5, in general, the following form can be written for the deformation of the local coordinate systems of the members of the regular structure:

$$\boldsymbol{\delta} = \mathbf{A}^t\mathbf{\Delta} = \mathbf{A}^t\mathbf{S}^{-1}\mathbf{P} \tag{9.6}$$

where **A** is the equilibrium matrix of the regular structure. Considering the equilibrium equations in the local coordinate system and Eq. 9.6, in general the vector of internal forces of the members of the regular structure under the action of an imaginary external unit load can be obtained as:

$$\mathbf{Q}_0 = \mathbf{s}\boldsymbol{\delta} = \mathbf{s}\left(\mathbf{A}^t\mathbf{S}^{-1}\mathbf{P}\right) = \left(\mathbf{s}\mathbf{A}^t\mathbf{S}^{-1}\right)\mathbf{P} \tag{9.7}$$

Here **s** is the block diagonal matrix containing the stiffness of the members of the regular part of the structure. Therefore the vector of internal forces of a regular structure can be obtained having the external forces in the following form:

$$\mathbf{Q}_0 = \mathbf{R}\mathbf{P} \quad ; \quad \mathbf{R} = \mathbf{s}\mathbf{A}^t\mathbf{S}^{-1} \tag{9.8}$$

If $\overline{\mathbf{X}}$ contains the internal forces of the excessive members in the global coordinate system and **P** is the external force vector of the structure, then the internal forces of the members of the regular structure when part of it is near-regular, can be obtained as:

$$\mathbf{Q}_1 = \mathbf{R}\mathbf{P} + \mathbf{R}\overline{\mathbf{X}} \tag{9.9}$$

Thus for the analysis of near-regular structure discussed in here, $\mathbf{Q}_1$ is the vector of internal forces of the regular structure. The vectors **P** and $\overline{\mathbf{X}}$ can be expressed as:

$$\overline{\mathbf{X}} = \mathbf{N}\mathbf{X} \quad ; \quad \mathbf{P} = \mathbf{I}\mathbf{P} \tag{9.10}$$

I is a unit matrix and **N** is a matrix for transforming the local coordinate system to the global coordinate system. **X** is the internal force vector of excessive members.

Here the method for the formation of **A** and **N** is explained. If $\mathbf{A}_T$ is the equilibrium matrix of a near-regular structure, then by partitioning according to the numbers of internal forces of the excessive members, the matrices **A** and **N** can be formed as:

$$\mathbf{A}_{\mathrm{T}} = [\mathbf{A}|\mathbf{N}] \tag{9.11}$$

where **A** is the equilibrium matrix of the regular structure. In Eqs. 9.9 and 9.10 by taking the common factor and extracting the vector of the assumed forces we have:

$$\mathbf{Q}_1 = \mathbf{R}[\mathbf{I} \quad \mathbf{N}]\begin{bmatrix}\mathbf{P}\\ \mathbf{X}\end{bmatrix} \tag{9.12}$$

In general case the internal forces of the near-regular structure will be as follows:

$$\mathbf{Q} = \begin{bmatrix}\mathbf{Q}_1\\ \mathbf{X}\end{bmatrix} \tag{9.13}$$

Therefore adding **X** to Eq. 9.12 the matrix **Q** can be written as

$$\mathbf{Q} = \begin{bmatrix}\mathbf{Q}_1\\ \mathbf{X}\end{bmatrix} = \left[\begin{array}{c|c}\mathbf{R} & \mathbf{RN}\\ \hline \mathbf{Z} & \mathbf{I}\end{array}\right]\begin{bmatrix}\mathbf{P}\\ \mathbf{X}\end{bmatrix} \tag{9.14}$$

Now the matrices $\mathbf{B}_0$ and $\mathbf{B}_1$ can be formed by partitioning the above matrix according to the numbering of the internal forces of the excessive members of the structure in the following form:

$$\mathbf{B}_0 = \begin{bmatrix}\mathbf{R}\\ \mathbf{Z}\end{bmatrix} \quad ; \quad \mathbf{B}_1 = \begin{bmatrix}\mathbf{RN}\\ \mathbf{I}\end{bmatrix} \tag{9.15}$$

In the above equations **Z** is a matrix of zeros with dimension $t \times k$ and **I** is a unit matrix of dimension $t \times t$. The matrices $\mathbf{B}_0$ and $\mathbf{B}_1$ have dimensions $(e + t) \times k$ and $(e + t) \times t$, respectively. t is the total number of internal forces of the excessive members, k is the DOFs of the near-regular structure in global coordinate system and e is the number of internal forces of the regular structure.

In this method we need to form the matrix $\mathbf{A}_{\mathrm{T}}$ and in the subsequent section a simple method will be presented for this formation.

The formation of the matrix **B** can be summarized as follows:

Step 1: Form the matrices $\mathbf{S}^{-1}$ and **s** for the regular structure.
Step 2: Form the matrix $\mathbf{A}_{\mathrm{T}}$ for all the members of the structure consisting of regular and excessive members.
Step 3: Partition $\mathbf{A}_{\mathrm{T}}$ using Eq. 9.11 and form the matrices **A** and **N**.
Step 4: Calculate the matrix **R** using Eq. 9.8.
Step 5: Calculate the matrices $\mathbf{B}_0$ and $\mathbf{B}_1$ using Eq. 9.15.

9.3.2 A Simple Method for the Formation of the Matrix A_T

A general method for the formation of the equilibrium matrix consists of writing equilibrium of the forces at the nodes of the structure. For a quick calculation of the matrix $\mathbf{A}_T$ one can assemble the rotation matrices of the members of the near-regular structure. Then it can be partitioned using the relationship presented in the previous section. In the following the approach for positioning the rotation matrices of the members in each column of the equilibrium matrix is illustrated. For the formation of the equilibrium matrix $\mathbf{A}_T$ of the near-regular matrix we perform the following process:

If we consider i as the nodal DOFs of the assumed member j in the local coordinate system, and r are the nodal DOFs of the assumed member j in the global coordinate system, then the columns corresponding to i in the matrix $\mathbf{A}_T$ will be as follows:

$$\mathbf{A}_T(\mathrm{r}, \mathrm{i}) = \mathbf{T}_j^t \tag{9.16}$$

The remaining rows of these columns are zero. We repeat this process for all the members of the near-regular structure. $\mathbf{T}_j$ is the modified rotation matrix of the jth member. This matrix can be represented as follows:

Space truss member

$$\mathbf{T}_j = [\mathbf{T}_1 \mid -\mathbf{T}_1] \;,\; \mathbf{T}_1 = [\mathrm{Cos}\alpha \quad \mathrm{Cos}\beta \quad \mathrm{Cos}\gamma]$$

Planar frame member

$$\mathbf{T}_j = [\mathbf{T}_1 \mid \mathbf{T}_3] \quad , \quad \mathbf{T}_3 = \mathbf{s}_1^{-1}\mathbf{s}_2\mathbf{T}_2 \quad , \quad \mathbf{s}_j = \left[\begin{array}{c|c} \mathbf{s}_1 & \mathbf{s}_2 \\ \hline \mathbf{s}_2 & \mathbf{s}_1 \end{array}\right] \tag{9.17}$$

$$\mathbf{T}_1 = \begin{bmatrix} \mathrm{Cos}\alpha & \mathrm{Sin}\alpha & 0 \\ -\mathrm{Sin}\alpha & \mathrm{Cos}\alpha & 0 \\ 0 & 0 & 1 \end{bmatrix} \;,\; \mathbf{T}_2 = \begin{bmatrix} -\mathrm{Cos}\alpha & -\mathrm{Sin}\alpha & 0 \\ \mathrm{Sin}\alpha & -\mathrm{Cos}\alpha & 0 \\ 0 & 0 & 1 \end{bmatrix}$$

s_j is the jth block of the stiffness matrix $\mathbf{s}$. Here, α, β and γ are the angles with the x, y and z axis, respectively.

Similar to Eq. 9.11 the above matrix can be transformed to $\mathbf{A}$ and $\mathbf{N}$ by partitioning and numbering the internal forces of the excessive members which are separated from the structure.

Algorithm
The algorithm can be summarized as:
Step 1: Formation of the matrices $\mathbf{T}_j$ for members of the near-regular structure.
Step 2: Formation of the equilibrium matrix $\mathbf{A}_T$ by assembling the rotation matrices of the near-regular matrix using Eq. 9.16.
Step 3: Formation of the matrices **A** and **N** by partitioning of the matrix $\mathbf{A}_T$.

9.4 Analysis of Regular Structures with Excessive Members

In this section the analysis of those structures for which the irregularity is produced by excessive members is studied. Here the force method is used for the analysis, with the only difference that instead of removing member to obtain a statically determinate structure, members are removed to transform the structure into a regular one. Here the relationships required for the force method are presented. Base on the concepts of the force method, the internal forces of the members of the near-regular structure can be expressed as:

$$[\mathbf{Q}] = [\mathbf{B}]\begin{bmatrix}\mathbf{P}\\ \mathbf{X}\end{bmatrix} \tag{9.18}$$

After the formation of the matrix **B** which was described in Sect. 9.2, one can calculate the internal forces of the excessive members using the following relationships:

$$\mathbf{D}_2 = \mathbf{B}_1{}^t\mathbf{F}\mathbf{B}_1 \quad ; \quad \mathbf{D}_1 = \mathbf{B}_1{}^t\mathbf{F}\mathbf{B}_0 \tag{9.19}$$

$$\mathbf{X} = -\mathbf{D}_2{}^{-1}\mathbf{D}_1\mathbf{P} \tag{9.20}$$

Here **F** is a block matrix of dimension (e + t) × (e + t) and contains all the flexibility matrices of the members of the near-regular structure. The matrix $\mathbf{D}_1$ is of dimension t × k and the matrix $\mathbf{D}_2$ is of dimension t × t. This means that for calculating the internal forces of the excessive members, only the inverse of a matrix of dimension t is needed.

At the end, the forces of **X** are added to the external force vector **P** denoted by $\mathbf{P}^*$ which is defined as the equivalent external load of the regular structure. According to this, the displacements of the structure can be obtained by the inverse of the stiffness matrix of the regular structure as follows:

$$\mathbf{P}^* = \mathbf{P} + \mathbf{N}\mathbf{X} \tag{9.21}$$

$$\boldsymbol{\Delta} = \mathbf{S}^{-1}\mathbf{P}^* \tag{9.22}$$

where $\mathbf{S}^{-1}$ is the inverse of the stiffness matrix of the DOFs of the regular structure and can be obtained using the existing methods, Kaveh [3]. The vector $\boldsymbol{\Delta}$ contains the displacements of the near-regular structure.

The matrix **N** is the transformation matrix of the internal forces in excessive members from local to global coordinate systems.

9.4.1 Summary of the Algorithm

Step 1: Numbering the DOFs, nodes and members of the near-regular structure and formation of the external force vector **P**.
Step 2: Formation of the matrices $\mathbf{S}^{-1}$ and **s** .
Step 3: Formation of the equilibrium matrix of the near-regular structure using Eq. 9.16.
Step 4: Calculation of the $\mathbf{B}_0$ and $\mathbf{B}_1$ matrices using Eq. 9.15.
Step 5: Formation of the flexibility matrix **F** in a block diagonal form for all the members of the near-regular structure.
Step 6: Calculation of the matrices $\mathbf{D}_1$ and $\mathbf{D}_2$ using Eq. 9.19.
Step 7: Calculation of the vector **X** using Eq. 9.20.
Step 8: Calculation of the equivalent external load of the regular structure using Eq. 9.21.
Step 9: Calculation of the nodal displacements of the near-regular structure using Eq. 9.21.

The above explanations are further explained through the following simple example.

9.4.2 Investigation of a Simple Example

For the 10-bar truss shown in Fig. 9.14, deleting member 10, the structures become regular. Here using the force method, the internal force of the member 10 is calculated and as an additional force it is added to the external forces. Then the regular structure is analyzed with the new loads.

It should be noted that in the standard force method the basic structure is selected for a redundant structure is often statically determinate. For the structure of this example we have four statical indeterminacy and four redundants should be chosen. However, in our approach the basic structure is selected as a regular structure which is not necessarily statically determinate.

In this example, EA is assumed to be unit for all the members and the external load vector is as follows:

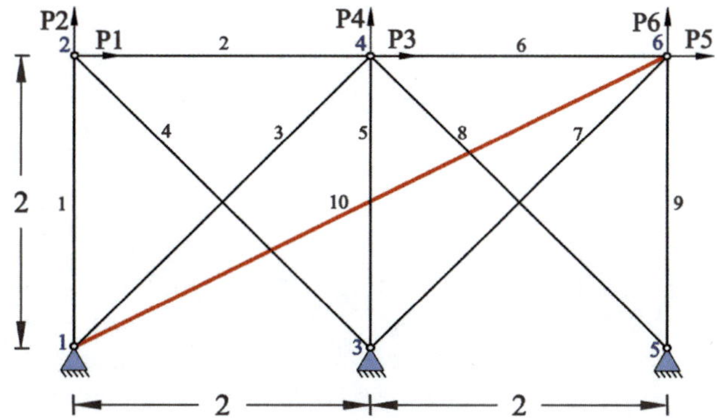

Fig. 9.14 A 10-bar truss transformable to a regular structure

$$\mathbf{P} = [\,10 \quad 0 \quad 0 \quad 20 \quad 0 \quad 0]^{t}$$

One of the methods for the formation of the equilibrium matrix is to use the equilibrium equations of forces at the nodes and formation of the matrix of the coefficients of the forces. In this example the equilibrium matrix of the near-regular structure is calculated using this approach. The obtained equilibrium matrix is partitioned into **A** and **N** using Eq. 9.11.

The force equilibrium equations will be as follows:

$$\begin{array}{ll} P_1 = -Q_2 - \sqrt{2}/2Q_4 & , \quad P_2 = Q_1 + \sqrt{2}/2Q_4 \\ P_3 = Q_2 + \sqrt{2}/2Q_3 - Q_6 - \sqrt{2}/2Q_8 & , \quad P_4 = \sqrt{2}/2Q_3 + Q_5 + \sqrt{2}/2Q_8 \\ P_5 = Q_6 + \sqrt{2}/2Q_7 + 2/\sqrt{5}Q_{10} & , \quad P_6 = \sqrt{2}/2Q_7 + Q_9 + 1/\sqrt{5}Q_{10} \end{array}$$

The relation between the equilibrium matrix **A** and the vector of external and internal forces of the structure can be written as:

$$\mathbf{P} = \mathbf{AQ} \tag{9.23}$$

In this way the matrix **A** and the partitioning considering the excessive member 10 will be as follows:

$$\mathbf{A}_{\mathrm{T}} = [\mathbf{A} \mid \mathbf{N}] = \left[\begin{array}{ccccccccc|c} 0 & -1 & 0 & -0.7071 & 0 & 0 & 0 & 0 & 0 & 0 \\ 1 & 0 & 0 & 0.7071 & 0 & 0 & 0 & 0 & 0 & 0 \\ 0 & 1 & 0.7071 & 0 & 0 & -1 & 0 & -0.7071 & 0 & 0 \\ 0 & 0 & 0.7071 & 0 & 1 & 0 & 0 & 0.7071 & 0 & 0 \\ 0 & 0 & 0 & 0 & 0 & 1 & 0.7071 & 0 & 0 & 0.9844 \\ 0 & 0 & 0 & 0 & 0 & 0 & 0.7071 & 0 & 1 & 0.4472 \end{array}\right]$$

Thus the matrix **N** will be as

$$\mathbf{N} = [0 \quad 0 \quad 0 \quad 0 \quad 0.9844 \quad 0.4472\,]^{t}$$

The stiffness matrix of the regular structure by elimination of the member 10 will become:

$$\mathbf{S} = \begin{bmatrix} 0.6767 & -0.1767 & -0.5 & 0 & 0 & 0 \\ -0.1767 & 0.6767 & 0 & 0 & 0 & 0 \\ -0.5 & 0 & 1.3535 & 0 & -0.5 & 0 \\ 0 & 0 & 0 & 0.8535 & 0 & 0 \\ 0 & 0 & -0.5 & 0 & 0.6767 & 0.1767 \\ 0 & 0 & 0 & 0 & 0.1767 & 0.6767 \end{bmatrix}$$

Replacing two columns 5 and 6 with columns 3 and 4, and also their corresponding rows, This matrix will get Form III pattern and calculating its eigenvalues leads to the formation of $\mathbf{S}^{-1}$. Here, **s** is a block diagonal matrix having the stiffness of the members of the regular structure. Since the structure is a truss, therefore this matrix becomes a diagonal one.

$$\mathbf{s} = \text{diag}\{0.5, 0.5, 0.3535, 0.3535, 0.5, 0.5, 0.3535, 0.3535, 0.5\}$$

In this relation diag represents a block diagonal matrix.

Substituting the above matrices in Eq. 9.15 leads to the formation of $\mathbf{B}_0$ and $\mathbf{B}_1$ matrices.

$$\mathbf{B}_0 = \begin{bmatrix} 0.3535 & 0.8311 & 0.1846 & 0 & 0.1464 & -0.0382 \\ -0.6464 & -0.1688 & 0.1846 & 0 & 0.1464 & -0.0382 \\ 0.3535 & 0.0923 & 0.4459 & 0.2928 & 0.3535 & -0.0923 \\ -0.5 & 0.2387 & -0.2612 & 0 & -0.2071 & 0.0540 \\ 0 & 0 & 0 & 0.5857 & 0 & 0 \\ -0.1464 & -0.0382 & -0.1847 & 0 & 0.6464 & -0.1688 \\ 0.2071 & 0.0540 & 0.2612 & 0 & 0.5 & 0.2387 \\ -0.3535 & 0.0923 & -0.4459 & 0.2928 & -0.3535 & 0.0923 \\ -0.1464 & -0.0382 & -0.1847 & 0 & -0.3535 & 0.8311 \\ 0 & 0 & 0 & 0 & 0 & 0 \end{bmatrix}$$

$$\mathbf{B}_1 = [-0.1138 \ -0.1138 \ -0.2749 \ 0.1610 \ 0 \ -0.5026 \ -0.5540 \ 0.2749 \ -0.0557 \ 1]^t$$

The flexibility matrix of the near-regular structure **F** in general is a block diagonal matrix. Since the considered structure is a truss, thus this matrix has numerical values in its diagonal.

$$\mathbf{F} = \text{diag}\left\{2, 2, 2\sqrt{2}, 2\sqrt{2}, 2, 2, 2\sqrt{2}, 2\sqrt{2}, 2, 2\sqrt{5}\right\}$$

Using Eq. 9.19 the matrices $\mathbf{D}_1$ and $\mathbf{D}_2$ are formed as follows:

$$\mathbf{D}_1 = [-0.8719 \quad -0.2277 \quad -1.0997 \quad 0 \quad -2.1050 \quad -0.1109];$$
$$\mathbf{D}_2 = [6.4045]$$

Now employing Eq. 9.20, the internal forces of the excessive members are calculated as:

$$\mathbf{X} = -\mathbf{D}_2{}^{-1}\mathbf{D}_1\,\mathbf{P}$$
$$= -\frac{1}{6.4045} \times \left[\overset{1}{-0.8719} \;\ldots\; \overset{4}{0} \;\ldots \right] \cdot [10\ 0\ 0\ 20\ 0\ 0]^{\mathrm{t}} = -1.3624$$

Substituting the values of $\mathbf{X}$ in Eq. 9.21, the vector $\mathbf{P}^*$ can be obtained. Now according to Eq. 9.22, multiplying this vector with $\mathbf{S}^{-1}$, the displacement vector of the nodal forces of the near-regular structure can be obtained.

$$\mathbf{P}^* = \{10, 0, 0, 20, 1.2177, 0.6089\}^{\mathrm{t}}$$
$$\boldsymbol{\Delta} = \{\,25.8839 \quad 6.7609 \quad 12.6449 \quad 23.4314 \quad 8.3473 \quad -3.0400\,\}^{\mathrm{t}}$$

It should be noted that the aim of this example was the explanation of the method by means of a simple example and for showing the capabilities of the presented method is not sufficient. The reduction in dimensions of the matrices achieved by the present method and the speed of calculation will be illustrated in Sect. 9.5.

9.5 Analysis of Regular Structures with Some Missing Members

In this section we consider those structures which need addition of some members to become a regular one. Obviously the method presented in the previous section can not be applied directly for these structures. However, for transforming this case to the previous one, a pair of members with equal modulus of elasticity having different signs, are added where we have lack of members for regularity. In the next step the members with negative modulus of elasticity are considered as excessive members and separated from the structure. The remaining process of the analysis is the same as the previous case. The internal forces of the members with negative modulus of elasticity are calculated and added to the external forces. Then the regular structure with the external loads together with the internal forces of the excessive members which are applied as the additional external forces, is analyzed. In the following a simple example is considered for further explanation.

9.5.1 Investigation of an Illustrative Simple Example

In this section a simple example is used to describe the process of the algorithm. In a 4-bar structure shown in Fig. 9.15a, it is obvious that if we add a member between the nodes 2 and 3, the structure will be transformed into Form II and one can easily calculate its inverse. Now we add two members 5 and 6 of identical properties have modulus of elasticity of different signs between the two nodes 2 and 3. This is

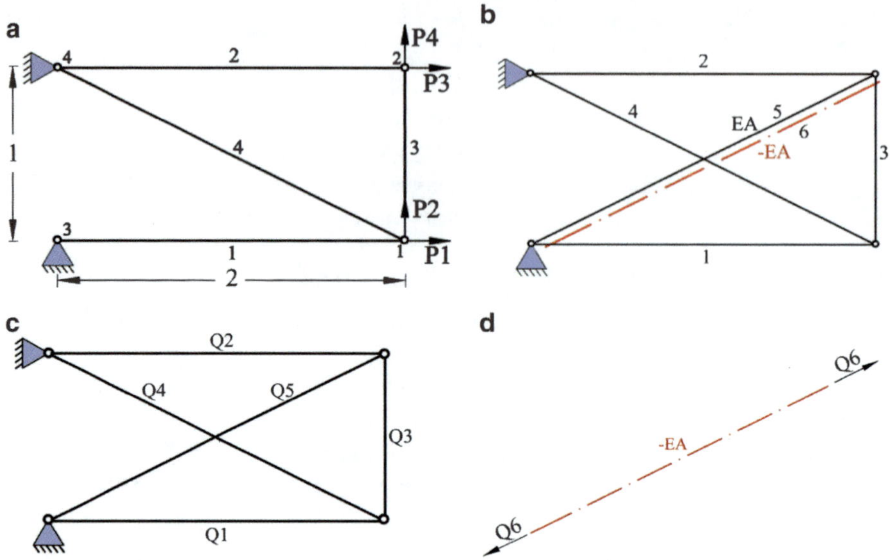

Fig. 9.15 (**a**) An irregular structure. (**b**) The irregular structure with a pair of members being added. (**c**) Representation of the internal forces in the regular structure. (**d**) The added member with negative modulus

logical assumption because the property of one member can be nullified by the other member.

Member 6 has negative modulus of elasticity and we consider it as a member separation of which transforms the structure into a regular one as shown in Fig. 9.15c. The structure obtained in this way is equivalent to the basic structure of the force method. From here onward all the previous steps can be employed. Figure 9.15d shows the excessive bar with negative modulus of elasticity which is separated from the truss shown in Fig. 9.15b.

The external force vector will be as follows:

$$\mathbf{P} = [0 \quad 0 \quad 0 \quad -10]^{\mathrm{t}}$$

For the formation of the equilibrium matrix of the structure shown in Fig. 9.15b, one can either use the equations corresponding to the equilibrium of the forces at the nodes, or alternatively use Eq. 9.16.

$$\mathbf{A}_\mathrm{T} = \left[\begin{array}{ccccc|c} 1 & 0 & 0 & 0.8944 & 0 & 0 \\ 0 & 0 & -1 & -0.4472 & 0 & 0 \\ 0 & 1 & 0 & 0 & 0.8944 & 0.8944 \\ 0 & 0 & 1 & 0 & 0.4472 & 0.4472 \end{array}\right]$$

The matrices **A** and **N** can be obtained by partitioning the above matrix according to Eq. 9.11 and numbering the internal force of the separated member.

$$\mathbf{A} = \begin{bmatrix} 1 & 0 & 0 & 0.8944 & 0 \\ 0 & 0 & -1 & -0.4472 & 0 \\ 0 & 1 & 0 & 0 & 0.8944 \\ 0 & 0 & 1 & 0 & 0.4472 \end{bmatrix} \quad ; \quad \mathrm{N} = \begin{bmatrix} 0 \\ 0 \\ 0.8944 \\ 0.4472 \end{bmatrix}$$

The matrix $\mathbf{S}^{-1}$ corresponding to the regular structure shown in Fig. 9.15c, and the matrix s are as follows:

$$\mathbf{S}^{-1} = \begin{bmatrix} 1.5935 & 2.0508 & -0.4065 & 1.9491 \\ 2.0508 & 9.8338 & -1.9491 & 9.3465 \\ -0.4065 & -1.9491 & 1.5935 & -2.0508 \\ 1.9491 & 9.3465 & -2.0508 & 9.8338 \end{bmatrix}$$

$$\mathbf{s} = \mathrm{diag}\{\, 0.5 \quad 0.5 \quad 1 \quad 0.4472 \quad 0.4472 \,\}$$

It can be seen that the matrix S can be transformed into Form II by multiplying the row and column 2 by -1. Using the above matrices and Eq. 9.8, the matrix **R** is obtained as:

$$\mathbf{R} = \begin{bmatrix} 0.7967 & 1.0254 & -0.2032 & 0.9745 \\ -0.2032 & -0.9745 & 0.7967 & -1.0254 \\ -0.1016 & -0.4873 & -0.1016 & 0.4872 \\ 0.2272 & -1.1464 & 0.2272 & -1.0896 \\ 0.2272 & 1.0896 & 0.2272 & 1.1464 \end{bmatrix}$$

Using Eq. 9.15, the matrices $\mathbf{B}_0$ and $\mathbf{B}_1$ are obtained as:

$$\mathbf{B}_0 = \begin{bmatrix} 0.7967 & 1.0254 & -0.2032 & 0.9745 \\ -0.2032 & -0.9745 & 0.7967 & -1.0254 \\ -1.1016 & -0.4873 & -0.1016 & 0.4872 \\ 0.2272 & -1.1464 & 0.2272 & -1.0896 \\ 0.2272 & 1.0896 & 0.2272 & 1.1464 \\ 0 & 0 & 0 & 0 \end{bmatrix} ; \quad \mathbf{B}_1 = \begin{bmatrix} 0.2540 \\ 0.2540 \\ 0.1270 \\ -0.2840 \\ 0.7159 \\ 1 \end{bmatrix}$$

The flexibility matrix **F** for the truss shown in Fig. 9.15b will be as follows:

$$\mathbf{F} = \mathrm{diag}\{2 \quad 2 \quad 1 \quad 2.2360 \quad 2.2360 \quad -2.2360\}$$

Using Eq. 9.19 the following matrices are obtained:

$$\mathbf{D}_1 = [0.5081 \quad 2.4364 \quad 0.5081 \quad 2.5635]; \ \mathbf{D}_2 = [-0.6351]$$

The matrix X is calculated from Eq. 9.20 as

$$\mathbf{X} = -\mathbf{D}_2{}^{-1}.\mathbf{D}_1\,\mathbf{P} = \frac{1}{0.6351} \times [0.5081 \ 2.4364 \ 0.5081 \ 2.5635].[0 \ 0 \ 0 \ -10]^t = -40.3607$$

Adding **X** to the vector of external loads according to Eq. 9.21 and multiplying the matrix $\mathbf{S}^{-1}$ employing Eq. 9.22 we will have:

$$\mathbf{P}^* = [0 \quad 0 \quad -36.0997 \quad -28.0498]^t$$
$$\boldsymbol{\Delta} = [-40 \quad -191.803 \quad 0 \quad -201.803]^t$$

Finally, using **X** and Eq. 9.14 one can find the internal forces of the structure shown in Fig. 9.15b as follows:

$$\mathbf{Q} = [-20 \quad 0 \quad -10 \quad 22.3607 \quad -40.3607 \quad -40.3607]^t$$

One can recognize the equality of the internal forces of the entries 5 and 6.

9.6 Practical Examples

Here four examples are presented. The first two examples correspond to Sect. 9.3 and the third example belongs to Sect. 9.4. The fourth example corresponds to the combination of the methods presented in Sects. 9.3 and 9.4. The latter example is chosen as a frame structure to showing the applicability of the presented method to other skeletal structures other than trusses.

Example 1. A truss with 47 members is considered in the form of a single layer rotational dome, having two members 26 and 27 making the truss a near-regular one, Fig. 9.16. If we remove these two members then the remaining regular structure can easily be solved using the method of Kaveh and Rahami [8]. The value of EA = 1 N is assumed to be identical for all the members and the force $P_z^2 = 10$ N and $P_y^8 = 20$ N are applied at nodes 2 and 8, in z direction and y direction, respectively.

For solution of this problem the members 26 and 27 are considered as excessive members. For the above near-regular structure the equilibrium matrix $\mathbf{A}_T$ has dimension 30 × 47 and by partitioning using Eq. 9.11, the matrices **A** and **N** with

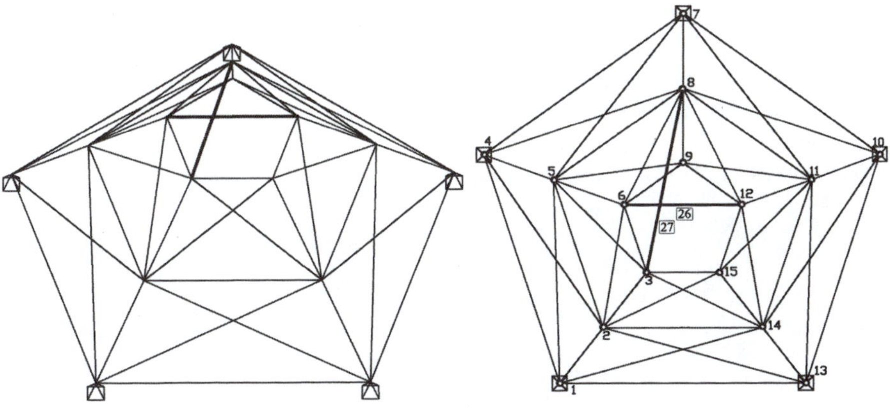

Fig. 9.16 The space dome of the Example 1 with 47 members

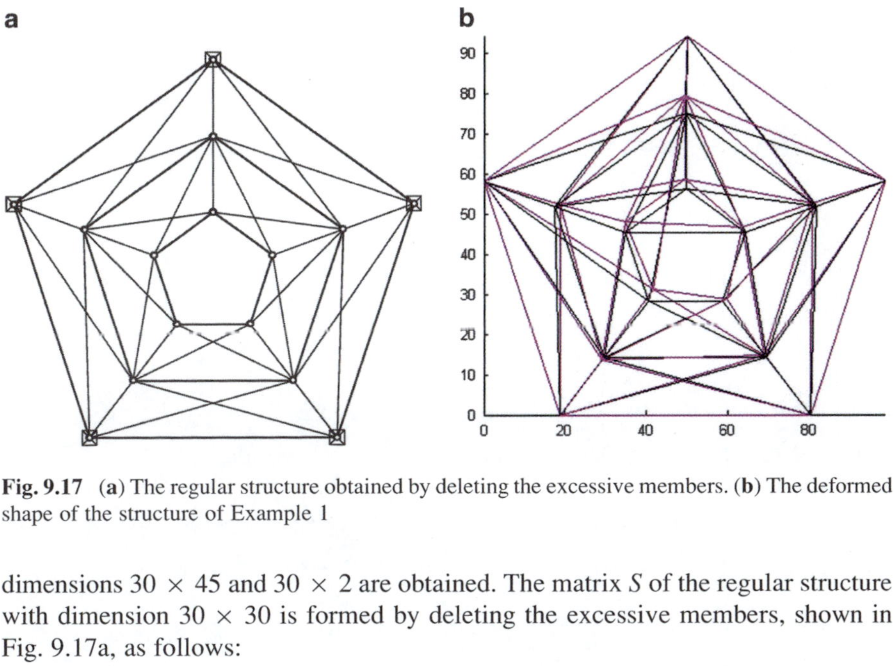

Fig. 9.17 (**a**) The regular structure obtained by deleting the excessive members. (**b**) The deformed shape of the structure of Example 1

dimensions 30 × 45 and 30 × 2 are obtained. The matrix S of the regular structure with dimension 30 × 30 is formed by deleting the excessive members, shown in Fig. 9.17a, as follows:

$$S = \sum_{i=1}^{5} (P_i \otimes A_i)$$

In this relation the matrices $\mathbf{A}_i$ and $\mathbf{P}_i$ are the submatrices constituting the matrix **S**.

Using the method presented in Kaveh and Rahami [8] the inverse of the matrix S is formed by using the eigenvalues and eigenvectors of five 6 × 6 matrices.

s is a diagonal matrix of dimension 45 × 45 containing the stiffness matrices of the members of the regular structure and **F** is a diagonal matrix of dimension 47 × 47 consisting of the flexibility matrices of the near-regular structure . Therefore the matrices $\mathbf{B}_0$ and $\mathbf{B}_1$ of dimension 47 × 30 and 47 × 2 can be obtained using the Eq. 9.15.

The matrices $\mathbf{D}_1$ and $\mathbf{D}_2$ of dimension 2 × 30 and 2 × 2 obtained by using Eq. 9.19 as follows:

$$\mathbf{D}_1 = \begin{bmatrix} \overset{1}{\cdots} & \overset{3}{5.6163} & \cdots & \overset{14}{-30.9438} & \overset{30}{\cdots} \\ \cdots & -62.3208 & \cdots & -29.6975 & \cdots \end{bmatrix}$$

$$\mathbf{D}_2 = \begin{bmatrix} 426.7122 & -156.126 \\ -156.126 & 267.2256 \end{bmatrix}$$

The vector of the internal forces of the excessive members **X** can be calculated from Eq. 9.20 as:

$$\mathbf{X}{=}-\mathrm{inv}\left(\begin{bmatrix} 426.7122 & -156.126 \\ -156.126 & 267.2256 \end{bmatrix}\right)\begin{bmatrix} \overset{1}{\ldots} & \overset{3}{5.6163} & \ldots & \overset{14}{-30.9438} & \overset{30}{\ldots} \\ \ldots & -62.3208 & \ldots & -29.6975 & \ldots \end{bmatrix}$$
$$\begin{bmatrix} \overset{1}{0} & \ldots & \overset{3}{10} & \ldots & \overset{14}{20} & \ldots & \overset{30}{0} \end{bmatrix}^t$$
$$\mathbf{X}{=}\{-1.9283{,}0.7148\}^t$$

The matrix of the internal forces of the near-regular structure can be obtained by substituting X in Eq. 9.14 as follows:

$$\mathbf{Q} = \begin{bmatrix} \overset{1}{8.7095} & \overset{2}{0.1002} & \overset{3}{-0.6626} & \overset{4}{-2.1497} & \ldots & \overset{44}{-0.6706} & \overset{45}{0.4741} & \overset{46}{-1.9283} & \overset{47}{0.7148} \end{bmatrix}^t$$

We substitute X in Eq. 9.21, and substitute the vector of the equivalent external forces of the regular structure in Eq. 9.22. The vector of the displacements for the near-regular structure can then be obtained using the inverse of the stiffness matrix of the regular structure.

$$\boldsymbol{\Delta} = \begin{bmatrix} \overset{1}{-70.6421} & \overset{2}{-47.0104} & \overset{3}{399.9523} & \overset{4}{117.5682} & \ldots & \overset{27}{-9.5185} & \overset{28}{110.8598} & \overset{29}{56.1357} & \overset{30}{90.5555} \end{bmatrix}^t$$

The deformed shape of the structure is shown in Fig. 9.17b.

If we solve the structure by a conventional method we have to find the inverse of a matrix of dimension 30 × 30, while the present approach requires the inverse of 5 matrices of dimension 6 × 6 and the inverse of the matrix $\mathbf{D}_1$ of dimension 2 × 2.

Example 2. A communication space tower studied in Kaveh and Rahami [8], is considered in here. This model can be expressed as the product two graphs. In practice the towers have horizontal belts employed in different heights. These belts make their model irregular. In here we want to analyze those towers which have

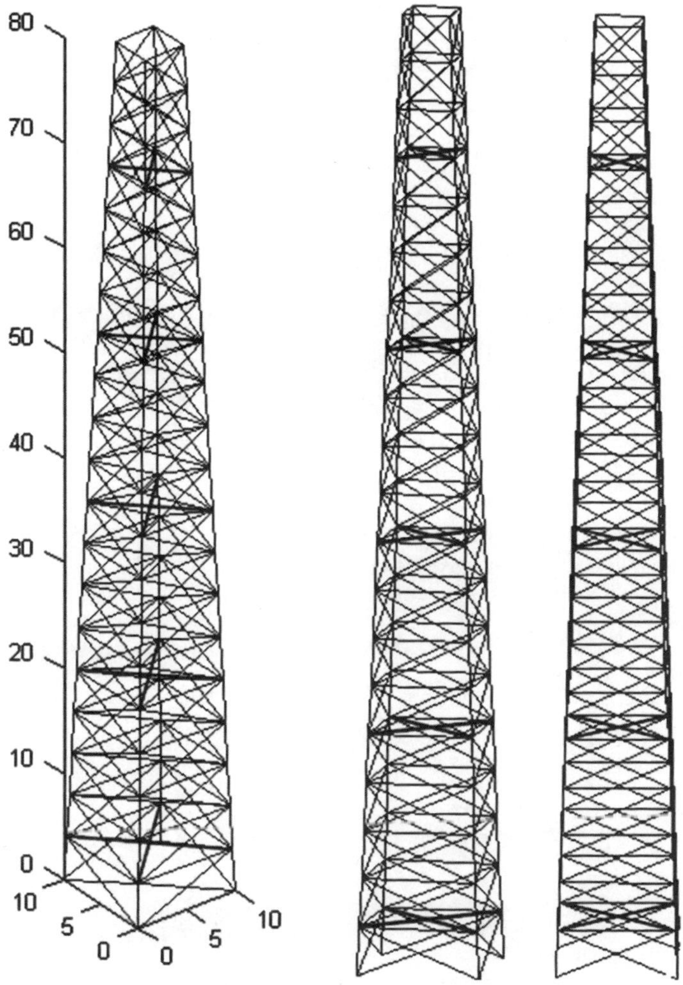

Fig. 9.18 A communication space tower with five pairs of excessive members being shown at the central core of the structure

belts at the central core of the structure (Fig. 9.18). If we consider these members as additional ones, we will obtain a regular structure. Such a regular structure can be generated by rotation of one of its faces. Using the force method the internal forces of the excessive members will be calculated and together with other external loads will be applied to the regular structure.

This tower has 84 nodes and 330 members. The load applied to the structure is P = 1 kN applied in all DOFs of the 4 upper nodes of the tower. The value of EA = 100 N for all members is considered to be identical.

The structure has 80 free nodes and 330 members and 10 members belong to the belt of the structure.

Forming the equilibrium matrix of the near-regular structure which is of dimension 240 × 330, and its partitioning by employing Eq. 9.11, the matrices A and N of dimensions 240 × 320 and 240 × 10 can be obtained.

The stiffness matrix $\mathbf{S}$ of the regular structure is of 240 × 240 and has the following form:

$$\mathbf{S} = \sum_{i=1}^{4} (\mathbf{P}_i \otimes \mathbf{A}_i)$$

Using the method presented in Kaveh and Rahami [8], the inverse of the stiffness matrix of the regular structure can be obtained using the eigenvalues and eigenvectors of four 60 × 60 matrices. In this example, since the structure is truss, s is a diagonal matrix of dimension 320 × 320 containing the stiffness matrices of the members of the regular structure. Having the above mentioned matrices and using Eq. 9.15 one can easily form the $\mathbf{B}_0$ and $\mathbf{B}_1$ matrices which are of dimension 330 × 240 and 330 × 10. For this structure, the flexibility matrix $\mathbf{F}$ of the near-regular structure has dimension 330 × 330. Having the matrices $\mathbf{B}_0$, $\mathbf{B}_1$ and $\mathbf{F}$, the matrices $\mathbf{D}_1$ and $\mathbf{D}_2$ of dimensions 10 × 320 and 10 × 10 are obtained from Eq. 9.19. Here $\mathbf{D}_2$ is as follows:

$$\mathbf{D}_2 = \begin{bmatrix} 0.2999 & -0.1033 & 0.2226 & -0.2225 & 0.3121 & -0.3121 & 0.4017 & -0.4017 & 0.4912 & -0.4912 \\ -0.1033 & 0.2999 & -0.2225 & 0.2226 & -0.3121 & 0.3121 & -0.4017 & 0.4017 & -04912 & 0.4912 \\ 0.2226 & -0.2225 & 2.9468 & -2.7704 & 5.5850 & -5.5850 & 8.3715 & -8.3715 & 11.1580 & -11.1581 \\ -0.2225 & 0.2226 & -2.7704 & 2.9468 & -5.5850 & 5.5850 & -8.3715 & 8.3715 & -11.158 & 11.1581 \\ 0.3121 & -0.3121 & 5.5850 & -5.5850 & 15.3371 & -15.1836 & 25.2968 & -25.296 & 35.3849 & -35.385 \\ -0.3121 & 0.3121 & -5.5850 & 5.5850 & -15.183 & 15.3371 & -25.296 & 25.2968 & -35.385 & 35.3849 \\ 0.4017 & -0.4017 & 8.3715 & -8.3715 & 25.2968 & -25.2968 & 48.2162 & -48.085 & 71.7608 & -71.760 \\ -0.4017 & 0.4017 & -8.3715 & 8.3715 & -25.296 & 25.2968 & -48.0854 & 48.2162 & -71.760 & 71.7608 \\ 0.4912 & -0.4912 & 11.1580 & -11.158 & 35.3849 & -35.385 & 71.7608 & -71.760 & 116.622 & -116.514 \\ -0.4912 & 0.4912 & -11.158 & 11.1580 & -35.385 & 35.3849 & -71.760 & 71.7608 & -116.514 & 116.622 \end{bmatrix}$$

In this way and using Eq. 9.20, the vector of internal forces of the excessive members can be obtained as

$$\mathbf{X} = \{-84.7178, -84.7178, -85.0153, -85.0153, -106.172, -106.172, -133.071, -133.071, -163.915, -163.915\}^t$$

Adding X to the external load vector using Eq. 9.21 and applying the load to the regular structure in Eq. 9.22, the vector displacements for the near-regular structure is obtained. The nodal displacements in some DOFs are as follows:

$$\boldsymbol{\Delta} = \left[\overset{1}{\ldots} \;\; \overset{61}{9513.052} \;\; \ldots \;\; \overset{121}{31468.4} \;\; \ldots \;\; \overset{181}{65318.71} \;\; \overset{240}{\ldots} \right]^t$$

The deformed shape of the structure is shown in Fig. 9.19.

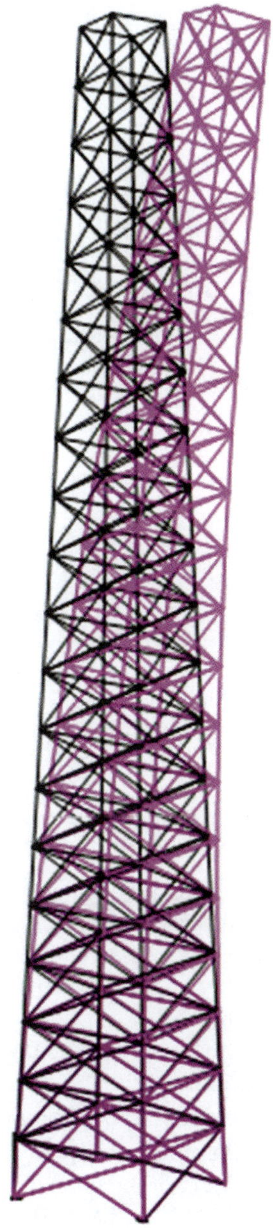

Fig. 9.19 A communication transmission tower together with its deformation

For solution of this structure using a conventional stiffness method we have to find the inverse of 240×240, while the present approach requires the inverse of four matrices of dimension 60×60 and the inverse of the matrix $\mathbf{D}_1$ of dimension 10×10 to complete the analysis of the near-regular structure.

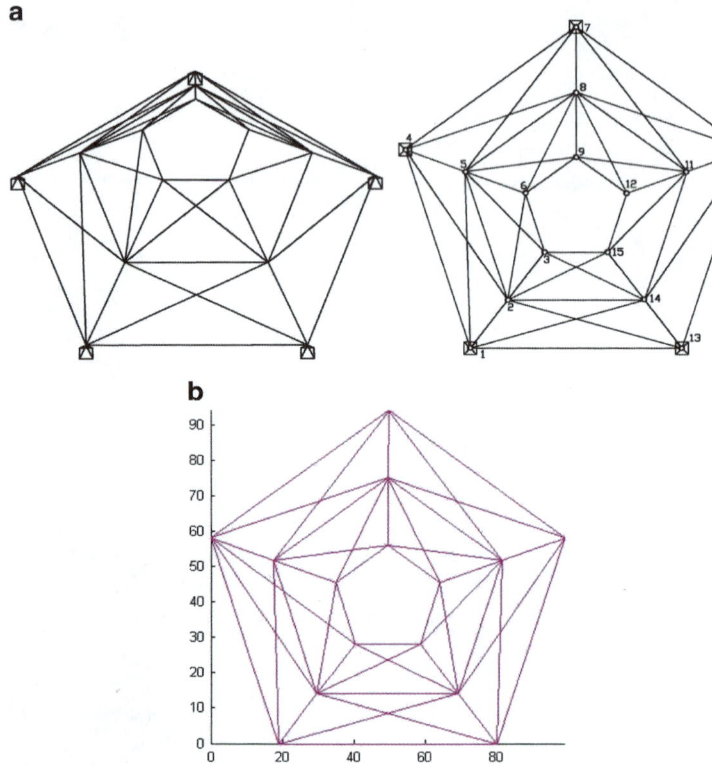

Fig. 9.20 (**a**) Two and three dimensional representations of the 43-bar structure. (**b**) The corresponding regular structure [1]

Example 3. Consider a 43-bar truss structure shown in Fig. 9.20a. This structure becomes a cyclically symmetric structure by addition of two members between the nodes 12 and 14, and nodes 11 and 13.

A pair of members with identical geometry and equal modulus of elasticity having different signs, are added where we have lack of members for regularity. In the next step the members with negative modulus of elasticity are considered as excessive members are separated from the structure. The external forces consist of $P_z^2 = P_y^8 = 10$ N. For all the member we consider EA = 1 N.

Forming the equilibrium matrix $\mathbf{A}_T$ of the near-regular structure according to Eq. 9.16, which is of dimension 30×47, and its partitioning by employing Eq. 9.11, the matrices **A** and **N** of dimensions 30×45 and 30×2 are obtained. It should be noted that the matrix $\mathbf{A}_T$ corresponds to the near-regular structure which has both members of positive and negative modulus of elasticity.

$$\mathrm{N}=\begin{bmatrix} \overset{1}{0} & \ldots & \overset{19}{0.0248} & 0.8650 & 0.5010 & 0 & 0 & 0 & 0 & 0 & \overset{27}{0} & \ldots & \overset{30}{0} \\ 0 & \ldots & 0 & 0 & 0 & -0.1375 & 0.8347 & 0.5332 & 0.1375 & -0.8347 & -0.5332 & \ldots & 0 \end{bmatrix}^{t}$$

The matrix S corresponding to Fig. 9.20b can be expressed as

$$\mathbf{S}=\sum_{i=1}^{5}(\mathbf{P}_i \otimes \mathbf{A}_i)$$

Thus as we observed in Example 1, the inverse of the stiffness matrix which is of dimension 30×30 can be calculated by evaluating the eigenvalues and eigenvectors of five matrices of dimension 6×6. The diagonal matrix **s** is of dimension 45×45 containing the stiffness matrices of the members of the structure shown in Fig. 9.20b. The diagonal matrix **F** is of dimension 47×47 containing the flexibility matrices of the members of the regular structure together with the excessive members.

Using Eq. 9.15 one can easily form the $\mathbf{B}_0$ and $\mathbf{B}_1$ matrices which are of dimension 47×30 and 47×2. Having the matrices $\mathbf{B}_0$, $\mathbf{B}_1$ and $\mathbf{F}$, the matrices $\mathbf{D}_1$ and $\mathbf{D}_2$ of dimensions 2×30 and 2×2 are obtained from Eq. 9.19. Here $\mathbf{D}_2$ is as follows:

$$\mathbf{D}_2=\begin{bmatrix} -19.4620 & 1.8186 \\ 1.8186 & -10.7629 \end{bmatrix}$$

Equation 9.20 can be employed to find the vector X as:

$$\mathbf{X}=-\mathbf{D}_2{}^{-1}\mathbf{D}_1\mathbf{P}$$
$$=-\begin{bmatrix} -19.4620 & 1.8186 \\ 1.8186 & -10.7629 \end{bmatrix}^{-1}\begin{bmatrix} \ldots & \overset{3}{0.3207} & \ldots & \overset{14}{1.3854} & \ldots \\ \ldots & -0.0640 & \ldots & -0.3018 & \ldots \end{bmatrix}\begin{bmatrix} \overset{1}{0} & \ldots & \overset{3}{10} & \ldots & \overset{14}{10} & \ldots & \overset{30}{0} \end{bmatrix}^{t}$$

$$\mathbf{X}=\begin{bmatrix} 0.8584 \\ -0.1949 \end{bmatrix}$$

Using Eq. 9.21 the equivalent external forces of the regular structure are obtained, and using the inverse of S one can easily find the nodal displacements of the near-regular structure by Eq. 9.22.

$$\mathbf{P}^*=\begin{bmatrix} \overset{1}{0} & \ldots & \overset{3}{10} & \ldots & \overset{14}{10} & \ldots & \overset{19}{0.0213} & 0.7426 & 0.4301 & 0.0268 & -0.1627 & -0.1039 & -0.0268 & 0.1627 & \overset{27}{0.1039} & \ldots & \overset{30}{0} \end{bmatrix}^{t}$$

$$\boldsymbol{\Delta}=\begin{bmatrix} \overset{1}{-105.512} & \overset{2}{-134.943} & 380.134 & 51.383 & \ldots & -18.489 & 50.4185 & \overset{29}{-75.205} & \overset{30}{116.529} \end{bmatrix}^{t}$$

The internal forces of the regular structure and the excessive members can be found using Eq. 9.14 as

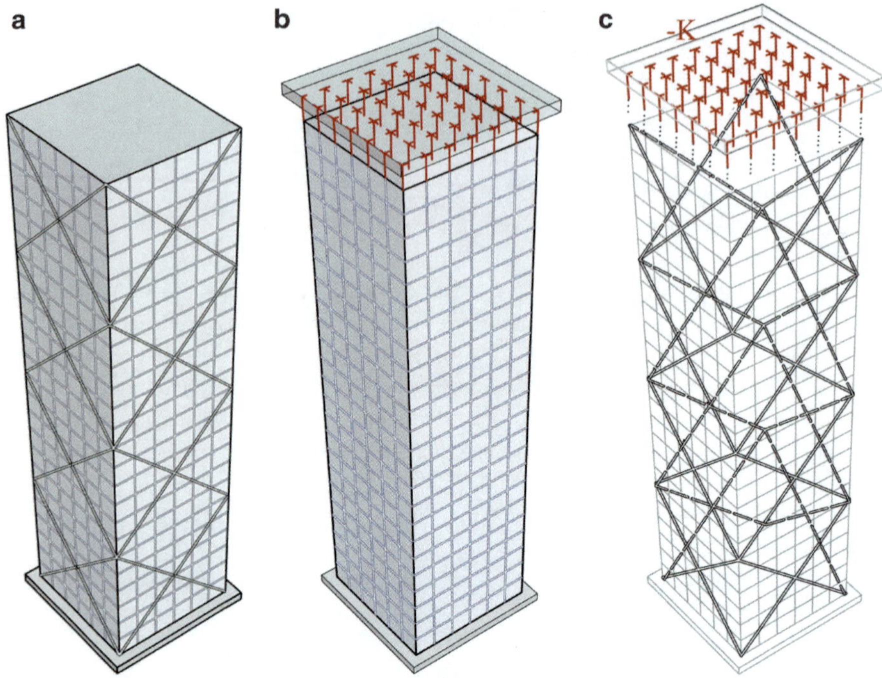

Fig. 9.21 (**a**) A 24-story irregular frame with bracing. (**b**) The regular part of the irregular frame with fictitious columns of positive modulus of elasticity being added. (**c**) The bracing part consisting of 32 bracing elements and 49 fictitious bending elements with negative modulus of elasticity being added as shown at the *top* of the structure [1]

$$\mathbf{Q} = \left[\overset{1}{6.8906}\ \ \overset{2}{-0.9281}\ \ \overset{3}{1.7904}\ \ \overset{5}{-0.950}\ \ \overset{6}{-5.60}\ \ \ldots\ \ \overset{43}{-0.0003}\ \ \overset{44}{0.8584}\ \ \overset{45}{-0.1949}\ \ \overset{46}{0.8584}\ \ \overset{47}{-0.1949}\right]^t$$

As it can be seen, the internal forces in members 44, 46 and 45, 47 which are the added pairs of members are the same as the entries of **X**.

For solution of this near-regular structure using a conventional stiffness method we have to find the inverse of 30 × 30, while the present approach requires the inverse of 5 matrices of dimension 6 × 6 and the inverse a matrix of dimension 2 × 2 to complete the analysis of the near-regular structure .

Example 4. A 24-story 3D frame is shown in Fig. 9.21a, with 49 columns and 84 beam in each story. The dimensions of all the beams and columns are assumed to be identical in all the stories. In each face of the building 8 bracing elements are added to increase the stiffness of the structure. Naturally these elements make the model irregular. Here using the presented method, the bracing elements are decomposed from the structure the analysis is performed for two separate parts, namely the regular bending frame and the excessive bracing elements.

If the top part of the structure is fixed similar to the bottom part, then the bending frame structure can be easily analyzed using the method presented in [3]. Therefore here we consider pairs of bending elements with positive and negative modulus of elasticity at end part of the structure, similar to the columns of the other stories, as illustrated in Fig. 9.21b. The columns are connected to the top part of the structure and fixed at the other ends. The elements with negative modulus of elasticity are considered as excessive members and are separated from the structure. Therefore the excessive members consist of 32 bracing elements and 49 bending elements with negative modulus of elasticity. These elements are highlighted in Fig. 9.21c. In this way, the regular structure consists of a 24 story frame together with additional bending elements with positive modulus of elasticity as illustrated in Fig. 9.21b.

The required parameters for the analysis are as $k = 7,056$, $t = 326$, and $e = 19,446$. Using Eqs. 9.16 and 9.17 and employing the rotation and stiffness matrices of the elements in their local coordinate systems, the matrix $\mathbf{A}_T$ of dimension 7056×19772 can be constructed. For the formation of the stiffness matrices of the elements, the local coordinate systems should be selected such that the form given in Eq. 9.17 is formed. For each bending elements, six internal forces, and for bracing elements only one axial force are assumed. It should be noted that the fictitious elements with $\pm$ modulus of elasticity contribute in the formation of this matrix.

By partitioning the matrix $\mathbf{A}_T$ we obtain two matrices $\mathbf{A}$ and $\mathbf{N}$ having dimensions 7056×19446 and 7056×326, respectively. Since the regular part contains 3,241 bending elements, thus the unassembled stiffness matrix s is of dimension 19446×19446. The assembled matrix $\mathbf{S}$ of the regular part has dimension 7056×7056. Utilizing the method of Ref. [3], the inverse of this matrix can easily be obtained calculating the eigenvalues of 24 matrices of dimension 294×294.

In this way forming the inverse of the stiffness matrix and using Eq. 9.8, the matrix $\mathbf{R}$ of dimension 19446×7056 can be obtained. Having this matrix the matrices $\mathbf{B}_0$ and $\mathbf{B}_1$ of dimensions 19772×7056 and 19772×326 will be formed using Eq. 9.15. The flexibility matrix $\mathbf{F}$ contains the flexibility of all the elements of the near-regular structure (bending elements, bracing elements, and pair of fictitious elements with + and − signs). This matrix is a block matrix such that for the bending members blocks are 6×6 and for the bracing members the blocks are 1×1. Thus the dimension of F is 19772×19772.

With help of Eq. 9.19 the matrices $\mathbf{D}_1$ and $\mathbf{D}_2$ of dimensions 326×7056 and 326×326 are obtained, respectively. Using Eq. 9.20 and finding the inverse of $\mathbf{D}_2$ leads to the vector of unknown $\mathbf{X}$ of dimension 326×1. Substituting this in Eq. 9.21, the equivalent external force vector of the regular part of the structure is obtained. Multiplying the inverse of the stiffness matrix of the regular part, the displacement vector of the near-regular structure of dimension 7056×1 is obtained.

It can be observed that the analysis of the problem with the help of this method for frame structures is the same as that of the trusses which were discussed in the previous examples, with the only difference that the rotation and stiffness matrices for the bending elements in the local coordinate systems should be defined according to the Eq. 9.17.

In this problem instead of inverting the stiffness matrix of dimension 7056×7056 in direct analysis of the near-regular structure, one needs to find the inverse of the matrix $\mathbf{D}_2$ of dimension 326×326, and calculate the eigenvalues of 24 matrices of dimension 294×294. This shows the efficiency of the present method. Obviously increasing the number of stories this efficiency will become more apparent. In other words in this method a matrix of dimension 7056 is decomposed into 24 matrices of dimension 294.

References

1. Kaveh A, Rahami H, Mirghaderi SR, Ardalan Asl M (2013) Analysis of near-regular structures using the force method. Eng Comput 30:21–48
2. Imrich W, Klavžar S (2009) Product graphs; structure and recognition. Wiley, New York
3. Kaveh A (2013) Optimal analysis of structures by concepts of symmetry and regularity. Springer/GmbH, Wien/New York
4. Sabidussi G (1960) Graph multiplication. Math Z 72:446–457
5. Gross J, Yellen J (1998) Graph theory and its applications. CRC Press, New York
6. Weichsel PM (1962) The Kronecker product of graphs. Proc Am Math Soc 13:47–52
7. Argyris JH, Kelsey S (1960) Energy theorems and structural analysis. Butterworth, London
8. Kaveh A, Rahami H (2011) Block circulant matrices and applications in free vibration analysis of cyclically repetitive structures. Acta Mech 217:51–62

Chapter 10
Simultaneous Analysis, Design and Optimization of Structures Using Force Method and Supervised Charged System Search

10.1 Introduction

Developing methods with higher computation efficiency is a crucial subject in advanced engineering problems of multi-physics nature. For instance, analyzing structures with larger number of members requires larger memory size and longer computation time. In addition, this costly computation has to be repeated many times, typically over 5,000 times, because the cross section size of the members is not determined in the early stages of designing such structures. Therefore, reducing the size of structural matrices and eliminating the unduly repetitions in the design and analysis procedures can lead to a considerable reduction in the computation efficiency [1, 2]. In this chapter, this goal is achieved utilizing meta-heuristics algorithms which minimize the energy function indirectly. Besides, design procedure and minimizing the weight of the structure is added to the analysis procedure. One of the most reliable meta-heuristic methods recently developed is Charged System Search (CSS) [3, 4], that is used in here. In this chapter, supervisor agents are considered to increase the exploration ability of the CSS algorithm. This method is called supervised CSS abbreviated as SCSS. Also a new formulation of the penalty function is made to improve the performance of the supervised CSS.

Designing structures with minimum weight can be achieved by using minimum energy methods, and members with pre-defined stress ratios [5], instead of the direct solution of classic equations. This results in avoiding not only the repetitive computations in the design and analysis, but also avoiding the computation of the solution of equations with large matrices. For this purposed, one needs to formulate the equations based on the minimum energy principle, and employ them in an efficient optimization algorithm. Combining the SCSS algorithm and the force method provides a suitable means for this purpose. The former is a suitable optimization algorithm and the latter can be used to derive the energy equations.

In the first part of this chapter, supervisor agents are introduced. In the second part energy formulation based on the force method is derived and the supervised SCSS algorithm is applied to the analysis procedure. In the third part, using the

A. Kaveh, *Computational Structural Analysis and Finite Element Methods*, 407
DOI 10.1007/978-3-319-02964-1_10,

SCSS and prescribed stress ratios, structures are analyzed and designed, and finally in the last part weight minimization is performed by imposing the analysis procedure as a constraint to the SCSS. In recent years the CSS has been applied successfully to many engineering optimization problems. For optimal design of structures, CSS has performed very well and improved all of the resulted design parameters and weights achieved by the other algorithms. Large-scale structures are analyzed and designed in this chapter in order to show the accuracy of the method when applied to different kinds of structures.

10.2 Supervised Charged System Search Algorithm

In the CSS algorithm, each vector of variables is an agent that moves through the search space and finds the minimal solutions [3, 4]. Throughout the search process, an agent might go to a coordinate in the search space that already has been searched by the same agent or another. If this coordinates have a good fitness, it will be saved in the Charged Memory [3] but if this coordinate does not have a good fitness, it will not be saved anywhere. Therefore, this step of the search process becomes redundant. This unnecessary step adversely affects the exploration ability of the algorithm. In this chapter, the supervisor agents are introduced to improve the exploration ability of the CSS algorithm. The supervisor agent is an independent agent of constant values that repels the agent if its coordinate has a bad fitness or attracts the agents if its coordinate has a good fitness. This procedure is repeated in all of the iterations and gives an overall view of the search space. The number of supervisor agents is selected at the beginning of the algorithm, and then their constant coordinates in the search space are determined as follows:

$$xs_{j,i} = \frac{(i-1)\left[x_{max,j} - x_{min,j}\right]}{NOSA - 1} + x_{min,j} \tag{10.1}$$

where NOSA is the number of supervisor agents, and $xs_{j,i}$ is the jth variable of the ith supervisor agent; $x_{min,j}$ and $x_{max,j}$ are the minimum and the maximum limits of the jth variable. The kind of the force for these agents is determined as

$$p = \log\left(\frac{\overline{fit}}{fit_i}\right) \tag{10.2}$$

where p is the same as the parameter in the original version of the CSS [3], fit_i is equal to the fitness value of the ith supervisor agent and $\overline{fit}$ is the average value of the fitness of the normal agents. Calculating other properties of the supervisor agents such as force and radius are similar to the standard CSS algorithm [3]. Supervisor agents do not move from their coordinate determined from Eq. 10.1, yet they apply additional forces on the normal agents. By doing so, they determine the fitness values of their fixed coordinate and its neighborhood, resulting in a better exploration ability of the CSS algorithm.

10.3 Analysis by Force Method and Charged System Search

In the presented approach, force method is applied to analyze structures. Since this method leads to less number of unknowns, it is preferred to displacement method. In the force method, the redundant forces are unknowns, whereas in the displacement method, the nodal displacements are unknowns. In this method [1, 2, 5], the energy relationships of the structure that satisfies the compatibility, force-displacement and equilibrium conditions are derived, and then, minimized using the SCSS. Suppose $\{\mathbf{p}\} = \{p_1,p_2,\ldots,p_n\}^{\mathbf{t}}$ is the vector of nodal forces, $\{\mathbf{q}\} = \{q_1, q_2,\ldots,q_n\}^{\mathbf{t}}$ is the vector of redundant forces, and $\{\mathbf{r}\} = \{s_1,s_2,\ldots,s_m\}^{\mathbf{t}}$ comprises of the internal forces of the members. Equilibrium condition results in the following equation [1, 2]:

$$\mathbf{r} = \mathbf{B_0 p} + \mathbf{B_1 q} = [\mathbf{B_0} \quad \mathbf{B_1}]\begin{bmatrix} \mathbf{p} \\ \mathbf{q} \end{bmatrix} \tag{10.3}$$

In addition, the complementary energy function is:

$$\mathrm{U^c} = \frac{1}{2}\,\mathbf{r^t F_m r} \tag{10.4}$$

where $[\mathbf{F_m}]$ is the unassembled flexibility matrix of the structure. According to the Castigliano's principle, a group of the redundant forces that minimize the complementary energy function is the exact solution that satisfies compatibility condition. By substituting $\{\mathbf{r}\}$ from Eq. 10.3 in Eq. 10.4, the following equation obtained:

$$\mathrm{U^c} = \frac{1}{2}[\mathbf{p^t} \quad \mathbf{q^t}][\mathbf{H}]\begin{bmatrix} \mathbf{p} \\ \mathbf{q} \end{bmatrix} \tag{10.5}$$

where $[\mathbf{H}] = [\mathbf{B_0} \quad \mathbf{B_1}]^{\mathbf{t}}[\mathbf{F_m}][\mathbf{B_0} \quad \mathbf{B_1}]$. Decomposing matrix $[\mathbf{H}]$ into four submatrices leads to:

$$\mathrm{U^c} = \frac{1}{2}\left(\{\mathbf{p}\}^{\mathbf{t}}[\mathbf{H_{pp}}]\{\mathbf{p}\} + \{\mathbf{p}\}^{\mathbf{t}}[\mathbf{H_{pq}}]\{\mathbf{q}\} + \{\mathbf{q}\}^{\mathbf{t}}[\mathbf{H_{qp}}]\{\mathbf{p}\} + \{\mathbf{q}\}^{\mathbf{t}}[\mathbf{H_{qq}}]\{\mathbf{p}\}\right) \tag{10.6}$$

In the classical method, the derivative of $\mathbf{U^c}$ in terms of $\{\mathbf{q}\}$ is calculated and is equated to zero leading to:

$$\{\mathbf{q}\} = -[\mathbf{H}_{qq}]^{-1}[\mathbf{H}_{qp}]\{\mathbf{p}\} \tag{10.7}$$

Since **[H]** is symmetric, $[\mathbf{H}_{qp}]^t = [\mathbf{H}_{pq}]$, Ref. [5].

Accordingly, in the classical method the inverse of $[\mathbf{H}_{qq}]$ needs to be calculated. This is a difficult task, and requires extensive computer memory, especially in the case of large scale structures. Therefore, finding **{q}** that minimizes the complementary energy without calculating the inverse of $[\mathbf{H}_{qp}]$ reduces the computation time and computer memory. The first term of Eq. 10.6 is constant and the second and third terms are equal. It can be shown that the third and fourth terms of $\mathbf{U}^c$ are symmetric. Therefore

$$F_u = \{\mathbf{q}\}^t[\mathbf{H}_{qp}]\{\mathbf{p}\} \tag{10.8}$$

is the equation that should be minimized [5].

Enhanced Charged System Search [4] is used to minimize Eq. 10.8. In this part, the force method analysis is applied to different types of structures to illustrate the performance of the method.

Case Study 1. The first example is an 11-member truss with three degrees of statical indeterminacy, as shown in Fig. 10.1. Consequently, the energy function includes three variables.

The classical method that calculates the exact and minimum amount of U^c leads to 419.8475, whereas, using the present approach with CSS, $U^c = 419.8476$ is obtained and **{q}** is calculated as:

$$\{\mathbf{q}\} = \{4.6394 - 3.7629\ 8.1900\}^t$$

The optimization history is shown in Fig. 10.2. The number of agents is selected as 20.

Case Study 2. The second example is an unbraced planar frame with constant EI having 36° of statical indeterminacy, as shown in Fig. 10.3. In this example, the axial force, shear and moment in the first node of the beams are considered as the redundant forces. As a result, the energy function includes 36 variables. Note that only the bending energy is considered as the energy of the frame. Loading condition is considered as:

1. A load −10 kN in the y-direction at nodes 8–11,
2. A load 10 kN in the x-direction at nodes 8–11,
3. A bending moment 10 kN.m in the x-y surface at nodes 8–11.

The exact calculation of U^c leads to 1,234.8; while it is $U^c = 1{,}249.2$ utilizing the CSS algorithm. Figure 10.4 shows the variation of $\mathbf{F}_U$ versus the number of iterations. As shown above, there is a very close agreement between the exact and the calculated value for the energy function, verifying the accuracy of the algorithm. In this case, the redundant forces are obtained as follows:

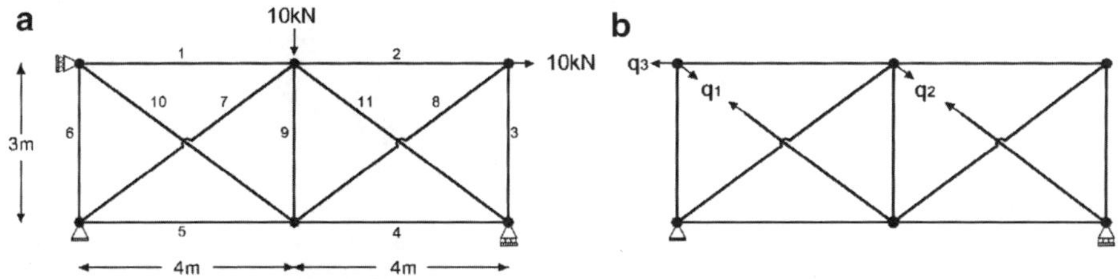

Fig. 10.1 A simple truss and the selected basic structure (Case Study 1): (**a**) A planar truss. (**b**) The selected basic structure

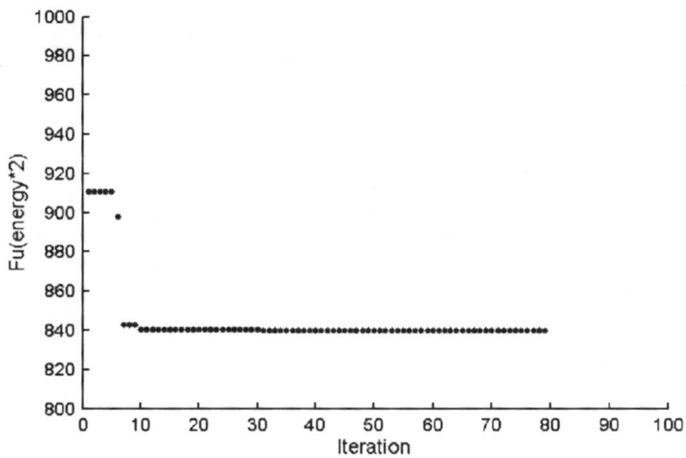

Fig. 10.2 Variation of F_U versus the number of iterations in the 11-member truss (Case Study 1)

$$\{\mathbf{q}\} = \{1.1275, 5.3155, 14.0096, 2.4854, 4.8316, 12.0549, 4.0405, 4.2845, 10.7913, -3.0551, 1.2459, 2.9740, -4.0016, 1.3874, 3.2303, 5.5762, 1.4122, 1.3221, 0.0660, 0.2315, 0.4707, 0.1680, 0.2155, 0.4678, 0.4265, 0.1987, 0.2503, -0.1444, 0.0425, -0.0728, 0.0540, 0.0052, 0.0351, 0.0373, 0.0847, 0.0901\}^t$$

Case Study 3. In the third example, a 40-element grilling system is considered to illustrate the accuracy of the force method and CSS in analyzing space frames. Geometry, nodal loads and basic structure are shown in Fig. 10.5. Torsion and shear in z direction, and moment around the axis with a greater moment of inertia in each member are considered as redundant forces.

Both the torsion and bending energies are considered as energy function in this structure. G, I and E are constant for members and the Poisson's ratio (υ) is considered 0.3. The cross-sections of members are considered to be 272 W-section as given in LRFD-AISC. Using the least square regression, the polar moment of inertia (J) is expressed as a function of the moment of inertia (I):

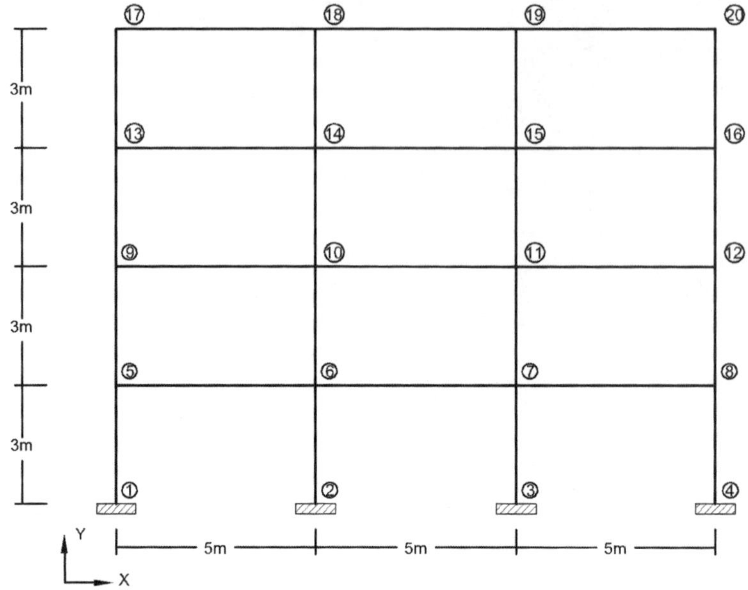

Fig. 10.3 An unbraced planar frame (Case Study 2)

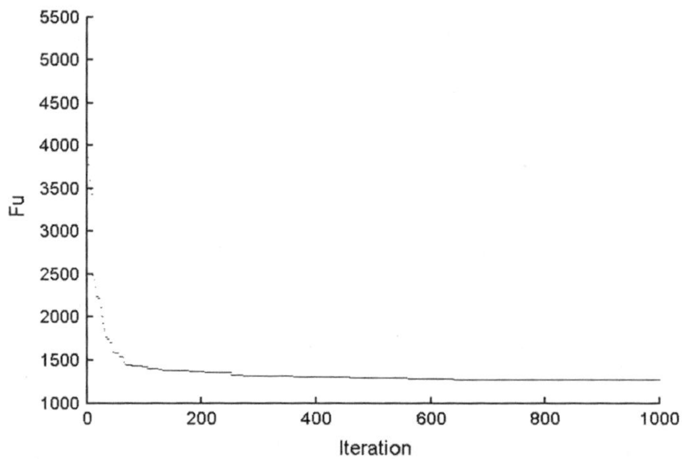

Fig. 10.4 Variation of $\mathbf{F}_U$ versus the number of iterations in the unbraced planar frame analysis (Case Study 2)

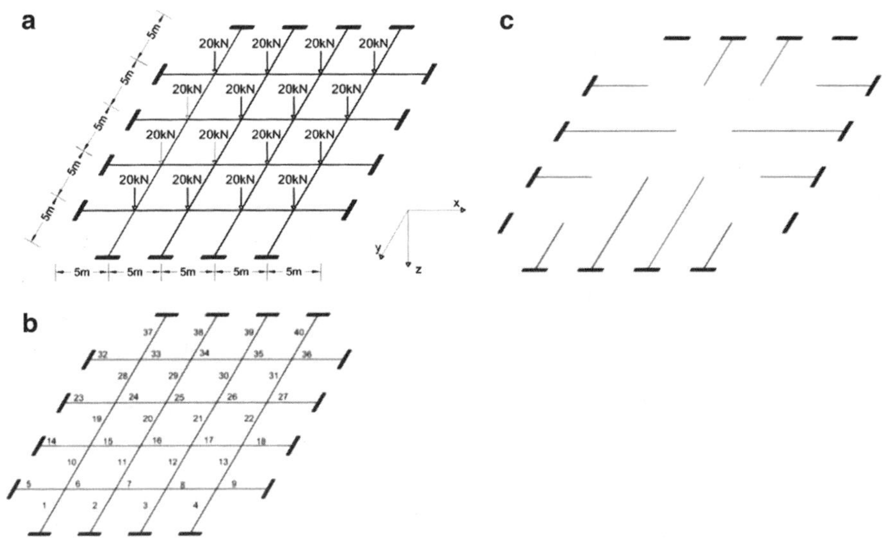

Fig. 10.5 A 40-element grillage (Case Study 3). (**a**) Geometry. (**b**) Node and element ordering. (**c**) Basic structure

$$J = 1.04I \tag{10.9}$$

Also

$$E = 2G(1 + \upsilon) \tag{10.10}$$

By substituting Eqs. 10.9 and 10.10 in $[\mathbf{F_m}]$, the energy function is derived. The exact calculation of energy using the classical method leads to 170,840, whereas, using the present approach $U^c = 177{,}460$ is obtained. The redundant forces, $\{\mathbf{q}\}$, are shown in Table 10.1.

Case Study 4. The Last example of this part is a 26-story tower with 246° of statical indeterminacy selected from Ref. [6], as shown in Fig. 10.6a, b. The energy function has 246 unknowns. The cross section and module of elasticity for all of the elements are considered constant and equal. Geometry and basic structure is shown in Fig. 10.6c.

The loading on the structure consists of:

1. The vertical load at each node in the first section is equal to −3 kips (−13.344 kN)
2. The vertical load at each node in the second section is equal to −6 kips (−26.688 kN)
3. The vertical load at each node in the third section is equal to −9 kips (−40.032 kN)

Table 10.1 The calculated redundant forces of 40-element grilling system (Case Study 3) × 10^4

q_1	−0.914	q_{19}	−0.0084	q_{37}	−0.0312	q_{55}	0.7119
q_2	0.2167	q_{20}	−1.8335	q_{38}	−5.1336	q_{56}	−4.0377
q_3	−3.9005	q_{21}	0.7346	q_{39}	−0.0287	q_{57}	−0.2541
q_4	−0.6323	q_{22}	−1.0314	q_{40}	0.5316	q_{58}	0.0398
q_5	0.314	q_{23}	3.6083	q_{41}	−1.9493	q_{59}	−6.1707
q_6	−0.3381	q_{24}	0.0769	q_{42}	0.0136	q_{60}	2.1362
q_7	0.1307	q_{25}	−0.0497	q_{43}	−0.0397	q_{61}	0.1051
q_8	−0.0469	q_{26}	0.0678	q_{44}	0.0061	q_{62}	−3.0445
q_9	2.8322	q_{27}	5.0685	q_{45}	4.5725	q_{63}	1.9832
q_{10}	0.4806	q_{28}	1.0572	q_{46}	−0.2432	q_{64}	−0.0718
q_{11}	−0.3335	q_{29}	−0.1714	q_{47}	−1.6436	q_{65}	0.2401
q_{12}	2.1219	q_{30}	5.5207	q_{48}	0.296	q_{66}	1.3579
q_{13}	−0.7939	q_{31}	0.4753	q_{49}	1.2002	q_{67}	0.0941
q_{14}	0.2277	q_{32}	−4.0345	q_{50}	−5.6626	q_{68}	−2.4965
q_{15}	3.3177	q_{33}	0.0442	q_{51}	0.1194	q_{69}	−0.2361
q_{16}	0.1725	q_{34}	−0.3564	q_{52}	1.1286	q_{70}	−0.8848
q_{17}	−1.4645	q_{35}	−3.7443	q_{53}	−5.547	q_{71}	−3.9475
q_{18}	−0.8168	q_{36}	0.055	q_{54}	−0.17	q_{72}	0.2642

4. The horizontal load at each node on the right side in the x direction is equal to −1 kips (−4.448 kN)
5. The horizontal load at each node on the left side in the x direction is equal to 1.5 kips (6.672 kN)
6. The horizontal load at each node on the front side in the y direction is equal to −1 kips (−4.448 kN)
7. The horizontal load at each node on the back side in the y direction is equal to 1 kips (4.448 kN)

In this example, the exact calculation of the energy function leads to 1.8008 × 107, and it is obtained as 1.8252 × 107 using the force method and CSS that is very close to the exact value.

10.4 Procedure of Structural Design Using Force Method and the CSS

In this section, design and optimization procedures are added to the analysis presented in the previous section. There are two major approaches to formulate the objective function in the simultaneous analysis and design of an optimal structure:

1. Using the pre-selected stress ratio.
2. Minimizing the structure weight.

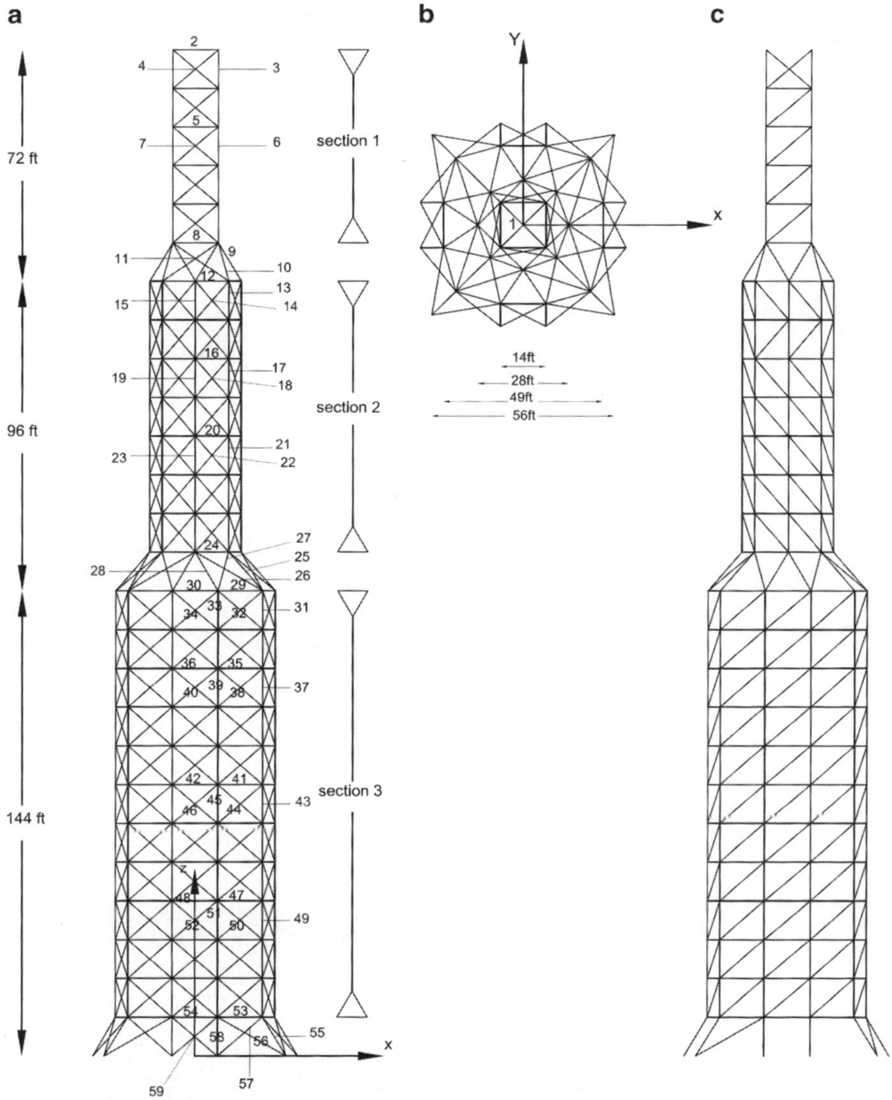

Fig. 10.6 A 26-story tower. (**a**) Geometry and grouping. (b) Top view. (**c**) Basic structure (Case Studies 4 and 10)

10.4.1 Pre-selected Stress Ratio

In this approach [5], a preselected stress ratio is assumed for each member, and then the complementary energy is minimized as the objective function. If the cross sections $\mathbf{A_i}$ (i = 1,. . .,m) are known, then the analysis can be performed using a meta-heuristics method such as CSS, described in the Sect. 3.

However, usually the cross sectional areas are not determined at the beginning of the design procedure. This problem leads to a new formulation of the complementary energy that eliminates $\mathbf{A_i}$ (i = 1,...,m) from the energy function [5].

Each agent in the CSS is a vector of redundant forces. Moreover, according to Eq. 10.3, the internal forces of members, $\{\mathbf{r}\}$, is obtained from the selected agents. The ratio between the stress in each member (σ_i) and its corresponding allowable stress (σ_a) is defined as **C**:

$$C = \frac{\sigma_i}{\sigma_a} \tag{10.11}$$

where $\sigma_i = \frac{r_i}{A_i}$. By substituting σ_i in Eq. 10.11, the cross section area of each member is obtained in terms of the internal force r_i, stress ratio **C**, and the allowable stress σ_a

$$A_i = \frac{r_i}{C\sigma_a} \tag{10.12}$$

Consequently, one can express the unassembled flexibility matrix of each member as a function of L, E, q and C as follows:

$$\mathbf{F_m} = \frac{L}{EA} = \frac{1}{Ef(\mathbf{r}, L, C)} = g(\mathbf{q}, C, L, E) \tag{10.13}$$

Substituting $\mathbf{F_m}$ in Eq. 10.4, leads to the elimination of $\mathbf{A_i}$ from the formulation of the complimentary energy:

$$\mathrm{Min}U^c = \frac{1}{2E}[\mathbf{p} \quad \mathbf{q}]^t[\mathbf{B_0} \quad \mathbf{B_1}]^t[\mathbf{g}(\mathbf{q}, \mathbf{C}, \mathbf{L})][\mathbf{B_0} \quad \mathbf{B_1}][\mathbf{p} \quad \mathbf{q}] \tag{10.14}$$

Pre-selected stress ratio is a parameter controlling the weight of the structure and stress constraint, simultaneously. Therefore, by minimizing the energy function in the analysis procedure, weight optimization and stress constraints satisfaction are fulfilled.

Case Study 5. As an example consider the truss shown in Fig. 10.7. This truss is designed with the constraints explained in Table 10.2 and using Eq. 10.14 as the objective function. In this example, two cases are considered. In case I, the stress ratios of the members is different, whereas in case II, it is assumed to be constant for all the members. For the sake of simplicity, the cross-sections are selected as hollow squares, as shown in Fig. 10.8. In this example, a population of 20 agents is considered in the CSS algorithm. The magnitude of A_i is determined considering the selected values of C_i. Enhanced CSS with supervisor agent is utilized in the simultaneous analysis and design of this structure and the results are shown in Tables 10.3 and 10.4. The convergence history is shown in Fig. 10.9. To verify the efficiency of the present method and combining the CSS algorithm and force

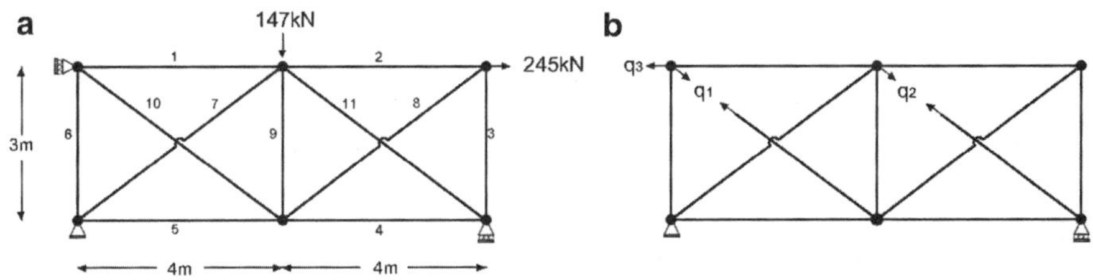

Fig. 10.7 A simple truss with pre-selected stress ratios (Case Study 5). (**a**) Geometry. (**b**) Basic structure

method in minimizing the structural weight, the design parameters and redundant forces obtained from CSS, are compared to those computed using the Genetic Algorithm (GA), reported by Kaveh and Rahami [5]. The comparison results are shown in Tables 10.3 and 10.4 for Case I and Case II, respectively.

In this example, the exact calculation of the energy function leads to 6.5989×10^5, and it is obtained as 6.6056×10^5 using the force method and CSS for case I. Besides, the exact calculation of the energy function leads to 7.5368140×10^5, and it is obtained as 7.5368147×10^5 using the force method and CSS for case II. The close agreement between these values verifies the accuracy of the calculated redundant forces shown in Tables 10.3 and 10.4 for case I and case II, respectively. Also variation of F_U versus the iteration is shown in Fig. 10.9.

10.4.1.1 Fully Stress Design (FSD) for Statically Indeterminate Structures

In this part, the presented CSS and force method is applied to an Optimally Criteria Method (OCM), namely Fully Stress Design (FSD). FSD leads to a correct optimal weight for statically determinate structures under a single load condition. In the FSD all the members are supposed to be subjected to their maximal allowable stresses [5]. Achieving such a design for an indeterminate structure with fixed geometry is not always possible. Even by changing the geometry, a FSD may not be achieved. Here a formulation presented by Kaveh and Rahami [5] is used to indirect analysis in the process of optimization. This formulation can be applied to all types of structures, however, a truss with the following strain energy is considered:

$$U^c = \sum \frac{P^2L}{EA} = \sum \frac{\gamma P^2LA}{\gamma EA^2} = \frac{1}{\gamma E}\sum \sigma_i^2 w_i \tag{10.15}$$

It should noted that for constant E and γ, the minimum weight can be achieved only when the stresses in all the members are identical. Therefore, in Eq. 10.15, the term corresponding with the stresses, i.e. σ_i^2, may be moved out of the summation. On the other hand, in the design procedure, one can consider the fully stress

Table 10.2 Design data for the 11-bar planar truss (Case Study 5)

Design variables

Redundant and size variables q_1; q_2; q_3; A_1; A_2; A_3; A_4; A_5; A_6; A_7; A_8; A_9; A_{10}; A_{11}

Material and section property

Young's modulus is assumed to be constant

Density of the material: $\rho = 0.00277\ \text{kg/cm}^3 = 0.1\ \text{lb/in}^3$

$A = 0.4h^2, r = \sqrt{0.4A}$, thicknesst $= 0.1h$.

Constraint data

Stress ratios

Case 1: $C = \{0.9, 0.8, 0.85, 0.8, 0.9, 0.85, 0.95, 0.9, 0.8, 0.9, 0.95\}$

Case 2: $c_i = 1; i = 1, \ldots, 11$

For tensile members

$F_a \leq 0.6\ F_y$ and $\lambda_i \leq 300$

For compressive members

$\lambda_i \leq 200$

$$F_a = \frac{\left[\left(1 - \frac{\lambda_i}{2C_c^2}\right)F_y\right]}{\left(\frac{5}{3} + \frac{3\lambda_i}{8C_c} - \frac{\lambda_i^3}{8C_c^3}\right)} \quad \text{for} \quad \lambda_i \leq C_c$$

$$F_a = \frac{12\pi^2 E}{23\lambda_i^2} \quad \text{for} \quad \lambda_i \leq C_c$$

Stress constraints

$\sigma_i < 234.43$ MPa; $i = 1, \ldots, 11$

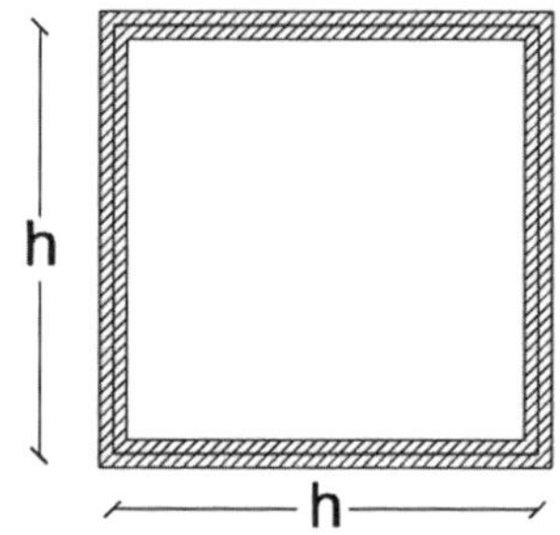

Fig. 10.8 A hollow square cross-section (Case Study 5)

Table 10.3 Optimal design comparison for the 11-bar truss (Case Study 5) (case 1)

Weight (N)

2,136.25

Size variable(cm^2)

A_1	A_2	A_3	A_4	A_5	A_6	A_7	A_8	A_9	A_{10}	A_{11}
11.55	13.36	41.20	4.44	4.44	42.51	6.94	9.15	61.02	9.71	17.51

Redundant variables $\times 10^3$ (N)

q_1	q_2	q_3
123.04	−5.04	244.69

Table 10.4 Optimal design comparison for the 11-bar truss (Case Study 5) (case 2)

Weight (N)										
1,914.84										
Size variable (cm²)										
A_1	A_2	A_3	A_4	A_5	A_6	A_7	A_8	A_9	A_{10}	A_{11}
11.55	13.36	41.20	4.44	4.44	42.51	6.94	9.15	61.02	9.71	17.51
Redundant variables ×10³ (N)										
q_1	q_2	q_3								
94.04	−0.0000541	198.66								

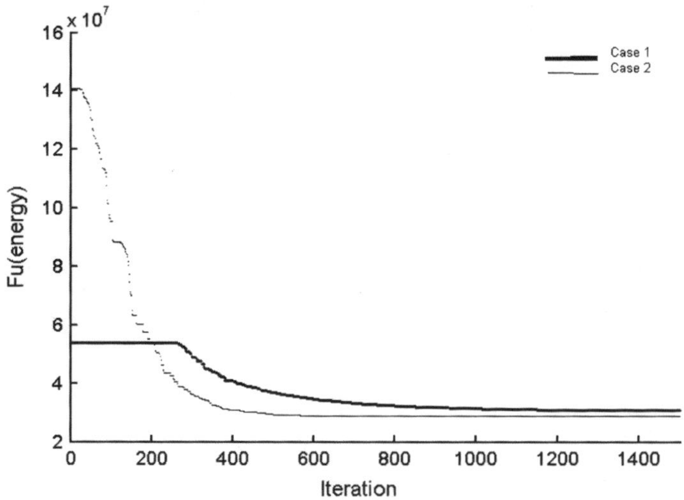

Fig. 10.9 Variation of F_U versus the iteration in the design procedure for the 11-member truss (Case Study 5)

constraint instead of minimum weight. This is because the minimum weight corresponds to a structure for that all the members are subjected to their maximum allowable stress.

Case Study 6. As an example, consider the structure shown in the Fig. 10.10, selected from Ref. [7]. The design and member size constraints are reported in Table 10.5. Redundant forces in this example are selected as internal forces in members 1 and 9. Twenty agents are selected in the CSS algorithm.

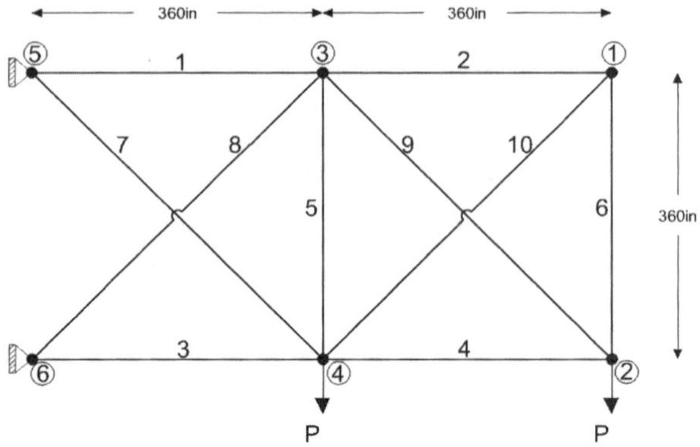

Fig. 10.10 A 10-bar truss example (Case Studies 6 and 7, Ref. [7])

Table 10.5 Design data for the 10-bar planar truss (Case Study 6)

Loading			
Node	Px: kips (kN)	Py: kips (kN)	Pz: kips (kN)
2	0	−100(−444.8)	0
4	0	100(−444.8)	0
Design variables			
Variables: q_1; q_2 (and A_1;A_2;A_3;A_4;A_5;A_6;A_7;A_8;A_9;A_{10} in case 3)			
Material property and constraint data			
Young's modulus: $E = 1e7$ psi $= 6.895e7$MPa			
Density of the material: $\rho = 0.1$ lb/in^3 $= 0.00277$ kg/cm^3			
For all members: $A_i \geq 0.1$ in^2; $i = 1, \ldots, 10$			
Stress constraints			
(a) FSD			
Case 1: $\lvert\sigma_i\rvert \leq 25$ ksi(172.375 MPa); $i = 1, \ldots, 10$			
Case 2: $\lvert\sigma_i\rvert \leq 25$ ksi; $i = 1, \ldots, 8, 10$ and $\lvert\sigma_9\rvert \leq 50$ ksi (344.75 MPa)			
(b) Weight minimization			
Case 3: $\lvert\sigma i\rvert \leq 25$ ksi; i $= 1, \ldots, 8, 10$ and $\lvert\sigma 9\rvert \leq 50$ ksi (344.75 MPa)			

10.5 Minimum Weight

In the second approach of simultaneous design and analysis of structures, the objective function is the weight of the structure, and the equilibrium, compatibility, and force/displacement conditions are the constraints. In summary, all these three conditions are called analysis criteria for simplicity. Other constraint such as stress, displacement, dynamical properties, and etc. can also be imposed to the fitness function. Penalty function is the most common approach to satisfying the constraints. The penalty function imposes a penalty to the fitness value of the solution, if the constraint is not satisfied:

$$\mathrm{f} = \mathrm{A} + \alpha B \tag{10.16}$$

In Eq. 10.16, f is the fitness value, A is the objective function and B is the penalty function and α is often selected as a big number. According to this equation, when B goes to zero and A goes to its minimum value, f goes to the minimum value of the fitness. However, since the minimum complementary energy is not zero, this form of penalty function cannot be used. In this case, W is minimum while the corresponding U^c is not minimum, i.e. the structure is not analyzed yet. Also a small value of α does not guarantee the minimum value of the B. On the other hand, in a structure that is in equilibrium and compatibility state, sum of the complementary energy U^c and the strain energy **U** is zero. Therefore, instead of the complementary energy, the sum of the complementary energy and the strain energy is used as the analysis criteria and is imposed to CSS as a constraint. The strain energy is a function of nodal displacements as follows:

$$\{\mathbf{d}\} = [\mathbf{B_0}]^{\mathbf{t}}[\mathbf{F_m}]([\mathbf{B_0}]\{\mathbf{p}\} + [\mathbf{B_1}]\{\mathbf{q}\}) \tag{10.17}$$

and

$$\mathrm{U} = \frac{1}{2}\{\mathbf{d}\}^{\mathbf{t}}[\mathbf{K}]\{\mathbf{d}\} - \{\mathbf{d}\}^{\mathbf{t}}\{\mathbf{F}\} \tag{10.18}$$

where **[K]** is the stiffness matrix and **{F}** is the nodal force vector. For equilibrium, U is negative and U + U^c is equal to zero. This formulation is used for the 10-bar truss example (Case Study 6) of Case III. Table 10.6 shows the results. Twenty agents are selected in the CSS algorithm. Also the resulting minimum weight is compared to the one obtained by Kaveh and Rahami in [5], and Kaveh and Hassani in [8] for the same example. The result of comparison is shown in Table 10.7. Similar to the other cases, CSS with supervisor agents have shown a better performance. Kaveh and Rahami in [5] used a different formulation to impose the analysis criteria as a constraint. In this method, using the derivative of U^c in Eq. 10.6 with respect to **{q}** leads to:

$$\frac{\partial \mathrm{U^c}}{\partial \mathrm{q}} = [\mathbf{H_{qp}}]\{\mathbf{p}\} + [\mathbf{H_{qq}}]\{\mathbf{q}\} = \mathbf{0} \tag{10.19}$$

Equation 10.19 indicates that the complementary energy of the structure is equal to its minimum value in the compatibility condition. Thus **{q}** should be selected such that Eq. 10.19 holds. The left hand of this equation is a zero vector and it should be changed to a scalar. The best way is calculation of the norm, because the norm of a vector is equal to zero when all the entries is equal to zero. Here, we use the equilibrium itself. For this purpose we can write

$$\mathrm{F}(\mathbf{q},\mathbf{A}) = \mathrm{W}(\mathbf{A})\left(1 + \alpha norm\left([\mathbf{H_{qp}}]\{\mathbf{p}\} + [\mathbf{H_{qq}}]\{\mathbf{q}\}\right)\right) \tag{10.20}$$

Having **{q}** and **{A},** the magnitude of **F** can be calculated from Eq. 10.20 and its minimum for a large value of α corresponds to minimum W. Other constraints such as stress constraints, displacement constraints or dynamical properties constraints

Table 10.6 Results of the 10-bar planar truss (Case Study 6) (case 1–3)

Case 1 (FSD)
A = {7.94 0.10 8.05 3.91 0.10 0.10 5.73 5.57 5.54 0.11}in^2, W = 1,591.8 lb
Case 2 (FSD)
A = {7.77 0.24 8.25 3.79 0.1011 0.22 5.97 5.41 3.67 0.31}in^2, W = 1,591.8 lb
Case 3 (weight minimization)
A = {7.77 0.24 8.25 3.79 0.1011 0.22 5.97 5.41 3.67 0.31}in^2, W = 1,516.2 lb

Table 10.7 Optimal design comparision for the 10-bar truss (Case Study 6)

Method	(GA) [5]	(ACO) [8]	Kaveh and Ahmadi [25]
Best weight (case 1) lb	1,593.5	1,593.5	1,591.8
Best weight (case 2) lb	1,723.5	1,723.5	1,724.6
Best weight (case 3) lb	1,519.2	1,519.2	1,516.2

can be applied to Eq. 10.20 after normalizing and selecting a penalty coefficient. Therefore, the final formulation will be as follow:

$$\text{Find}\rightarrow q,A;A\in\{S_d \text{ or } S_c\}$$
$$\text{Min}F(q,A)=\sum_{i=1}^{ne} A_i l_i \rho_i \left(1+\alpha norm\left(\left[\mathbf{H_{qp}}\right]\{\mathbf{p}\}+\left[\mathbf{H_{qq}}\right]\{\mathbf{q}\}\right)\right)+\sum_{m=1}^{nc}\max(0,g_m(A)) \tag{10.21}$$

where S_d and S_c are the discrete and continuous sections, respectively. $g_m(A)$ corresponds to violation of the constraints. Because of indirect analysis, internal forces in earlier iterations are not reliable. In other words, since the redundant forces are not exact, the calculated constraints are not exact either, and cannot be relied on. Reliability criteria can be norm($[\mathbf{H_{qp}}]\{\mathbf{p}\}$ + $[\mathbf{H_{qq}}]\{\mathbf{q}\}$). Accordingly, the design constraints penalty function can be altered to:

$$F(q,A)=\sum_{i=1}^{ne} A_i l_i \rho_i \left(1+\alpha norm\left(\left[\mathbf{H_{qp}}\right]\{\mathbf{p}\}+\left[\mathbf{H_{qq}}\right]\{\mathbf{q}\}\right)\right)+\sum_{m=1}^{nc}\max(0,g_m(A))^{R(norm)} \tag{10.22}$$

where R(norm) is a function of norm($[\mathbf{H_{qp}}]\{\mathbf{p}\}$ + $[\mathbf{H_{qq}}]\{\mathbf{q}\}$). This function can be considered as follows:

$$R(norm) = \log(10 + NORM) \tag{10.23}$$

where NORM is equal to norm($[\mathbf{H_{qp}}]\{\mathbf{p}\}$ + $[\mathbf{H_{qq}}]\{\mathbf{q}\}$). In all of the examples studied in the following, Eq. 10.22 has been used in the CSS algorithm.

Table 10.8 Design data for the 10-bar planar truss (Case Study 7)

Material property and constraint data
Young's modulus: E = 1e7 psi = 6.895e7MPa
Density of the material: $\rho = 0.1$ lb/in^3 = 0.00277 kg/cm^3
Stress constraints
$\lvert\sigma_i\rvert \leq 25$ ksi(172.375 MPa); i =1, . . ., 10
Nodal displacement constraint in all directions of the co-ordinate system
$\lvert\Delta_i\rvert \leq 2$ in (5.08 cm); i = 1, . . ., 4
List of the available profiles
Case 1: (Discrete sections)
A_i = {1.62, 1.80, 1.99, 2.13, 2.38, 2.62, 2.63, 2.88, 2.93, 3.09, 3.13, 3.38, 3.47, 3.55, 3.63, 3.84, 3.87, 3.88, 4.18, 4.22, 4.49, 4.59, 4.80, 4.97, 5.12, 5.74, 7.22, 7.97, 11.5, 13.5, 13.9, 14.2, 15.5, 16.0, 16.9, 18.8, 19.9, 22.0, 22.9, 26.5, 30.0, 33.5} in^2
A_i = {10.4516, 11.6129, 12.8387, 13.7419, 15.3548, 16.9032, 16.9677, 18.5806, 18.9032, 19.9354, 20.1935, 21.8064, 22.3871, 22.9032, 23.4193, 24.7741, 24.9677, 25.0322, 26.9677, 27.2258, 28.9677, 29.6128, 30.9677, 32.0645, 33.0322, 37.0322, 46.5806, 51.4193, 74.1934, 87.0966, 89.6772, 91.6127, 99.9998, 103.2256, 109.0320, 121.2901, 128.3868, 141.9352, 147.7416, 170.9674, 193.5480, 216.1286} cm^2
Case 2: (Continuous sections)
$0.1 \leq A_i \leq 35$ in^2 (225.8960) cm^2; i = 1, . . ., 10

Case Study 7: A 10-bar Planar Truss. The 10-bar truss as shown in Fig. 10.10 is considered for optimal design. Table 10.8 contains the necessary data. As seen displacement constraint is added to the design procedure. Two cases are considered, the first is optimal design using discrete sections and the second corresponds to continuous sections. Equation 10.22 is used as the objective function in the CSS, where a population of 20 CPs is used. In both cases, Λ and $\mathbf{q}$ are variables. In discrete case a code is utilized that moves the section between two available sections to one of them based of a probabilistic function. Results are obtained in Tables 10.9 and 10.10 for discrete and continuous sections, respectively.

Case Study 8: A 25-bar Space Truss. Geometry, nodal ordering and grouping of members are sown in Fig. 10.11 and Table 10.11, respectively. Table 10.12 contains the necessary data for design. Table 10.13 contains the results and shows the efficiency of this method and combining the CSS and force method compared to the other algorithms.

In this example, the calculated maximum displacement in case 1 and case 2, using exact displacement method, are equal to 0.3482 in and 0.3503 in. and those of the present method are 0.3496 in and 0.3498 in. respectively. There is another set of areas for case 2 as A = {0.10, 0.10, 3.7598, 0.10, 1.8932, 0.7755, 0.1408, 3.8460} and the corresponding weight is equal to 468.1998. Maximum displacement of this set of areas leads to 0.3497 in.

Table 10.9 Optimal design comparison for the 10-bar planar truss (Case Study 7) (discrete)

Method	Weight lb(kN)	A_1	A_2	A_3	A_4	A_5	A_6	A_7	A_8	A_9	A_{10}
Kaveh and Rahami [5]	5,490.738 (24.4228)	33.50	1.62	22.90	14.20	1.62	1.62	7.97	22.9	22.00	1.62
Shih [9]	5,491.71 (24.4271)	33.50	1.62	22.90	15.50	1.62	1.62	7.97	22.00	22.00	1.62
Rajeev [10]	5,613.84 (24.9704)	33.50	1.62	22.90	15.50	1.62	1.62	14.20	19.90	19.90	2.62
Kaveh and Hassani [8]	5,517.72 (24.5702)	33.50	1.62	22.90	14.20	1.62	1.62	11.50	22.00	19.90	1.62
Kaveh and Ahmadi [25]	5,475.40 (24.3817)	30.00	1.62	22.90	16.00	1.62	1.62	7.97	22.90	22.90	1.62

Table 10.10 Optimal design comparison for the 10-bar planar truss (Case study 8) (continuous)

Method	Weight lb (kN)	A_1	A_2	A_3	A_4	A_5	A_6	A_7	A_8	A_9	A_{10}
Kaveh and Rahami [5]	5,061.90 (22.5153)	30.67	0.1	22.87	15.34	0.1	0.46	7.48	20.96	21.70	0.1
Schmit and Farshi [11]	5,089.0 (22.6359)	33.43	0.1	24.26	14.26	0.1	0.1	8.39	20.74	19.69	0.1
Schmit and Miura [12]	5,076.85 (22.5818)	30.67	0.1	23.76	14.59	0.1	0.1	8.59	21.07	20.96	0.1
Schmit and Miura [12]	5,107.3 (22.7173)	30.57	0.37	23.97	14.73	0.1	0.36	8.55	21.11	20.77	0.32
Venkayya [13]	5,084.9 (22.6176)	30.42	0.13	23.41	14.91	0.10	0.10	8.70	21.08	21.08	0.19
Gellatly and Berke [14]	5,112.0 (22.7382)	31.35	0.1	20.03	15.60	0.14	0.24	8.350	22.21	22.06	0.1
Dobbs and Nelson [15]	5,080.0 (22.5958)	30.50	0.1	23.29	15.43	0.1	0.21	7.65	20.98	21.82	0.1
Rizzi [16]	5,076.66 (22.5810)	30.73	0.1	23.93	14.73	0.1	0.1	8.54	20.95	21.84	0.1
Khan and Willmert [17]	5,066.98 (22.5379)	30.98	0.1	24.17	14.81	0.1	0.41	7.547	21.05	20.94	0.1
Kaveh and Hassani [8]	5,095.46 (22.6899)	30.86	0.1	23.55	15.01	0.1	0.22	7.63	21.65	21.32	0.1
Kaveh and Ahmadi [25]	5,059.39 (22.5041)	30.5	0.1	21.99	15.70	0.1	0.5	7.55	21	22	0.1

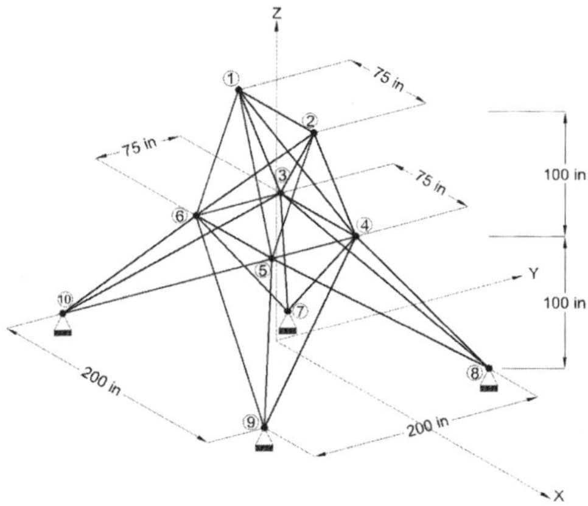

Fig. 10.11 Geometry of a 25-bar space truss (Case Study 8)

Table 10.11 Member grouping

Group number	Members
1	1–2
2	1–4,2–3,1–5,2–6
3	2–5,2–4,1–3,2–6
4	3–6,4–5
5	3–4,5–6
6	3–10,6–7,4–9,5–8
7	3–8,4–7,6–9,5–10
8	3–7,4–8,5–9,6–10

Table 10.12 Design data for a 25-bar space truss (Case Study 8)

Design variables

Size variables $A_1;A_2;A_3;A_4;A_5;A_6;A_7;A_8;\ q_1;\ q_2\ q_3;\ q_4;\ q_5;\ q_6;\ q_7$

Material property and constraint data

Young's modulus: $E =$ 1e7 psi

Density of the material: $\rho = 0.1$ lb/in^3 = 0.00277 kg/cm^3

Stress constraints

$|\sigma_i| \leq 40$ ksi (275.8 MPa); $i = 1, \ldots, 25$

Displacement constraint in the directions of X and Y in the co-ordinate system

$|\Delta_i| \leq 0.35$ in *(0.8890 cm)*; $i = 1, 2$

List of the available profiles

Case 1: (Discrete sections)

Ai = {0.1, 0.5 × I (I = 1,2,. . .,76), 39.81, 40} in^2

Ai = {0.6452, 3.2258 × I (I = 1, 2, . . ., 76), 256.8382, 258.0640} cm^2

Case 2: (Continuous sections)

Ai ≥0.1 in^2 (0.6452)

Loading data

Node	Px: kips (kN)	Py: kips (kN)	Pz: kips (kN)
1	−10 (44.48)	−10 (44.48)	−10 (44.48)
2	0	−10 (44.48)	−10 (44.48)
3	0.5 (2.224)	0	0
6	0.5 (2.224)	0	0

Table 10.13 Optimal design comparision for the 25-bar space truss (Case Study 8)

Method	Weight lb (kN)	A_1	A_2	A_3	A_4	A_5	A_6	A_7	A_8
Rajeev [10]	546.01 (2.4287)	0.1	1.8	2.3	0.2	0.1	0.8	1.8	3.0
Erbatur [18]	493.80 (2.1964)	0.1	1.2	3.2	0.1	1.1	0.9	0.4	3.4
Kaveh and Kalatjari [19]	480.23 (2.1361)	0.1	0.1	3.5	0.1	2.0	1.0	0.1	4.0
Kaveh and Rahami (C.1) [5]	479.75 (2.1340)	0.1	0.5	3.0	0.1	2.0	1.0	0.1	4.0
Kaveh and Rahami (C.2) [5]	467.6293 (2.0800)	0.1	0.1	3.7598	0.1	1.8552	0.7755	0.1408	3.8460
Kaveh and Ahmadi [25] (**C.1**)	479.75 (2.1340)	0.1	0.5	3.0	0.1	2.0	1.0	0.1	4.0
Kaveh and Ahmadi [25] (**C.2**)	467.0660 (2.0774)	0.1	0.1001	3.6979	0.1005	1.8687	0.7888	0.1420	3.8612

Table 10.14 Design data for a 120-bar space dome (Case Study 9)

Design variables
Vaariables: $A_1;A_2;A_3;A_4;A_5;A_6;A_7$; q_1; q_2; q_3; q_4; q_5; q_6; q_7; q_8; q_9
Material property and constraint data
Young's modulus: E = 30,450 ksi = 210,000 MPa
Density of the material: $\rho = 0.288\ \text{lb/in}^3 = 7971.810\ \text{kg/cm}^3$
For all members: $0.775 \leq A_i \leq 20\ \text{in}^2$; i =1, ..., 120
Constraints
$\lambda_i = \frac{l_i}{r}$ $r = \sqrt{0.4 \times A}$
Stress constraints
For tensile members
$F_a \leq 0.6\ F_y$ and $\lambda_i \leq 300$
For compressive members
$\lambda_i \leq 200$
$F_a = \left[\left(1 - \frac{\lambda_i}{2C_c^2}\right)F_y\right] / \left(\frac{5}{3} + \frac{3\lambda_i}{8C_c} - \frac{\lambda_i^3}{8C_c^3}\right)$ for $\lambda_i \leq C_c$
$F_a = \frac{12\pi^2 E}{23\lambda_i^2}$ for $\lambda_i \leq C_c$
$\sigma_i < 58.0\ ksi$ (400 MPa); i =1, ..., 120
Displacement constraint in the directions of X, Y anc Z in all unsupported nodes
$|\Delta_i| \leq 0.1969 in$

Case Study 9: A 120-bar Dome. A 120-bar dome structure is considered in this example. This structure has 9° of statical indeterminacy. The necessary data for design, and constraints are shown in Table 10.14. Optimal design comparison for the 26-story tower is obtained in Table 10.15. Geometry, ordering and member grouping structure are shown in Fig. 10.12. Loading condition is considered as:

1. A vertical load at node 1 equal to −13.49 kips (−60 kN)
2. Vertical loads at node 2 through 14 equal to −6.744 kips (−30 kN)
3. Vertical loads at the rest of the nodes equal to −2.248 kips (−10 kN)

Redundant forces are considered as the reactions at nodes 39, 43 and 47.

For the present approach the maximum stress ratio is equal to 0.9552 and the maximum displacement using the exact displacement method is equal 0.17335 in, and the maximum displacement using the present method is calculated as 0.17339 in.

In this example, when the displacement method is utilized as an analysis procedure, the unknowns change from redundant forces to nodal displacements. Then number of unknowns drastically increase from 9 redundant forces to 111 nodal displacements. This imposes a highly computational cost on the optimization procedure. Equation 10.24 will be used to analysis using displacement method.

Table 10.15 Optimal design comparison for the 120-bar dome (Case Study 10)

	Optimal cross-sectional areas (in^2)							
Element group	A_1	A_2	A_3	A_4	A_5	A_6	A_7	Best weight(lb)
Kaveh et al. (IACS) [20]	3.026	15.060	4.707	3.100	8.513	3.694	2.503	33,320.52
Kaveh and Talatahari (PSOPC) [21]	3.040	13.149	5.646	3.143	8.759	3.758	2.502	33,481.20
Kaveh and Talatahari (PSACO) [21]	3.026	15.222	4.904	3.123	8.341	3.418	2.498	33,263.90
Kaveh and Talatahari (HPSACO) [21]	3.095	14.405	5.020	3.352	8.631	3.432	2.499	33,248.90
Kaveh and Talatahari (HBB-BC) [22]	3.037	14.431	5.130	3.134	8.591	3.377	2.500	33,287.90
Kaveh and Talatahari (CSS) [6]	3.027	14.606	5.044	3.139	8.543	3.367	2.497	33,251.90
Kaveh and Ahmadi [25]	4.795	5.153	4.527	4.353	1.650	3.533	3.888	28,712.69

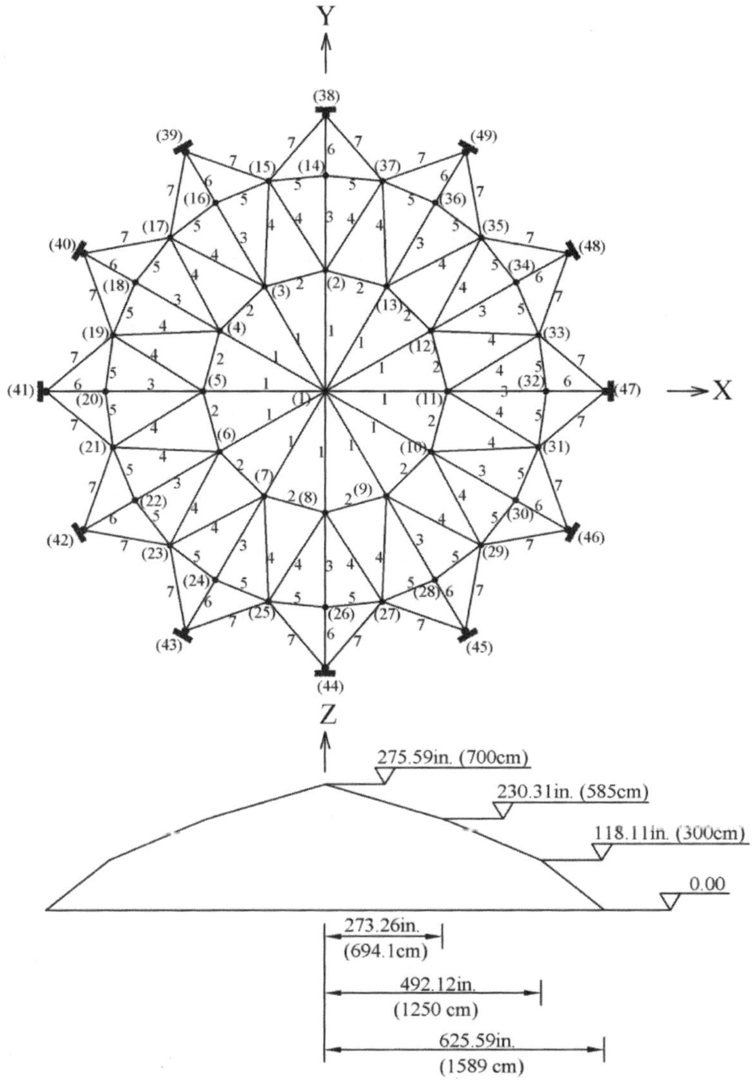

Fig. 10.12 A 120-bar dome (Case Study 9)

$$\mathrm{norm}([\mathbf{K}]\{\mathbf{X}\} - \{\mathbf{F}\}) = 0 \tag{10.24}$$

where **K** is considered as the stiffness matrices of the structure. **X** is considered as the nodal displacement vector and **F** is the nodal forces vector.

Case Study 10: A 26-story Tower. The main aim of the present method is to avoid the computation of the inverse of the large-scale structures matrices. This method must be applied to the large-scale structures to show the superiority of the present

Table 10.16 Design data and constraints of 26-story truss (Case Study 10)

Design variables

Variables $A_1;A_2;A_3;...;A_{59}; q_1; q_2;...; q_{246}$

Material property and constraint data

Young's modulus: $E = 1e7$ psi

Density of the material: $\rho = 0.1\ \text{lb/in}^3 = 0.00277\ \text{kg/cm}^3$

Stress constraints

$|\sigma_i| \leq 25$ ksi(172.375 MPa); $i = 1, \ldots, 942$

Displacement constraint in the directions of X and Y in the co-ordinate system

$|\Delta_i| \leq 15$ in *(about 1/250 of the total height of the tower) for the four nodes of the top level in the x, y and z directions*

List of the available profiles

Case 1: Ai $\geq$1 in^2 (6.452) cm^2 Ai <200 in^2 (1,290.32) cm^2

Table 10.17 Design comparison for the 26-story truss (Case Study 10)

Variable (in^2)	Erbatur and Hasançebi [23]	Rahami et al. [24]	Kaveh and Talatahari [6]	Kaveh and Ahmadi [25]
A1	1	2.7859	0.962	1.0376
A2	1	1.3572	2.557	2.0424
A3	3	5.0362	1.65	1.6003
A4	1	2.2398	0.402	1.0113
A5	1	1.2226	0.657	1.0033
A6	17	14.9575	18.309	2.5260
A7	3	2.9568	0.346	1.0001
A8	7	10.9038	3.076	1.0981
A9	20	14.4177	2.235	2.4705
A10	1	3.709	3.813	1.0222
A11	8	5.7076	0.856	1.2531
A12	7	4.9264	1.138	1.0024
A13	19	14.1751	3.374	1.8253
A14	2	1.9043	0.573	1.0463
A15	5	2.8101	19.53	1.6020
A16	1	1	1.512	1.0760
A17	22	18.807	2.667	2.2508
A18	3	2.6151	0.478	1.0177
A19	9	12.5328	17.873	3.4032
A20	1	1.1314	0.335	1.0012
A21	34	30.5122	2.78	5.3252
A22	3	3.346	0.43	1.0003
A23	19	17.045	3.048	4.4083
A24	27	18.0785	5.112	10.7550
A25	42	39.2717	19.352	5.0916
A26	1	2.6062	0.476	1.0029
A27	12	9.8303	2.887	5.5097
A28	16	13.1126	19.5	7.9683
A29	19	13.6897	4.772	4.4314
A30	14	16.9776	5.063	5.3373

(continued)

Table 10.17 (continued)

Variable (in^2)	Erbatur and Hasançebi [23]	Rahami et al. [24]	Kaveh and Talatahari [6]	Kaveh and Ahmadi [25]
A31	42	37.6006	15.175	6.7094
A32	4	3.0602	1.176	1.6518
A33	4	5.5106	0.839	3.1108
A34	4	1.8014	1.394	1.0434
A35	1	1.1568	0.153	1.2485
A36	1	1.2423	0.247	1.0746
A37	62	62.7741	18.673	6.8163
A38	3	3.3276	0.696	1.2514
A39	2	4.2369	1.395	5.4658
A40	4	1.7202	0.422	1.1308
A41	1	1.0148	0.417	1.3079
A42	2	5.6428	0.679	1.0063
A43	77	78.0094	19.584	9.9490
A44	3	3.2206	0.533	1.1061
A45	2	3.5934	1.64	7.3345
A46	3	4.7668	0.618	2.3035
A47	2	1.1531	0.531	2.3722
A48	3	2.1698	1.374	1.0706
A49	100	99.6406	19.656	13.9159
A50	4	4.1469	0.888	2.7680
A51	1	2.16	4.456	5.2249
A52	4	4.1499	0.386	1.0024
A53	6	11.207	10.398	11.7689
A54	3	11.0904	18.834	12.1676
A55	49	35.9499	18.147	19.9929
A56	1	2.1937	3.28	9.2241
A57	62	66.1705	2.972	1.0313
A58	1	3.3402	4.927	8.1362
A59	3	4.0525	0.288	1.0025
Weight (lb)	143,436	142,295.75	47,370.8412	47,108.4972

method. For this purpose, a 26-story tower as shown in Fig. 10.6 is considered. Loading condition is defined in Case Study 4. Design data and constraints are maintained in the Table 10.16. This structure has 246° of statical indeterminacy. The member grouping has 59 groups as shown in Fig. 10.6. The simultaneous analysis, design and optimization of this structure have 305 variables. A population of 100 CPs is considered in the CSS algorithm. Equation 10.22 is taken as the objective function in the CSS algorithm with supervisor agents. Optimal design comparison for the 26-story tower is provided in Table 10.17.

In this example, the exact maximum nodal displacement calculated for the four top nodes, using the displacement method, is 14.3442 in. The present method leads to 14.7688 in. The maximum stress ratio is equal to 94.90 %. According to the above table the efficiency of the CSS and especially the present method in analysis, design and optimization of large-scale structures in comparison to other methods become apparent.

References

1. Kaveh A (2006) Optimal structural analysis, 2nd edn. Wiley, Somerset
2. Kaveh A (2004) Structural mechanics: graph and matrix methods, 3rd edn. Research Studies Press, Baldock
3. Kaveh A, Talatahari S (2010) A novel heuristic optimization method: charged system search. Acta Mech 213(3–4):267–286
4. Kaveh A, Talatahari S (2010) An enhanced charged system search for configuration optimization using the concept of fields of forces. Struct Multidiscip Optim 43:339–351
5. Kaveh A, Rahami H (2006) Analysis, design and optimization of structures using force method and genetic algorithm. Int J Numer Methods Eng 65:1570–1584
6. Kaveh A, Talatahari S (2010) Optimal design of skeletal structures via the charged system search algorithm. Struct Multidiscip Optim 41(10):893–911
7. Kirsch U (1981) Optimum structural design, concepts, methods and applications. McGraw-Hill, New York
8. Kaveh A, Hassani M (2009) Simultaneous analysis, design and optimization of structures using force method and ant colony algorithms. Asian J Civil Eng 10(4):381–396
9. Shih CJ, Yang YC (2002) Generalized Hopfield network based structural optimization using sequential unconstrained minimization technique with additional penalty strategy. Adv Eng Softw 33:721–729
10. Rajeev S, Krishnamoorthy CS (1992) Discrete optimization of structures using genetic algorithms. J Struct Eng ASCE 118:1233–1250
11. Schmit LA Jr, Farshi B (1974) Some approximation concepts for structural synthesis. AIAA J 12:692–699
12. Schmit LA Jr, Miura H (1976) Approximation concepts for efficient structural synthesis, NASA CR-2552. NASA, Washington, DC
13. Venkayya VB (1971) Design of optimum structures. Comput Struct 1:265–309
14. Gellatly RA, Berke L (1971) Optimal structural design, AFFDLTR-70-165. Air Force Flight Dynamics Laboratory, Wright-Patterson AFB
15. Dobbs MW, Nelson RB (1976) Application of optimality criteria to automated structural design. AIAA J 14:1436–1443
16. Rizzi P (1976) Optimization of multiconstrained structures based on optimality criteria. In: AIAA/ASME/SEA 17th structures, structural dynamics and material conference, King of Prussia
17. Khan MR, Willmert KD, Thoronton WA (1979) An optimality criterion method for large-scale structures. AIAA J 17:735–761
18. Erbatur F, Hasançebi O, Tutuncil I, Kihc H (2000) Optimal design of planar and structures with genetic algorithms. Comput Struct 75:209–224
19. Kaveh A, Kalatjari V (2002) Genetic algorithm for discrete sizing optimal design of trusses using the force method. Int J Numer Methods Eng 84:55–72
20. Kaveh A, Farahmand Azar B, Talatahari S (2008) Ant colony optimization for design of space trusses. Int J Space Struct 23(3):167–181
21. Kaveh A, Talatahari S (2009) Particle swarm optimizer, ant colony strategy and harmony search scheme hybridized for optimization of truss structures. Comput Struct 87(5–6):267–283
22. Kaveh A, Talatahari S (2009) Size optimization of space trusses using Big-Bang Big Crunch algorithm. Comput Struct 87:1129–1140
23. Erbatur F, Hasançebi O (2000) Optimal design of planar and space structures with genetic algorithms. Comput Struct 75:209–224
24. Rahami H, Kaveh A, Aslani M, Najian Asl R (2011) A hybrid modified Genetic-Nelder Mead simplex algorithm for large-scale truss optimization. Int J Optim Civil Eng 1:29–46
25. Kaveh A, Ahmadi B (2013) Simultaneous analysis, design and optimization of structures using force method and supervised CSS algorithm. Scientia Iranica 20:65–76

Printed by Printforce, the Netherlands